Dynamical Systems and Population Persistence

Hal L. Smith
Horst R. Thieme

Graduate Studies in Mathematics
Volume 118

American Mathematical Society
Providence, Rhode Island

2010 *Mathematics Subject Classification.* Primary 37N25, 92D25, 92D30; Secondary 37B25, 37Lxx.

For additional information and updates on this book, visit
www.ams.org/bookpages/gsm-118

Library of Congress Cataloging-in-Publication Data

Smith, Hal L.
Dynamical systems and poplulation persistence / Hal L. Smith, Horst R. Thieme.
p. cm. – (Graduate studies in mathematics ; v. 118)
Includes bibliographical references and index.
ISBN 978-0-8218-4945-3 (alk. paper)
1. Biology–Mathematical models. 2. Population biology. I. Thieme, Horst R., 1948– II. Title.

QH323.5.S58 2011
577.8′80151539–dc22 2010033476

∞ The paper used in this book is acid-free and falls within the guidelines
established to ensure permanence and durability.
Visit the AMS home page at `http://www.ams.org/`

10 9 8 7 6 5 4 3 2 1   16 15 14 13 12 11

To our doctoral advisors
Willi Jäger (HRT) and Paul Waltman (HLS)

Contents

Preface

The primary focus of this book is the mathematical theory of persistence. The theory is designed to provide an answer to such questions as which species, in a mathematical model of interacting species, will survive over the long term. In a mathematical model of an epidemic, will the disease drive a host population to extinction or will the host persist? Can a disease remain endemic in a population? Persistence theory can give a mathematically rigorous answer to the question of persistence by establishing an initial-condition-independent positive lower bound for the long-term value of a component of a dynamical system such as population size or disease prevalence.

Mathematically speaking, in its simplest formulation for systems of ordinary or delay differential equations, and for a suitably prescribed subset I of components of the system, persistence ensures the existence of $\epsilon > 0$ such that $\liminf_{t\to\infty} x_i(t) > \epsilon,\ i \in I$ provided $x_i(0) > 0,\ i \in I$. We say that these components persist uniformly strongly, or, more precisely, that the system is uniformly strongly ρ-persistent for the persistence function $\rho(x) = \min_{i\in I} x_i$. This *persistence function* $\rho(x)$ may be viewed as the distance of state x to a portion of the boundary of the state-space $\mathbb{R}^n_+$, namely the states where one or more of species $i \in I$ are extinct.

The adjective "strong" is often omitted; uniform weak ρ-persistence is defined similarly but with limit superior in place of limit inferior. The adjective "uniform" emphasizes that the lower bound ϵ is independent of initial data satisfying the restriction $x_i(0) > 0,\ i \in I$. Similarly, as in the definition of Lyapunov stability, the precise value of ϵ is unspecified and usually difficult to estimate. Uniform persistence is a qualitative notion, not a quantitative one. However, in rare cases, ϵ can be related to system parameters; this ideal situation is referred to as "practical persistence" [**24, 25, 26, 28, 27, 35**].

Weaker notions of weak and strong persistence drop the "uniformity with respect to initial data" (i.e., ϵ), requiring merely positivity of the limit superior, respectively, the limit inferior.

The definition of persistence and the related concept of permanence (uniform persistence plus an upper bound on limit superior of all components) evolved in the late 1970s from the work of Freedman and Waltman [**75**], Gard [**80, 81**] Gard and Hallam [**82**], Hallam [**96**], and Schuster, Sigmund, Wolff [**196**]. Most of these early papers show weak persistence, but Schuster, Sigmund, and Wolff [**196**] prove uniform strong ρ-persistence for the hypercycle equation in the n-simplex with $\rho(x) = x_1 \cdots x_n$ as persistence function.

The notion of a persistence function seems to have been introduced by Gard and Hallam [**82, 80**], though with a more technical intention than here. It was later superseded by a more general concept which combines the usual Lyapunov function methods with time averages [**104**] and became known as average Lyapunov function [**109**]. "Persistence function" (together with the ρ-symbol [**80, 82**]) is revived here as a means to make precise which parts of a system persist; in applications ρ has a very concrete and intuitive interpretation like the number of infected individuals to describe disease persistence in an epidemic model. Such a "hands on" interpretation would be lacking for a typical average Lyapunov function like $x_1^{p_1} \cdots x_n^{p_n}$ [**81, 82**]. Zhao [**238**] uses the notion *generalized distance function* to stress the idea that ρ measures the distance to the brink of extinction.

Persistence theory developed rapidly in the 1980s because the necessary machinery from dynamical systems, a theory of attractors and repellers, was already in place. Early work focused on persistence of components of systems of ordinary differential equations. Later, this was extended to discrete time or difference equations, and then to infinite dimensional dynamical systems generated by delay differential equations and partial differential equations. Application of the theory was initially slow to catch on in the applied literature on population biology and epidemiology, but it has more recently become an accepted tool in theoretical population dynamics.

Although the theory is now quite "user friendly" in the sense that a user does not need to be an expert to use it, it is a mathematically sophisticated theory. Our motivation for writing this monograph grew out of the problem of teaching the theory to our graduate students. There are very few sources where one can find self-contained treatments that are accessible to graduate students. The survey articles by Waltman [**231**] and by Hutson and Schmitt [**110**] remain useful, although they do not contain more recent refinements of the theory which are scattered in the literature. Recent monographs by Cantrell and Cosner [**25**], Hofbauer and Sigmund [**106**], Thieme [**217**], and Zhao [**238**] are good sources, but their focus is broader than persistence theory.

This monograph began as a set of lecture notes for graduate students in a team-taught course on Dynamical Systems in Biology offered by the authors in fall 2005, spring 2007, and spring 2009 at Arizona State University. It contains a large number of homework exercises. A description of the contents of the chapters follows.

Chapters 1 through 8 contain our main results on persistence theory for autonomous dynamical systems.

Chapter 1 begins with a review of metric spaces, the natural abstract setting or "state space" for finite and infinite dimensional dynamics. The notion of a semiflow on a metric space is developed; it gives the dynamics. We distinguish discrete and continuous time semiflows simply by the time set: nonnegative integers in the former case, the nonnegative reals in the latter case. The basic properties of a semiflow are independent of the time set. This unified treatment of discrete time and continuous time semiflows allows us to unify the later treatment of persistence theory for discrete time and continuous time dynamics. In the literature, the theory developed separately for discrete and continuous time systems, but we have been largely successful in our attempt to present a unified treatment, avoiding as much as possible separate approaches. Persistence theory often requires that the underlying dynamics are dissipative in some sense. The strongest sense is that there is a compact attractor of bounded subsets. Although the trend of recent work, and one of our goals here, is to weaken these compactness requirements where possible, we present the theory of attractors in Chapter 2.

Chapter 3 begins with the definitions of persistence, both uniform weak and uniform strong persistence relative to a persistence function. However, its main focus is on uniform weak persistence and on elementary methods for establishing it. Several examples illustrating such methods are introduced. These include the continuous time model of an SEIRS infectious disease in a meta-population with host travel between patches, the classical May-Leonard system of three competing populations, and discrete time nonlinear matrix models of population dynamics. The latter include the LPA model of flour beetle dynamics and nonlinear versions of the well-studied Leslie-type demographic models.

Uniform strong persistence is the desired conclusion; uniform weak persistence is more easily obtainable. It has long been known that uniform weak persistence plus suitable compactness properties of the dynamical system give uniform strong persistence. See Freedman and Moson [**72**] for flows on locally compact metric spaces, Freedman, Ruan, and Tang [**73**] for flows, and Thieme [**215**] for semiflows on general metric spaces. In Chapter 4, we present a number of such results. Some of these have relatively weak compactness assumptions at the expense of lengthy and seemingly technical hypotheses. Others require more compactness assumptions but are more easily and concisely formulated. A number of applications are treated in detail, including those introduced in the previous chapter.

The choice of a persistence function may be not obvious; several different choices may be appropriate. The question then naturally arises as to whether, and how to prove that, persistence with respect to one such function implies persistence with respect to another persistence function. This issue is treated in Chapter 4.6.

For semiflows that are dissipative in a suitable strong sense and that are uniformly ρ-persistent, there is an elegant decomposition of the attractor into an extinction attractor, a persistence attractor, and a family of total trajectories whose α limit sets are contained in the extinction attractor and whose ω limit sets are contained in the persistence attractor. This result, versions of which were first proved by Hale and Waltman [**95**] and later by Zhao [**238**] and Magal and Zhao [**158**], is proved in Chapter 5.

The brief Chapter 6 explores various scenarios whereby one may establish that persistence implies the existence of a "persistence equilibrium", that is, an equilibrium x^* for which $\rho(x^*) > 0$ where ρ denotes the persistence function. This provides an extra incentive for taking the trouble to establish persistence. The monograph by X.-Q. Zhao [**238**] contains a nice summary of the history of results in this direction. See the notes to Chapter 1 of [**238**]. Newer results also appear in the recent paper by Magal and Zhao [**158**].

Nonlinear matrix models, such as those introduced in Chapter 3, are increasingly being used in population modeling as indicated by the recent monographs [**29, 44, 39**]. Therefore, we devote Chapter 7 to applying the results of the previous chapters to them.

Chapter 8 treats the mathematically more sophisticated topological approach to persistence and its consequences. As Josef Hofbauer [**79, 104**] has repeatedly pointed out, the theory of attractors and repellers, formulated by several mathematicians including Zubov, Ura, Kimura, and Conley, lead directly to proofs of many of the results of persistence theory. See [**79**] for many historical references to the work of Zubov and Ura and Kimura; the monograph of Bhatia and Szegő [**16**] contains some of this work. The notion of chain recurrence and chain transitivity of Conley [**33**] has also proved to be very useful. These notions were originally established for flows on locally compact spaces but are needed for semiflows on potentially infinite dimensional Banach spaces for persistence theory. We give here a self-contained treatment of these ideas and how they are used in the theory. Most notable among the results implied by these ideas are the Butler-McGehee Theorem and the acyclicity theorem establishing uniform weak persistence.

These were first formulated by Butler, Freedman and Waltman [**22, 23**] for flows on locally compact spaces, extended to discrete time systems by Freedman and So [**74**] and Hofbauer and So [**107**]. They were later generalized to semiflows on infinite dimensional spaces by Hale and Waltman [**95**] assuming the existence of a compact attracting set, an assumption that was relaxed in [**215**].

Several more or less straightforward applications of the acyclicity approach to persistence are included. These include the classical three-level (ODE) food chain model considered by Hastings and collaborators, nonlinear matrix models for biennial species, and a metered epidemic model — a hybrid of both discrete and continuous time.

Finally, we show how classical Lyapunov exponents may be used to establish one of the key hypotheses in the acyclicity theorem, namely, that a compact invariant set, belonging to the "extinction set", is uniformly weakly repelling in directions normal to the extinction set. The use of Lyapunov exponents in the study of biological models was pioneered by Metz et al. [**168**], who proposed that the dominant Lyapunov exponent gives the best measure of invasion fitness, and by Rand et al. [**180**] who used it to characterize the invasion "speed" of a rare species. Roughly, a positive dominant Lyapunov exponent corresponding to a potential invading species in the environment set by a resident species attractor implies that the invader can successfully invade. Our treatment is patterned after the approach taken by Paul Salceanu in his thesis [**186**] and [**187, 188, 189, 190**].

Chapter 9 focuses on an SI epidemic model where infectives are structured by age since infection and where the force of infection depends on an age-since-infection weighted average of current infectives. The model can be reduced to a system of integral equations; existence and uniqueness of solutions, and boundedness of solutions are proved. The host is shown to (uniformly) persist, the basic replacement number $\mathcal{R}_0$ is identified, disease extinction is shown to occur if $\mathcal{R}_0 < 1$, and uniform weak persistence of the disease is shown if $\mathcal{R}_0 > 1$. In order to obtain uniform persistence of the disease, it is useful to reformulate the dynamics as a semiflow on a suitable Banach space. This is done by showing that solutions satisfy a weakly formulated semilinear Cauchy problem. One can then show the existence of a compact attractor of bounded sets under suitable restrictions. This in turn facilitates the argument for uniform persistence of the disease when $\mathcal{R}_0 > 1$. The existence of an endemic equilibrium is also established, and rather unrestrictive conditions for its global stability are derived. It should be noted that persistence is indispensable for doing the latter because the Lyapunov function that is used is not defined on the whole state space but only on the persistence attractor (see also [**155**]).

Chapter 10 is devoted to a brief treatment of the semilinear Cauchy problem $u' = Au + F(u)$, $u(0) = u_0$ in a Banach space setting. Here A is a closed linear operator and F is a nonlinear map. Notions of classical, integral, and mild solutions are defined, and the equivalence of mild and integral formulation is shown. Globally defined integral solutions are shown to define a semiflow, and local existence is established by the contraction mapping principle when F satisfies a Lipschitz condition. If F is suitably bounded, global in time existence is also shown. Conditions for the induced semiflow to be asymptotically smooth, a key requirement for showing the existence of a compact global attractor, are identified. As we have biological examples in mind, positivity of solutions must be satisfied. Conditions which ensure positivity of solutions are formulated.

Chapter 11 treats microbial growth on a growth-limiting nutrient in a tubular bioreactor. Fresh nutrient enters the left side of the tube, and unused nutrient and microbes leak out the right side of the tube in proportion to their concentration. Both nutrient and microbes are assumed to diffuse throughout the tube. The issue is whether or not the influx of nutrient is sufficient to allow the microbes to persist in the bioreactor. Relying heavily on the machinery of Chapter 10, we show that the system of reaction-diffusion equations generates a dissipative semiflow. Linearized stability analysis of the so-called washout equilibrium solution (no microbes) leads to a basic reproduction number $\mathcal{R}_0$. If $\mathcal{R}_0 < 1$, the microbes are "washed out" of the bioreactor, and, if $\mathcal{R}_0 > 1$, they uniformly persist and there is a unique colonization equilibrium.

Chapter 12 considers a model of microbial growth in a chemostat where microbial cells of different age take up nutrient at differing rates and divide at an age-dependent rate. Ignoring growth and uptake, focusing only on demographics of cell division, we begin by obtaining a renewal equation for cell population division rate and showing that it has a unique solution. This leads to the definition of a semigroup of operators and ultimately to a formulation of the full model, including growth and uptake, as an abstract ODE in a Banach space setting. Its mild solutions are shown to generate a semiflow. Consideration of the "washout state", absent microbes, allows identification of the basic "biomass production number" for the model. When it is less than one, and an additional condition satisfied, the microbes are washed out;

when it exceeds one, the cell population persists uniformly weakly. Proofs of these results make use of the Laplace transform. In fact, uniform persistence of the cell population holds when the basic production number exceeds one under additional assumptions, but the proof is deferred to a later chapter. The problem is in establishing sufficient compactness of the semiflow. A different approach provides another route from uniform weak to uniform strong persistence, which succeeds for this model.

Chapter 13 is devoted to persistence for nonautonomous systems. Practical persistence is established, under suitable conditions, for a population of micro-organisms growing in a chemostat with time-dependent dilution rate using elementary arguments. It is also established that all positive solutions are asymptotic to each other. The abstract notion of a nonautonomous semiflow is introduced, corresponding definitions of persistence are given, and several results giving conditions under which uniform weak persistence implies uniform strong persistence are proved. Special attention is devoted to the case of periodic nonautonomous semiflows and nonautonomous semiflows that are asymptotic to such semiflows. The implication that uniform weak persistence implies uniform persistence for these cases is specialized. Finally, uniform persistence is established for the (autonomous) cell division model treated in Chapter 12 by using the methods developed for nonautonomous semiflows.

As noted in our description of Chapter 3 above, persistence functions were introduced early on in the history of persistence theory as a means to obtain uniform persistence (permanence) in much the same way that they are used in Lyapunov stability theory to obtain stability results for equilibria of dynamical systems [**82**], and there is now a well-developed approach to establishing persistence using so-called average Lyapunov functions (a generalization of persistence functions in the sense of Gard and Hallam [**82**]). The works of Fonda [**71**], Gard [**80**], Hofbauer [**103**], Hutson [**109**], and Schuster, Sigmund and Wolff [**196**] have been very influential. Some of their ideas, as well as later work, have been reviewed in the paper of Hutson and Schmitt [**110**] and the monograph of Sigmund and Hofbauer [**106**]. However, so far in this work, we have used persistence functions primarily as a means to precisely define what is meant by persistence, not as a tool with which to establish it.

In Chapter 15, taking inspiration from this large literature, we formulate some general results which yield persistence using the average Lyapunov function approach. The adjective "average" in the terminology signifies that a time-average of the function over a sufficiently large interval should be positive. We formulate an approach which works for nonautonomous semiflows and, as usual, seeks to minimize compactness requirements. These goals force rather technical statements, but the main results are simple: The existence of a weak average Lyapunov function ρ implies weak ρ-persistence; the existence of a strong average Lyapunov function implies uniform ρ-persistence. As an application, the hypercycle equation, treated in Chapter 12 of [**106**] in the autonomous case, is extended to the case where replication rates may be time-dependent.

The book ends with two appendices. The first, Appendix A, covers some useful techniques in differential equations which are not usually covered in a basic course. Chief among these are differential inequalities, a key tool in applied dynamics. Here, we mean Kamke's comparison theorem for ODEs and the strong maximum principle for PDEs. The former result is proved, the latter is merely stated and references are given. Dynamical systems in biology typically deal with nonnegative quantities, and therefore one needs to establish that solutions that begin nonnegative, remain so in the future. Another essential tool for dealing with positivity and stability is the Perron-Frobenius theory which we state but do not prove. Finally, an elementary but powerful method which can sometimes establish persistence is the method of fluctuation. It provides the means to explicitly estimate the limit inferior and limit superior of bounded components of solutions of systems of ordinary and delay differential equations.

Appendix B introduces selected useful tools from functional analysis. Among them are compactness criteria in L^p spaces, inequalities for Volterra integral equations, proof of the equivalence of integral and mild solutions of linear differential equations in Banach spaces, and Fourier transform methods for integro-differential equations. The latter leads to conditions implying that any bounded solution of a class of integro-differential inequalities or equations vanishes identically, and this result may be used to establish global stability results. These tools are used in Chapter 9 and Chapter 12.

One should also disclose what is not in this book that a reader might expect given the title. One such omission is the notion of robust persistence, more precisely, the reasonable expectation that the notion of uniform persistence should be structurally stable to small changes in system dynamics in some topology. For example, if the dissipative system takes the Kolmogorov form $x_i' = x_i f_i(x)$ on $\mathbb{R}_+^n$, then small perturbations should mean small changes in the per capita growth rates f_i, say in the C^r-topology. Robust (C^r) ρ-persistence with $\rho(x) = \min_i x_i$ for this system would mean the existence of $\epsilon, \delta > 0$ such that $\liminf_{t\to\infty} x_i(t) > \epsilon$, $\forall i$ provided $x(t)$ satisfies $x_i' = x_i g_i(x)$ where $\|f - g\|_{C^r} < \delta$ and $x_i(0) > 0$, $\forall i$. Such results were first established by Schreiber [**193**]. See also Hirsch et al. [**101**]. We do not include these results since they are partly covered in the monograph of Zhao [**238**].

Finally, we have not included recent work on persistence for stochastic systems [**11, 105, 194**] or for skew-product semiflows [**169, 238**].

There is a huge body of literature on persistence theory, and this book does not span nearly all of it. We ask the forgiveness of our valued fellow scholars whose works we have failed to reference.

We would like to acknowledge the many students, especially Thanate Dhirasakdanon, who have contributed to this work through their questions, suggestions, and their homework solutions.

We thank our wives, Kathryn Smith and Adelheid Thieme, for their unwavering support, though this endeavor must have been shrouded in mystery for them.

As much as any other science, mathematics takes place in a tapestry of teachers, peers, and students; we gratefully dedicate this monograph to our Ph.D. advisors Willi Jäger (HRT) and Paul Waltman (HLS).

Hal Smith was supported in part by NSF Grant DMS-0918440.
Horst Thieme was supported in part by NSF Grant DMS-0715451.

Introduction

The temporal development of a natural or artificial system can conveniently be modeled by a semiflow. A semiflow consists of a state space, X, a time-set, J, and a map, Φ.

The state space X comprehends all possible states of the system: the amounts or densities of the system parts and, if there are one or several system structures, their structural distributions.

According to the interests of the authors, this book concentrates on biological, ecological, and epidemiological systems. For the last, for example, the state space typically contains the amounts or densities of susceptible and infective and possibly exposed and removed individuals. For spatial spread, spatial distributions are included in the state space. If age-structure is thought to be important, age-distributions are included as well.

Time can be considered as a continuum or in discrete units; the most common choices for the time set J are the nonnegative reals or the nonnegative integers, $\mathbb{R}_+ = [0, \infty)$ and $\mathbb{Z}_+ = \mathbb{N} \cup \{0\} = \{0, 1, \dots\}$. Depending on the model, the time unit can be a year, month, or day.

The most important ingredient of a semiflow is the semiflow map $\Phi : J \times X \to X$. Often Φ itself is called the semiflow. If $x \in X$ is the initial state of the system (at time 0), then $\Phi(t, x)$ is the state at time t. This interpretation immediately leads to the identity

$$\Phi(0, x) = x, \qquad x \in X.$$

Further, semiflows are characterized by the semiflow property:

$$\Phi(t + r, x) = \Phi(t, \Phi(r, x)), \qquad r, t \in J, \quad x \in X.$$

This property has the following interpretation: If x is the initial state and the system develops for a time r and if the state $\Phi(r, x)$ is taken as a new initial state and the system develops for another time t, then the resulting state is the same as if the system develops from x for the time $t + r$.

It may be that a system does not last for all future times. Then the semiflow map Φ is only defined on a subset of $J \times X$ and the semiflow property must be appropriately supplemented. If seasonal influences are important, nonautonomous semiflows need to be considered (Chapter 13).

Semiflows are induced by differential equations of all kind (ordinary, partial, functional, and combinations of these): in the case of an ordinary differential equation, $\Phi(t, x)$ is the solution at time t when x is the initial datum (at time 0).

To be more concrete, consider the following endemic model for a fertility reducing infectious disease caused by a viral, bacterial, or fungal parasite. S and I denote the respective numbers of susceptible and infective hosts,

$$S' = (\beta - \mu)S + q\beta I - \kappa SI, \qquad I' = \kappa SI - (\mu + \alpha)I. \tag{0.1}$$

Here $\beta > \mu > 0$ are the per capita birth and death rates, $\kappa > 0$ is the per capita infection rate and $\alpha \geq 0$ the additional death rate due to the disease. The factor $q \in [0, 1]$ expresses the reduction of fertility for an infective individual.

It follows from standard arguments in ordinary differential equations that, for each pair $S_0, I_0 \geq 0$, there exist unique solutions $S, I : \mathbb{R}_+ \to \mathbb{R}_+$ with $S(0) = S_0$, $I(0) = I_0$, where $\mathbb{R}_+ = [0, \infty)$ is the set of nonnegative real numbers. Then

$$\Phi(t, (S_0, I_0)) = (S(t), I(t))$$

defines a semiflow with state space $\mathbb{R}^2_+$ and time-set $\mathbb{R}_+$. The semiflow property follows from the uniqueness of solutions.

It is an important question whether the dynamical system persists (remains safely away from extinction) as a whole or at least in parts (which parts?). This question can be mathematically formulated and addressed by using a *persistence function*

$$\rho : X \to \mathbb{R}_+. \tag{0.2}$$

For $x \in X$, $\rho(x)$ is the amount of the part of the system that is of particular interest. For the model of a fertility-reducing infectious disease, if emphasis is on whether the disease becomes endemic or can be eradicated, $\rho(S, I) = I$ is the number or density of infective (or infected) individuals. If emphasis is on whether the disease threatens to drive the host population into extinction, then $\rho(S, I) = S + I$ is the total number of hosts.

The semiflow Φ is called *uniformly ρ-persistent* if there exists some $\epsilon > 0$ such that

$$(0.3) \qquad \liminf_{t\to\infty} \rho(\Phi(t,x)) \geq \epsilon \quad \text{whenever } x \in X, \rho(x) > 0.$$

A stepping-stone to uniform persistence is uniform weak persistence: Φ is *uniformly weakly ρ-persistent* if (0.3) holds with lim sup replacing lim inf. If Φ is uniformly ρ-persistent, the amount of the system part of interest is eventually bounded away from 0 with the bound being independent of the initial state (as long as the interesting part is present initially). If Φ is only uniformly weakly ρ-persistent, then the amount can come arbitrarily close to 0 but always bounces back.

In terms of the infectious disease, we talk about uniform (weak) host persistence if $\rho(S,I) = S + I$ and about uniform (weak) parasite persistence (or disease endemicity) if $\rho(S,I) = I$.

This book embarks on the strategy of establishing uniform weak persistence first and then deriving uniform persistence.

From uniform weak to uniform persistence. Trivially, uniform persistence implies uniform weak persistence; the converse is not always true. Let us return to the example of the fertility-reducing disease. If $q = 0$ (i.e., the disease sterilizes), the host-parasite system becomes a special case of the Lotka-Volterra prey-predator system for which the whole first quadrant is filled with periodic orbits. This means that the system is uniformly weakly persistent for both host and parasite, but uniformly persistent for neither.

Strangely enough, the lack of uniform persistence concurs with a lack of the solutions to be eventually uniformly bounded: there is no $c > 0$ such that $\limsup_{t\to\infty}(S(t) + I(t)) < c$ for all nonnegative solutions.

To continue this line of thought in more generality, let us assume that the state space X is a metric space (though persistence theory is also possible without a topology on X).

The preliminary insight (which will turn out to be not completely correct) that some boundedness is needed for proceeding from uniform weak to uniform persistence can now be formulated in various ways. The strongest such formulation assumes the existence of a compact global attractor [**91**], and it becomes a natural question under which conditions a compact global attractor exists (Chapter 2). The presentation in this book differs from others [**91, 197, 238**] as an approach is chosen that deals with continuous and discrete time in a unified way.

Beyond that, two opposite directions are pursued: on the one hand, harvest the full fruit of assuming a compact global attractor (persistence à la Caesar) and, on the other hand, relax the assumption of a compact attractor as much as possible (persistence à la Münchhausen, persistence via Arzela-Ascoli, and persistence via Laplace transform).

Persistence à la Caesar (*Attractor est omnis divisus in partes tres*, cf. De bello gallico) divides the compact global attractor, A, in three invariant parts: the extinction attractor on which the persistence functional ρ is zero, the persistence attractor, A_1, on which ρ is strictly positive, and a set of orbits connecting the extinction attractor to the persistence attractor. Every compact set on which ρ is strictly positive has a neighborhood U that is attracted by A_1:

$$d(\Phi(t,x), A_1) \to 0, \quad t \to \infty, \qquad \text{uniformly for } x \in U.$$

Here $d(y, A_1)$ is the distance from the point y to the set A_1. This convergence result implies that A_1 is stable.

In several examples (Chapter 8.7 and Chapter 9), we will find conditions under which the persistence attractor is a singleton set and thus, automatically, a locally asymptotically stable equilibrium. The techniques used for this involve Lyapunov type functionals [**163**] and Fourier transforms [**151**].

Persistence à la Münchhausen (Chapter 4.5) tries to get away with as few compactness assumptions as possible. It is called that way as it has some resemblance to the feat of Lügenbaron (lying baron) Karl Friedrich Hieronymus Freiherr von Münchhausen (1720-1797) who, in one of the tales told about him, escapes from a swamp lifting himself and his horse up pulling at his own hair [**17**] (see the drawing by Theodor Hosemann (1807-1875), Figure 4.1). While compactifying properties of the semiflow cannot be completely eliminated as assumptions (recall the Volterra predator-prey model), one can get quite far in this direction, though, at the expense of a considerable increase in technicality. For illustration, let us return to the model (0.1) for a fertility-reducing infectious disease.

Recall that, for $q = 0$ (sterilizing disease), this is a Lotka-Volterra predator-prey model which is known to be uniformly weakly persistent, but not uniformly persistent.

If $q \in (0,1]$, one can show that both the susceptible and the infective part of the population persist uniformly, i.e., both the host and the parasite persist uniformly, though their numbers can grow without bound for certain parameter values (if $q\beta > \mu + \alpha$).

The reason for the different persistence scenarios becomes a little clearer when we reformulate the equations in terms of the total host population size $N = S + I$ and the fraction of infective hosts $y = I/N$,

$$\begin{aligned} N' &= N\Big(\beta(1-y) - \mu + (q\beta - \alpha)y\Big), \\ y' &= y\Big((\kappa N - \alpha - \beta)(1-y) - q\beta y\Big). \end{aligned} \tag{0.4}$$

Note the change in state space that becomes $X = \mathbb{R}_+ \times [0,1]$. Whatever q, as long as $0 \le q \le 1$, this system has the invariant set $\{(N,0); N > 0\}$ on which host population grows exponentially. This lack of boundedness does not necessarily impede uniform host or parasite persistence. Notice that

$$N' \ge N(q\beta - \mu - \alpha).$$

Assume that $q\beta - \mu - \alpha > 0$. Then $N(t)$ grows exponentially if $N(0) > 0$. This has the consequence that $y(t) \to 1$ as $t \to \infty$ if $0 < y(0) \le 1$, i.e., the disease pervades the host population, and both host and disease persistence are uniform in the strongest possible way.

For $q = 0$, the case of a sterilizing disease, the state space has another invariant set, $\{(N,1); N \ge 0\}$, on which $N(t) \to 0$ as $t \to \infty$. This set has an attracting part where N is large and a repelling part where N is small.

Uniform weak host persistence can be shown whether or not $q = 0$. However, if $q > 0$, it can be shown on the state space

$$X = \{(N,y); N > 0, 0 \le y \le 1\},$$

while for $q = 0$ it can only be shown on the state space

$$\tilde{X} = \{(N,y); N > 0, 0 \le y < 1\}.$$

See Theorem 3.3 for details. For $q = 0$, it is not the lack of boundedness in the host component that is an impediment for showing uniform host persistence, but the loss of completeness of the state space by the necessary exclusion of $y = 1$. So uniform host persistence only holds for $q > 0$ (Theorem 4.14).

Using uniform host persistence, uniform persistence of the disease can now be established, but it requires the full Baron von Münchhausen hair-pulling stunt which we cannot explain here (see Theorem 4.17 and the subsequent application).

Interestingly enough, if $0 < q\beta < \mu + \alpha$, the ideas of ρ-persistence can be used to show that the host population size is eventually uniformly bounded: there is some $c > 0$ such that $\limsup_{t\to\infty} N(t) < c$ for all solutions N and y with $y(0) > 0$. Simply use $\rho(N,y) = \frac{1}{1+N}$ as persistence function. But uniform disease persistence must be established first before it can be established that the disease imposes a bound on the host population size that is eventually uniform (Exercise 4.9).

Since our ODE model is two-dimensional, one can alternatively use phase-plane methods. This way, for $0 < q\beta < \mu + \alpha$, one can directly (without using persistence theory) obtain a compact host and disease persistence attractor that attracts all compact sets in $\{(N,y); N > 0, 0 < y \le 1\}$.

The use of persistence theory becomes unavoidable, however, if one considers several competing parasite strains which provide complete cross-protection [**218**] or several stages of infection (Section 4.7).

The existence of an attractor with host and disease persistence concurs with the existence of an equilibrium in $(0, \infty) \times (0, 1)$ which is called either an *endemic equilibrium* or a *host-parasite-coexistence equilibrium* depending on whether the view point is mainly epidemiological or ecologic. By the Poincaré-Bendixson limit set trichotomy, any solution in the host and disease persistence attractor is either this equilibrium or a periodic orbit or a homoclinic orbit connecting the equilibrium to itself. The Bendixson-Dulac criterion (use the Dulac function $\frac{1}{Ny(1-y)}$) rules the second and third possibility out, and the host and parasite persistence attractor is the singleton set consisting of the endemic equilibrium. Notice that the stability of the endemic equilibrium follows without a linearized stability analysis. Of course, the latter would be easy for a small system like this, but could be quite harrowing for a large system.

If the state space is infinite dimensional, a compact attractor can be elusive for other reasons than the lack of eventual uniform boundedness or completeness as in the case of a model for cells with age-dependent division growing in a chemostat (Chapters 12 and 13.13.4). In this case, the Arzela-Ascoli theorem can come to the rescue and imply that, for certain sequences (x_j) that may have no convergent subsequences, $\rho(\Phi(t, x_j))$ has a subsequence that converges as $j \to \infty$ uniformly for t in compact subsets of $\mathbb{R}_+$. Notice that this remedy relies on the persistence function ρ and even works without a topology on the state space. It extends to persistence for nonautonomous semiflows (nonlinear evolutionary systems) (Chapter 13).

Another line of keeping assumptions at minimum concerns the continuity of the semiflow if time is a continuous variable. In certain models of physiologically structured populations, it is either necessary or convenient to choose the space of measures as a state space: either the solutions become measure-valued even for smooth-initial data by some mild form of shock-formation [**3, 213**], or there are equilibria to be taken care of that are measures [**2, 227**]. For a state space of measures, the semiflow is typically continuous in the space variable but sometimes not in the time variable (at least not in the same, the strong, topology). To illustrate this phenomenon and various ways of overcoming the associated difficulties, we consider the endemic model with variable infectivity also for infection-age distributions that are measures (Section 9.9.10).

How to get uniform weak persistence. Historically, there have been two main approaches to persistence theory. The first considers the semiflow on the "extinction boundary" of the state space, which can be facilitated by

Morse decompositions or acyclic decompositions of its point attractor (Chapter 8). This approach again relies on compactness assumptions and has a topological character (repeller-attractor pairs, chain-recurrence). The second uses so-called average Lyapunov functions. The concept of ρ-persistence can reconcile both approaches by either letting $\rho(x)$ be the distance from x to the boundary of extinction or by identifying ρ with the average Lyapunov function (Chapter 15). It also applies to the persistence of nonautonomous semiflows [**216**] (Chapters 13 to 15).

By example, we also present various ad hoc methods for proving uniform weak persistence. Typically, they work by contradiction. One is the method of fluctuations [**102**], together with differential inequalities and the Perron-Frobenius theory of quasipositive matrices (Appendix A and Chapter 3); another is the use of the Laplace transform. The resurgence of this classical tool is not so surprising as many semiflows are generated by nonlinear perturbations of linear semiflows (alias operator semigroups, Chapter 10) for which the Laplace transform is a major tool of investigation [**5**]. The Laplace transform can be quite effective (Chapter 5.7, Chapter 9) and also works in cases where the existence of a compact attractor cannot be established (Chapter 12).

Chapter 1

Semiflows on Metric Spaces

Since metric spaces are natural state-spaces of semiflows, a brief introduction of metric spaces is given below, mainly to fix notation. Otherwise the reader is supposed to be familiar with metric spaces and continuity of functions as they are taught in an introductory graduate course in analysis.

1.1. Metric spaces

Definition 1.1. A *metric space* (X, d) is a nonempty set X together with a function $d : X \times X \to \mathbb{R}$ which satisfies the following axioms:

(1) $d(x, y) = d(y, x)$ for all $x, y \in X$. [symmetry]

(2) $d(x, z) \leq d(x, y) + d(y, z)$ for all $x, y, z \in X$. [triangle inequality]

(3) $d(x, x) = 0$ for all $x \in X$.

(4) $d(x, y) \neq 0$ for all $x, y \in X$ with $x \neq y$.

$d(x, y)$ measures the distance between x and y. d is called a *metric* on X. If (4) does not hold, d is called a *semimetric*.

Example 1.2 (trivial metric). Let X be an arbitrary nonempty set. Define

$$d(x, y) = \begin{cases} 0, & x = y, \\ 1, & x \neq y. \end{cases}$$

Then d is a metric on X, called the *trivial metric* or the *discrete metric*.

Proof. The only nontrivial property is the triangle inequality.

Case 1: $x = z$.

Then $d(x, z) = 0 = 0 + 0 \leq d(x, y) + d(y, x)$.

Case 2: $x \neq z$.

Then $x \neq y$ or $y \neq z$. So $d(x, y) + d(y, z) \geq 1 = d(x, z)$. □

This metric is useless for applications but great for finding counterexamples.

Proposition 1.3. *Let d be a semimetric on the nonempty set X. Then*

$$|d(x, y) - d(y, z)| \leq d(x, z) \qquad \forall x, y, z \in X.$$

In particular, $d(x, z) \geq 0$ for all $x, z \in X$.

Proof. Obviously, the first statement implies the second. Let $x, y, z \in X$. Then, by the triangle inequality,

$$d(x, y) \leq d(x, z) + d(y, z).$$

By subtraction,

$$d(x, y) - d(y, z) \leq d(x, z).$$

Again, by the triangle inequality,

$$d(y, z) \leq d(y, x) + d(x, z).$$

By subtraction and symmetry of the semimetric,

$$-(d(x, y) - d(y, z)) = d(y, z) - d(y, x) \leq d(x, y).$$

We combine the two inequalities and obtain the first statement. Recall that the absolute value $|\alpha|$ of a real number α is either α or $-\alpha$. □

Definition 1.4. (convergence) Let (X, d) be a metric space and (x_j) a sequence in X, $x \in X$. We say that (x_j) *converges to* x if, for any $\epsilon > 0$, there exists an $N \in \mathbb{N}$ such that

$$d(x_j, x) < \epsilon \qquad \forall j > N.$$

The point $x \in X$ to which the sequence (x_j) converges is called the *limit* of the sequence, and we write

$$x = \lim_{j \to \infty} x_j$$

or

$$x_j \to x, \quad j \to \infty.$$

Remark 1.5. A sequence (x_j) converges towards a point x in a metric space (X, d) if and only if $d(x_j, x) \to 0$ as $j \to \infty$ in $\mathbb{R}$ with the absolute value.

The following relation will be needed in the future.

Lemma 1.6. *Let X be a metric space and d its metric.*

If (x_n), (y_n) are sequences in X and $x_n \to x \in X$, $y_n \to y \in X$ as $n \to \infty$, then $d(x_n, y_n) \to d(x, y)$ as $n \to \infty$.

Proof. By Proposition 1.3,

$$\big|d(x,y) - d(y,z)\big| \leq d(x,z) \qquad \forall x, y, z \in X.$$

Let (x_n), (y_n) be sequences in X and $x, y \in X$. Then

$$\big|d(x_n, y_n) - d(x,y)\big| = \big|d(x_n, y_n) - d(y_n, x) + d(y_n, x) - d(x,y)\big|.$$

By the triangle inequality for the absolute value and the inequality above,

$$\begin{aligned} \big|d(x_m, y_n) - d(x,y)\big| &\leq \big|d(x_n, y_n) - d(y_n, x)\big| + \big|d(y_n, x) - d(x,y)\big| \\ &\leq d(x_n, x) + d(y_n, y). \end{aligned}$$

If $y_n \to y$ and $x_n \to x$, then $d(y_n, y) \to 0$ and $d(x_n, x) \to 0$, so $d(x_n, y_n) - d(x,y) \to 0$ as $n \to \infty$. □

Definition 1.7 (distance from a point to a set). Let $\emptyset \neq B \subset X$ and $x \in X$. Then the *distance from the point x to the set B* is defined as

$$d(x, B) := \inf\{d(x,y); y \in B\}.$$

Definition 1.8. The subset S of a metric space (X, d) is said to be *compact* if every infinite sequence (x_n) in S has a subsequence which has a limit in S.

In a topological Hausdorff space, this property is called *sequential compactness.*

Lemma 1.9. *Let $B \subset X$, $x \in X$, and let B be nonempty and compact. Then*

$$d(x, B) = \min\{d(x,y); y \in B\}.$$

Proof. By definition of the infimum, there exists a sequence (y_n) in B such that $d(x, y_n) \to d(x, B)$ as $n \to \infty$. Since B is compact, there exists a subsequence (y_{n_j}) of (y_n) and some $y \in B$ such that $y_{n_j} \to y$ as $j \to \infty$. By Lemma 1.6, $d(x, y_{n_j}) \to d(x, y)$ as $j \to \infty$; but also $d(x, y_{n_j}) \to d(x, B)$ as $j \to \infty$. Hence $d(x, B) = d(x, y)$ with $y \in B$. □

Lemma 1.10. *Let $\emptyset \neq B \subset X$. Then*

$$|d(x, B) - d(z, B)| \leq d(x, z) \qquad \forall x, z \in X.$$

Proof. Let $x, z \in X$ and $y \in B$. By the triangle inequality,

$$d(x, y) \leq d(x, z) + d(z, y).$$

By definition of the infimum,

$$d(x, B) \leq d(x, z) + d(z, y).$$

By subtraction

$$d(x, B) - d(x, z) \leq d(z, y).$$

Since $y \in B$ was arbitrary, this inequality holds for all $y \in B$ and so, by definition of the infimum,

$$d(x, B) - d(x, z) \leq d(z, B).$$

We reorganize,

$$d(x, B) - d(z, B) \leq d(x, z).$$

By symmetry, this inequality also holds, if x and z are interchanged, and the inequality follows. □

Definition 1.11. Let S be a subset of a metric space (X, d).

- A point $x \in X$ is called a *limit point* of S if there is a sequence $(x_n)_{n=1}^{\infty}$ in S such that $x_n \to x$ as $n \to \infty$.
  Notice that every $x \in S$ is a limit point of S.
- A point $x \in X$ is called an *accumulation point* of S, if x is a limit point of S and the sequence (x_n) can be chosen such that $x_n \neq x$ for all $n \in \mathbb{N}$.

Definition 1.12. Let S be the subset of a metric space (X, d).

- The *closure* of S, denoted by $\bar{S}$, is the set of all limit points of S.
  $\bar{S}$ is the union of S and the set of accumulation points of S. Since every $x \in S$ is a limit point of S, $S \subset \bar{S}$.
- S is called *closed* if $S = \bar{S}$, i.e., if S contains all its limit points.

Definition 1.13. A subset S of a metric space X is called *precompact* if its closure $\bar{S}$ is compact.

The *open neighborhood* of x, or open ball centered at x, with radius r is defined as

$$B_r(x) = \{y \in X; d(y, x) < r\} = U_r(x), \tag{1.1}$$

while the *closed neighborhood (or ball)* is defined as

$$\bar{B}_r(x) = \{y \in X; d(y, x) \leq r\} = \bar{U}_r(x). \tag{1.2}$$

Definition 1.14. Let S be the subset of a metric space X. A point $x \in S$ is called an *interior point* of S if there exists some $\epsilon > 0$ such that $B_\epsilon(x) \subset S$. The set of interior points of S is called the *interior* of S and is denoted by $\breve{S}$ or Int S.

If $x \in X$ and $S \subset X$, then S is called a *neighborhood of* x if x is an interior point of S.

Definition 1.15. Let S be the subset of a metric space X with metric d. S is called *open* if $S = \breve{S}$, i.e., every point in S is an interior point of S.

Warning: While the closed ball is the closure of the open ball and the open ball is the interior of the closed ball in a normed vector space, this is not necessarily true in an arbitrary metric space.

Many metric spaces are subsets of normed vectors spaces over a field $\mathbb{K}$, $\mathbb{K} = \mathbb{R}$ or $\mathbb{K} = \mathbb{C}$.

Definition 1.16. Let X be a vector space over $\mathbb{K}$. A function $\xi : X \to \mathbb{R}$ is called a *seminorm* on X, if the following hold:

(1) $\xi(x + y) \leq \xi(x) + \xi(y)$ for all $x, y \in X$. [triangle inequality]

(2) $\xi(\alpha x) = |\alpha|\xi(x)$ for all $x \in X$, $\alpha \in \mathbb{K}$. [homogeneity]

ξ is called a *norm* on X if, in addition,

(3) $\xi(x) \neq 0$ if $x \in X$, $x \neq 0$.

In this case, we often write $\|x\|$ instead of $\xi(x)$ and call $(X, \|\cdot\|)$ a normed vector space. The norm induces a metric on X by $d(x, y) = \|x - y\|$. If the vector space is finite-dimensional, we will sometimes use $|\cdot|$ for the norm.

Remark 1.17. (a) Let (X, d) be a metric space, $x, y \in X$, $r > 0$. Then $d(y, \bar{B}_r(x)) \geq d(y, x) - r$.

(b) Let $(V, \|\cdot\|)$ be a normed vector space and d the metric induced by $\|\cdot\|$. Then $d(y, B_r(x)) = d(y, \bar{B}_r(x)) = d(y, x) - r$ for all $x \in X$, $r > 0$ and $y \in X \setminus B_r(x)$.

(c) Give an example of a metric space where $d(y, \bar{B}_r(x)) > d(y, x) - r$ for appropriate $x \in X$, $r > 0$, $y \in X \setminus \bar{B}_r(x)$.

Definition 1.18 (diameter, bounded set)**.** The *diameter* of a subset of the metric space X is defined as

$$\text{diam } B = \sup\{d(x, y); x, y \in B\}. \tag{1.3}$$

A subset of X is called *bounded* if it has a finite diameter.

Definition 1.19 (distance from one set to another)**.** Let A, B be nonempty subsets of X. We define the *distance from the set A to the set B* as

$$d(A, B) = \sup\{d(x, B); x \in A\}.$$

In this generality, it is possible that $d(A,B) = \infty$.

Lemma 1.20. *Let A, B be nonempty subsets of X.*

(a) *If A is bounded, $d(A,B) < \infty$.*

(b) *If A is compact,*

$$d(A,B) = \max\{d(x,B); x \in A\}.$$

(c) *If A and B are compact, then there exist $x \in A$ and $y \in B$ such that $d(A,B) = d(x,y)$.*

Proof. (a) Let A be bounded. By Lemma 1.10, for all $x, y \in A$,

$$d(x,B) - d(y,B) \le d(x,y) \le \text{ diam } A.$$

We reorganize,

$$d(x,B) \le \text{ diam } A + d(y,B) \qquad \forall x, y \in A.$$

By definition of the supremum,

$$d(A,B) \le \text{ diam } A + d(y,B) < \infty.$$

(b) By definition of the supremum, there exists a sequence (x_n) in A such that $d(x_n, B) \to d(A,B)$ as $n \to \infty$. Since A is compact, after choosing a subsequence, $x_n \to x$ as $n \to \infty$ for some $x \in A$. By Lemma 1.10,

$$|d(x_n,B) - d(x,B)| \le d(x_n,x) \to 0, \qquad n \to \infty,$$

so

$$d(A,B) = \lim_{n\to\infty} d(x_n,B) = d(x,B) < \infty.$$

(c) If both A and B are compact, by part (b) and Lemma 1.9, it follows that $d(A,B) = d(x,y)$ for some $y \in B$. □

Let B be a subset of a metric space X and $\epsilon > 0$. The ϵ-neighborhood of B is defined as the set of points whose distance to B is less than ϵ,

$$U_\epsilon(B) := \{x \in X; d(x,B) < \epsilon\}. \tag{1.4}$$

Notice that

$$U_\epsilon(B) = \{x \in X; \exists b \in B \text{ such that } d(x,b) < \epsilon\} = \bigcup_{x\in B} B_\epsilon(x). \tag{1.5}$$

The letter U traditionally reminds us of the German word "Umgebung" which means "neighborhood".

Remark 1.21. In general, $d(A,B) \neq d(B,A)$. Choose A and B such that $A \subset \bar{B}$, but $B \not\subset \bar{A}$. Then $d(A,B) = 0$, but $d(B,A) > 0$.

Lemma 1.22. *Let A, B, C be nonempty subsets of X. Then*

$$d(A,B) \le d(A,C) + d(C,B).$$

Proof. Let $x \in A$ and $z \in C$. By Lemma 1.10,

$$d(x, B) \leq d(z, B) + d(x, z).$$

By definition of the supremum,

$$d(x, B) \leq d(C, B) + d(x, z).$$

If $d(C, B) = \infty$, the desired inequality is automatically satisfied. So we assume that $d(C, B) < \infty$. By subtraction,

$$d(x, B) - d(C, B) \leq d(x, z).$$

Since $z \in C$ is arbitrary, this inequality holds for all $z \in C$. By definition of the infimum,

$$d(x, B) - d(C, B) \leq d(x, C).$$

By definition of the supremum,

$$d(x, B) - d(C, B) \leq d(A, C).$$

Hence,

$$d(x, B) \leq d(A, C) + d(C, B).$$

Since $x \in A$ is arbitrary, this inequality holds for all $x \in A$. By definition of the supremum,

$$d(A, B) \leq d(A, C) + d(C, B). \qquad \square$$

Let A and B be nonempty compact subsets of X. The *Hausdorff distance* between A and B is defined by

$$d_H(A, B) := \max\left(d(A, B), d(B, A)\right).$$

Remark 1.23. The reader is asked to prove the following in Exercise 1.9:

(a) The Hausdorff distance defines a metric on the set $K(X)$ consisting of all nonempty compact subsets of X.

(b) Let $\epsilon > 0$ and $A, B \in K(X)$. Then $d_H(A, B) < \epsilon \Leftrightarrow A \subset U_\epsilon(B)$ and $B \subset U_\epsilon(A)$.

(c) Let $\{A_n\}$ be a sequence in $K(X)$ where X is a compact metric space and $B \in K(X)$. Then $A_n \to B$ in $K(X)$ if and only if $B = C$ where C consists of all points $c \in X$ that are subsequential limit points of some sequence $\{a_n\}_n$ with $a_n \in A_n$.

The following result can be found in [**10**] (see Chapt. II, Theorem 7.1 and Exercise 7.6) .

Proposition 1.24. *If X is a compact metric space, then $K(X)$ with d_H is a compact metric space.*

Remark 1.25. Let B be a subset of a metric space X and $\epsilon > 0$. Recall that the ϵ-neighborhood of B is defined as the set

$$U_\epsilon(B) := \{x \in X; d(x, B) < \epsilon\}.$$

The following hold (see Exercise 1.8):

(a) $U_\epsilon(B)$ is an open set.

(b) If C is a compact subset of X and D a closed subset of X with $C \cap D = \emptyset$, then there exists some $\epsilon > 0$ such that $U_\epsilon(C) \cap D = \emptyset$.

For checking that a set is compact (in Proposition 1.24, e.g.) the following concept can be useful. Proof of the upcoming results can be found in many analysis books ([**164**, Chap. 7.9], e.g.).

Definition 1.26. Let S be the subset of a metric space X with metric d. S is called *totally bounded* if, for each $\epsilon > 0$, there are finitely many points $x_1, \ldots, x_n \in X$ such that

$$S \subset \bigcup_{j=1}^{n} B_\epsilon(x_j).$$

In other words, for each $x \in S$, there exists some $j \in \{1, \ldots, n\}$ such that $d(x, x_j) < \epsilon$. The set $\{x_1, \ldots, x_n\}$ is called an *ϵ-net.*

Remark 1.27. (a) It is immediate that every subset of a totally bounded set is totally bounded itself.

(b) The ϵ-net in Definition 1.26 can be chosen as a subset of S.

Proposition 1.28. *Let S be the subset of a metric space (X, d). If S is compact, then S is totally bounded.*

To obtain a converse we need the following concepts.

Definition 1.29. A sequence (x_j) in a metric space (X, d) is called *a Cauchy sequence* (or fundamental) if, for every $\epsilon > 0$, there exists some $N \in \mathbb{N}$ such that

$$d(x_j, x_k) < \epsilon \qquad \forall j, k > N.$$

Definition 1.30. A subset S of metric space (X, d) is *complete* if every Cauchy sequence in S has a limit in S. A normed vector space that is complete under the metric induced by the norm is called a *Banach space.*

Theorem 1.31. *Let S be the subset of a metric space (X, d). Then S is compact if and only if S is complete and S is a totally bounded subset of X.*

1.2. Semiflows

In order to have a unified approach to continuous and discrete semiflows, we introduce the concept of a time-set.

Definition 1.32. A subset $J \subset [0, \infty)$ is called *a time-set* if it has the following properties:

(1) $0 \in J$ and $1 \in J$.

(2) If $s, t \in J$, then $s + t \in J$.

(3) If $s, t \in J$, and $s < t$, then $t - s \in J$.

A time-set J is called a *closed time-set* if J is a closed subset of $[0, \infty)$.

It is not hard to see that J is a time-set if and only if $J \cup (-J)$ is a subgroup of $(\mathbb{R}, +)$ containing $\mathbb{Z}$ and that, if J is closed, then either $J = [0, \infty)$ or $J = \{mh; m \in \mathbb{Z}_+\}$ where $1 = nh$ for some natural number n (Proposition 4.1).

The two standard examples of time-set are $\mathbb{R}_+ = [0, \infty)$ itself and the nonnegative integers, $\mathbb{Z}_+$. Obviously, every time-set contains $\mathbb{Z}_+$. It also contains $mJ = \{mt; t \in J\}$ for every $m \in \mathbb{N}$.

Definition 1.33. Let J be a time-set and X be a nonempty set (the "state space"). A map $\Phi : J \times X \to X$ is called a *(global autonomous) semiflow* if:

(1) $\Phi(0, x) = x$ for all $x \in X$.

(2) $\Phi(t + s, x) = \Phi(t, \Phi(s, x))$ for all $t, s \in J$ and $x \in X$.

For a map $\Phi : J \times X \to X$, $\Phi(t, \cdot)$ is a map from X to X which is denoted by Φ_t,

$$\Phi_t(x) = \Phi(t, x).$$

The semiflow property in (2) can then be written as $\Phi_t \circ \Phi_s = \Phi_{t+s}$ and (1) takes the form $\Phi_0 =$ identity map.

Let X be a metric space and d the metric. A map $\Phi : J \times X \to X$ is called *state-continuous*, if all maps $\Phi_t : X \to X$, $t \in J$, are continuous.

Φ is called *state-continuous, uniformly in finite time,* if for every $x \in X$, $t \in J$, $\epsilon > 0$ there exists exists some $\delta > 0$ such that $d(\Phi(s, y), \Phi(s, x)) < \epsilon$ whenever $s \in [0, t] \cap J$, $d(y, x) < \delta$.

Φ is called *time-continuous*, if for every $x \in X$, $\Phi(t, x)$ is a continuous function of $t \geq 0$.

Φ is called *continuous*, if Φ is continuous as a map from $J \times X$ to X with $J \times X$ carrying the product topology.

Φ is called a *discrete semiflow*, if $J = \mathbb{Z}_+$.

Φ is called an *injective semiflow* if all maps Φ_t, $t \in J$, are one-to-one.

$\Phi : \hat{J} \times X \to X$ is called a *flow*, if $\hat{J}$ is a subgroup of $(\mathbb{R}, +)$ containing $\mathbb{Z}$ and $\Phi(0, x) = x$ for all $x \in X$ and $\Phi(t + s, x) = \Phi(t, \Phi(s, x))$ for all $t, s \in \hat{J}$, $x \in X$.

The notions of state-continuity, state-continuity uniform in finite time, continuity, and discreteness are defined for flows in analogy. With obvious modifications they also apply to maps from $J \times X$ into a metric space Y.

Lemma 1.34. *Let the time-set J be closed.*

(a) *Every continuous semiflow is state-continuous, uniformly in finite time.*

(b) *Let $J = \mathbb{Z}_+$ and $\Phi : J \times X \to X$ be a discrete semiflow such that Φ_1 is continuous. Then Φ is a continuous semiflow.*

(c) *Every semiflow that is state-continuous, uniformly in finite time, and time-continuous is continuous.*

Proof. (a) Let $\Phi : J \times X \to X$ be a continuous semiflow. Suppose the claim is not true. Then there are $t \in J$, $x \in X$, and $\varepsilon > 0$ such that for each $n \in \mathbb{N}$, there are $s_n \in J \cap [0, t]$ and $y_n \in X$ with $d\big(\Phi(s_n, y_n), \Phi(s_n, x)\big) \geq \varepsilon$ and $d(y_n, x) < \frac{1}{n}$. Since J is closed, $J \cap [0, t]$ is compact and so there is a subsequence $\{s_{n_k}\}$ of $\{s_n\}$ and $s \in J \cap [0, t]$ such that $s_{n_k} \to s$ as $k \to \infty$. We also have $y_{n_k} \to x$ as $k \to \infty$. Since Φ is continuous, we have $d\big(\Phi(s_{n_k}, y_{n_k}), \Phi(s, x)\big) \to 0$ as $k \to \infty$. Hence we have

$$
\begin{aligned}
d\big(\Phi(s_{n_k}, y_{n_k}), \Phi(s_{n_k}, x)\big) \leq\ & d\big(\Phi(s_{n_k}, y_{n_k}), \Phi(s, x)\big) \\
& + d\big(\Phi(s, x), \Phi(s_{n_k}, x)\big) \to 0 \text{ as } k \to \infty,
\end{aligned}
$$

which contradicts $d\big(\Phi(s_n, y_n), \Phi(s_n, x)\big) \geq \varepsilon$ for all $n \in \mathbb{N}$.

(b) Since $J = \mathbb{Z}_+$, Φ is time-continuous. To show that Φ is continuous, by part (c) it suffices to show that Φ is state-continuous. Now, $x \mapsto \Phi(0, x) = x$ is continuous and for each $m \in \mathbb{N}$,

$$
x \mapsto \Phi(m, x) = \underbrace{\Phi_1 \circ \cdots \circ \Phi_1}_{m \text{ times}}(x)
$$

is continuous. Hence, the claim follows.

(c) Let $\{(t_n, x_n)\}_{n \in \mathbb{N}} \subset J \times X$ be a sequence converging to $(t, x) \in J \times X$. Let $T \in J$ be such that $t_n \leq T$ for all $n \in \mathbb{N}$. Let $\varepsilon > 0$. Since Φ is state-continuous uniformly in $J \cap [0, T]$, there is $N_1 \in \mathbb{N}$ such that $d\big(\Phi(t_n, x_n), \Phi(t_n, x)\big) < \varepsilon/2$ for all $n \geq N_1$. Since Φ is time-continuous, there is $N_2 \geq N_1$ such that $d\big(\Phi(t_n, x), \Phi(t, x)\big) < \varepsilon/2$ for all $n \geq N_2$. Hence, for all $n \geq N_2$, we have $d\big(\Phi(t_n, x_n), \Phi(t, x)\big) \leq d\big(\Phi(t_n, x_n), \Phi(t_n, x)\big) + d\big(\Phi(t_n, x), \Phi(t, x)\big) < \varepsilon$. Therefore, $\Phi(t_n, x_n) \to \Phi(t, x)$ as $n \to \infty$. Since this is true for any convergent sequence in $J \times X$ and $J \times X$ is a metric space, Φ is a continuous semiflow. □

Remark 1.35. (a) Global semiflows induced by autonomous ordinary differential equations are continuous.

Here $\Phi(t, x_0) = x(t)$ where $x' = f(x)$ and $x(0) = x_0$ and f is a locally Lipschitz continuous vector field. The semiflow property follows from the uniqueness of solutions. Φ is obviously time-continuous. State-continuity, uniformly in finite time, follows from the continuous dependence of solutions on initial data.

(b) If Φ is a discrete semiflow,

$$\Phi_m = \Phi_1^m = \underbrace{\Phi_1 \circ \cdots \circ \Phi_1}_{m \text{ times}} \qquad \forall m \in \mathbb{N}.$$

Discrete semiflows are associated with periodic ordinary differential equations, difference equations, and metered models.

(c) J is a time-set if and only if $\hat{J} = (-J) \cup J$ is a subgroup of $(\mathbb{R}, +)$ which contains $\mathbb{Z}$.

1.3. Invariant sets

Throughout this section, let J be a time-set and let $\Phi : J \times X \to X$ be a (global autonomous) semiflow.

Definition 1.36. Let K be a nonempty subset of X.

- K is called *forward invariant (under Φ)*, if $\Phi_t(K) \subset K$ for all $t \in J$.
- K is called *backward invariant (under Φ)*, if $\Phi_t(K) \supseteq K$ for all $t \in J$.
- K is called *invariant (under Φ)* if K is both forward and backward invariant, i.e., if $\Phi_t(K) = K$ for all $t \in J$.
- A subset K of X is called a *minimal set*, if K is a nonempty, closed, invariant set and contains no set with these three properties other than itself.

Remark 1.37. (a) If K is forward invariant, then $\Phi_t(K)$ is forward invariant for all $t \in J$ because

$$\Phi_s(\Phi_t(K)) = \Phi_{t+s}(K) = \Phi_t(\Phi_s(K)) \subset \Phi_t(K) \qquad \forall s, t \in J.$$

(b) If the semiflow is induced by an ordinary differential equation $x' = f(x)$ with forward uniqueness of solutions and continuous f, then K is forward invariant if and only if all solutions that start in K stay in K thereafter, $x(0) \in K$ implies $x(t) \in K$ for all $t \geq 0$. If $K \subset X \subset \mathbb{R}^n$ is a closed subset of $\mathbb{R}^n$, K is invariant if and only if it satisfies the subtangential condition

$$\frac{1}{h} d(x + hf(x), K) \to 0, \qquad h \searrow 0, x \in K. \tag{1.6}$$

See Theorem 3.1, Chapter VI of [**159**] and Remark 2.3 or Theorem 2.2.28 in [**1**].

Theorem 1.38. *Let Φ be state-continuous and let K be a compact forward invariant subset of X. Then K contains a compact minimal set K_0.*

Proof. Let $\mathcal{K}$ be the collection of all nonempty subsets of K which are forward invariant and compact. $\mathcal{K}$ is a partially ordered set via the superset relation, $K_1 \leq K_2$ if $K_2 \subset K_1$. We apply Zorn's Lemma [**164**, Ch.1.2].

Let $\tilde{\mathcal{K}}$ be a totally ordered set (a chain) in $\mathcal{K}$. Set $\tilde{K} = \bigcap\{M; M \in \tilde{\mathcal{K}}\}$. Obviously, $\tilde{K}$ is an upper bound, provided it is nonempty. Since $\tilde{\mathcal{K}}$ is a totally ordered set, every finite intersection of sets in $\tilde{\mathcal{K}}$ is nonempty. Since K is compact, $\tilde{K}$ is nonempty. By Zorn's Lemma, $\mathcal{K}$ has a maximal element, K_0. Let $t > 0$. Then $\Phi_t(K_0)$ is a nonempty, forward invariant set contained in K_0. Since K_0 is compact, and Φ_t continuous, $\Phi_t(K_0)$ is compact. Since K_0 is maximal with respect to the superset relation, $\Phi_t(K_0) = K_0$. This shows that K_0 is invariant and K_0 is a minimal set. □

Invariant sets have the surprising property that each of their elements lies on a total trajectory. We make this precise:

Definition 1.39. A function $\phi : J \cup (-J) \to X$ is called a *total trajectory* (of Φ), if $\Phi(t, \phi(s)) = \phi(t + s)$ for all $t \in J$, $s \in J \cup (-J)$. If $t_0 \in J, x_0 \in X$ and $\phi(t_0) = x_0$, then ϕ is called a *total trajectory through* (t_0, x_0).

The *orbit* of a total trajectory ϕ is defined as its range, $\phi(J \cup (-J))$, and called a *total orbit.*

In general, there may be several total trajectories through (t_0, x_0). Total orbits are invariant.

If Φ is induced by the solutions of $x' = f(x)$, a total trajectory through (t_0, x_0) is a solution x on $\mathbb{R}$ with $x(t_0) = x_0$.

Observe that the following result involves no topology; Φ is not assumed to be continuous nor is K assumed to have special topological properties.

Theorem 1.40. *A set K is invariant if and only if, for every element $x_0 \in K$, there is a total trajectory through $(0, x_0)$ with values in K.*

Proof. "If": Let $x \in K$. Then there exists a total trajectory ϕ through $(0, x)$ with values in K. Let $t \in J$. Then $\Phi_t(x) = \phi(t) \in K$. Further,

$$x = \phi(0) = \phi(t - t) = \Phi_t(\phi(-t)) \in \Phi_t(K).$$

Since $x \in K$ is arbitrary, $\Phi_t(K) = K$.

"Only if": Let $x_0 \in K$. Since K is invariant, for every $y \in K$, the set $X_y = \{x \in K; \Phi(1, x) = y\}$ is nonempty. We successively choose elements x_{-n}, $n \in \mathbb{Z}_+$ such that $\Phi_1(x_{-n-1}) = x_{-n}$. (Rigorously, we may need the Axiom of Choice to do that). By construction and the semiflow property, $\Phi_k(x_{-m}) = x_{k-m}$ for $m, k \in \mathbb{Z}_+$ with $m \geq k$. Let $-t \in J$. Choose the largest nonnegative integer m such that $m \leq -t$. Since J is a time-set, there exists some $s \in J$ such that $-t = m + s$, $s < 1$, and $1 - s \in J$. Then $-t = m + 1 - (1 - s)$. This way we have shown: For any $t \in -J$, there exist $k \in \mathbb{Z}_+$ and $r \in J$ such that $t = -k + r$. This is also trivially true for $t \in J$ with $k = 0$ and $r = t$. We define

$$\phi(t) = \Phi_r(x_{-k}), \qquad t = -k + r, \quad t \in J \cup (-J), r \in J, k \in \mathbb{Z}_+.$$

Since K is forward invariant, $\phi(t) \in K$. We show that the definition is independent of the choice of k and r. Let $t = -k + r = -m + s$ with $k, m \in \mathbb{Z}_+$, $r, s \in J$. Let $-m < -k$. Then $s > r$ and $s - r = m - k \in J$. Now

$$\Phi_s(x_{-m}) = \Phi_r(\Phi_{s-r}(x_{-m})) = \Phi_r(\Phi_{m-k}(x_{-m})) = \Phi_r(x_{-k}).$$

To check that ϕ is a total trajectory, let $t \in J \cup (-J)$ and choose $k \in \mathbb{Z}_+$ and $r \in J$ such that $t = -k + r$. Then, for $s \in J$,

$$\Phi_s(\phi(t)) = \Phi_s(\Phi_r(x_{-k})) = \Phi_{s+r}(x_{-k}) = \phi(s + r - k) = \phi(s + t).$$

Further, since $0 = 0 + 0$, $\phi(0) = \Phi_0(x_0) = x_0$. Hence, ϕ is a total trajectory through $(0, x_0)$. □

The total trajectories in a compact invariant set form an equicontinuous family.

Proposition 1.41. *Let the semiflow $\Phi : J \times X \to X$ be continuous, J a closed time-set, and K a compact invariant subset of X. Then the total trajectories in K are equicontinuous, uniformly on $\mathbb{R}$. More precisely: For every $\epsilon > 0$, there exists $\delta > 0$ such that $d(\phi(t), \phi(s)) \leq \epsilon$ for all total trajectories ϕ in K and all $t, s \in \mathbb{R}$, $|t - s| \leq \delta$.*

Proof. Since Φ is continuous, Φ is uniformly continuous on the compact set $(J \cap [0, 1]) \times K$. Let $\epsilon > 0$. Choose $\delta \in (0, 1)$ such that $d(\Phi(s, x), x) \leq \epsilon$ for $s \in [0, \delta) \cap J$,$x \in K$. Now let ϕ be a total trajectory in K, $t, r \in J$, $|t - r| \leq \delta$. Without loss of generality, we can assume $r \leq t \leq r + \delta$. Then $d(\phi(t), \phi(r)) = d(\Phi(t - r, \phi(r)), \phi(r)) \leq \epsilon$ because $\phi(r) \in K$. □

The total trajectories in a compact invariant set form a compact family in the topology of uniform convergence on compact sets. We formulate this result as follows.

Proposition 1.42. *Let the semiflow $\Phi : J \times X \to X$ be continuous, J be a closed time-set, and K a compact invariant subset of X. Let (x_j) be a sequence in K converging to some element $x_\infty \in K$. Then, after choosing a subsequence, for all $j \in \mathbb{N} \cup \{\infty\}$, there exist total trajectories ϕ_j in K such that $\phi_j(0) = x_j$ and $\phi_j \to \phi_\infty$ as $j \to \infty$ uniformly on compact subsets of $\hat{J} = J \cup (-J)$.*

Proof. By Theorem 1.40, there exist total trajectories $\phi_j : \hat{J} \to K$ with $\phi_j(0) = x_j$ for all $j \in \mathbb{N}$. Since K is compact, the set $\{\phi_j(t); j \in \mathbb{N}\}$ is compact for each $t \in \hat{J}$. By Proposition 1.41, $\{\phi_j; j \in \mathbb{N}\}$ is an equicontinuous family in $C(\hat{J}, K)$. By the Arzela-Ascoli Theorem [**164**, Ch.8.3], there exists a subsequence of ϕ_j, which we again call (ϕ_j), and a function $\phi_\infty : \hat{J} \to K$ such that $\phi_j(t) \to \phi_\infty(t)$ as $j \to \infty$, uniformly on compact subsets of $\hat{J}$. For $t \in J$, $s \in \hat{J}$, we have

$$\phi_\infty(t+s) = \lim_{j\to\infty} \phi_j(t+s) = \lim_{j\to\infty} \Phi(t, \phi_j(s)) = \Phi(t, \phi_\infty(s)).$$

Hence, ϕ_∞ is a total trajectory. Obviously, $\phi_\infty(0) = x_\infty$. □

Theorem 1.43. *Let the time-set J be $\mathbb{Z}_+$ or $\mathbb{R}_+$, $\Phi : J \times X \to X$ be a semiflow and K be an invariant compact subset of X. Let $(\phi_k)_{k\in\mathbb{N}}$ be a sequence of total Φ-trajectories with range in K. Then, after choosing a subsequence, $(\phi_k(j))$ converges as $k \to \infty$ for every $j \in \mathbb{Z}$; if Φ is state-continuous, $(\phi_k(t))$ converges as $k \to \infty$ for every $t \in \hat{J}$; if Φ is state-continuous, uniformly in finite time, the convergence is uniform on compact subsets of $\hat{J}$.*

Proof. Since K is compact, we successively find subsequences $(\phi_k^\ell)_k$, $\ell \in \mathbb{N}$, of (ϕ_k) with the following properties:

$(\phi_k^{\ell+1})$ is a subsequence of (ϕ_k^ℓ).

$\phi_k^\ell(j)$ converges as $k \to \infty$ for $j = -\ell, \dots, 0, \dots, \ell$.

We define a new sequence,$\psi^k = \phi_k^k$ for $k \in \mathbb{N}$. Then (ψ_k) is a subsequence of (ϕ_k) and $\psi_k(j)$ converges as $k \to \infty$ for all $j \in \mathbb{Z}_+$. Now let $t \in \hat{J}$. Choose $j \in \mathbb{Z}$ such that $t \geq j$. If Φ is state-continuous, $\phi_k(t) = \Phi(t-j, \phi_k(j))$ converges as $k \to \infty$. If $\tilde{J}$ is a compact subset of $\hat{J}$, then choose $j \in \mathbb{Z}$ such that $t \geq j$ for all $t \in \tilde{J}$. By the same argument, if Φ is state-continuous, uniformly in finite time, the convergence is uniform for $t \in \tilde{J}$. □

Theorem 1.44. *Let $\Phi : J \times X \to X$ be a semiflow, K an invariant subset of X, and suppose that $\Phi_t|_K$ is injective for each $t > 0$ in J. Then $\Phi|_K$ can* *be extended to a flow $\hat{\Phi} : \hat{J} \times K \to K$ with $\hat{J} = J \cup (-J)$. If K is compact and Φ is continuous (state-continuous), so is $\hat{\Phi}$.*

Proof. Let $\hat{\Phi}_t$ be the restriction of Φ_t to K for $t \in J$. Since K is invariant and Φ is injective, $\hat{\Phi}_t$ is bijective and its inverse relation $\hat{\Phi}_t^{-1}$ is a bijective map. We set $\hat{\Phi}_{-t} = \hat{\Phi}_t^{-1}$, $t \in J$. Let $t, r \in J$. Then $\hat{\Phi}_{t+r} = \hat{\Phi}_t \circ \hat{\Phi}_r$. Taking inverses,

$$\hat{\Phi}_{-t-r} = \hat{\Phi}_{-r} \circ \hat{\Phi}_{-t}.$$

If $t - r \in J$,

$$\hat{\Phi}_{-r} \circ \hat{\Phi}_t = \hat{\Phi}_r^{-1} \circ \hat{\Phi}_r \circ \hat{\Phi}_{t-r} = \hat{\Phi}_{t-r}.$$

Similarly, $\hat{\Phi}_t \circ \hat{\Phi}_{-r} = \hat{\Phi}_{t-r}$. If $r - t \in J$, then

$$\hat{\Phi}_{r-t} = \hat{\Phi}_r \circ \hat{\Phi}_{-t} = \hat{\Phi}_{-t}\hat{\Phi}_r.$$

Taking inverses,

$$\hat{\Phi}_{t-r} = \hat{\Phi}_t \circ \Phi_{-r} = \hat{\Phi}_{-r} \circ \Phi_t.$$

Thus this formula holds for all $t, r \in J$. Taking all cases together, we have shown that $\hat{\Phi}_{t+r} = \hat{\Phi}_t \circ \hat{\Phi}_r$ for all $r, t \in \hat{J}$.

We now assume that K is compact and Φ is continuous. Assume that $\hat{\Phi}$ is not continuous. Then there exist $\epsilon > 0$, $t \in J$, $x \in X$ and sequences (t_k) in J and (x_k) in X such that $t_k \to t$, $x_k \to x$, $k \to \infty$, but $d(y_k, \hat{\Phi}(-t, x)) \geq \epsilon$ for all $k \in \mathbf{N}$, $y_k := \Phi(-t_k, x_k)$. Since K is compact, after choosing a subsequence, $y_k \to y$ as $k \to \infty$ and $y \neq \hat{\Phi}(-t, x)$. Since Φ is continuous, $x_k = \Phi(t_k, y_k) \to \Phi(t, y)$ as $k \to \infty$. So $x = \hat{\Phi}(t, y)$, a contradiction.

The proof that $\hat{\Phi}$ inherits state-continuity from Φ is similar. □

Theorem 1.45. *Every compact minimal set of a state-continuous semiflow is the closure of a total orbit.*

Proof. By Theorem 1.38, K contains the orbit of a total trajectory. One easily sees that the orbit of a total trajectory is invariant. By Exercise 1.12, its closure is also invariant. Since K is minimal, K equals this closure. □

The converse is not true in general. There exist compact closures of total orbits which are not minimal sets, e.g. an orbit connecting two equilibria.

We introduce some special total orbits which are compact minimal sets.

Definition 1.46. A point $x^* \in X$ is called a *fixed point* or an equilibrium (point) of the semiflow Φ, if $\Phi_t(x^*) = x^*$ for all $t \in J$.

A total trajectory ϕ is called *periodic* with period $p \in J$ if $\phi(t+p) = \phi(t)$ for all $t \in J \cup (-J)$ and $\phi(t)$ is not an equilibrium for all $t \in J \cup (-J)$. A subset K is called a *periodic orbit* if it is the orbit of a periodic trajectory.

Obviously, if x^* is an equilibrium, $\phi(t) = x^*$ for all $t \in J$ defines a constant total trajectory. Equilibria and periodic orbits are compact minimal sets.

Definition 1.47. Let $\Phi : \mathbb{R}_+ \times X \to X$ be a semiflow. Define the *continuity space* of Φ as

$$X_\circ = \{x \in X; \Phi(t,x) \to x, t \to 0\}.$$

Proposition 1.48. *Let $\Phi : \mathbb{R}_+ \times X \to X$ be a semiflow which is state-continuous, uniformly in finite time. Then:*

(a) *$X_\circ$ is a closed subset of X which is forward invariant under Φ, and*

(b) *the restriction of Φ to $\mathbb{R}_+ \times X_\circ$ is a continuous semiflow from $\mathbb{R}_+ \times X_\circ$ to $X_\circ$.*

(c) *Actually, if $x \in X_\circ$, $t \geq 0$, then $\Phi(s,y) \to \Phi(t,x)$ as $y \in X, y \to x$ and $s \to t$.*

(d) *If $r > 0$, $x \in X$ and $\Phi(t,x)$ is a continuous function of $t \geq r$, then $\Phi_s(x) \in X_\circ$ for all $s \geq r$.*

Proof. (a) Let $t \geq 0$, $x \in X_\circ$. To show that $\Phi_t(x) \in X_\circ$, we observe that

$$\Phi(s, \Phi_t(x)) = \Phi(t+s, x) = \Phi(t, \Phi_s(x)).$$

Since $x \in X_\circ$, as $s \to 0$, $\Phi_s(x) \to x$ and, since Φ is state continuous, $\Phi(t, \Phi_s(x)) \to \Phi(t,x)$. So, as $s \to 0$, $\Phi(s, \Phi_t(x)) \to \Phi_t(x)$ which shows that $\Phi_t(x) \in X_\circ$, and $\Phi(t+s, x) \to \Phi(t,x)$ which shows that $\Phi(\cdot, x)$ is right-continuous.

To show that $X_\circ$ is closed, let $x_n \in X_\circ$, $x \in X$ and $x_n \to x$ as $n \to \infty$. By the triangle inequality, for each $n \in \mathbb{N}$,

$$d(\Phi(s,x), x) \leq d(\Phi(s,x), \Phi(s,x_n)) + d(\Phi(s,x_n), x_n) + d(x_n, x).$$

Let $\epsilon > 0$. Since Φ is state-continuous, uniformly in finite time, there exists some $n \in \mathbb{N}$ such that $d(\Phi(s,x), \Phi(s,x_n)) < \epsilon/3$ for all $s \in [0,1]$ and $d(x_n, x) < \epsilon/3$. Since $x_n \in X_\circ$, there exists some $\delta \in (0,1)$ such that $d(\Phi(s,x_n), x_n) < \epsilon/3$ for $s \in [0,\delta]$. So $d(\Phi(s,x), x) < \epsilon$ for all $s \in [0,\delta]$.

Part (b) follows from (a) and (c).

(c) Now let $x \in X_\circ$, $y \in X$ and $t \geq 0$, $h \geq 0$. By the triangle inequality,

$$d(\Phi(t+h,y), \Phi(t,x)) \leq d(\Phi(t+h,y), \Phi(t+h,x)) + d(\Phi(t+h,x), \Phi(t,x)).$$

Let $\epsilon > 0$. Since Φ is state-continuous, uniformly in finite time, there exists some $\delta_1 > 0$ such that $d(\Phi(t+h,y), \Phi(t+h,x)) < \epsilon/2$ whenever $d(x,y) < \delta_1$ and $h \in [0,1]$. As we have already seen above, $\Phi(\cdot, x)$ is right-continuous. So there exists some $\delta_2 \in (0,1]$ such that $d(\Phi(t+h,x), \Phi(t,x)) < \epsilon/2$ if $h \in [0,\delta_2]$. So $d(\Phi(t+h,y), \Phi(t,x)) < \epsilon$ whenever $d(x,y) < \delta_1$ and $h \in [0,\delta_2]$.

Now let $x \in X_\circ$, $y \in X$ and $t > 0$, $h \in [0, t]$. Then

$$\begin{aligned} d(\Phi(t-h,y),\Phi(t,x)) =& d(\Phi(t-h,y),\Phi(t-h,x)) \\ &+ d(\Phi(t-h,x),\Phi(t-h,\Phi_h(x))). \end{aligned}$$

Let $\epsilon > 0$. Since Φ is state-continuous uniformly in finite time, there exists some $\delta_1 > 0$ such that $d(\Phi(s,y),\Phi(s,x)) < \epsilon/2$ whenever $d(y,x) < \delta_1$ and $s \in [0,t]$. Since $\Phi_h(x) \to x$ as $h \to 0$, there exists some $\delta_2 > 0$ such that, if $h \in [0,\delta_2]$, then $d(\Phi_h(x),x) < \delta_1$ and $d(\Phi(s,x),\Phi(s,\Phi_h(x))) < \epsilon/2$ for all $s \in [0,t]$. So $d(\Phi(t-h,y),\Phi(t,x)) < \epsilon$ whenever $d(x,y) < \delta_1$ and $h \in [0,\delta_2]$.

(d) This is an obvious consequence of the semiflow property. □

1.4. Exercises

Exercise 1.1. Verify Remark 1.17.

Exercise 1.2. A subset U of the metric space X is closed if and only if its complement $X \setminus U$ is open.

Exercise 1.3. Let B be a nonempty subset of X and $x \in X$. Show $d(x,B) = 0$ if and only if $x \in \bar{B}$.

Exercise 1.4. Let S be a subset of a metric space X. Then S is precompact if and only if every sequence in S has a subsequence which has a limit in X.

Exercise 1.5. What is the diameter of $B_r(x)$? Warning: This may depend on the metric space.

Exercise 1.6. Let A, B be nonempty subsets of X with A compact and let $\epsilon > 0$. Show that $d(A,B) < \epsilon$ if and only if $A \subset U_\epsilon(B)$.

Exercise 1.7. Let A, B be nonempty subsets of X. Show $A \subset \bar{B}$ if and only if $d(A,B) = 0$.

Exercise 1.8. Verify Remark 1.25.

Exercise 1.9. Verify Remark 1.23.

Exercise 1.10. Let X be a normed vector space and d the induced metric. Let B be a convex subset of X: $x, \tilde{x} \in B$ and $\xi \in [0,1]$ implies $\xi x + (1-\xi)\tilde{x} \in B$.

Then the function $\rho : X \to \mathbb{R}_+$ defined by $\rho(x) = d(x,B)$ is convex: $x, \tilde{x} \in X$ and $\xi \in [0,1]$, implies

$$\rho(\xi x + (1-\xi)\tilde{x}) \le \xi \rho(x) + (1-\xi)\rho(\tilde{x}).$$

Exercise 1.11. Let $J \subset [0,\infty)$ and $\hat{J} = J \cup (-J)$. Show J is a time-set if and only if $\hat{J}$ is a subgroup of $(\mathbb{R},+)$ which contains $\mathbb{Z}$.

Exercise 1.12. Let Φ be a state-continuous semiflow. Show :

(a) The closure of a forward invariant set is forward invariant.

(b) If the closure of a backward invariant set is compact, then it is also backward invariant.

Exercise 1.13. Show that every periodic orbit is a minimal set.

Exercise 1.14. Show the necessity of (1.6) in Remark 1.37. Show that (1.6) is equivalent to the condition of Proposition A.17 for $D = \mathbb{R}^n_+$ when $f(t,x) = f(x)$.

Exercise 1.15. Let Φ be a semiflow on X and $M \subset X$.

Show :

(a) M is forward invariant if and only if $M \subset \Phi_t^{-1}(M)$ for all $t \in J$.

(b) If X invariant and $\Phi_t^{-1}(M) \subset M$ for all $t \in J$, then M is backward invariant.

(c) If M is backward invariant and all maps Φ_t are injective, then $\Phi_t^{-1}(M) \subset M$ for all $t \in J$.

Exercise 1.16. Let Φ be a semiflow on X and M a subset of X. Define

$$\bigcup \Phi^{-1}(M) = \bigcup_{t \in J} \Phi_t^{-1}(M) = \{x \in X;\ \exists t \in J : \Phi(t,x) \in M\}.$$

Show:

(a) $\Phi_r^{-1}(\bigcup \Phi^{-1}(M)) \subset \bigcup \Phi^{-1}(M)$ for all $r \in J$.

(b) If X is invariant, $\bigcup \Phi^{-1}(M)$ is backward invariant.

(c) If M is forward invariant, so is $\bigcup \Phi^{-1}(M)$.

Exercise 1.17. Let Φ be a semiflow on X and M a subset of X. Define

$$\bigcap \Phi^{-1}(M) = \bigcap_{t \in J} \Phi_t^{-1}(M) = \{x \in X;\ \forall t \in J : \Phi(t,x) \in M\}. \tag{1.7}$$

Notice that

$$\bigcap \Phi^{-1}(X \setminus M) = X \setminus \bigcup \Phi^{-1}(M) \text{ and } \bigcup \Phi^{-1}(X \setminus M) = X \setminus \bigcap \Phi^{-1}(M).$$

Show:

(a) M is forward invariant if and only if $M \subset \bigcap \Phi^{-1}(M)$.

(b) $\bigcap \Phi^{-1}(M)$ is forward invariant provided that it is nonempty.

(c) $X \setminus \bigcap \Phi^{-1}(M)$ is backward invariant if X is invariant.

Assume in addition that $\Phi_s^{-1}(M) \subset M$ for all $s \in J$.

(d) $\Phi_s^{-1}(\bigcap \Phi^{-1}(M)) = \bigcap \Phi^{-1}(M)$ for all $s \in J$. If X is invariant, $\bigcap \Phi^{-1}(M)$ is invariant provided that it is nonempty.

(d) $X \setminus \bigcap \Phi^{-1}(M)$ is forward invariant. If X is invariant, so is $X \setminus \bigcap \Phi^{-1}(M)$.

Exercise 1.18. The translation of a total trajectory is again a total trajectory: if ϕ is a total trajectory and $\phi_r(t) := \phi(r+t)$, $r, t \in \hat{J} = J \cup (-J)$, then ϕ_r is a total trajectory.

Exercise 1.19. Total orbits (i.e., ranges of total trajectories) are invariant.

Chapter 2

Compact Attractors

The study of the dynamics of a semiflow (inertial manifolds, persistence) is largely facilitated if there is a global compact attractor, i.e. a compact invariant subset which attracts a sufficiently broad class of subsets of X.

Recall Definition 1.7 of the distance $d(x, B)$ from the point x to the set B and Definition 1.19 of the distance $d(A, B)$ from the set A to the set B.

If $Y \subset X$ and $\{Y_t\}_{t \in J}$ a family of subsets of X indexed over an unbounded subset J of $\mathbb{R}_+$, there are different ways in which Y_t can approach Y as $t \to \infty$. Recall $U_\epsilon(Y) = \{x \in X; d(x, Y) < \epsilon\}$. Let

$$J_r = J \cap [r, \infty) \text{ for } r \in J. \tag{2.1}$$

Definition 2.1. We say that Y_t *converges to* Y as $t \to \infty$, symbolically $Y_t \to Y$, if for any open subset U of X with $Y \subset U$, there exists some $r \in J$ such that $Y_t \subset U$ for all $t \in J_r$.

We say that Y_t *is attracted to* Y as $t \to \infty$, if for every $\epsilon > 0$ there is some $r \in J$ such that $Y_t \subset U_\epsilon(Y)$ for all $t \in J_r$.

Remark 2.2. Y_t is attracted to Y as $t \to \infty$ if and only if $d(Y_t, Y) \to 0$ as $t \to \infty$.

Since $U_\epsilon(Y)$ is open, convergence of sets is a stronger concept than attraction of sets.

Lemma 2.3. *Let $\{Y_t; t \in J\}$ be a family of nonempty sets in the metric space X, $\emptyset \neq Y \subset X$.*

(a) *If $Y_t \to Y$ as $t \to \infty$, then $d(Y_t, Y) \to 0$ as $t \to \infty$.*

(b) *If Y is compact and $d(Y_t, Y) \to 0$ as $t \to \infty$, then $Y_t \to Y$ as $t \to \infty$.*

(c) *If* Y *is open, then* $Y_t \to Y$ *as* $t \to \infty$ *if and only if there exists some* $r \in J$ *such that* $Y_t \subseteq Y$ *for all* $t \in J_r$.

Part (b) may be false if Y is closed rather than compact. Choose X as the real numbers with the usual metric, $Y = \mathbb{N}$ and $y_n = n + (1/n)$ for $n \in \mathbb{N}$. Then $d(y_n, Y) \leq 1/n$ but $y_n \not\to Y$ as can be seen by choosing

$$U = \bigcup_{n \in \mathbb{N}} \left(n - \frac{1}{n^2}, n + \frac{1}{n^2}\right).$$

2.1. Compact attractors of individual sets

Definition 2.4. Let J be a time-set and $\Phi : J \times X \to X$ a semiflow.

- A set $K \subset X$ is said to *attract* a set $M \subset X$, if $K \neq \emptyset$ and $d(\Phi_t(M), K) \to 0$ as $t \to \infty$.

  We also say that M is attracted by K.

- K is called an *attractor* of M, if K is invariant and attracts M. In this situation, we also say that M has the attractor K. K is called a *compact attractor* of M if K is compact in addition.

Remark 2.5. The first definition also makes sense if Φ is not necessarily a semiflow.

In this section we are going to clarify the connection between these two concepts, namely that a set which is attracted by a compact set also has a compact attractor. We need some technical results and another concept, ω-limit sets. Also, the concept of asymptotic compactness will be useful. In the next section, we will give sufficient conditions for the existence of compact attractors of classes of sets.

Lemma 2.6. *Let* K *and* M *be subsets of* X. *Then the following statements are equivalent:*

(1) K *attracts* M.

(2) $d(\Phi_t(x), K) \to 0$, $t \to \infty$, *uniformly for* $x \in M$.

(3) *All sequences* (t_j) *in* J, $t_j \to \infty$ *as* $j \to \infty$, *and* (x_j) *in* M *satisfy*

$$d(\Phi(t_j, x_j), K) \to 0, \qquad j \to \infty.$$

Definition 2.7. The ω*-limit set* of a subset M of X is defined as

$$\omega(M) = \bigcap_{t \in J} \overline{\Phi(J_t \times M)}, \qquad J_t = J \cap [t, \infty).$$

Obviously, $\omega(M)$ is a (possibly empty) closed set. Some special cases warrant special attention. If M is forward invariant, then $\Phi(J_t \times M) = \Phi(t, M)$, so

$$\omega(M) = \cap_{t\in J}\overline{\Phi(t,M)} \subset \overline{\Phi(s,M)} \subset \overline{M}, \ s \in J.$$

If M is invariant, then $\Phi(t, M) = M$ and $\omega(M) = \overline{M}$.

The following alternative characterization will be very useful.

Lemma 2.8. *An element x in X satisfies $x \in \omega(M)$ if and only if there are sequences (t_j) in J, $t_j \to \infty$ as $j \to \infty$, and (x_j) in M such that $\Phi(t_j, x_j) \to x$ as $j \to \infty$.*

The subsequent definition is taken from [**197**].

Definition 2.9. Let J be a time-set and $\Phi : J \times X \to X$ a map, $M \subset X$. Φ is called *asymptotically compact on* M, if, for any sequences (t_i) in J, $t_i \to \infty$ as $i \to \infty$, and (x_i) in M, $(\Phi(t_i, x_i))$ has a convergent subsequence.

The following equivalencies even hold if Φ is not necessarily a semiflow.

Proposition 2.10. *Let J be a time-set and $\Phi : J \times X \to X$ a map, $\emptyset \neq M \subset X$. Then the following statements are equivalent:*

(1) *Φ is asymptotically compact on M.*

(2) *M is attracted by a nonempty compact set $K \subset X$.*

(3) *$\omega(M)$ is nonempty, compact, and attracts M.*

If one and then all of these three statements hold, $\omega(M) \subset K$ for every compact $K \subset X$ that attracts M.

Proof. "(1) ⇒(3)": Since $M \neq \emptyset$, we can pick some $x \in M$ and consider the sequence $(\Phi(k, x))$. Since Φ is asymptotically compact on M, a convergent subsequence exists, the limit of which is an element in $\omega(M)$ by Lemma 2.8. To show that $\omega(M)$ is compact, let (z_k) be a sequence in $\omega(M)$. By Lemma 2.8, for every $k \in \mathbb{N}$ there exist some $x_k \in M$ and $t_k \in J$, $t_k > k$, such that $d(z_k, \Phi(t_k, x_k)) < \frac{1}{k}$. Since Φ is asymptotically compact on M, after choosing a subsequence, there is some $x \in X$ such that $\Phi(t_k, x_k) \to x$ as $k \to \infty$. Hence, (z_k) has a convergent subsequence with limit $x \in \omega(M)$. To show that $d(\Phi_t(M), \omega(M)) \to 0$ as $t \to \infty$, assume the contrary. Then, by Lemma 2.6, there exist some $\epsilon > 0$ and sequences $J \ni t_j \to \infty, j \to \infty$, and (x_j) in M such that $d(\Phi(t_j, x_j), \omega(M)) \geq \epsilon$ for all $j \in \mathbb{N}$. Since Φ is asymptotically compact on M, after choosing a subsequence, $\Phi(t_j, x_j) \to z$ as $j \to \infty$ with some $z \in X$. By Lemma 2.8, $z \in \omega(M)$, a contradiction.

Obviously, (3) implies (2) with $K = \omega(M)$.

"(2) $\Rightarrow$(1)": Let $K \subset X$ be a nonempty compact set that attracts M, $J \ni t_j \to \infty$, $j \to \infty$, and (x_j) a sequence in M. By Lemma 2.6, $d(\Phi(t_j, x_j), K) \to 0$ as $j \to \infty$. Then we find a sequence (z_j) in K such that $d(\Phi(t_j, x_j), z_j) \to 0$ as $j \to \infty$. Since K is compact, (z_j) has a convergent subsequence with limit in K and so has $(\Phi(t_j, x_j))$. Hence Φ is asymptotically compact on M.

A similar argument shows that $\omega(M) \subset K$. Thus $\omega(M)$ is the smallest compact set that attracts M. □

If Φ is a state-continuous semiflow, we get stronger results.

Theorem 2.11. *Let J be a time-set and $\Phi : J \times X \to X$ a state-continuous semiflow, $\emptyset \neq M \subset X$. Then $\omega(M)$ is forward invariant and contains every backward invariant subset of $\overline{\Phi(J \times M)}$.*

Assume, in addition, that M is attracted by a compact set or, equivalently, Φ is asymptotically compact on M.

Then $\omega(M)$ is nonempty and invariant and M has a compact attractor, namely $\omega(M)$. In particular, $\omega(M)$ is the unique compact attractor of M contained in $\overline{\Phi(J \times M)}$.

Further, if C is a subset of X and attracts $\omega(M)$, then C attracts M.

In summary, $\omega(M)$ contains every backward invariant set belonging to the closure of its orbit $\Phi(J \times M)$, but it may not attract M or be compact. If, in addition, Φ is asymptotically compact on M, then $\omega(M)$ is nonempty and a compact attractor of M. In this case, it is the largest invariant set contained in the closure of the orbit of M and the smallest compact set attracting M.

Proof. (Theorem 2.11) Let K be a backward invariant subset of $\overline{\Phi(J \times M)}$ and $x \in K$. Let $k \in \mathbb{N}$. Then there exists some $x_k \in K$ such that $x = \Phi_k(x_k)$. Since Φ_k is continuous, we find some $y_k \in \Phi(J \times M)$ such that $d(\Phi_k(x_k), \Phi_k(y_k)) < \frac{1}{k}$. Further we find $s_k \in J$, $z_k \in M$ such that $y_k = \Phi(s_k, z_k)$. Hence

$$x = \Phi_k(x_k) = \lim_{k \to \infty} \Phi_k(y_k) = \lim_{k \to \infty} \Phi_k(\Phi(s_k, z_k)) = \lim_{k \to \infty} \Phi(k + s_k, z_k).$$

Since $k + s_k \to \infty$ as $k \to \infty$ and $z_k \in M$, $x \in \omega(M)$. Since $x \in K$ was arbitrary, $K \subset \omega(M)$.

To show that $\omega(M)$ is forward invariant, let $t \in J$ and $x \in \omega(M)$. By Lemma 2.8, there exist sequences $J \ni t_j \to \infty$, $j \to \infty$, and (x_j) in M such that $\Phi(t_j, x_j) \to x$ as $j \to \infty$. Since Φ_t is continuous and Φ a semiflow,

$$\Phi_t(x) = \lim_{j \to \infty} \Phi_t(\Phi(t_j, x_j)) = \lim_{j \to \infty} \Phi(t + t_j, x_j).$$

By Lemma 2.8, $\Phi_t(x) \in \omega(M)$. Since $x \in \omega(M)$, was arbitrary, $\Phi_t(\omega(M)) \subset \omega(M)$.

Now assume that Φ is asymptotically compact on M. By Proposition 2.10, $\omega(M)$ is nonempty, compact, and attracts M.

For the same t and x as above, we now have also

$$x = \lim_{j\to\infty} \Phi(t_j, x_j) = \lim_{j\to\infty} \Phi_t(\Phi(t_j - t, x_j)).$$

After choosing a subsequence, $\Phi(t_j - t, x_j) \to y$ for some $y \in \omega(M)$. Since Φ_t is continuous, $x = \Phi_t(y) \in \Phi_t\big(\omega(M)\big)$. Since $x \in \omega(M)$ has been arbitrary, $\omega(M) \subset \Phi_t(\omega(M))$.

If K is a compact attractor contained in $\overline{\Phi(J \times M)}$, we also have $\omega(M) \subset K$ by Proposition 2.10. Since $K \subset \omega(M)$ by the first assertion, $\omega(M)$ is the unique compact attractor of M that is contained in $\overline{\Phi(J \times M)}$.

Finally let C be a subset of X that attracts $\omega(M)$. By Lemma 1.22,

$$d(\Phi_t(M), C) \le d(\Phi_t(M), \omega(M)) + d(\omega(M), C).$$

Since $\omega(M)$ is invariant,

$$d(\Phi_t(M), C) \le d(\Phi_t(M), \omega(M)) + d(\Phi_t(\omega(M)), C).$$

Since $\omega(M)$ attracts M and C attracts $\omega(M)$, the right-hand side of this inequality converges to 0 as $t \to \infty$. So C attracts M. □

Recall the definition of a connected metric space [**61**, Ch.V.1].

Definition 2.12. A metric space X is called *connected* if the only sets in X that are both open and closed are the empty set and X itself.

A subset K of X is called connected if it is connected when considered as a metric space itself.

Remark 2.13. A subset K of a metric space X is not connected (*disconnected*) if there are two open sets U and V such that

- $U \cap V \cap K = \emptyset$,
- $U \cap K \neq \emptyset, \qquad V \cap K \neq \emptyset$,
- $K \subset U \cup V$.

Equivalently, K is not connected if there are two closed sets U and V such that these properties are satisfied.

This implies that a compact set is disconnected if it is the disjoint union of compact sets: there exist nonempty compact sets K_1 and K_2 such that $K = K_1 \cup K_2$ and $K_1 \cap K_2 = \emptyset$.

Analogous statements hold for open and closed sets.

Definition 2.14. A compact invariant set K is called *flow-connected* if the following holds: Whenever K_1 and K_2 are nonempty compact sets such that $K = K_1 \cup K_2$ and $K_1 \cap K_2 = \emptyset$, then neither K_1 nor K_2 are forward invariant.

The notion of a flow-connected set is slightly stronger than the notion of an *invariantly connected* set [**137, 93**].

Obviously every compact invariant set that is connected is also flow-connected. The opposite is not necessarily the case: If Φ is a discrete semiflow, and $\phi : \mathbb{Z} \to X$ a periodic trajectory, then the range of ϕ, $\{\phi(0), \dots, \phi(p)\}$ with minimum period $p \in \mathbb{N}$, is flow-connected, but not connected.

Corollary 2.15. *Let the assumptions of Theorem 2.11 be satisfied. If M is connected, the following hold in addition:*

(1) *If Φ is state-continuous, uniformly in finite time, then $\omega(M)$ is flow-connected.*
(2) *If $J = \mathbb{R}$ and Φ is a continuous semiflow, then $\omega(M)$ is connected.*

Proof. (1) To see that $\omega(M)$ is flow-connected if M is connected and Φ state-continuous, uniformly in finite time, let

$$\omega(M) = K_1 \cup K_2, \qquad K_1 \cap K_2 = \emptyset,$$

with nonempty compact sets K_1 and K_2. Then there exists some $\delta > 0$ such that

$$U_\delta(\omega(M)) = U_\delta(K_1) \cup U_\delta(K_2), \qquad U_\delta(K_1) \cap U_\delta(K_2) = \emptyset.$$

Since $\omega(M)$ attracts M, there exists some $t \in J$ such that

$$\Phi_s(M) \subset U_\delta(\omega(M)) \qquad \forall s \in J,\ s \geq t.$$

Since M is connected and Φ_s is continuous, $\Phi_s(M)$ is connected. Hence $\Phi_s(M) \subset U_\delta(K_1)$ or $\Phi_s(M) \subset U_\delta(K_2)$ for every $s \in J, s \geq t$. Since K_1 and K_2 are nonempty, by Lemma 2.8 there are arbitrarily large numbers $s \in J$ such that $\Phi_s(M) \subset U_\delta(K_1)$ and arbitrarily large numbers $r \in J$ such that $\Phi_r(M) \subset U_\delta(K_2)$.

To show that neither K_1 nor K_2 are forward invariant, without loss of generality we suppose that K_1 is forward invariant and derive a contradiction.

We can find increasing sequences $(s_j), (t_j)$ in J such that $s_{j+1} \geq 1 + t_j$ and

$$\Phi_{s_j}(M) \subset U_\delta(K_1), \qquad \Phi_{t_j}(M) \subset U_\delta(K_2), \qquad 0 < t_j - s_j \leq 1.$$

Notice the asymmetry in the sequences. This is to obtain a contraction resulting from the assumption that K_1 is forward invariant.

To construct the sequences, assume that $s_j, t_j \in J$ have been found with the desired properties for $j = 1, \dots, n$. Then $\Phi_s(M) \subset U_\delta(K_1)$ for some $s \geq 2 + t_n$. Further there is some $r \in J$, $r > s$, such that $\Phi_r(M) \in U_\delta(K_2)$.

Let $\tilde{s}$ be the supremum of those $s \in [2 + t_n, r]$, $s \in J$, with $\Phi_s(M) \subset U_\delta(K_1)$.

There exists some $s \in [\tilde{s} - (1/2), \tilde{s}] \cap J$ such that $\Phi_s(M) \subset U_\delta(K_1)$. Choose $s_{n+1} = s$ and $t_{n+1} = \min\{s+1, r\}$. Then $s_{n+1} \in J$ and $s_{n+1} \geq 1+t_n$. Further $t_{n+1} \in J$ and $0 \leq t_{n+1} - s_{n+1} \leq 1$. If $s + 1 \geq r$, $t_{n+1} = r$ and $\Phi_{t_{n+1}}(M) \subset U_\delta(K_2)$. If $s + 1 \leq r$, then $\tilde{s} < r$ and $\Phi_{\tilde{r}}(M) \subset U_\delta(K_2)$ for all $\tilde{r} \in (\tilde{s}, r] \ni s + 1 = t_{n+1}$. Since $U_\delta(K_1) \cap U_\delta(K_2) = \emptyset$, $s_{n+1} < t_{n+1}$.

This finishes the recursive construction of the sequences (s_j) and (t_j).

In particular $s_j, t_j \to \infty$ as $j \to \infty$. Let $x \in M$. Then, after choosing subsequences,

$$\Phi_{s_j}(x) \to y \in \omega(M), \qquad \Phi_{t_j}(x) \to z \in \omega(M), \qquad j \to \infty.$$

Actually $y \in K_1$ and $z \in K_2$. Now

$$\Phi_{t_j}(x) = \Phi_{\epsilon_j}(\Phi_{s_j}(x))$$

with $\epsilon_j = t_j - s_j \in (0, 1] \cap J$. Since Φ is state-continuous, uniformly in finite time,

$$d(\Phi_{t_j}(x), \Phi_{\epsilon_j}(y)) = d\big(\Phi_{\epsilon_j}(\Phi_{s_j}(x)), \Phi_{\epsilon_j}(y)\big) \to 0, \qquad j \to \infty.$$

By the triangle inequality,

$$\Phi_{\epsilon_j}(y) \to z, \qquad j \to \infty.$$

Since we assumed that K_1 is forward invariant and since $y \in K_1$, $\phi_{\epsilon_j}(y) \in K_1$ for all $j \in J$ and so $z \in K_1 \cap K_2$. This contradiction implies that K_1 is not forward invariant. Since the scenario is symmetric for K_1 and K_2, K_2 is not forward invariant either.

(2) Let now $J = \mathbb{R}$ and Φ be a continuous semiflow, and let M be connected. As a Cartesian product of two connected sets, $[t, \infty) \times M$ is connected in the product space $[0, \infty) \times X$ [**61**, Thm.V.1.7]. Since Φ is continuous, $M_t = \Phi([t, \infty) \times M)$ is connected in X [**61**, Thm.V.1.4]. Suppose that $\omega(M)$ is not connected. Then there exist nonempty compact sets K_1 and K_2 such that $\omega(M) = K_1 \cup K_2$ and $K_1 \cap K_2 = \emptyset$. As before, there exists some $\delta > 0$ such that

$$U_\delta(\omega(M)) = U_\delta(K_1) \cup U_\delta(K_2), \qquad U_\delta(K_1) \cap U_\delta(K_2) = \emptyset.$$

Since $\omega(M)$ attracts M, there exists some $t > 0$ such that

$$\Phi_s(M) \subset U_\delta(\omega(M)) \qquad \forall s \geq t.$$

In other words, the connected set $M_t = \Phi([t, \infty) \times M)$ is a subset of $U_\delta(K_1) \cup U_\delta(K_2)$.

By Remark 2.13, $M_t \cap U_\delta(K_1) = \emptyset$ or $M_t \cap U_\delta(K_2) = \emptyset$. Without loss of generality let us assume that $M_t \cap U_\delta(K_1) = \emptyset$. Since $K_1 \neq \emptyset$, there exists some $x \in K_1 \subset \omega(M)$. By Lemma 2.8, there exist sequences (t_j) in J and (x_j) in M such that $t_j \to \infty$ as $j \to \infty$ and $\Phi(t_j, x_j) \to x$ as $j \to \infty$. For sufficiently large j, $\Phi(t_j, x_j) \in U_\delta(K_1)$ and $\Phi(t_j, x_j) \in M_t$, a contradiction. □

2.2. Compact attractors of classes of sets

So far we have considered compact attractors of individual sets. We expand our view to compact attractors of classes of sets. As before, let X be a metric space with metric d, J a time-set, and $\Phi : J \times X \to X$ a (global autonomous) semiflow. Moreover we assume that Φ is state-continuous.

Definition 2.16. Let $\mathcal{C}$ denote a class of subsets of X (e.g., singleton sets, compact sets, bounded sets). A nonempty, compact, invariant set $K \subset X$ is called a *compact attractor of* $\mathcal{C}$ if K attracts all sets in $\mathcal{C}$.

The following condition for the existence of a compact attractor of $\mathcal{C}$ is sufficient and necessary, but perhaps not very practical.

Theorem 2.17. *The following are equivalent for a class $\mathcal{C}$ of subsets of X.*

(a) *There exists a compact attractor of $\mathcal{C}$.*

(b) *Φ is asymptotically compact on every set $M \in \mathcal{C}$ and $\bigcup_{M \in \mathcal{C}} \omega(M)$ has compact closure in X.*

(c) *There is a compact set in X that attracts every set $M \in \mathcal{C}$.*

If (a), (b) or (c) and then all of them hold, the closure of $\bigcup_{M \in \mathcal{C}} \omega(M)$ is the smallest compact attractor of $\mathcal{C}$.

Proof. By definition, (a) $\Rightarrow$ (c).

(c) $\Rightarrow$ (b). Let A be a compact subset of X that attracts each $M \in \mathcal{C}$. Then Φ is asymptotically compact on each $M \in \mathcal{C}$ and $\omega(M) \subset A$ by Proposition 2.10. Thus $\bigcup_{M \in \mathcal{C}} \omega(M)$ is contained in A, and so is its closure which is compact because A is compact.

(b) $\Rightarrow$ (a). For each $M \in \mathcal{C}$, $\omega(M)$ attracts M and is compact and nonempty by Proposition 2.10 and a compact attractor of M by Theorem 2.11. Then $\bigcup_{M \in \mathcal{C}} \omega(M)$ attracts every set in $\mathcal{C}$ and is invariant. Its closure also attracts every set in $\mathcal{C}$ and, since it is compact, is also invariant (Exercise 1.12). □

Lemma 2.18. *Let $\mathcal{C}$ be a class of subsets of X and A a compact attractor of $\mathcal{C}$. Then A contains every backward invariant set in $\mathcal{C}$.*

Proof. Let $B \in \mathcal{C}$. Since A attracts B, for every open set $U \supseteq A$ there exists some $t \in J$ such that $\Phi_t(B) \subset U$. If B is backward invariant, $B \subset U$. Suppose $B \not\subset A$. Then there exists some $x \in B$ such that $x \notin A$. Since A is compact, there exists some open set U such that $x \notin U$, $A \subset U$. This contradicts $B \subset U$. □

The following uniqueness result for compact attractors follows from Lemma 2.18.

Theorem 2.19. *Let $\mathcal{C}$ be a class of subsets of X. There exists at most one compact attractor of $\mathcal{C}$ that is itself an element of $\mathcal{C}$. It would contain all other invariant sets in $\mathcal{C}$.*

The next result states that a compact attractor of $\mathcal{C}$ is the union of total orbits that are contained in some set in $\mathcal{C}$.

Theorem 2.20. *Let $\mathcal{C}$ be a class of subsets of X and A a compact attractor of $\mathcal{C}$, $A \in \mathcal{C}$. Then $x \in X$ is an element of A if and only if there exist a set $M \in \mathcal{C}$ and a total Φ-trajectory through x with range in M.*

Proof. $\Rightarrow$. By definition of an attractor, A is invariant. By Theorem 1.40, for each $x \in A$ there exists some total Φ-trajectory through x with range in $A \in \mathcal{C}$.

$\Leftarrow$. Let $x \in X$ such that there are a set $M \in \mathcal{C}$ and a total Φ-trajectory ϕ with $x = \phi(0) \in M$. Since the range of ϕ is invariant and M attracted by A, the range of ϕ is a subset of A. So $x \in A$. □

We will use the notion of a global attractor as little as possible because there is no agreement in the literature about this term.

Definition 2.21. Instead we use the following terminology for a nonempty compact invariant set A:

If $\mathcal{C}$ is the class of singleton sets in X and A attracts $\mathcal{C}$, A is called a *compact attractor of points.*

If $\mathcal{C}$ is the class of bounded (compact) sets in X and A attracts $\mathcal{C}$, A is called an (actually the) *compact attractor of bounded (compact) sets.* The following somewhat loose language needs further explanation:

A is a (the) *compact attractor of neighborhoods of compact sets* if every compact set in X has a neighborhood that is attracted by A.

The term *global compact attractor* has been used in various ways in the literature: compact attractor of points [**134**], compact attractor of neighborhoods of compact sets [**101, 158, 197, 238**] (see Corollary 2.32), and compact attractor of bounded sets [**60, 91, 181, 210**].

Hale [**92**] shows that on every infinite dimensional Hilbert space one can construct a continuous semiflow with time-set $\mathbb{Z}_+$, induced by a linear map, where $\{0\}$ is the compact attractor of compact sets but not a compact attractor of neighborhoods of compact sets. See Example 2.45.

Magal and Zhao [**158**] construct several examples of continuous semiflows on complete metric spaces, both with time-set $\mathbb{Z}_+$ and time-set $\mathbb{R}_+$, that have a compact attractor of neighborhoods of compact sets but no compact attractor of bounded sets.

2.2.1. Compact attractors of compact sets.

Definition 2.22. A nonempty compact invariant set A in X is called a *local compact attractor of compact sets* if it attracts all compact subsets in a neighborhood of itself.

Theorem 2.23. *Let A be a compact attractor of points in X and a local compact attractor of compact sets. Then A is the compact attractor of compact sets in X.*

Proof. Let U be an open neighborhood of A such that A attracts all compact subsets of U. Let C be a compact subset of X and $x \in C$. Since A is an attractor of points, there exists some $t \in J$ such that $\Phi_t(x) \in U$. Since Φ_t is continuous, there exists some closed neighborhood V of x such that $\Phi_t(V) \subset U$. $C \cap V$ is compact, and so is $\Phi_t(C \cap V) \subset U$. By assumption, A attracts $\Phi_t(C \cap V)$ and thus $C \cap V$. Since C is compact, we find closed sets $V_1, \ldots, V_n$ with nonempty interior such $C \subset V_1 \cup \cdots \cup V_n$ and $C \cap V_k$ is attracted by A. Let W be a neighborhood of A. Then there exists $t_k \in J$ such that $\Phi_t(C \cap V_k) \subset W$ for all $t \in J$, $t \geq t_k$. Set $r = \max_{k=1}^n t_k$. Then $\Phi_t(C) \subset W$ for all $t \in J_r$. □

We resume our discussion of when compact attractors are connected (cf. [**91**, L.2.4.1]). Unlike in Corollary 2.15, it will not be required that Φ is a continuous semiflow.

Proposition 2.24. (a) *If K is the compact attractor of a connected set B, $K \subset B \subset X$, then K is connected.*

(b) *If X is the closed convex subset of a Banach space and $K \subset X$ a compact attractor of compact sets in X, then K is connected.*

If X is just a connected metric space and $J = \mathbb{Z}_+$, a continuous semiflow can have a compact attractor of bounded sets that is not connected [**84**].

Proof. (a) Suppose that the compact set K is not connected. Then there exist disjoint open sets U and V in X such that $K \subset U \cup V$ and $K \cap U \neq \emptyset$,

$K \cap V \neq \emptyset$. Since K attracts B, $\Phi_t(B) \to K$ as $t \to \infty$. As K is invariant, there exist some $t \in J$ such that

$$K = \Phi_t(K) \subset \Phi_t(B) \subset U \cup V.$$

Since Φ_t is continuous, $\Phi_t(B)$ is connected, but $\Phi_t(B) \cap U \neq \emptyset$ and $\Phi_t(B) \cap V \neq \emptyset$, a contradiction.

(b) By one of Mazur's theorems [**62**, Theorem V.2.6], there exists a compact convex set $\tilde{K}$ in X that contains K. In particular, $\tilde{K}$ is connected and attracted by K, and the statement follows from part (a). □

We are not aware of any conditions in terms of the semiflow Φ that are both sufficient and necessary for the existence of a compact attractor of compact sets. The sufficient conditions we present below will imply that A is a compact attractor in a stronger sense. Example 2.45 will show that asymptotic smoothness is not necessary for a compact attractor of compact sets to exist.

The following concepts are taken from [**91**] (the first two) and [**197**] (the third).

Definition 2.25. Let $\Phi : J \times X \to X$ be a state-continuous semiflow.

- Φ is called *point-dissipative* (or *ultimately bounded*) if there exists a bounded subset B of X which attracts all points in X.
- Φ is called *asymptotically smooth* if Φ is asymptotically compact on every forward invariant bounded closed set.
- Φ is called *eventually bounded* on a set $M \subset X$ if $\Phi(J_r \times M)$, $J_r = J \cap [r, \infty)$, is bounded for some $r \in J$.

Remark 2.26. (a) By Exercise 1.12, an asymptotically smooth semiflow is asymptotically compact on every bounded forward invariant set.

(b) If X is a closed subset of $\mathbb{R}^n$, Φ is asymptotically smooth. More generally, if Φ_r is compact on the metric space X for some $r \in J$, Φ is asymptotically smooth. See Exercise 2.9.

The relevance of these concepts lies in their applicability (see Section 2.3 for asymptotic smoothness).

Proposition 2.27. *Let Φ be asymptotically smooth, $M \subset X$. Then M has a compact attractor, $\omega(M)$, if and only if Φ is eventually bounded on M.*

Proof. One direction is obvious. Assume that Φ is eventually bounded on M. By definition, there exists some $r \in J$ such that $\Phi(J_r \times M)$ is bounded, $J_r = J \cap [r, \infty)$. $\Phi(J_r \times M)$ is also forward invariant. Since Φ is asymptotically smooth, Φ is asymptotically compact on $\Phi(J_r \times M)$. To show that Φ is asymptotically compact on M, consider sequences $J \ni t_j \to \infty$, $j \to \infty$, and (x_j) in M. Set $s_j = t_j - r$, $y_j = \Phi_r(x_j)$. For large j, $s_j \in J_r$; further $s_j \to \infty$ as $j \to \infty$. By definition, $y_j \in \Phi(J_r \times M)$ and $\Phi(s_j, y_j) = \Phi(s_j, \Phi_r(x_j)) = \Phi(t_j, x_j)$. Since Φ is asymptotically compact on $\Phi(J_r \times M)$, $(\Phi(t_j, x_j))$ has a convergent subsequence. This shows that Φ is asymptotically compact on M. By Theorem 2.11, M has a compact attractor, $\omega(M)$. □

The following condition for the existence of compact attractor of points is more practical than the specialization of Theorem 2.17 to attractors of points.

Theorem 2.28. *Let $\Phi : J \times X \to X$ be a state-continuous semiflow. Assume that Φ is point-dissipative and asymptotically smooth. Then there exists a compact attractor of points, namely the closure $\bar{\Omega}(X)$ of $\Omega(X) = \bigcup_{x \in X} \omega(x)$.*

Proof. Since Φ is point-dissipative, there is a bounded set B that attracts all points in X. We can assume that B is closed. Then, for every $x \in X$, Φ is eventually bounded on $\{x\}$. By Proposition 2.27, every $x \in X$ has a compact attractor, $\omega(x)$, because Φ is asymptotically smooth. Set $K = \bigcup_{x \in X} \omega(x)$. Then $K \subset B$, K is invariant, and K attracts all points in X. Since Φ is asymptotically smooth, Φ is asymptotically compact on K. To show that K has compact closure, let (x_j) be a sequence in K. Since K is invariant, for every $j \in \mathbb{N}$, there exists $y_j \in K$ such that $\Phi_j(y_j) = x_j$. Since Φ is asymptotically compact on K, (x_j) has a convergent subsequence. By Exercise 1.12, $\bar{K}$ is invariant. So $\bar{K}$ is a compact attractor of every point in X. □

Cholewa and Hale [**31**] give an example of a continuous semiflow with time-set $\mathbb{Z}_+$ that has a compact attractor of points and is asymptotically smooth, but not eventually bounded on bounded sets. It is not even eventually bounded on arbitrarily small neighborhoods of some point.

2.2.2. Compact attractors of neighborhoods of compact sets. This type of compact attractor has been called global compact attractor in [**101, 158, 197, 238**]. See the forthcoming Corollary 2.32.

The next result still has a local flavor, but will be helpful in obtaining global attraction results.

Proposition 2.29. *Let Φ be state-continuous, U an open subset of X, and Φ asymptotically compact on U. Then U has a compact attractor, $\omega(U)$, which attracts a neighborhood of every compact set in $\bigcup \Phi^{-1}(U)$.*

Here

$$\bigcup \Phi^{-1}(U) = \bigcup_{t \in J} \Phi_t^{-1}(U) = \{x \in X; \exists t \in J : \Phi(t,x) \in U\}. \tag{2.2}$$

See Exercise 1.16.

Proof. By Theorem 2.11, U has a compact attractor, $\omega(U)$.

> **Step 1:** For each $x \in \bigcup \Phi^{-1}(U)$, there exists an open neighborhood U_x of x such that U_x is attracted by $\omega(U)$.

Proof of Step 1: Let $x \in \bigcup \Phi^{-1}(U)$. Then there exists some $t \in J$ such that $\Phi_t(x) \in U$. Since U is open and Φ_t is continuous, there exists an open neighborhood U_x of x such that $\Phi_t(U_x) \subset U$. In order to show that $\omega(U)$ attracts U_x, let V be an open neighborhood of $\omega(U)$. Since $\omega(U)$ attracts U, there exists some $r \in J$ such that $\Phi_s(U) \subset V$ for all $s \in J_r$. If $s \in J_r$,

$$\Phi_{s+t}(U_x) = \Phi_s(\Phi_t(U_x)) \subset \Phi_s(U) \subset V.$$

In other words, $\Phi_u(U_x) \subset V$ for all $u \in J_{r+t}$.

> **Step 2:** Finale.

Let K be a compact subset of $\bigcup \Phi^{-1}(U)$. Then, by step 1, for each $x \in K$, there exists an open neighborhood U_x such that U_x is attracted by $\omega(U)$. Since $K \subset \bigcup_{x \in K} U_x$, there exist finitely many open sets $U_1, \dots, U_k$ such that $K \subset \bigcup_{j=1}^k U_j$ and U_j is attracted by $\omega(U)$ for $j = 1, \dots, k$. Let V be an open neighborhood of $\omega(U)$. By Definition 2.1, for each $j = 1, \dots, k$ there exists $r_j \in J$ such that $\Phi_t(U_j) \subset V$ for all $t \in J_{r_j}$. Define $r = \max_{j=1}^k r_j$ and $W = \bigcup_{j=1}^k U_j$. Then W is an open neighborhood of K and $\Phi_t(W) \subset V$ for all $t \in J_r$. Hence $\omega(U)$ attracts W. □

The following result has been inspired by [**158**, Thm.2.6]. See also Theorems 2.4.2 and 3.4.2 in [**91**] and [**93**, Thm.8.7].

Theorem 2.30. *The following are equivalent for a state-continuous semiflow $\Phi : J \times X \to X$.*

(a) *Φ is point-dissipative, asymptotically smooth, and eventually bounded on every compact subset K of X.*

(b) *There exists a compact attractor A of neighborhoods of compact sets in X; further, A attracts every subset of X on which Φ is eventually bounded.*

Proof. (b) $\Rightarrow$ (a). The only part which may not be immediate is asymptotic smoothness. Let B be a forward invariant bounded set. Then Φ is eventually bounded on B, and B is attracted by the compact set A. By Proposition 2.10, Φ is asymptotically compact on B.

(a) $\Rightarrow$ (b). Let $\tilde{A}$ be a compact attractor of points which exists according to Theorem 2.28.

Step 1: If $\delta > 0$ is sufficiently small, Φ is asymptotically compact on $U = U_\delta(\tilde{A})$.

By Proposition 2.27, it is sufficient to show that Φ is eventually bounded on $U_\delta(\tilde{A})$. Suppose the latter is false. Then there exist a sequence (x_n) in X and a sequence (t_n) in J such that $x_n \to \tilde{A}$, $t_n \to \infty$, and the elements $\Phi(t_n, x_n)$ form an unbounded sequence. By this, we mean $d(y, \Phi(t_n, x_n)) \to \infty$ for some $y \in X$. After choosing a subsequence, we can also assume that $x_n \to x$ for some $x \in \tilde{A}$. Then $K = \{x_n; n \in \mathbb{N}\} \cup \{x\}$ is a compact set. By assumption, Φ is eventually bounded on K, a contradiction.

Step 2: If K is a compact subset of X, $A = \omega(U)$ attracts a neighborhood of K.

By Step 1, Φ is asymptotically compact on the open set $U = U_\delta(\tilde{A})$. By Proposition 2.29, $\omega(U)$ attracts a neighborhood of every compact set in $\bigcup \Phi^{-1}(U) = X$. The last equality holds because $\tilde{A}$ attracts all points in X and U is a neighborhood of $\tilde{A}$.

Step 3: If B is a subset of X and Φ is eventually bounded on B, then B is attracted by A.

Let B be a subset of X on which Φ is eventually bounded. Since Φ is asymptotically smooth, by Proposition 2.27, B has a compact attractor, namely $\omega(B)$. By Step 2, A attracts $\omega(B)$. By Theorem 2.11, A attracts B. □

When we drop asymptotic smoothness, we still have the following.

Theorem 2.31. *Let Φ be state-continuous. Then the following are equivalent:*

(a) *Φ has a compact attractor, A, of neighborhoods of compact sets in X.*

(b) *There exists an open set U with the following properties:*
   (i) *Φ is asymptotically compact on U,*
   (ii) *for every $x \in X$, there exists some $t \in J$ such that $\Phi(t, x) \in U$.*

(c) *Φ has a compact attractor $\tilde{A}$ of points in X that attracts a neighborhood of itself.*

If one and then all of (a), (b), (c) hold, $A = \omega(U) = \tilde{A}$.

Proof. Obviously (a) implies (c).

(c)⇒(b). Assume that there exists a compact attractor $\tilde{A}$ with the properties described in (c). Then $\tilde{A}$ attracts an open neighborhood U of itself. U has the properties (i) and (ii).

(b)⇒(a). Assume that there exists an open set U satisfying (i) and (ii). By (ii), $\bigcup \Phi^{-1}(U) = X$. By Proposition 2.29, $A := \omega(U)$ is the compact attractor with the desired properties.

Combining (c)⇒(b) and (b)⇒(a), we see that the attractor in (c) has the properties of the attractor in (a). Since the properties in (a) uniquely determine the attractor (Theorem 2.19), $A = \tilde{A}$. □

Corollary 2.32. *Let A be a compact attractor of points in X that attracts a neighborhood of itself. Then A is the compact attractor of neighborhoods of compact sets in X.*

In Chapter 8, a nonempty subset that attracts a neighborhood of itself will be called *locally attracting* (Definition 8.1). In the present context, this terminology does not seem particularly effective.

We have no answer to the following question: Is asymptotic smoothness necessary for the existence of a compact attractor of neighborhoods of compact sets?

2.2.3. Compact attractors of bounded sets. This type of compact attractor has been called global compact attractor in [**60, 91, 181, 210**]. Magal and Zhao [**158**] present examples of continuous semiflows (both with time-sets $\mathbb{Z}_+$ and $\mathbb{R}_+$) that are asymptotically smooth and have a compact attractor of neighborhoods of compact sets that is not a compact attractor of bounded sets.

The following result (cf. [**91**, Thm.2.4.6]) is immediate from Theorem 2.30.

Theorem 2.33. *The following statements are equivalent:*

(1) *There exists a compact attractor of bounded sets (which is unique and contains every bounded backward invariant set).*

(2) *Φ is point-dissipative, asymptotically smooth, and eventually bounded on every bounded set in X.*

The next result says that the compact attractor of bounded sets is the union of all bounded total trajectories. In particular, it contains all unstable manifolds of equilibria and periodic orbits (Exercise 2.18).

Proposition 2.34. *Let A be a compact attractor of bounded sets for the state-continuous semiflow Φ on X. Then A contains every bounded backward* *invariant set and*

$$A = \{x \in X : \exists \text{ bounded total trajectory } \phi \text{ through } x\}.$$

This is a special case of Lemma 2.18 and Theorem 2.20 where $\mathcal{C}$ is the class of bounded subsets of $\mathcal{C}$.

2.2.4. Elementary examples.

Example 2.35. Consider the scalar ordinary equation $x' = x^2(1-x)$ on $\mathbb{R}$. Every solution starting in $(-\infty, 0]$ converges to 0, and every solution starting in $(0, \infty)$ converges to 1. So the set $\{0, 1\}$ is a compact attractor of points. The compact attractor of bounded sets (which, in this case, coincides with the compact attractor of compact sets) is the interval $[0, 1]$, which is also a compact attractor of points. In particular, a compact attractor of points is not uniquely determined.

Example 2.36. The Lotka-Volterra competition model (LVCM):

$$\begin{aligned} x' &= rx(1 - x - ay), \\ y' &= sy(1 - bx - y), \end{aligned}$$

where $a, b, r, s > 0$, defines a continuous semiflow on $X = \mathbb{R}^2_+ = \mathbb{R}_+ \times \mathbb{R}_+$. By Remark 2.26 and Theorem 2.33, to establish the existence of the compact attractor of bounded sets, we need only show that the system is point dissipative and eventually bounded on every bounded set. Both of these properties can be verified using the differential inequalities

$$\begin{aligned} x' &\leq rx(1 - x), \\ y' &\leq sy(1 - y) \end{aligned}$$

satisfied by solutions of (LVCM). Since $x(t) \leq u(t)$ and $y(t) \leq v(t)$, where $u(0) = x(0)$ and $u(t)$ satisfies the differential equation $u' = ru(1-u)$ and $v(0) = y(0)$ and $v(t)$ satisfies the differential equation $v' = sv(1-v)$, it follows that:

(a) $B = [0, 1]^2$ attracts all points.

(b) $sB = [0, s]^2$, $s \geq 1$, is forward invariant.

Then (a) implies point dissipativity and (b) implies that the semiflow is eventually bounded on every bounded set. For if M is a bounded set in X, then $M \subset sB$ for some large enough s. Since sB is forward invariant, it contains the forward orbit of M. See Exercise 2.16 below for a precise description of the attractor of bounded sets.

Example 2.37. We return to our red thread, the SI type model of an infectious disease that reduces fertility, (0.1), rewritten in (0.4),

$$N' = N\Big(\beta(1-y) - \mu + (p\beta - \alpha)y\Big),$$
$$y' = y\Big((\kappa N - \alpha - \beta)(1-y) - p\beta y\Big).$$

Here $\beta > \mu > 0$, $\kappa > 0$ and $\alpha \geq 0$ and $p \in (0,1]$. Assume that $0 < p\beta < \mu + \alpha$. We consider the solution semiflow that is induced on the forward invariant set $X = \{(N,y), N > 0, 0 < y \leq 1\}$.

We will show that Φ is not eventually bounded on all bounded sets and so has no compact attractor of bounded set in X, but that Φ has a compact attractor of compact sets in X. Actually this compact attractor is the singleton set formed by the endemic equilibrium.

Notice that the state space in the previous example is not complete. Infinite-dimensional examples of semiflows on complete metric spaces that have a compact attractor of neighborhoods of compact sets, but not of bounded sets are given in [**158**] and the references therein.

Proof. It is useful to consider the solution semiflow induced on $\mathbb{R}_+ \times [0,1]$ as well, which has X as a forward invariant subset. There are two invariant subsets which are not equilibria: $M_1 = \{(0,y); 0 \leq y \leq 1\}$ and $M_2 = \{(N,0); N \geq 0\}$. On M_1, $y(t) \to 0$ as $t \to \infty$, exponentially fast while $N(t)$ grows exponentially on M_2. In particular, the origin is a saddle with M_1 being the stable and M_2 being the unstable manifold.

There is an interior equilibrium, (N^*, y^*), also called the endemic equilibrium in view of the epidemiological interpretation,

$$\begin{aligned} y^* &= \frac{\beta - \mu}{\beta(1-p) + \alpha}, \\ N^* &= \frac{(\alpha + \beta)(1 - y^*) + p\beta y^*}{1 - y^*}. \end{aligned} \tag{2.3}$$

Notice that $y^* < 1$ if and only if $\beta < \mu + \alpha$. We will show that Φ is not eventually bounded on the bounded set $B = \{(N^*, y); 0 < y \leq 1\}$. Suppose it is. Then there exists some $r > 0$ such that $\Phi([r,\infty) \times B)$ is bounded; in particular, there exists some $c > 0$ such that $N(t) \leq c$ for all $t \geq r$ and all solutions with initial data in B. For such solutions, $N(t) \leq N^* e^{(\beta-\mu)r}$ for all $t \in [0,r]$. So by choosing $c > 0$ large enough, $N(t) \leq c$ for all $t \geq 0$ and all solutions starting in B. For such solutions, $y' \leq \kappa c y$ for all $t \geq 0$ and $y(t) \leq y(0)e^{\kappa c t}$. So

$$N' \geq N(\beta - \mu - (\beta + \alpha)y(0)e^{\kappa c t}).$$

For $n \in \mathbb{N}$, choose

$$y(0) = \frac{\beta - \mu}{2(\beta + \alpha)} e^{-\kappa c n} =: \breve{y}_n$$

and let (N_n, y_n) be the solutions with initial data $(N^*, \breve{y}_n)$. Then

$$N_n' \geq N_n (1/2)(\beta - \mu), \qquad 0 \leq t \leq n$$

and $N_n(n) \geq N^* e^{(\beta - \mu)n/2} \to \infty$ as $n \to \infty$, a contradiction.

In order to show that Φ has a compact attractor on X, we will construct a forward invariant open set U that attracts all point in X and whose closure in $\mathbb{R}^2$ is contained in X. Then Φ is asymptotically compact on U and the desired attractor exists by Theorem 2.31. For the construction we need the isoclines of this planar system. The N-isocline is given by $y = y^*$; more precisely, $N' > 0$ if $y < y^*$ and $N' < 0$ if $y > y^*$. To determine the y-isocline, we solve the equation $0 = y'$,

$$y = \frac{\kappa N - \alpha - \beta}{\kappa N - \alpha - \beta + p\beta} =: \phi(N). \tag{2.4}$$

Notice that $\phi(N) \in [0, 1]$ if and only if $N \geq \frac{\alpha+\beta}{\kappa} = N_\diamond$. Further $\phi : [N_\diamond, \infty) \to \mathbb{R}$ is strictly increasing and concave, $\phi(N_\diamond) = 0$, $\phi(N) \to 1$ as $N \to \infty$. Further $y' > 0$ if $y < \phi(N)$ and $N' < 0$ if $y > \phi(N)$. We construct a closed curve in $(0, \infty) \times (0, 1]$ as follows: The first part is the solution curve of the system starting at $(N^*, 1)$. For increasing t, both $N(t)$ and $y(t)$ decrease until the solution curve intersects the N-isocline $y = y^*$. Then $N(t)$ increases and $y(t)$ decreases until the solution curve intersects the y isocline $y = \phi(N)$ at some point (N_1, y_1) with $0 < N_1 < N^*$ and $0 < y_1 < y^*$. This is the first part of the curve we are constructing. The second part is the horizontal line segment from (N_1, y_1) to (N_1, y^*). The third part is the solution curve that starts at (N_1, y^*). Both $N(t)$ and $y(t)$ increase until the solution curve intersects the N-isocline $y = y^*$ from where $N(t)$ decreases and $y(t)$ increases until the solution curves hits the N-isocline $y = \phi(N)$ at some point (N_2, y_2) with $N_2 > N^*$ and $y^* < y_2 < 1$. This is the third part of the curve. The forth part is the vertical line segment from (N_2, y_2) to $(N_2, 1)$, and the fifth part the horizontal line segment between $(N_2, 1)$ and $(N^*, 2)$. Let U be the interior of this closed curve. U is forward invariant because its boundary, the closed curve, either consists of solution curves which cannot be crossed by uniqueness or of line segments on which the direction field points into U. All solutions starting outside of U eventually enter U through one of the two line segments lying in U.

This concludes the proof that Φ has a compact attractor in X that attracts all compact subsets of X. As already explained in the introduction, the Poincaré-Bendixson theorem together with the Bendixson-Dulac criterion imply that the attractor is the singleton set $\{(N^*, y^*)\}$. □

2.2.5. Compact attractors and stability.

Definition 2.38. A forward invariant subset M of X is called *stable* if, for any neighborhood U of M there exists some neighborhood V of M such that $\Phi(J \times V) \subset U$.

A stable subset M of X is called *locally asymptotically stable* if there exists a neighborhood V of M such that M attracts all points in V.

The subsequent results are related to those in Sections 2.2 and 3.3 in [**91**] (see also [**158**]).

Theorem 2.39. *Let the semiflow Φ be state-continuous, uniformly in finite time, and A be a compact forward invariant subset of X that attracts all compact subsets of a neighborhood of itself (e.g. a local compact attractor of compact sets). Then A is locally asymptotically stable.*

Proof. We only need to prove the stability. Assume that A attracts all compact subsets of a neighborhood W of itself but that A is not stable. Then there exists a neighborhood U of A and sequences (x_k) in X and (t_k) in J such that $d(x_k, A) \to 0$ and $\Phi(t_k, x_k) \notin U$ for all $k \in \mathbb{N}$.

There exists a sequence (y_k) in A such that $d(x_k, y_k) \to 0$ as $k \to \infty$. After choosing subsequences, $y_k \to y$ as $k \to \infty$ for some $y \in A$ and $x_k \in W$ for all $k \in \mathbb{N}$. Then $x_k \to y$ as $k \to \infty$.

After choosing subsequences again, we can assume that (t_k) is bounded or $t_k \to \infty$ as $k \to \infty$.

Case 1: $t_k \to \infty$ as $k \to \infty$

Since $\{x_k; k \in \mathbb{N}\} \cup \{y\}$ is a compact subset of W, it is attracted by A. So $d(\Phi(t_k, x_k), A) \to 0$ as $k \to \infty$. This contradicts $\Phi(t_k, x_k) \notin U$ for all $k \in \mathbb{N}$.

Case 2: (t_k) is bounded.

Since Φ is state-continuous, uniformly in finite time,

$$d(\Phi(t_k, x_k), \Phi(t_k, y)) \to 0, \qquad k \to \infty.$$

Since A is forward invariant and $y \in A$,

$$d(\Phi(t_k, x_k), A) \to 0, \qquad k \to \infty.$$

Again, this contradicts $\Phi(t_k, x_k) \notin U$ for all $k \in \mathbb{N}$. $\square$

If Φ is asymptotically smooth and point-dissipative, there exists a compact attractor of points by Theorem 2.28, which is not necessarily unique (Example 2.35). This changes if this compact attractor of points is stable.

Theorem 2.40. *Let Φ be state-continuous and asymptotically smooth. Let A be a compact attractor of points that is stable. Then A is a compact attractor of neighborhoods of compact sets.*

Proof. Let A be a compact attractor of points that is stable. Let $\epsilon > 0$. Then $U_\epsilon(A) = \{x \in X; d(x, A) < \epsilon\}$ is an open bounded neighborhood of A (Exercise 1.8). Since A is stable, there exists a neighborhood V_ϵ of A such that $\Phi(J \times V_\epsilon) \subset U_\epsilon(A)$. So Φ is eventually bounded on V_ϵ. Since Φ is asymptotically smooth, $\omega(V_\epsilon)$ is a compact attractor of V_ϵ (Proposition 2.27). Since A is an invariant set contained in V_ϵ, $\omega(V_\epsilon)$ contains A and thus attracts all points in X. By Proposition 2.29, $\omega(V_\epsilon)$ attracts neighborhoods of compact sets in $\bigcup \Phi^{-1}(V_\epsilon) = \{x \in X; \exists t \in J : \Phi(t, x) \in V_\epsilon\}$. Since V_ϵ is a neighborhood of A and A attracts all points in X, $\bigcup \Phi^{-1}(V_\epsilon) = X$. So $\omega(V_\epsilon)$ is a compact attractor of neighborhoods of compact sets in X and as such unique. In particular, $\tilde{A} = \omega(V_\epsilon)$ does not depend on $\epsilon > 0$. Since $\tilde{A}$ is contained in the closure of $\Phi(J \times V_\epsilon) \subset U_\epsilon(A)$ for all $\epsilon > 0$, $\tilde{A} \subset A$ and so $\tilde{A} = A$. This implies that A is the compact attractor of neighborhoods of compact sets in X. □

We combine Theorem 2.39 and Theorem 2.40 (cf. [**93**, Cor.8.4]).

Corollary 2.41. *Let Φ be asymptotically smooth and state-continuous, uniformly in finite time. Then any compact attractor of compact sets is also a compact attractor of neighborhoods of compact sets.*

Example 2.45 below shows that asymptotic smoothness is necessary in Theorem 2.40 and Corollary 2.41. It presents a compact attractor of compact sets (of a continuous semiflow) that is, in particularly, stable but not a compact attractor of neighborhoods of compact sets.

We first show that, for *Lipschitz* semiflows, each compact attractor of points is also a compact attractor of compact sets. A semiflow $\Phi : J \times X \to X$ on a metric X is called a *Lipschitz semiflow* if there exists some $L > 0$ such that $d(\Phi(t, x), \Phi(t, y)) \le L d(x, y)$ for all $t \in J$, $x, y \in X$.

Theorem 2.42. *Let Φ be a Lipschitz semiflow on a metric space X. Then the following holds:*

(a) *Every compact invariant set is stable.*

(b) *If $A \subset X$ is a compact attractor of points, then it is a compact attractor of compact sets.*

Proof. (a) Let $x \in X$ and A a compact invariant set. Then

$$\begin{aligned} d(\Phi(t, x), A) =& d(\Phi(t, x), \Phi_t(A)) = \inf\{d(\Phi(t, x), \Phi(t, y)); y \in A\} \\ \le& \inf\{L d(x, y); y \in A\} = L d(x, A). \end{aligned}$$

Recall $U_\epsilon(A) = \{x \in X; d(x, A) < \epsilon\}$, the open ϵ-neighborhood of A. For each $\epsilon > 0$, $\Phi_t(U_\epsilon(A)) \subset U_{\epsilon L}(A)$. Now let U be a neighborhood of A. Since A is compact, there exists some $\epsilon > 0$ such that $U_{\epsilon L}(A) \subset U$. Then $\Phi_t(U_\epsilon(A)) \subset U$. This implies that A is stable.

(b) We adapt ideas from the proof of [**31**, Prop.4.1]. Let A be a compact attractor of points and suppose that A is not a compact attractor of compact sets. Then there exists a compact C such that A does not attract C: there exist $\epsilon > 0$, a sequence (t_n) in J, $t_n \to \infty$ as $n \to \infty$, and a sequence (x_n) in C such that $d(\Phi(t_n, x_n), A) > \epsilon$. After taking a subsequence we can assume that $x_n \to x$ as $n \to \infty$ for some $x \in C$. By Lemma 1.10,

$$\begin{aligned} d(\Phi(t_n, x_n), A)) \leq& d(\Phi(t_n, x), A) + d(\Phi(t_n, x_n), \Phi(t_n, x)) \\ \leq& d(\Phi(t_n, x), A) + Ld(x_n, x) \to 0 \end{aligned}$$

because A attracts $\{x\}$. This contradiction shows that A is a compact attractor of compact sets. □

We take a look at state-continuous semiflows on Banach spaces where the maps Φ_t interact is some way with the linear structure of the space. We write $U_\epsilon = U_\epsilon(0)$ for the ϵ-neighborhood of 0. The following theorem is inspired by Jack Hale's results for linear semiflows [**93**, Thm.6.1].

Theorem 2.43. *Let Φ be a state-continuous semiflow on a Banach space X with time-set $J = \mathbb{Z}_+$ or $J = \mathbb{R}_+$. Assume Φ is positively homogeneous of degree one:*

$$\Phi_t(\alpha x) = \alpha \Phi_t(x), \quad x \in X, t \in J, \alpha \geq 0. \tag{2.5}$$

Then 0 is a fixed point of Φ and the following hold:

(a) *$\{0\}$ is a compact attractor of bounded sets if and only if it attracts a neighborhood of itself.*

(b) *The following are equivalent:*

- *There exists a compact attractor of points.*
- *Φ is point-dissipative.*
- *$\{0\}$ is a compact attractor of points.*

Proof. (a) Since U_1 is bounded and a neighborhood of 0, 0 attracts a neighborhood of itself if it is the compact attractor of bounded sets. In turn, assume that 0 attracts a neighborhood U of itself. There exists some $\ell > 0$ such that $U_{1/\ell} \subset U$. So $\{0\}$ attracts $U_{1/\ell}$. Further let B be a bounded set. There exists some $m \in \mathbb{N}$ such that $B \subset U_m$. Notice that $U_m = \{m\ell y; y \in U_{1/\ell}\}$. By (2.5),

$$\begin{aligned} d(\Phi_t(U_m), 0) =& \sup\{\|\Phi_t(x)\|; x \in U_m\} \\ =& \sup\{\|\Phi_t(m\ell y)\|; y \in U_{1/\ell}\} \leq m\ell\, d(\Phi_t(U_{1/\ell}), 0). \end{aligned}$$

Since $\{0\}$ attracts $U_{1/\ell}$, it also attracts U_m and thus B.

(b) If there is a compact attractor of points, then Φ is obviously point-dissipative. In turn, assume that Φ is point-dissipative. Then there exists a bounded set that attracts all points. In particular, there exists some $r > 0$ such that the r-neighborhood of 0, U_r, attracts all points. Let $x \in X$. For all $n \in \mathbb{N}$, there exists some $t_n \in J$ such that $\Phi(t, nx) \in U_{r+1}$ for all $t \in J$, $t \geq t_n$. By (2.5),

$$\|\Phi(t,x)\| = \frac{1}{n}\|\Phi(t,nx)\| \leq \frac{r+1}{n}, \quad t \geq t_n.$$

This implies that $\Phi(t,x) \to 0$ as $t \to \infty$. □

Theorem 2.44 ([**93**, Thm.6.1]). *Let Φ be a point-dissipative continuous semiflow on a Banach space X with time-set $J = \mathbb{Z}_+$ or $J = \mathbb{R}_+$. Assume that all maps Φ_t, $t \in J$, are linear.*

Then $\{0\}$ is the compact attractor of compact sets. Further $\{0\}$ is the compact attractor of bounded sets if and only if it attracts a neighborhood of itself.

Proof. Since Φ is point-dissipative, there exists some $t \in J$ such that $\Phi(\cdot, x)$ is bounded on $J \cap [t, \infty)$. Since Φ is continuous, $\Phi(\cdot, x)$ is bounded on the compact set $J \cap [0, t]$ and thus bounded on J. By the uniform bounded principle [**100**, Thm.2.5.4] and linearity, Φ is a Lipschitz semiflow. The statements now follow from Theorem 2.43 and Theorem 2.42. □

The next example, for a compact attractor of compact sets that does not attract a neighborhood of itself, is due to Hale [**92**] and can also be found in [**31**].

Example 2.45. Let X be any of the sequence spaces ℓ^p with $1 \leq p < \infty$, $\|(x_n)\| = \left(\sum_{n=1}^{\infty} |x_n|^p\right)^{1/p}$, or c_0 of sequences (x_n) with $x_n \to 0$ as $n \to \infty$ $\|(x_n)\| = \sup_n |x_n|$. Let $J = \mathbb{Z}_+$ or $J = \mathbb{R}_+$. For $x = (x_n)$ define the multiplication semigroup $\Phi(t,x) = (e^{-\lambda_n t}x_n)_{n\in\mathbb{N}}$ where $\lambda_n > 0$ and $\lambda_n \to 0$ as $n \to \infty$. Notice that $T(t) = \Phi_t$ is a bounded linear operator and $\|T(t)\| \leq 1$ for the operator norm of $T(t)$. This implies that Φ is a Lipschitz semiflow (actually a contraction semigroup). The zero sequence forms a compact invariant set A in X. For each $m \in \mathbb{N}$,

$$\|\Phi(t,x)\| \leq \sup_{n\leq m} e^{-\lambda_n t}\|x\| + \|(x_{n+m})_{n\in\mathbb{N}}\|.$$

Since $\lambda_n > 0$, for every $m \in \mathbb{N}$, $\limsup_{t\to\infty} \|\Phi(t,x)\| \leq \|(x_{n+m})_{n\in\mathbb{N}}\|$. As $\|(x_{n+m})_{n\in\mathbb{N}}\| \to 0$ as $m \to \infty$, $\limsup_{t\to\infty} \|\Phi(t,x)\| = 0$. This shows that A is a compact attractor of points. By Theorem 2.42 or Theorem 2.44, A is a compact attractor of compact sets and stable.

To show that $\{0\}$ does not attract a neighborhood of itself, let e^k denote the sequence where the kth term is 1 and all other terms are 0. Then $\|e^k\| = 1$ and $\Phi(t, e^k) = e^{-\lambda_k t}$. Thus $d(\Phi_t(U_1), 0) \geq \sup\{e^{-\lambda_k t}; k \in \mathbb{N}\} = 1$ because $\lambda_k \to 0$ as $k \to \infty$. So 0 does not attract the bounded set U_1. By Theorem 2.43, $\{0\}$ does not attract a neighborhood of itself. In particular, it is not an attractor of neighborhoods of compact sets.

2.3. A sufficient condition for asymptotic smoothness

There are various conditions for asymptotic smoothness that can be verified in applications [**91**]. The following one will be used in Chapter 9.5. Recall the definition of the diameter of a set.

Theorem 2.46 (cf. [**91**, L.2.3.2]). *Let X be a closed subset of a Banach space E, J a time set and $\Phi : J \times X \to X$ a state-continuous semiflow. Then Φ is asymptotically smooth if there are maps $\Psi, \Theta : J \times X \to X$ such that*

$$\Phi(t, x) = \Theta(t, x) + \Psi(t, x)$$

and the following hold for any bounded closed set C that is forward invariant under Φ:

- $\liminf_{J \ni t \to \infty} \operatorname{diam} \Theta_t(C) = 0,$

and

- •• *there is some $r_C \in J$ such that $\Psi_t(C)$ has compact closure for all $t \in J$, $t \geq r_C$.*

Proof. Let C be a bounded closed set that is forward invariant under Φ. We will show that Φ is asymptotically compact on C. To this purpose let (t_k) and (x_k) be sequences in J and C respectively, $t_k \to \infty$ as $k \to \infty$. To show that $(\Phi(t_k, x_k))$ has a convergent subsequence, we show that $\{\Phi(t_k, x_k); k \in \mathbb{N}\}$ is totally bounded (Theorem 1.31). Let $\epsilon > 0$. Choose $r_C \in J$ according to ••. By •, we can choose some $s \in J$, $s \geq r_C$, such that $\operatorname{diam} \Theta_s(C) < \epsilon/2$ and $\Theta_s(C)$ is contained in a ball with radius $\epsilon/2$. Choose some $\ell \in \mathbb{N}$ such that $t_k \geq s$ for all $k > \ell$. The finite set $\{\Phi(t_k, x_k); k \leq \ell\}$ can be covered by finitely many balls of radius ϵ. Let $k > \ell$. Since Φ is a semiflow and C forward invariant,

$$\Phi(t_k, x_k) = \Phi(s, y_k), \qquad y_k = \Phi(t_k - s, x_k) \in C.$$

By assumption,

$$\Phi(t_k, x_k) = \Psi(s, y_k) + \Theta(s, y_k).$$

Recall $s \geq r_C$. Since $\Psi_s(C)$ has compact closure by ••, the set $\{\Psi(s, y_k); k \geq \ell\}$ can be covered by finitely many balls of radius $\epsilon/2$. Since $\{\Theta(s, y_k); k \geq \ell\} \subset \Theta_s(C)$ is contained in a ball with radius $\epsilon/2$, the set $\{\Phi(t_k, x_k); k \geq$

$\ell\}$ can be covered by finitely many balls of radius ϵ and so can the set $\{\Phi(t_k, x_k); k \in \mathbb{N}\}$. This means that the set $\{\Phi(t_k, x_k); k \in \mathbb{N}\}$ is totally bounded and has compact closure. □

2.4. α-limit sets of total trajectories

Let J be a time-set of $[0, \infty)$, $\hat{J} = J \cup (-J)$, $\Phi : J \times X \to X$ a state-continuous semiflow, and $\phi : \hat{J} \to X$ a total trajectory. If Φ is continuous (e.g. $J = \mathbb{Z}_+$ and Φ state-continuous, or $J = \mathbb{R}_+$ and Φ continuous), then $\phi : \hat{J} \to X$ is continuous.

The α-limit set of a total trajectory ϕ is defined as

$$\alpha(\phi) = \bigcap_{t \in J} \overline{\phi(-J_t)} \qquad -J_t = (-\infty, -t] \cap (-J). \tag{2.6}$$

We define the ω-limit set of ϕ as $\omega(\phi)$ as $\omega(\phi(0))$. It is easy to show that $\omega(\phi) = \omega(\phi(r))$ for all $r \in \hat{J}$.

The following alternative characterization is helpful.

Lemma 2.47. *Let ϕ be a total trajectory, $x \in X$. Then $x \in \alpha(\phi)$ if and only if there exists a sequence $-J \ni t_j \to -\infty$, $\phi(t_j) \to x$ as $j \to \infty$.*

Theorem 2.48. *Let $\phi : \hat{J} \to X$ be a total trajectory for a state-continuous semiflow Φ. Assume that $\phi(-J)$ has compact closure. Then the following hold:*

(1) *$\alpha(\phi)$ is nonempty, compact and invariant.*

(2) *$\phi(t) \to \alpha(\phi)$ as $t \to -\infty$.*

(3) *If Φ is a continuous semiflow ($J = [0, \infty)$), then $\alpha(\phi)$ is connected.*

(4) *If Φ is state-continuous, uniformly in finite time, then $\alpha(\phi)$ is flow-connected.*

The proof is left as an exercise (Exercise 2.11).

2.5. Invariant sets identified through Lyapunov functions

Let J be a time-set, $\hat{J} = J \cup (-J)$, X a metric space, and $\Phi : J \times X \to X$ a state-continuous semiflow.

Definition 2.49. If Y is a forward invariant subset of X, a continuous function $L : Y \to \mathbb{R}$ is called a *Lyapunov function* on Y if $L(\Phi(t, x)) \leq L(x)$ for all $x \in Y$ and $t \in J$.

Recall that for every element x in an invariant set A there exists a total Φ-trajectory with $\phi(0) = x$ (Theorem 1.40). The following result is then readily shown.

Lemma 2.50. *Let A be an invariant subset of X and let $L : A \to \mathbb{R}$ be continuous. Then L is a Lyapunov function on A if $L \circ \phi$ is decreasing on $\hat{J}$ for every total Φ-trajectory $\phi : \hat{J} \to A$.*

A function $f : D \to \mathbb{R}$, $D \subset \mathbb{R}$, is *decreasing* if $f(t) \leq f(s)$ whenever $s, t \in D$ and $s \leq t$. f is increasing if $-f$ is *decreasing.*

Proposition 2.51. *Let L be a Lyapunov function on the forward invariant subset Y of X.*

If $x \in Y$, then L is constant on $\omega(x)$, $L(y) = \lim_{t\to\infty} L(\Phi(t, x)) \leq L(x)$ for all $y \in \omega(x)$.

If $\phi : \ddot{J} \to Y$ is a total trajectory, then L is constant on $\alpha(\phi)$, $L(z) = \lim_{t\to-\infty} L(\phi(t)) \geq L(\phi(0))$ for all $z \in \alpha(x)$.

Proof. Let $x \in Y$. Since L is a Lyapunov function, and Φ a semiflow, $L(\Phi(t, x))$ is a decreasing function of $t \in J$. The assertion is trivially satisfied for $\omega(x)$ if this set is empty. Let $y \in \omega(x)$. By Lemma 2.8, there exists a sequences (t_j) in J such that $t_j \to \infty$ and $\Phi(t_j, x) \to y$. Since L is continuous and a Lyapunov function,

$$L(y) = \lim_{j\to\infty} L(\Phi(t_j, x)) = \lim_{t\to\infty} L(\Phi(t, x)) \leq L(x).$$

The proof for the α-limit set is similar. □

The next result takes the role of the LaSalle invariance principle in the context of compact attractors [**137**, Thm.6.3].

Theorem 2.52. *Let L be a Lyapunov function on A and $\tilde{A} \subset A$. Assume that L is constant on $\tilde{A}$. Further assume that, if $\phi : \hat{J} \to A$ is a total Φ-trajectory and $L \circ \phi$ is constant, then $\phi(\hat{J}) \subset \tilde{A}$.*

Then $A = \tilde{A}$.

Proof. Let $x \in A$. Let $\phi : \hat{J} \to A$ be a total Φ-trajectory with $\phi(0) = x$ which exists according to Theorem 1.40. Then $L \circ \phi : \hat{J} \to \mathbb{R}$ is decreasing and bounded. Since A is compact, $\omega(x)$ and $\alpha(x)$ are nonempty, compact, and invariant subsets of A. By Proposition 2.51, for all $y \in \omega(x)$ and all $z \in \alpha(x)$,

$$L(y) = \lim_{t\to\infty} L(\phi(t)), \qquad L(z) = \lim_{t\to-\infty} L(\phi(t)),$$

and L is constant on $\omega(x)$ and on $\alpha(x)$. To show that $\omega(x)$ and $\alpha(x)$ are subsets of $\tilde{A}$, let $y \in \omega(x)$. By Theorem 1.40, there exists a total Φ-trajectory $\psi : \hat{J} \to \omega(x)$ such that $\psi(0) = y$. Then $L \circ \psi$ is constant and $y \in \psi(\hat{J}) \subset \tilde{A}$

by assumption. This implies $\omega(x) \subset \tilde{A}$. By the same token, $\alpha(x) \in \tilde{A}$. Since L is constant on $\tilde{A}$, $\lim_{t\to\infty} L(\phi(t)) = \lim_{t\to-\infty} L(\phi(t))$. Since $L \circ \phi$ is decreasing, $L \circ \phi$ is constant. By assumption, $\phi(\hat{J}) \subset \tilde{A}$ and, in particular, $x = \phi(0) \in \tilde{A}$. Since x has been an arbitrary element in A, $A \subset \tilde{A}$. □

The next theorem is an immediate consequence of Theorem 2.52. Here, $\hat{J} = \mathbb{R}$.

Theorem 2.53. *Let $\Phi : \mathbb{R}_+ \times X \to X$ be a state-continuous semiflow, A a compact invariant set in X, and let $L : A \to \mathbb{R}$ be continuous. Assume that, for all total trajectories $\phi : \mathbb{R} \to A$, either $L \circ \phi$ is locally absolutely continuous on $\mathbb{R}$ and $(L\circ\phi)' \le 0$ a.e. or $(L\circ\phi)'_+ \le 0$ on $\mathbb{R}$. Further assume that $\tilde{A} \subset A$, L is constant on $\tilde{A}$ and that $\phi(\mathbb{R}) \subset \tilde{A}$ for any total Φ-trajectory with $(L \circ \phi)' \equiv 0$.*

Then $A = \tilde{A}$.

See Sections A.A.1 and A.A.2 for notations and concepts.

Proof. Obviously L is a Lyapunov function. Now let $\phi : \mathbb{R} \to \tilde{A}$ be a total trajectory such that $L \circ \phi$ is constant on $\mathbb{R}$. Then $(L \circ \phi)' \equiv 0$ on $\mathbb{R}$. By assumption $\phi(\mathbb{R}) \subset \tilde{A}$. The assumptions of Theorem 2.52 are satisfied. □

This theorem will be used for analyzing the dynamics of the cyclic competition model by May and Leonard (Section 5.4) and an endemic model with class-age dependent infectivity (Ch. 9, Sec. 9.9). Similar ideas have been used in [**155**].

2.6. Discrete semiflows induced by weak contractions

Let X be a metric space with metric d and $f : X \to X$ a weak contraction (or nonexpansive): $d(f(x), f(y)) \le d(x, y)$ for all $x, y \in X$. We consider the semiflow Φ with $\Phi_n = f^n$, $n \in \mathbb{Z}_+$, given by the n-fold iterate of f. Then $L : X \to \mathbb{R}_+$, $L(x) = d(x, f(x))$, is a Lyapunov function for the semiflow Φ as $L(f(x)) \le L(x)$. The next result follows from Proposition 2.51.

Lemma 2.54. *Let $x \in X$. Then there exists some $c_\omega \in [0, d(x, f(x))]$ such that $d(y, f(y)) = c_\omega$ for all $y \in \omega(x)$. If $\phi : \mathbb{Z} \to X$ is a total Φ-trajectory, $\phi(0) = x$, then there exists a constant $c_\alpha \ge d(x, f(x))$ such that $d(z, f(z)) = c_\alpha$ for all $z \in \alpha(\phi)$.*

Proposition 2.55. *Let $Y \subset X$ and $f(Y) \subset Y$. Assume that there exists some $c \ge 0$ such that $d(y, f(y)) = c$ for all $y \in Y$. Further assume that for any pair $y, z \in Y$ with $z \ne y$ there exists some $k \in \mathbb{N}$ such that $d(f^k(y), f^k(z)) < d(y, z)$. Then $c = 0$ and Y is a singleton set formed by a fixed point of f.*

Proof. It is enough to show that $c = 0$. Then every point in Y is a fixed point of f. By our extra assumption, fixed points in Y are unique.

So we suppose that $c > 0$. Pick some $y \in Y$. Then $y \neq f(y)$. By assumption, there exists some $k \in \mathbb{N}$ such that

$$d(y, f(y)) > d(f^k(y), f^k(f(y)) = d(f^k(y), f(f^k(y))).$$

Since Y is forward invariant, $f^k(y) \in Y$, so $d(f^k(y), f(f^k(y))) = c$ and $d(y, f(y)) > c$, a contradiction. This implies the assertion. □

Theorem 2.56. *Let K be a nonempty compact subset of X, $f(K) = K$. Assume that for each pair $y, z \in K$, $y \neq x$, there exists some $k \in \mathbb{N}$ such that $d(f^k(y), f^k(z)) < d(y, z)$. Then K is a singleton set formed by a fixed point of f.*

Proof. Let K be a compact invariant set and $x \in K$. By Theorem 1.40, there exists a total orbit $(x_n)_{n\in\mathbb{Z}}$ through x, $x_0 = x$. Since K is compact and invariant, $\omega(x)$ and $\alpha(x)$ are nonempty, compact and invariant. By Lemma 2.54, $d(y, f(y)) = c_\omega$ for all $y \in \omega(x)$ and $d(y, f(y)) = c_\alpha$ for all $y \in \alpha(x)$, $c_\omega \leq d(x, f(x)) \leq c_\alpha$. By Proposition 2.55, we have $c_\alpha = 0 = c_\omega$, so $d(x, f(x)) = 0$ and x is the unique fixed point of f in K. □

Theorem 2.57. *Let f be asymptotically smooth and let $\{f^n(x); n \in \mathbb{N}\}$ be bounded for some $x \in X$. Assume that for any compact invariant set K and for any $z, y \in K$ with $z \neq y$ there exists some $n \in \mathbb{N}$ such that $d(f^n(z), f^n(y)) < d(z, y)$. Then there exists a unique fixed point that attracts all bounded sets.*

Proof. Since $M_0 = \{f^n(x); n \in \mathbb{N}\}$ is bounded and forward invariant and f is asymptotically smooth, f is asymptotically compact on M_0. So $\omega(x)$ is nonempty, compact, and invariant. By Theorem 2.56, $\omega(x)$ is a singleton set that is formed by a fixed point $\tilde{x}$ of f.

Let B be a bounded subset of X. Then $B \subset U_r(\tilde{x})$ for some $r > 0$. $U_r(\tilde{x})$ is bounded and forward invariant because f is a weak contraction. Since f is asymptotically smooth, f is asymptotically compact on $U_r(\tilde{x})$. By Theorem 2.11, $\omega(U_r(\tilde{x}))$ is nonempty, compact, invariant and attracts $U_r(\tilde{x})$. By Theorem 2.56, $\omega(U_r(\tilde{x}))$ is a singleton set formed by a fixed point of f. Since $\omega(U_r(\tilde{x}))$ contains the fixed point $\tilde{x}$, $\omega(U_r(\tilde{x})) = \{\tilde{x}\}$. So $\tilde{x}$ attracts $U_r(\tilde{x}))$ and B as well. □

Example 2.58. The set $(0, \infty)$ is a complete metric space under the metric $d(x, y) = |\ln x - \ln y|$. This metric is a special case of the *part metric* (also called *Thompson metric* [**228**]). The metric spaces $((0, \infty), |\cdot|)$ and $((0, \infty), d)$ are topologically equivalent, i.e., they have the same open, closed, and compact sets, but the metrics are not equivalent. For instance, the set

$(0,1]$ is bounded in the first, but not the second metric space. Every interval $[a,b]$ with $0 < a < b < \infty$ is compact with respect to the metric d.

Let $f : (0,\infty) \to (0,\infty)$. If $f(x)/x$ decreases and $xf(x)$ increases in $x > 0$, then f is a weak contraction. If both the decrease and the increase are strict, f is a strict contraction: $d(f(x), f(y)) < d(x,y)$ if $x,y > 0$, $x \neq y$.

Assume that

$$\lim_{x\to\infty} f(x)/x < 1 < \lim_{x\to 0} f(x)/x,$$

with the letter possibly infinite. Then there exists a unique fixed point of x that attracts all compact subsets of $(0,\infty)$.

Proof. We show the strict statement. Let $x \neq y$. We can assume that $0 < x < y$. Then $d(x,y) = \ln y - \ln x$ and $\frac{f(x)}{x} > \frac{f(y)}{y}$. Since the logarithm is strictly increasing,

$$\ln f(x) - \ln x > \ln f(y) - \ln y.$$

So

$$\ln f(y) - \ln f(x) < \ln y - \ln x = d(y,x).$$

Further, $yf(y) > xf(x)$. Since the logarithm is strictly increasing,

$$\ln y + \ln f(y) > \ln x + \ln f(x),$$

so

$$\ln f(y) - \ln f(x) > \ln x - \ln y = -d(y,x).$$

We combine the two inequalities and obtain

$$|\ln f(y) - \ln f(x)| < d(y,x).$$

The other assumptions imply that each orbit is contained in a compact subset of $(0,\infty)$. The assertion now follows from Theorem 2.57. □

Discrete systems involving nonlinearities with properties like those in Example 2.58 have been considered in [**212**].

For generalizations of the part metric to higher finite and infinite dimensions see [**129, 209**].

Weak contractions do not need to have bounded orbits. Consider $f : (0,\infty) \to (\infty)$, $f(x) = \frac{x(2+x)}{1+x}$. Then f is strictly increasing and so $xf(x)$ is strictly increasing; further $f(x)/x$ is strictly decreasing. By the previous example, $d(f(x), f(y)) < d(x,y)$ for all $x,y \in (0,\infty)$, $x \neq y$, with the part metric d. However, since $f(x)/x > 1$ for all $x > 0$ and f is strictly increasing, $f^n(x) \to \infty$ as $n \to \infty$, $x > 0$.

2.7. Exercises

Exercise 2.1. Let $f : \mathbb{C} \to \mathbb{C}$ be given by $f(z) = z^2$ and let Φ be the induced discrete time semiflow on $\mathbb{C}$.

(a) Show that $K_1 = \{z : 1 < |z| < 2\}$ is backward invariant, $K_2 = \{z : |z| < 1/2\}$ is forward invariant, and $K = \{z : |z| = 1\}$ is invariant.

(b) Find several different total trajectories $\phi : \mathbb{Z} \to K$ with $\phi(0) = 1$.

(c) Determine $\omega(M)$ where $M = \{z : |z| = 1/2\}$ and prove your assertion. Do the same for $L = \{z : |z| \leq 1\}$.

Exercise 2.2. Prove Lemma 2.3.

Exercise 2.3. Prove Lemma 2.6.

Exercise 2.4. Prove Lemma 2.8.

Exercise 2.5. Let M be a subset of X and Φ a continuous semiflow on X. Show that the following two statements are equivalent:

(1) M has compact closure and Φ is asymptotically compact on M.

(2) $\Phi([0, \infty) \times M)$ has compact closure.

Exercise 2.6. Let A, B and C be nonempty subsets of a metric space X and let B be backward invariant or C attract a neighborhood of B.

Show: If B attracts A and C attracts B, then C attracts A.

Hint: If B is backward invariant, use $d(\Phi_t(A), C) \leq d(\Phi_t(A), B) + d(B, C)$ and $B \subset \Phi_t(B)$.

Exercise 2.7. Let Φ be a continuous semiflow on X and asymptotically compact on $M \subset X$. Show the following:

(a) Φ is asymptotically uniformly time-continuous on M in the following sense: For every $\epsilon > 0$ there exist $r > 0, \delta > 0$ such that $d(\Phi(t, x), \Phi(s, x)) < \epsilon$ for all $x \in M$, $t, s \geq r$, with $|t - s| < \delta$.

(b) Φ is eventually bounded on M, i.e., there exists some $r > 0$ such that $\Phi(([r, \infty) \cap J) \times M)$ is bounded.

Exercise 2.8. Let $\Phi : \mathbb{R}_+ \times X \to X$ be a semiflow, $M \subset X$. Show:

(1) If Φ is asymptotically compact on M, then, for every $r \in J$, $r > 0$, the discrete semiflow $(\Phi_{mr})_{m \in \mathbb{Z}_+}$ is asymptotically compact on M.

(2) If Φ is continuous and there is some $r \in J$, $r > 0$, such that the discrete semiflow $(\Phi_{mr})_{m \in \mathbb{N}}$ is asymptotically compact on M, then Φ is asymptotically compact on M.

Exercise 2.9. Let Φ be eventually bounded on $M \subset X$ and let Φ_r be a compact map for some $r \in J, r > 0$. Then Φ is asymptotically compact on M.

Recall: A compact map maps bounded sets into pre-compact sets.

Exercise 2.10. Prove Lemma 2.47 .

Exercise 2.11. Prove Theorem 2.48.

Exercise 2.12. Let $x \in X$, $r \in J$, $r > 0$, and $\Phi(r, x) = x$. Then there exists a periodic trajectory with $\phi(t) = \Phi(t, x)$ for $t \in [0, r]$ and $\phi(t+r) = \phi(t)$ for all $t \in \hat{J}$.

Hint: You may want to look at the proof of Theorem 1.40.

Exercise 2.13. (a) Let M be an invariant subset of X which consists of a single element x^*. Then x^* is an equilibrium of the semiflow, i.e., $\Phi(t, x^*) = x^*$ for all $t \geq 0$.

(b) Let Φ be asymptotically compact on $M \subset X$ and $\omega(M)$ consist of a single element x^*. Then x^* is an equilibrium and $\Phi(t, x) \to x^*$ as $t \to \infty$, uniformly in $x \in M$.

(c) Assume that the solutions of $x' = f(x)$ are uniquely determined by their initial data $x(0) = x_0$ in forward time. Further assume that the semiflow induced by $\Phi(t, x_0) = x(t)$ is a continuous semiflow. Then x^* is an equilibrium of Φ if and only if $f(x^*) = 0$.

Exercise 2.14. Let $f : \mathbb{R}^n_+ \to \mathbb{R}^n$ be locally Lipschitz continuous and satisfy the following conditions:

- For all $j = 1, \dots, n$, $f_j(x) \geq 0$ whenever $x \geq 0, x_j = 0$.

  Here f_j denotes the jth component of f, i.e., $f(x) = (f_1(x), \dots, f_n(x))$.
- There exist $b \in (0, \infty)^n$ and numbers $c_1, c_2 > 0$ such that

(2.7) $$\langle b, f(x)\rangle \leq c_1 + c_2\langle b, x\rangle \qquad \forall x \in \mathbb{R}^n_+.$$

(a) Show: For every $x_0 \in \mathbb{R}^n_+$, there exists a unique solution x of $x' = f(x)$, $x(0) = x_0$, which is defined and nonnegative for all nonnegative times. The semiflow Φ defined by $\Phi(t, x_0) = x(t)$ is a continuous semiflow from $[0, \infty) \times \mathbb{R}^n_+$ to $\mathbb{R}^n_+$.

Hint: Derive a differential inequality for $\langle b, x(t)\rangle$ where $b = (b_1, \dots, b_n)$.

Replace (2.7) by the following assumption:

- There exist numbers $c_3, \epsilon > 0$ such that

(2.8) $$\langle b, f(x)\rangle \leq c_3 - \epsilon\langle b, x\rangle \qquad \forall x \in \mathbb{R}^n_+.$$

(b) Show: Φ is point-dissipative.

(c) Show: Φ is eventually bounded on every bounded set.

(d) Show: There exists a compact attractor of bounded sets.

Exercise 2.15. Consider the quasilinear system

$$x_j' = \sum_{k=1}^{n} a_{jk}(x)x_k, \qquad j = 1, \dots, n, \tag{2.9}$$

with locally Lipschitz continuous functions $a_{jk} : [0,\infty)^n \to \mathbb{R}$. Further assume:

- If $x \in [0,\infty)^n$, $j \in \{1,\dots,n\}$, and $x_j = 0$, then $a_{jk}(x) \geq 0$ for all $k = 1,\dots,n$, $k \neq j$.
- There exist numbers $b_j, M_j > 0$, $j = 1,\dots,n$, such that

$$\sum_{j-1}^{n} b_j a_{jk}(x) \leq M_k \qquad \forall x \in \mathbb{R}_+^n, k = 1,\dots,n. \tag{2.10}$$

(a) Show: For every $x_0 \in \mathbb{R}_+^n$, there exists a unique solution x of (2.9), $x(0) = x_0$, which is defined and nonnegative for all nonnegative times. The semiflow Φ defined $\Phi(t,x_0) = x(t)$ is a continuous semiflow from $[0,\infty) \times \mathbb{R}_+^n$ to $\mathbb{R}_+^n$.

Make the following additional assumption:

- There exist numbers $c_j, \epsilon_j > 0$, $j = 1,\dots,n$, such that for every $k = 1,\dots,n$,

$$\sum_{j=1}^{n} b_j a_{jk}(x) \leq -\epsilon_k \qquad \text{whenever } x \in \mathbb{R}_+^n, x_k \geq c_k. \tag{2.11}$$

$b_1,\dots,b_n$ are the same as in (2.10).

(b) Show: There exists a compact attractor of bounded sets.

Exercise 2.16. The symmetric Lotka-Volterra competition system

$$\begin{aligned} x' &= x(1 - x - 3y), \\ y' &= y(1 - 3x - y), \end{aligned}$$

which you may assume generates a continuous semiflow on $\mathbb{R}_+^2$, has equilibria

$$E_0 = (0,0), \quad E_1 = (1,0), \quad E_2 = (0,1), \quad E_{12} = (1/4,1/4),$$

with E_1 and E_2 being the only asymptotically stable ones. By symmetry, the line $y = x$ is invariant, so the system has no periodic orbits.

(a) Describe the behavior of solutions starting on $y = x$.

(b) Describe all heteroclinic orbits (total trajectories) $E_i \to E_j$, $i,j \in \{0,1,2,12\}$ connecting different equilibria.

(c) Using Proposition 2.34, describe precisely the largest compact invariant set and the compact attractor of bounded sets.

Exercise 2.17. Let A be the compact attractor of compact sets for the state-continuous semiflow Φ on X. Then A contains every compact backward invariant set and

$$A = \{x \in X : \exists \text{ total trajectory } \phi \text{ through } x \text{ with pre-compact range}\}.$$

Hint: Apply Lemma 2.18 and Theorem 2.20.

Exercise 2.18. Let Φ be a continuous semiflow with a compact attractor A of compact sets. Let M be a compact invariant subset of X and let $W^u(M)$ be its unstable set consisting of all points $x \in X$ such that there is a total trajectory $\phi : J \cup (-J) \to X$, $\phi(0) = x$, with $\overline{\phi(-J)}$ compact and $\alpha(\phi) \subset M$. Show that $M \subset W^u(M) \subset A$. A compact attractor of compact sets contains the unstable sets of its compact invariant sets. What about the stable set of M?

Hint: Show that $\phi(-J \cup J)$ is pre-compact and invariant, then use the previous exercise. For pre-compactness, you will need time-continuity of Φ.

Chapter 3

Uniform Weak Persistence

3.1. Persistence definitions

Let X be an arbitrary nonempty set and $\rho : X \to \mathbb{R}_+$.

Definition 3.1. A semiflow $\Phi : J \times X \to X$ is called *weakly ρ-persistent*, if

$$\limsup_{t\to\infty} \rho(\Phi(t,x)) > 0 \qquad \forall x \in X, \rho(x) > 0.$$

Φ is called *strongly ρ-persistent*, if

$$\liminf_{t\to\infty} \rho(\Phi(t,x)) > 0 \qquad \forall x \in X, \rho(x) > 0.$$

A semiflow $\Phi : J \times X \to X$ is called *uniformly weakly ρ-persistent*, if there exists some $\epsilon > 0$ such that

$$\limsup_{t\to\infty} \rho(\Phi(t,x)) > \epsilon \qquad \forall x \in X, \rho(x) > 0.$$

Φ is called *uniformly (strongly) ρ-persistent*, if there exists some $\epsilon > 0$ such that

$$\liminf_{t\to\infty} \rho(\Phi(t,x)) > \epsilon \qquad \forall x \in X, \rho(x) > 0.$$

It is convenient to introduce the following concepts as well.

Definition 3.2. A semiflow $\Phi : J \times X \to X$ is called *ρ-dissipative*, if there exists some $c > 0$ such that

$$\limsup_{t\to\infty} \rho(\Phi(t,x)) < c \qquad \forall x \in X.$$

Φ is called *weakly ρ-dissipative*, if there exists some $c > 0$ such that

$$\liminf_{t\to\infty} \rho(\Phi(t,x)) < c \qquad \forall x \in X.$$

Φ is called *ρ-permanent*, if Φ is both ρ-dissipative and uniformly ρ-persistent.

Mathematically, there is a connection between persistence and dissipativity: Φ is ρ-dissipative if and only if it is uniformly $\tilde{\rho}$-persistent with $\tilde{\rho}(x) = \frac{1}{1+\rho(x)}$.

In order to get a feel for these definitions we revisit the endemic model already considered in the Introduction.

3.1.1. An SI endemic model for a fertility reducing infectious disease. Recall the following endemic model for a fertility reducing infectious disease caused by a viral, bacterial, or fungal parasite. S and I denote the respective numbers of susceptible and infective hosts,

$$S' = (\beta - \mu)S + q\beta I - \kappa SI, \qquad I' = \kappa SI - (\mu + \alpha)I. \tag{3.1}$$

Here $\beta > \mu > 0$ are the per capita birth and death rates, $\kappa > 0$ is the per capita infection rate and $\alpha \geq 0$ the additional death rate due to the disease. The factor $q \in [0,1]$ expresses the reduction of fertility for an infective individual.

It follows from standard arguments in ordinary differential equations that, for each pair $S_0, I_0 \geq 0$, there exist unique solutions $S, I : \mathbb{R}_+ \to \mathbb{R}_+$ with $S(0) = S_0$, $I(0) = I_0$, where $\mathbb{R}_+ = [0,\infty)$ is the set of nonnegative real numbers. Then

$$\Phi(t, (S_0, I_0)) = (S(t), I(t))$$

defines a semiflow with state space $\mathbb{R}^2_+$ and time-set $\mathbb{R}_+$. The semiflow property follows from the uniqueness of solutions. If Φ is induced by the solutions of a differential equation as in this case, it is called the *solution semiflow.*

If $q = 0$ (i.e., the disease sterilizes), the host-parasite system becomes a special case of the Lotka-Volterra prey-predator system for which the whole first quadrant is filled with periodic orbits. This means that the system is uniformly weakly persistent for both host and parasite ($\rho(S,I) = S + I$ and $\rho(S,I) = I$, respectively), but uniformly persistent for neither. Concurrently, the system is weakly ρ-dissipative, but not dissipative for $\rho(S,I) = S + I$ or $\rho(S,I) = I$.

For a clearer distinction between host and parasite persistence, we write the system in terms of the total size of the host population, $N = S + I$, and the fraction of infected hosts, $y = I/N$. Since $S = N - I$ and $I = yN$,

$$N' = N\Big(\beta(1-y) - \mu + (q\beta - \alpha)y\Big) \tag{3.2}$$

and

$$y' = y\Big((\kappa N - \alpha - \beta)(1-y) - q\beta y\Big). \tag{3.3}$$

Note the change in state space that becomes $X = \mathbb{R}_+ \times [0,1]$. We explore persistence and dissipativity (eventual boundedness) of the host population.

Theorem 3.3. *Let $\rho(N, f) = N$. If $\beta > \mu$ and $q \in (0,1]$, then the solution semiflow of (3.2) and (3.3) is uniformly weakly ρ-persistent on the state space $X = \mathbb{R}_+ \times [0,1]$.*

If $\beta > \mu$ and $q = 0$, then the solution semiflow of (3.2) and (3.3) is uniformly weakly ρ-persistent on the state space $\check{X} = \mathbb{R}_+ \times [0,1)$.

If $q\beta < \mu + \alpha$, the solution semiflow is weakly ρ-dissipative on the state space $\tilde{X} = \mathbb{R}_+ \times (0,1]$.

If $q = 0$, uniform weak host persistence does not hold on the state space $X = \mathbb{R}_+ \times [0,1]$ because it contains the invariant set $\mathbb{R}_+ \times \{1\}$ on which the host is driven to extinction by the disease. Since $\check{X} = \mathbb{R}_+ \times [0,1)$ is not complete, our forthcoming methods will fail to provide uniform host persistence for $q = 0$ as they should because uniform host persistence does not hold in that case.

Proof. We argue by contradiction. Suppose that $\beta > \mu$ and Φ is not weakly ρ-persistent. Then, for any arbitrary $\epsilon > 0$ (which we will choose later), there exists a solution with $N(0) > 0$ and $\limsup_{t\to\infty} N(t) < \epsilon$. This implies $y(0) > 0$, otherwise $N(t) \to \infty$ as $t \to \infty$. Then $N(t) > 0$ and $y(t) > 0$ for all $t \geq 0$. Further there exists some $t_0 > 0$ such that $N(t) < \epsilon$ for all $t \geq t_0$. Also, for $t \geq t_0$, by (3.3),

$$\frac{y'}{y} \leq (\kappa\epsilon - \alpha - \beta)(1-y) - q\beta y.$$

Choose $\epsilon > 0$ small enough such that the right-hand side is negative or zero. Then y is monotone decreasing and $y(t) \to \tilde{y}$ for some $\tilde{y} \in [0,1]$. Assume $\tilde{y} > 0$.

If $q > 0$, the right-hand side of the inequality is bounded away from 0 and $y(t) \to 0$. If $q = 0$, we consider the state space $\check{X}$, so $y(0) < 1$. Then $y(t) \leq y(0) < 1$ for all $t \geq 0$ and also $\tilde{y} < 1$. Again the right-hand side is negative and bounded away from 0, and we obtain that $y(t) \to 0$. In either case, $y(t) \to 0$ as $t \to \infty$.

By (3.2),

$$\liminf_{t\to\infty} \frac{N'(t)}{N(t)} \geq \beta - \mu > 0,$$

and N increases exponentially, a contradiction.

Now assume that $q\beta \leq \mu + \alpha$ and that the semiflow is not weakly ρ-dissipative on $\tilde{X}$. Then, for any arbitrary $c > 0$, there exists some solution with $y(0) > 0$ such that $\liminf_{t\to\infty} N(t) > c$. Then $y(t) > 0$ for all $t \geq 0$. If $c > 0$ is chosen so large that $\kappa c > \alpha + \beta$, y remains bounded away from 0 because $y' > 0$ if y is close to 0. This implies that $y_\infty = \liminf_{t\to\infty} y(t) > 0$. By the fluctuation method (Theorem A.14), there exists a sequence $t_k \to \infty$ such that $y(t_k) \to y_\infty$ and $y'(t_k) \to 0$ as $t \to \infty$. From (3.3),

$$\begin{aligned} 0 = &\lim_{k\to\infty} \Big((\kappa N(t_k) - \alpha - \beta)(1 - y(t_k)) - q\beta y(t_k) \Big) \\ \geq &(\kappa c - \alpha - \beta)(1 - y_\infty) - q\beta y_\infty. \end{aligned}$$

We solve this inequality for y_∞,

$$y_\infty \geq \frac{\kappa c - \alpha - \beta}{\kappa c - \alpha - \beta + q\beta}.$$

Notice that y_∞ converges to 1 as $c \to \infty$. From (3.2), since its right-hand side decreases in y,

$$\limsup_{t\to\infty} \frac{N'(t)}{N(t)} \leq \Big(\beta(1 - y_\infty) - \mu + (q\beta - \alpha) y_\infty \Big).$$

The right-hand side of this inequality is negative for $y_\infty = 1$. Since $y_\infty \to 1$ as $c \to \infty$, the right-hand side is also negative if c is sufficiently large. Then $N(t) \to 0$ as $t \to 0$, a contradiction. □

After having shown some of the ad hoc methods used for establishing uniform weak persistence without hopefully having delved too far into technicalities, we add more disease stages and spatial heterogeneity. We will also use another type of disease incidence and replace the infection term κSI by $\kappa SI/N$.

3.2. An SEIRS epidemic model in patchy host populations

Motivated by [**185**] we consider a host population which is geographically distributed over n patches (regions, countries). (For more general results see [**54**].) The disease divides each patch population into four epidemiological classes (or compartments): susceptible individuals (represented by letter S), exposed individuals (infected but are not yet infective, represented by letter E), infective individuals (represented by letter I, and removed (or recovered) individuals (represented by letter R). We include that recovered individuals can become susceptible again, so our model will be of SEIRS type. We will use C $\in \{S, E, I, R\}$ as generic letter for an unspecified epidemiological class. We assume that disease transmission and transition between epidemiological classes only occurs on patches and not during travel.

Let $S_i(t), E_i(t), I_i(t), R_i(t)$ denote the respective numbers of susceptible, exposed, infective, and removed individuals in patch i at time t and $N_i(t)$ the total population size in patch i at time t. We introduce the following vector notation for the various epidemiological classes,

$$
\begin{aligned}
S = & \ (S_1, \ldots, S_n), & E = & \ (E_1, \ldots, E_n), \\
I = & \ (I_1, \ldots, I_n), & R = & \ (R_1, \ldots, R_n).
\end{aligned} \tag{3.4}
$$

The dynamics of the host population and the disease are described by the following system,

$$
\begin{aligned}
\frac{dS_i}{dt} &= \Lambda_i - \kappa_i \frac{S_i I_i}{N_i} - \mu_i^S S_i + \gamma_i^{\mathrm{R}} R_i + \sum_{k=1}^{n} (m_{ik}^S S_k - m_{ki}^S S_i), \\
\frac{dE_i}{dt} &= \kappa_i \frac{S_i I_i}{N_i} - (\gamma_i^E + \mu_i^E) E_i + \sum_{k=1}^{n} (m_{ik}^E E_k - m_{ki}^E E_i), \\
\frac{dI_i}{dt} &= \gamma_i^E E_i - (\gamma_i^I + \mu_i^I) I_i + \sum_{k=1}^{n} (m_{ik}^I I_k - m_{ki}^I I_i), \\
\frac{dR_i}{dt} &= \gamma_i^I I_i - (\gamma_i^R + \mu_i^R) R_i + \sum_{k=1}^{n} (m_{ik}^R R_k - m_{ki}^R R_i), \\
N_i &= S_i + E_i + I_i + R_i.
\end{aligned} \tag{3.5}
$$

Individuals are recruited into the local population of patch i at a rate Λ_i. This may happen by immigration or birth, or, in case of a sexually transmitted disease, by entering the sexually active part of the patch population. Individuals on patch i and in the epidemiological class $C \in \{S, E, I, R\}$ die at a per capita rate μ_i^C and move to the next epidemiological class at a per capita rate γ_i^C. Disease transmission is modeled by frequency-dependent (or standard) incidence with κ_i being the per capita infection rate on patch i. Individuals in epidemiological class C migrate from patch k to patch i at a per capita rate m_{ik}^C. Without loss of generality we can assume that $m_{kk}^C = 0,\ 1 \leq k \leq n,\ C = S, E, I, R$.

The differential equations above suggest that it will be useful to introduce matrices $M^C = (M_{ij}^C)$, $C = S, E, I, R$, the migration matrices, defined by $M_{ij}^C = m_{ij}^C$, $i \neq j$ and $M_{ii}^C = -\sum_k m_{ik}^C$, $1 \leq i \leq n$. Then M^C is quasi-positive (meaning that its off-diagonal entries are nonnegative, see Section A.A.5), and it has zero column sums. Let γ^C, μ^C, and κ denote diagonal matrices with the obvious diagonal entries taken from the equations (3.5).

The following assumptions are made throughout the section.

ASSUMPTION 3.1. *All parameter constants are nonnegative;*

(a) *for all* i*:* $\mu_i^C > 0,\ \kappa_i > 0,\ \gamma_i^E > 0,$

(b) $\sum_i \Lambda_i > 0,$

(c) *the matrix* $M^S = (m_{ik}^S)$ *is irreducible.*

Hypothesis (a) just says that death rates are positive, the disease contact rate is positive, and disease progression does not stall in the exposed class. As we assume positive death rates, we must have some influx into the population to prevent population collapse. This is the point of hypothesis (b). Hypothesis (c) is important since it says that susceptible individuals get around to all the patches and thus will come into contact with infectives if there are any. Note that infectives are not assumed able to get around to all patches.

We write $v \geq u$ ($v \gg u$) for vectors u, v when their components satisfy $v_i \geq u_i$ ($v_i > u_i$); similar notation is used for matrices, e.g., $A \gg 0$ means all entries are positive.

For $N_i = 0$, we define $\frac{S_i I_i}{N_i} = 0$. The expression $\frac{S_i I_i}{N_i}$ is then a locally Lipschitz continuous function of $(S_i, E_i, I_i, R_i) \in \mathbb{R}_+^4$. We define the total population size as

$$\bar{N}(t) = \sum_{i=1}^n N_i. \tag{3.6}$$

Theorem 3.4. *For any* $S^\circ, E^\circ, I^\circ, R^\circ \in \mathbb{R}_+^n$ *there exists a unique solution* $(S(t), E(t), I(t), R(t))$ *of (3.5) with initial data* $(S^\circ, E^\circ, I^\circ, R^\circ)$ *which is defined for all* $t \geq 0$ *and takes values in* $\mathbb{R}_+^{4n}$. *There exists some* $c > 0$ *such that, for all nonnegative solutions of (3.5),* $\limsup_{t\to\infty} \bar{N}(t) \leq c$ *and* $\bar{N}(t) \leq \max\{c, \bar{N}(0)\}$ *for all* $t \geq 0$.

Proof. By Proposition A.17, there exists a solution with values in $\mathbb{R}_+^{4n}$ which is defined on some interval $[0, b)$ with $b \in (0, \infty]$. If $b < \infty$, then $\limsup_{t\to b-} \bar{N}(t) = \infty$. We add all equations in system (3.5),

$$\frac{d\bar{N}}{dt} = \sum_{i=1}^n \Big(\Lambda_i - \mu_i^S S_i - \mu_i^E E_i - \mu_i^I I_i - \mu_i^R R_i\Big). \tag{3.7}$$

Set $\mu_m = \min_{i=1}^n \{\mu_i^C; C = S, E, I, R\}$, then $\mu_m > 0$ and

$$\bar{N}' \leq \sum_{i=1}^n \Lambda_i - \mu_m \bar{N}.$$

Now our first assertion follows by arguments given following Theorem A.19. □

Remark 3.5. We will often use the fact that a quasipositive matrix A with negative column sums satisfies $\mathbf{s}(A) < 0$ where $\mathbf{s}(A)$ denotes the stability modulus of A, the maximum of the real parts of the eigenvalues. See Exercise 3.2. If the quasipositive matrix A satisfies $\mathbf{s}(A) < 0$, then $-A^{-1} \geq 0$; if, in addition, it is irreducible, then $-A^{-1} \gg 0$. See Exercise A.6.

We make heavy use of the Perron-Frobenius theory in this section so the reader is advised to review Appendix A.A.5. We use that $e^{At} \geq 0,\ t \geq 0$ for a quasipositive matrix and that $e^{At} \gg 0,\ t > 0$ if it is also irreducible. See (A.12).

Differential inequalities for systems will also be used extensively. See especially the comparison Theorem A.19 in Appendix A.A.4.

The next result gives conditions guaranteeing that the population on each (or some) patch uniformly persists.

Proposition 3.6. *$S(t) \gg 0$ for $t > 0$. The population of every patch persists: $\exists\ c > 0$ such that for all solutions and all i, $\liminf_{t\to\infty} S_i(t) > c$.*

Proof. S satisfies

$$S' \geq \Lambda + (M^S - \mu^S - \kappa^S)S,$$

so

$$S(t) \geq e^{(M^S-\mu^S-\kappa^S)t}S(0) + \int_0^t e^{(M^S-\mu^S-\kappa^S)(t-s)}\Lambda ds.$$

This inequality can be seen either by applying the Comparison Theorem A.19 or directly by multiplying both sides of the differential inequality by the positive matrix $e^{(M^S-\mu^S-\kappa^S)t}$ and integrating. As $\Lambda > 0$ and the irreducibility of $M^S - \mu^S - \kappa^S$ ensures that the matrix exponential has all positive entries, $S(t) > 0$ for $t > 0$. Moreover,

$$e^{(M^S-\mu^S-\kappa^S)t}S(0) + \int_0^t e^{(M^S-\mu^S-\kappa^S)(t-s)}\Lambda ds \longrightarrow -(M^S - \mu^S - \kappa^S)^{-1}\Lambda,$$

because $\mathbf{s}(M^S - \mu^S - \kappa^S) < 0$ by Remark 3.5. Note the irreducibility of $M^S - \mu^S - \kappa^S$ ensures that $-(M^S - \mu^S - \kappa^S)^{-1} \gg 0$. □

3.2.1. Stability of the disease-free state. The subspace $E = I = R = 0$ is invariant and gives the disease-free dynamics on which $S = N$ and

$$\frac{dN_i}{dt} = \Lambda_i - \mu_i^S N_i + \sum_{k=1}^n \left(m_{ik}^S N_k - m_{ki}^S N_i\right)$$

or in matrix-vector notation

$$N' = \Lambda + (M^S - \mu^S)N.$$

As the sum of the entries of the i-th column of $M^S - \mu^S$ is negative, $\mathbf{s}(M^S - \mu^S) < 0$, $-(M^S - \mu^S)^{-1} \gg 0$, the equilibrium solution of the disease-free system is given by

$$N = \hat{S} = -(M^S - \mu^S)^{-1}\Lambda \gg 0.$$

The disease-free equilibrium (DFE) of (3.5) is given by

$$S_i = \hat{S}_i, \;\; E_i = I_i = R_i = 0, \;\; 1 \le i \le n \tag{3.8}$$

We can linearize at (DFE). The $4n \times 4n$ Jacobian matrix, variables ordered as (S, E, I, R), takes the form below where each entry represents an $n \times n$ matrix:

$$J = \begin{pmatrix} M^S - \mu^S & 0 & -\kappa & \gamma^R \\ 0 & M^E - \mu^E - \gamma^E & \kappa & 0 \\ 0 & \gamma^E & M^I - \mu^I - \gamma^I & 0 \\ 0 & 0 & \gamma^I & M^R - \mu^R - \gamma^R \end{pmatrix}.$$

The spectrum of J consists of the spectrum of $M^S - \mu^S$ and the spectrum of the quasipositive matrix

$$\begin{pmatrix} M^E - \mu^E - \gamma^E & \kappa & 0 \\ \gamma^E & M^I - \mu^I - \gamma^I & 0 \\ 0 & \gamma^I & M^R - \mu^R - \gamma^R \end{pmatrix},$$

which in turn consists of the spectrum of $M^R - \mu^R - \gamma^R$ and that of the quasipositive matrix

$$B = \begin{pmatrix} M^E - \mu^E - \gamma^E & \kappa \\ \gamma^E & M^I - \mu^I - \gamma^I \end{pmatrix}. \tag{3.9}$$

Proposition 3.7. *(DFE) is stable if* $\mathbf{s}(B) < 0$ *and unstable if* $\mathbf{s}(B) > 0$.

The proof is left to Exercise 3.5. Matrix B is irreducible if $\kappa_i, \gamma_i^E > 0$ for all i and either m^E or m^I is irreducible. See Exercise 3.6.

Theorem 3.8. *If* $\mathbf{s}(B) < 0$, *then* $(E, I) \to 0$ *as* $t \to \infty$ *for all solutions of (3.5). (DFE) is globally asymptotically stable.*

The proof is left to Exercise 3.7.

So far it is not clear how to estimate $\mathbf{s}(B)$ or even whether $\mathbf{s}(B) > 0$ is possible! The reader may find nice estimates of a quantity equivalent to $\mathbf{s}(B)$, namely the basic reproduction number $\mathcal{R}_0$, in [**185**]. It is useful to think of B as a function of one of its parameters, say κ, which we will view both as a diagonal matrix and as a vector: $B = B(\kappa)$. It is well known that $\mathbf{s}(B)$ is increasing in each of its entries b_{ij} [**15**]. Therefore, $\mathbf{s}(B(\kappa))$ is increasing in the variables κ_i, and when $\kappa = 0$, then $\mathbf{s}(B(0)) < 0$. We show that $\mathbf{s}(B(\kappa)) > 0$ for large κ by finding a positive vector $v = (v_1, v_2)^*$ and $r > 0$ such that $Bv \geq rv$. It then follows that $\mathbf{s}(B) \geq r > 0$ (see appendix [**217**]). Now $Bv = (Cv_1 + \kappa v_2, \gamma^E v_2 + Dv_1)$ where $C = M^E - \mu^E - \gamma^E$ and $D = M^I - \mu^I - \gamma^I$. Assume M^E and M^I are irreducible, so the same holds for C, D. Then there exists $v_1, v_2 \gg 0$ such that $Cv_1 = r_1 v_1$, $Dv_2 = r_2 v_2$ where $r_1 = \mathbf{s}(C) < 0$ and $r_2 = \mathbf{s}(D) < 0$. Now $Bv = (r_1 v_1 + \kappa v_2, \gamma^E v_1 + r_2 v_2)$. We may replace v_1 by a positive multiple tv_1, which we rename as v_1, where t is so large that $\gamma^E v_1 + r_2 v_2 \geq v_2$. Now choose κ sufficiently large that $r_1 v_1 + \kappa v_2 \geq v_1$. Then for such κ, $\mathbf{s}(B(\kappa)) \geq 1$.

3.2.2. Weak uniform persistence of the disease. We show below that the disease is endemic in the population, in the sense that the flow Φ is uniformly weakly ρ-persistent, with $\rho(S, E, I, R) = \sum_i I_i$, if $\mathbf{s}(B) > 0$. The same hypotheses are shown to imply that the disease persists in the stronger sense of uniform ρ-persistence in Theorem 4.6 and this is later extended in Theorem 4.23 to uniform ρ-persistence for $\rho(S, E, I, R) = \min_i \min\{E_i, I_i\}$.

Proposition 3.9. *Let B be irreducible. Then*

$$\sum_i (I_i(0) + E_i(0)) > 0 \implies E_j(t) > 0, I_j(t) > 0, \ 1 \leq j \leq n, \ t > 0. \tag{3.10}$$

If, in addition, $\mathbf{s}(B) > 0$, then there exists $\epsilon > 0$ such that

$$\limsup_{t\to\infty} \sum_i I_i(t) > \epsilon$$

for all solutions of (3.5) satisfying $\sum_i (I_i(0) + E_i(0)) > 0$.

Proof. Let $v(t)$ denote the column vector $(E(t), I(t))$ and observe that it satisfies the linear differential equation

$$v'(t) = \tilde{B}(t)v(t)$$

where $\tilde{B}(t)$ has the same $n \times n$ blocks as B in (3.9) except that its upper-right off-diagonal block is the diagonal matrix with diagonal entries $\kappa_i S_i(t)/N_i(t)$. These entries are positive by Proposition 3.6. Consequently, the irreducibility of $\tilde{B}(t)$ follows from that of B. Hence, (3.10) follows from Proposition A.28.

If $\mathbf{s}(B) > 0$, there exists $\delta > 0$ satisfying $\min_i \kappa_i > \delta > 0$ such that $\mathbf{s}(\hat{B}) > 0$ where $\hat{B}$ is obtained from B by subtracting δ from each κ_i.

Suppose the assertion is false. Then for arbitrarily small $\epsilon > 0$, there exists a solution satisfying $\sum_i (I_i(0) + E_i(0)) > 0$ and $\sum_i I_i(t) < \epsilon$ for all large t. As noted above, $\sum_i (I_i(0) + E_i(0)) > 0$ implies that $\sum_i E_i(t) > 0$ and $\sum_i I_i(t) > 0$ for all $t > 0$.

Then, from the differential inequality

$$E' \leq \kappa I + (M^E - \mu^E - \gamma^E)E,$$

we deduce that $\sum E_i(t) = O(\epsilon)$ for all large t. By this, we mean that there exists $\epsilon_0 > 0$ and $c > 0$ such that $|\sum E_i(t)| \leq c\epsilon$ for $0 < \epsilon \leq \epsilon_0$ and all large t. Similarly, from

$$R' = \gamma^I I + (M^R - \mu^R - \gamma^R)R$$

we find that $\sum_i R_i(t) = O(\epsilon)$ for all large t. Now, using Proposition 3.6, we have

$$\frac{S_i(t)}{N_i(t)} = \frac{S_i(t)}{S_i(t) + E_i(t) + I_i(t) + R_i(t)} = \frac{S_i(t)}{S_i(t) + O(\epsilon)} = 1 - O(\epsilon)$$

for large t. Therefore, by choosing $\epsilon > 0$ sufficiently small, we find such a solution such that $z' \geq \hat{B}z$ for all large t where $z = (E, I)^*$ and $z(t) \neq 0$. As $\hat{B}$ is irreducible, there exists $w \gg 0$ such that $\hat{B}^* w = \mathbf{s}(\hat{B})w$. Let $u(t) = w^* z(t)$. Then $u(t) > 0$. Taking the product of the differential inequality above with w^*, we get $u'(t) \geq w^* \hat{B} z = \mathbf{s}(\hat{B})u(t)$ for all large t. See Corollary A.27. As $\mathbf{s}(\hat{B}) > 0$, we conclude that $u(t) \to \infty$, $t \to \infty$, contradicting our conclusions above. □

Remark 3.10. Suppose we modify equations (3.5) by adding $\eta \geq 0$ to each equation for C_i, $C = E, I, R$, $1 \leq i \leq n$; think of this as allowing nonsusceptible immigrants. Then a slight modification of the proof above shows that the same conclusion holds for all solutions of the modified system with $\eta \in [0, \epsilon]$ and satisfying $\sum_i (I_i(0) + E_i(0)) > 0$. Indeed, $\sum_i (I_i(0) + E_i(0)) > 0$ is unnecessary when $\eta > 0$.

Theorem 3.11. *If B is irreducible and $\mathbf{s}(B) > 0$, then there exists an equilibrium (S^*, E^*, I^*, R^*) satisfying $C^* > 0$, $C = S, E, I, R$.*

Proof. Modify equations (3.5) as in Remark 3.10 by adding $\eta \in [0, 1]$ to each equation for C_i, $C = E, I, R$, $1 \leq i \leq n$. Let $N = \sum_C \sum_i C_i$ and add the equations as before, we find that $N' \leq \sum_i \Lambda_i + 3n - \mu N$ uniformly for $\eta \in [0, 1]$. Therefore,

$$P = \left\{(S, E, I, R) \in \mathbb{R}_+^{4n} : N \leq \left(\sum_i \Lambda_i + 3n\right)/\mu\right\}$$

is forward invariant for our perturbed system. As P is compact and convex, for each $\eta \in (0, 1]$ there exists one equilibrium solution $(S_\eta, E_\eta, I_\eta, R_\eta) \in P$ (see Hale, Theorem I.8.2 [**90**]). By Remark 3.10, there is an $\epsilon > 0$ such that $\sum_i (I_\eta)_i > \epsilon$ provided $0 < \eta < \epsilon$. Now take a positive sequence $\eta_n \to 0$ and denote by (S_n, E_n, I_n, R_n) a sequence of equilibria corresponding to $\eta = \eta_n$. It has a convergent subsequence converging to a point $x^* = (S^*, E^*, I^*, R^*)$ of P which by continuity must be an equilibrium of (3.5). Since $\sum_i (I_{\eta_n})_i > \epsilon$

for all large n it follows that $\sum_i I_i^* > \epsilon$. Consequently, $E^* \gg 0$ and $I^* \gg 0$ by (3.10). This implies that $R^* \gg 0$, $S^* \gg 0$ by Proposition 3.6. $\square$

In Chapter 6, we provide general results implying the existence of ρ-positive equilibria of systems that are weakly ρ-persistent. Using these methods, the reader is asked in Exercise 6.1 to give another proof of Theorem 3.11.

3.3. Nonlinear matrix models: Prolog

Consider the nonlinear matrix model

$$\begin{aligned} x(n+1) =& G(x(n)), \quad x(0) \in \mathbb{R}_+^m, \\ G(x) =& A(x)x, \end{aligned} \tag{3.11}$$

where $A : \mathbb{R}_+^m \to \mathbb{R}_+^{m\times m}$ is a continuous mapping into the $m \times m$ matrices with nonnegative entries. Such models arise frequently in demography and populations biology. The monographs of Caswell [**29**] and Cushing [**39**] are especially recommended. The components of the vector x may represent age classes or life-cycle stages of a population.

In the following, we employ vector inequalities $a \le b$ ($a \ll b$) if $a_i \le b_i$ ($a_i < b_i$) for all i; further, $a < b$ if $a \le b \ne a$; similarly $b \ge a$, $b > a$ and $b \gg a$. The same conventions hold for matrices. Norms on $\mathbb{R}^m$ are denoted by $|\cdot|$.

If $x \ge 0$, then also $A(x) \ge 0$ and $G(x) \ge 0$; so $\mathbb{R}_+^m$ is forward invariant for this discrete dynamical system. The semiflow is given by

$$\Phi(j, x) = G^j(x)$$

where G^j denotes the j-fold composition of G with itself.

For example, the LPA model of Costantino et al. [**36**] is greatly celebrated in population biology due to the close correspondence between its predictions and observed features in controlled laboratory experiments. It accounts for *larval, pupal, and adult* insect densities at 14-day time (census) units (length of the maturation interval of feeding larvae). The population dynamics of the flour beetle *Tribolium castaneum* have the remarkable aspect that adults cannibalize on both pupae and eggs and larvae cannibalize on eggs; this results in strong nonlinearities in the mathematical model. With x_1, x_2, x_3 representing the life-cycle stages of (feeding) larva, pupa and adult, respectively, the LPA model takes the form

$$\begin{aligned} x_1(n+1) =& d\, x_3(n) \exp(-ax_1(n) - bx_3(n)), \\ x_2(n+1) =& p\, x_1(n), \\ x_3(n+1) =& q\, x_2(n) \exp(-cx_3(n)) + rx_3(n). \end{aligned} \tag{3.12}$$

Here r is the adult survival probability, p is the transition/survival probability from the larval to the pupal stage, and q the transition/survival probability from the pupal to the adult stage. The coefficients a, b, and c are related to cannibalism and d to fecundity. See [**36, 43, 44**] for details of the modeling; we have renamed their parameters for mathematical convenience.

The LPA model can be rewritten in the form of (3.11) with

$$x = (x_1, x_2, x_3) \in \mathbb{R}^3_+,$$

written as column vectors, and

$$A(x) = \begin{pmatrix} 0 & 0 & d\exp(-ax_1 - bx_3) \\ p & 0 & 0 \\ 0 & q\exp(-cx_3) & r \end{pmatrix}. \tag{3.13}$$

Age-structured populations are often modeled by (3.11), where x_i denotes the number of individuals in age class $i = 1, 2, \dots, m$ and the matrices A are of Leslie form [**141**]. Following [**39**], we partition the matrices A,

$$A(x) = B(x) + T(x), \tag{3.14}$$

into *birth (or fertility) matrices* $B(x)$ and *transition/survival* matrices $T(x)$,

$$B = \begin{pmatrix} \beta_1 & \cdots & \beta_m \\ 0 & \cdots & 0 \\ \vdots & \ddots & \vdots \\ 0 & \cdots & 0 \end{pmatrix}, \quad T = \begin{pmatrix} 0 & 0 & \cdots & 0 \\ \tau_1 & 0 & \cdots & 0 \\ \vdots & \ddots & \ddots & \vdots \\ 0 & \cdots & \tau_{m-1} & 0 \end{pmatrix}, \tag{3.15}$$

where B and T and the β_j and τ_j are functions of x. We call demographic models of this kind *Leslie models* [**41**].

3.3.1. Stability of the extinction equilibrium. With $G(x) = A(x)x$,

$$G(x) - G(0) - A(0)x = [A(x) - A(0)]x.$$

Since A is continuous, $G'(0) = A(0)$. The stability of the $x = 0$ equilibrium depends on the spectral radius of $A(0)$, denoted by $\mathbf{r}(A(0))$.

Proposition 3.12. *If $\mathbf{r}(A(0)) < 1$, then the zero equilibrium is locally asymptotically stable. If, in addition,*

$$A(x) \le A(0), \quad x \ge 0, \tag{3.16}$$

then $x(n) \to 0$ for every $x(0) \ge 0$ (even uniformly for $x(0)$ in bounded sets).

Proof. The first assertion is standard. See Theorem V.6.1 in [**183**]. If the additional condition (3.16) holds, then

$$x(n+1) \le A(0)x(n) \le A(0)^{n+1}x(0).$$

Let $|\cdot|$ be any of the standard norms on $\mathbb{R}^m$, which are monotone. Then

$$|x(n+1)| \leq |A(0)^{n+1}|\,|x(0)|.$$

Since $|A(0)^n| \to 0$ by (A.9), the statement follows. □

Note that we do not assert that $\mathbf{r}(A(0)) > 1$ implies that $x = 0$ is unstable. This assertion would follow if the zero point were an interior point of a domain of interest and the Jacobian matrix were nonsingular there (see Corollary V.6.3 [**183**]). Below, we will obtain a stronger result than instability of $x = 0$ when $\mathbf{r}(A(0)) > 1$ under slightly stronger assumptions.

The LPA model obviously satisfies the sublinearity condition (3.16). So does the age-structured population model if both fertilities and survival probabilities decline, or do not increase, with increasing age-class densities.

For the LPA model one can derive the basic reproduction number $\mathcal{R}_0$ as follows. If we added a single adult to a batch of flour, it would survive for an average of $1/(1-r)$ census periods, producing a total of $d/(1-r)$ eggs of which $pqd/(1-r)$ survive to become adults. Thus

$$\mathcal{R}_0 = \frac{pqd}{1-r} \tag{3.17}$$

gives the basic reproduction number.

Remark 3.13. Let $r_0 = \mathbf{r}(A(0))$. The reader is asked to verify that $r_0 < 1$ ($r_0 > 1$) for the LPA model (3.13) if and only if $\mathcal{R}_0 < 1$ ($\mathcal{R}_0 > 1$) in Exercise 3.8.

Remark 3.14. In general, if there is an additive decomposition $A(0) = B(0) + T(0)$ and $B(0)$ and $T(0)$ have convincing interpretations as birth (fertility) and transition/survival matrices, then $\mathcal{R}_0$ is defined as the spectral radius of $B(0)(\mathbb{I} - T(0))^{-1}$. With this definition, Remark 3.13 remains valid [**239**, Thm.3]. This even holds if matrices are replaced by bounded linear operators [**220**, Thm.3.10]. If $\mathcal{R}_0 > 0$, it satisfies the equation [**142**]

$$\mathbf{r}(T(0) + \mathcal{R}_0^{-1} B(0)) = 1.$$

Example 3.15. Consider the Leslie model and let $r_0 = \mathbf{r}(A(0))$. Assume that $\tau_i(0) > 0$ for $i = 1, \ldots, m-1$. By the Perron Frobenius Theorem A.25, $A(0)w = r_0 w$ for some eigenvector $w \in \mathbb{R}^m_+$. Then

$$r_0 w_1 = \sum_{i=1}^{m} \beta_i(0) w_i, \qquad r_0 w_i = \tau_{i-1}(0) w_{i-1}, \quad i = 2, \ldots, m.$$

Assume $r_0 > 0$. Then

$$w_j = \prod_{i=1}^{j-1} \tau_i(0) r_0^{1-j} w_1. \tag{3.18}$$

This implies that $w_1 > 0$ and r_0 satisfies the equation

$$1 = \sum_{j=1}^{m} \beta_j(0) \prod_{i=1}^{j-1} \tau_i(0) r_0^{-j}. \tag{3.19}$$

This calculation also shows that if $A(0)\tilde{w} = \lambda\tilde{w}$ with $0 \neq \tilde{w} \in \mathbb{R}_+^m$, $0 \neq \lambda \in \mathbb{R}$, then $\lambda = r_0$ and $\tilde{w} = \alpha w$ for some $\alpha \in \mathbb{R}$.

Notice that $r_0 - 1$ has the same sign as $\mathcal{R}_0 - 1$,

$$\begin{aligned}\mathcal{R}_0 &= \sum_{i=1}^{m} \beta_i(0) \prod_{j=1}^{i-1} \tau_j(0) \\ &= \beta_1(0) + \tau_1(0)\beta_2(0) + \tau_1(0)\tau_2(0)\beta_3(0) + \cdots + \beta_m(0)\tau_1(0)\cdots\tau_{m-1}(0).\end{aligned}$$

$\mathcal{R}_0$ is the basic reproduction number for the Leslie model. See [**29, 39**]. The reasoning is easier starting with a single newborn. It will have $\beta_1(0)$ offspring and with probability $\tau_1(0)$ it will survive one census to leave $\tau_1(0)\beta_2(0)$ offspring and so on.

In turn, if $\mathcal{R}_0 > 0$, there exists a unique $r_0 > 0$ that satisfies (3.19). Setting $w_1 = 1$ and defining w_j by (3.18) produces a strictly positive eigenvector w associated with r_0.

Notice that the sign relation between r_0 and $\mathcal{R}_0$ holds and that the eigenvector w has all components positive whether or not $A(0)$ is irreducible. By the Perron-Frobenius theorem, there is also an eigenvector v of A_0^* associated with r_0,

$$r_0 v_i = \beta_i(0) v_1 + \tau_i v_{i+1}, \quad i = 1, \ldots, m-1, \qquad r_0 v_m = \beta_m(0) v_1.$$

Observe that $v_m = 0$ if $\beta_m(0) = 0$.

3.3.2. Uniform weak persistence.

Let X be a nonempty set, $F : X \to X$ be a map generating the discrete dynamical system

$$x(n+1) = F(x(n)), \qquad x(0) \in X,$$

and let $\rho : X \to [0, \infty)$.

Proposition 3.16. *Let $k \in \mathbb{N}$ and $\delta > 0$ be such that*

(a) $\rho(x) > 0 \implies \rho(F^k(x)) > 0$.

(b) $\inf\left\{\dfrac{\rho(F^k(x))}{\rho(x)};\ 0 < \rho(x) < \delta\right\} > 1$.

Then

$$\rho(x(0)) > 0 \implies \limsup_{n\to\infty} \rho(x(n)) \geq \delta.$$

Proof. Replacing F by F^k, we can assume $k = 1$ because

$$\limsup_{n\to\infty} \rho(F^n(x)) \geq \limsup_{n\to\infty} \rho(F^{nk}(x)).$$

By (b), there exist $\delta > 0$ and $\gamma > 1$ such that

$$0 < \rho(x) < \delta \implies \rho(F(x)) > \gamma\rho(x). \tag{3.20}$$

Suppose $\rho(x(0)) > 0$ and

$$\limsup_{n\to\infty} \rho(x(n)) < \delta.$$

Then there exists a natural number N such that $\rho(x(n)) < \delta$, $n \geq N$. By (3.20),

$$\rho(x(n)) = \rho(F(x(n-1))) \geq \gamma\rho(x(n-1)), \qquad n > N.$$

We repeat this step and use assumption (a),

$$\rho(x(n)) \geq \gamma^{n-N}\rho(x(N)) > 0, \qquad n \geq N.$$

This yields the contradiction $\lim_{n\to\infty} \rho(x(n)) = \infty$. □

Let $X = \mathbb{R}^m_+$ and let F be differentiable at 0 and $F'(0)$ be represented by a nonnegative Jacobian matrix A_0. A natural choice of ρ is

$$\rho(x) = \sum_i v_i x_i = \langle v, x\rangle, \tag{3.21}$$

where v is a nonnegative eigenvector of the transpose A_0^*:

$$\rho(A_0 x) = \langle v, A_0 x\rangle = \langle A_0^* v, x\rangle = r_0\langle v, x\rangle = r_0\rho(x), \qquad x \in \mathbb{R}^m_+,$$

with the associated eigenvalue r_0. In the following, we employ vector inequalities $a \leq b$ ($a \ll b$) if $a_i \leq b_i$ ($a_i < b_i$) for all i; further, $a < b$ if $a \leq b \neq a$; similarly, $b \geq a$, $b > a$ and $b \gg a$. The same conventions hold for matrices.

Theorem 3.17. *Suppose that*

(a) $\mathbb{R}^m_+ \setminus \{0\}$ *is forward invariant under* F.

(b) *There exist* $r_0 > 1$ *and* $v > 0$ *such that* $A_0^* v \geq r_0 v$, $A_0 = F'(0)$.

Then the semiflow induced by F *is uniformly weakly* ρ*-persistent for* $\rho(x) = \langle v, x\rangle$.

Proof. Hypothesis (a) of Proposition 3.16 holds with $k = 1$ by (a) above because ρ is the restriction of the norm $|x| = \sum_i v_i|x_i|$ to $\mathbb{R}^m_+$.

As for Hypothesis (b) of Proposition 3.16, let $|\cdot|_\infty$ denote the maximum norm on $\mathbb{R}^m$ and let $\mathbf{1}$ be the vector with all coordinates 1. Notice the following property: For any $c \geq 0$,

$$|x|_\infty \leq c \iff -c\mathbf{1} \leq x \leq c\mathbf{1}. \tag{3.22}$$

Let $\epsilon > 0$. Since F is differentiable at 0, $A_0 = F'(0)$, and all norms are equivalent on $\mathbb{R}^m$, there exists some $\delta > 0$ such that

$$|x| \le \delta \implies |F(x) - F(0) - A_0 x|_\infty \le \epsilon |x|.$$

By (3.22),

$$|x| \le \delta \implies F(x) - F(0) - A_0 x \ge -\epsilon |x| \mathbf{1}.$$

Recall that, for $x \ge 0$, $\rho(x) = \langle v, x\rangle = |x|$. Thus for $x \ge 0$, $\rho(x) \le \delta$,

$$\begin{aligned}\rho(F(x)) \ge& \langle v, A_0 x - \epsilon|x|\mathbf{1}\rangle = \langle A_0^* v, x\rangle - \epsilon\rho(x)\langle v, \mathbf{1}\rangle \\ =& r_0 \rho(x) - \epsilon \rho(x) \langle v, \mathbf{1}\rangle = \gamma\rho(x),\end{aligned}$$

with $\gamma > 1$ if $\epsilon > 0$ and $\delta > 0$ are chosen sufficiently small. □

We return to $F(x) = A(x)x$. Assume that the spectral radius of $A(0)$ exceeds unity: $r_0 = \mathbf{r}(A(0)) > 1$. Let $v \ge 0$ be an eigenvector corresponding to r_0 for the transpose of $A(0)$ provided by the Perron-Frobenius theory (Theorem A.25). If $A(0)$ is irreducible, then $v > 0$.

Corollary 3.18. *Suppose that*

(a) $\mathbb{R}_+^m \setminus \{0\}$ *is forward invariant for (3.11).*

(b) $\mathbf{r}(A(0)) > 1$ *and* $A(0)$ *is irreducible.*

Then (3.11) is uniformly weakly ρ*-persistent for* $\rho(x) = |x|$ *with any norm on* $\mathbb{R}^m$.

Corollary 3.18 applies directly to both the LPA model and the Leslie model when $A(0)$ is irreducible and $\mathcal{R}_0 > 1$. $A(0)$ is irreducible for the LPA model precisely when $pqd > 0$ and for the demographic model when $\beta_m(0)\prod_{i=1}^{m-1} \tau_i(0) > 0$.

Corollary 3.18 does not apply to Leslie models in which the highest age classes do not reproduce (see Example 3.15) making $A(0)$ reducible. Indeed, if every individual in the population is beyond reproductive age, the population will die out in a couple of years. Individuals beyond reproductive age cannot be ignored, however, because they may impact the rest of the population by using resources on the one hand and assisting in the bringing up of offspring on the other hand. We address Leslie models of this type as follows.

Definition 3.19. If $Q = (q_{ij})$ is a matrix, define its *support* as

$$\operatorname{supp}(Q) = \{(i,j); q_{ij} \ne 0\}. \tag{3.23}$$

A similar notion applies to vectors giving the indices of nonzero components.

Lemma 3.20. *Let* F *be of nonlinear matrix form* $F(x) = A(x)x$ *with* $A(x)$ *continuously depending on* $x \in \mathbb{R}_+^m$. *Then, for any* $\eta \in (0,1)$, *there exists some* $\delta > 0$ *such that* $A(x) \ge \eta A(0)$ *whenever* $|x| \le \delta$.

Proof. Let $\eta \in (0,1)$. Let $\epsilon > 0$, to be chosen later. Then there exists some $\delta > 0$ such that $|a_{ij}(x) - a_{ij}(0)| \le \epsilon$ whenever $(i,j) \in \text{supp}(A(0))$ and $|x| \le \delta$. If the latter is the case,

$$\frac{a_{ij}(x)}{a_{ij}(0)} = 1 - \frac{a_{ij}(x) - a_{ij}(0)}{a_{ij}(0)} \ge 1 - \frac{\epsilon}{a_{ij}(0)} \ge 1 - \frac{\epsilon}{\alpha}$$

with $\alpha = \min\{a_{ij}(0); (i,j) \in \text{supp}(A(0))\} > 0$. Choose $\epsilon > 0$ small enough such that $1 - \frac{\epsilon}{\alpha} > \eta$. Then $a_{ij}(x) \ge \eta a_{ij}(0)$ whenever $(i,j) \in \text{supp}(A(0))$ and $|x| \le \delta$. If $(i,j) \notin \text{supp}(A(0))$, this inequality is trivially satisfied. □

Theorem 3.21. *Let F be of nonlinear matrix form $F(x) = A(x)x$, and $A_0 = A(0)$. Assume:*

(a) *There exist some $v > 0$ and $r_0 > 1$ such that $A_0^* v \ge r_0 v$.*

(b) *If $\langle x, v\rangle > 0$, then $\langle F(x), v\rangle > 0$.*

(c) *There exists some $n \in \mathbb{N}$ and $c > 0$ such that $|F^n(x)| \le c\langle x, v\rangle$ for all $x \in \mathbb{R}^m_+$.*

Then the semiflow induced by F is uniformly weakly ρ-persistent for $\rho = \langle \cdot, v\rangle$.

Proof. Let $x \in \mathbb{R}^m_+$, $\langle x, v\rangle > 0$. Set $x(n) = F^n(x)$, $n \in \mathbb{Z}_+$. By assumption (a), $\langle x(n), v\rangle > 0$ for all $n \in \mathbb{Z}_+$. Let $\epsilon > 0$, to be determined later. Suppose that the semiflow induced by F is not uniformly weakly ρ-persistent. Then after a shift in time, $0 < \langle x(n), v\rangle < \epsilon$ for all $n \in \mathbb{Z}_+$. By (c), after another shift in time, $|x(n)| \le c\epsilon$ for all $n \in \mathbb{Z}_+$. Choose $\eta \in (0,1)$ such that $r_0\eta > 1$. By Lemma 3.20, there exists some $\delta > 0$ such that $F(x) \ge \eta A(0)x$ whenever $x \in \mathbb{R}^m_+$, $|x| \le \delta$. Choose $\epsilon > 0$ such that $c\epsilon < \delta$. Then $|x(n)| \le \delta$ for all $n \in \mathbb{N}$, and

$$\begin{aligned}\rho(x(n+1)) =& \langle F(x(n)), v\rangle \ge \langle \eta A(0)x(n), v\rangle = \eta\langle x(n), A_0^* v\rangle \\ \ge& \eta\langle x(n), r_0 v\rangle = \eta r_0 \rho(x(n)) > 0.\end{aligned}$$

Since $\eta r_0 > 1$, $\rho(x_n) \to \infty$ as $n \to \infty$, a contradiction. □

Example 3.22. Consider the Leslie model with $\tau_i(x) > 0$ for $i = 1, \dots, m-1$, $x \in \mathbb{R}^m_+$. Let $\mathcal{R}_0 > 1$ and $j \in \{1, \dots, m-1\}$ such that $\beta_j(0) > 0$ (such a j exists) and $\beta_j(x) > 0$ but $\beta_k(x) = 0$, $k = j+1, \dots, m$, $x \in \mathbb{R}^m_+$.

We check the assumptions of Theorem 3.21.

(a) The $j \times j$ Leslie matrix $\tilde{A}_0$ that results from $A_0 = A(0)$ by deleting the rows and columns $k = j+1, \dots, m$ is irreducible, $\mathbf{r}(\tilde{A}_0) = \mathbf{r}(A_0) =: r_0$. Since $\mathcal{R}_0 > 1$, also $r_0 > 1$ by Example 3.15. Let $v = (v_1, \dots, v_j)$ be the nonnegative eigenvector of $\tilde{A}_0^*$ associated with r_0 according the Perron-Frobenius Theorem A.25. Then $v_i > 0$, $i = 1, \dots, j$. Extend v to a vector in $\mathbb{R}_+^m$ by setting $v_k = 0$ for $k = j+1, \dots, m$. Then $A_0^* v = r_0 v$, and (a) is proved.

(b) If $x \in \mathbb{R}_+^m$ and $\langle x, v \rangle > 0$, then $x_i > 0$ for some $i \in \{1, \dots, j\}$. So $F_i(x) > 0$ and $\langle F(x), v \rangle > 0$.

(c) Let $\tilde{\beta} = \sup\{\beta_i(x); x \in \mathbb{R}_+^m, i = 1, \dots, j\} < \infty$ and $\tilde{\tau} = \sup\{\tau_i(x); x \in \mathbb{R}_+^m, i = 1, \dots, m-1\} < \infty$. Then

$$\sum_{i=1}^m F_i(x) \leq \tilde{\beta} \sum_{i=1}^j x_j + \tilde{\tau} \sum_{i=1}^{m-1} x_i.$$

By iteration, we find some $\tilde{c} > 0$ such that

$$\sum_{i=1}^m (F^m)_i(x) \leq \tilde{c} \sum_{i=1}^j x_i \leq \frac{\tilde{c}}{\min_1^j v_j} \langle x, v \rangle.$$

So the assumptions of Theorem 3.21 are satisfied and the semiflow induced by F is uniformly weakly $\langle \cdot, v \rangle$-persistent if $\mathcal{R}_0 > 1$. Then the semiflow is also uniformly weakly ρ-persistent for $\rho(x) = \sum_{i=1}^j x_i$.

Uniform persistence for nonlinear matrix models is taken up in Chapter 7.

3.4. The May-Leonard example of cyclic competition

May and Leonard [**162**] discovered an interesting phenomena in the dynamics of a symmetric Lotka-Volterra competition model of three species which has since become a standard counterexample in persistence theory. The system is given by

$$\begin{aligned} N_1' &= N_1[1 - N_1 - \alpha N_2 - \beta N_3], \\ N_2' &= N_2[1 - \beta N_1 - N_2 - \alpha N_3], \\ N_3' &= N_3[1 - \alpha N_1 - \beta N_2 - N_3]. \end{aligned} \tag{3.24}$$

The system has obvious cyclic symmetry which is formalized in the following result.

Lemma 3.23. *The vector field F given by the right-hand side of (3.24) satisfies $QF(N) = F(QN)$, where $N = (N_1, N_2, N_3)^T$, for the cyclic permutation matrix*

$$Q = \begin{pmatrix} 0 & 1 & 0 \\ 0 & 0 & 1 \\ 1 & 0 & 0 \end{pmatrix}.$$

Consequently, $QN(t) = (N_2(t), N_3(t), N_1(t))^T$ is a solution whenever $N(t) = (N_1(t), N_2(t), N_3(t))^T$ is a solution.

Proof. Notice that $Q(N_1, N_2, N_3) = (N_2, N_3, N_1)$. It is now readily seen that

$$[QN(t)]' = QN'(t) = QF(N(t)) = F(QN(t)). \qquad \square$$

It is readily observed that the line $N_1 = N_2 = N_3$ is invariant under the dynamics of (3.24). On that line, the dynamics are given by the logistic equation

$$\tilde{N}' = \tilde{N}[1 - (1 + \alpha + \beta)\tilde{N}] \tag{3.25}$$

with $N_i = \tilde{N}$. Therefore, the positive portion of this line belongs to the stable manifold of the positive (coexistence) equilibrium point:

$$E^* = \frac{1}{1 + \alpha + \beta}(1, 1, 1). \tag{3.26}$$

In addition, there are the trivial equilibrium $E_0 = (0, 0, 0)$ and the single-species equilibria $E_i = Q^{i-1}(1, 0, 0)$, $i = 1, 2, 3$.

We assume

$$0 < \alpha < 1 < \beta \tag{3.27}$$

which implies that there are no two-population equilibria; the inequality $0 < \beta < 1 < \alpha$ also gives this result; similar results may be obtained if the latter holds. Together with the cyclic symmetry, this means that there is no most harmful competitor. Species one suffers the most from species three, species two the most from species one, and species three the most from species two.

Linearizing around the equilibrium E_1, one finds that (3.27) implies that N_2 cannot invade E_1 but N_3 can, so by symmetry, N_1 cannot invade E_3 but N_2 can, and N_3 cannot invade E_2 but N_1 can. More precisely, we have the following.

Lemma 3.24. (a) *E_1 is globally asymptotically stable in the invariant hyperplane $N_3 = 0, N_1 > 0, N_2 \geq 0$. However, it is a uniform weak repeller for solutions with $N_3(0) > 0$: there exists some $\epsilon > 0$ such that $\limsup_{t\to\infty} |N(t) - E_1| \geq \epsilon$ for any solution with $N_3(0) > 0$.*

(b) *E_2 is globally asymptotically stable in the invariant hyperplane $N_1 = 0, N_2 > 0, N_3 \geq 0$, but is a uniform weak repeller for solutions with $N_1(0) > 0$.*

(c) *E_3 is globally asymptotically stable in the invariant hyperplane $N_2 = 0, N_2 \geq 0, N_3 > 0$, but is a uniform weak repeller for solutions with $N_2(0) > 0$.*

Proof. Because of the symmetry established in Lemma 3.23, it is enough to show (a). Local asymptotic stability in the hyperplane $N_3 = 0$ follows from linearization about the equilibrium E_1. Global asymptotic stability is shown using the Poincaré-Bendixson theory and the Dulac function $\frac{1}{N_1 N_2}$.

Now assume that E_1 is not a uniform weak repeller for solutions with $N_3(0) > 0$. Then, for every $\epsilon > 0$, there exists a solution with $N_3(0) > 0$ and $\limsup_{t\to\infty} |N(t) - E_1| < \epsilon$. Since $N_3(t) > 0$ for all $t \geq 0$, after a forward translation in time we can assume that $|N(t) - E_1| < \epsilon$ for all $t \geq 0$. By (3.24),

$$\frac{N_3'}{N_3} \geq 1 - \alpha(1+\epsilon) - \beta\epsilon - \epsilon.$$

Since $\alpha < 1$ we can choose $\epsilon > 0$ so small that $\inf_{t\geq 0} \frac{N_3'}{N_3} > 0$. Since N_3 is positive, this implies that $N_3(t) \to \infty$ as $t \to \infty$, a contradiction. □

It follows from the Poincaré-Bendixson theory and the properties of saddle points that there is a heteroclinic cycle, HC, more precisely an invariant set consisting of the three equilibria E_i, $i = 1, 2, 3$, together with three planar total trajectories. One total trajectory lies in $N_2 = 0$ with α-limit set E_1 and ω-limit set E_3; a second one belongs to $N_1 = 0$, with α-limit set E_3 and ω-limit set E_2; the third belongs to $N_3 = 0$ with α-limit set E_2 and ω-limit set E_1; symbolically

$$E_1 \mapsto E_3 \mapsto E_2 \mapsto E_1.$$

We will give sufficient conditions ensuring that HC is the ω-limit set of every trajectory with positive initial conditions that does not belong to the line $N_1 = N_2 = N_3$.

The Jacobian matrix for E^* is given by

$$J_0 := -\frac{1}{1+\alpha+\beta}\begin{pmatrix} 1 & \alpha & \beta \\ \beta & 1 & \alpha \\ \alpha & \beta & 1 \end{pmatrix}. \tag{3.28}$$

One can verify (see [**162, 106**]) that the eigenvalues are given by

$$\begin{aligned} \kappa_1 &= -1, \\ \kappa_{2,3} &= \frac{(\alpha+\beta-2) \pm i\sqrt{3}(\beta-\alpha)}{2(1+\alpha+\beta)}. \end{aligned}$$

The first eigenvalue can be guessed, it is associated with the eigenvector $(1,1,1)^T$. Alternatively, (3.25) has the equilibrium $\tilde{N}^* = \frac{1}{1+\alpha+\beta}$, which corresponds to E^*, and the eigenvalue of the associated (degenerate) Jacobian matrix is -1.

We assume that

$$\alpha + \beta > 2. \tag{3.29}$$

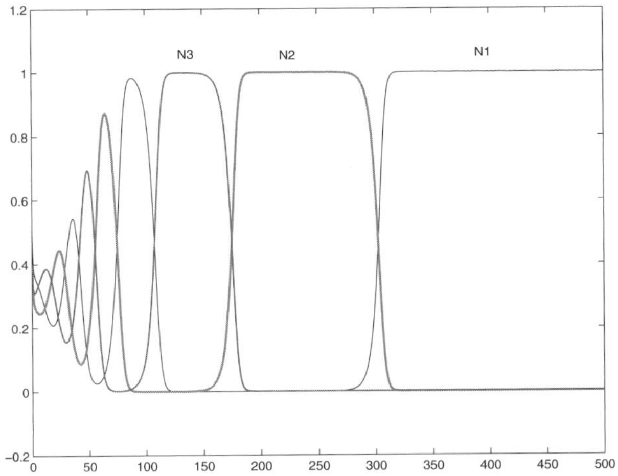

Figure 3.1. Time series for May-Leonard equation with $\alpha = 0.8$ and $\beta = 1.5$

Then E^* is an unstable saddle point with a two-dimensional unstable manifold and a one-dimensional stable manifold which coincides with the line $\{N_1 = N_2 = N_3\}$.

Remark 3.25. Let $S = N_1 + N_2 + N_3$. Then $\frac{dS}{dt} \leq S(1 - S)$ and therefore all solutions are bounded. Also, $\frac{dS}{dt} \geq S - \frac{\alpha+\beta}{2}S^2$ so all nonzero solutions are attracted by the set

$$T = \left\{N : N_i \geq 0,\ \frac{2}{\alpha + \beta} \leq N_1 + N_2 + N_2 \leq 1\right\}.$$

In particular, (3.24) is uniformly ρ-persistent where $\rho(N) = N_1 + N_2 + N_3$. The reader is asked to prove these assertions in Exercise 3.11.

The next result was first stated in [**162**] but was not proved. The proof below follows [**106**]. Our proof below uses the Butler-McGehee Theorem 8.8.

Proposition 3.26. *Let (3.27) and (3.29) hold. The ω-limit set of a positive solution not on the invariant line $N_1 = N_2 = N_3$ is the heteroclinic cycle HC.*

Proof. We first note that the limit set, denoted here by ω, cannot coincide with one of the equilibria E_i or E^* as they are saddle points and our initial data does not belong to the stable manifolds of any one of these equilibria.

Let $P = N_1N_2N_3$ and observe that

$$\frac{dP}{dt} = P[3 - (1 + \alpha + \beta)S], \qquad S = N_1 + N_2 + N_3. \tag{3.30}$$

Let $V = PS^{-3}$ and compute

$$\frac{dV}{dt} = PS^{-3}[3-(1+\alpha+\beta)S] - 3PS^{-4}\frac{dS}{dt},$$

$$\begin{aligned}\frac{dS}{dt} =& N_1[1-N_1-\alpha N_2-\beta N_3] \\ &+N_2[1-\beta N_1-N_2-\alpha N_3] \\ &+N_3[1-\alpha N_1-\beta N_2-N_3].\end{aligned}$$

We substitute the second equation into the first and simplify,

$$\frac{dV}{dt} = -PS^{-4}W$$

with

$$\begin{aligned}W =& (1+\alpha+\beta)S^2 - 3N_1[N_1+\alpha N_2+\beta N_3] \\ &- 3N_2[\beta N_1+N_2+\alpha N_3] - 3N_3[\alpha N_1+\beta N_2+N_3].\end{aligned}$$

We expand,

$$\begin{aligned}W =& (1+\alpha+\beta)(N_1^2+N_2^2+N_3^2+2N_1N_2+2N_1N_3+2N_2N_3) \\ &- 3N_1[N_1+\alpha N_2+\beta N_3] \\ &- 3N_2[\beta N_1+N_2+\alpha N_3] - 3N_3[\alpha N_1+\beta N_2+N_3].\end{aligned}$$

We start to simplify,

$$\begin{aligned}W =& (\alpha+\beta-2)(N_1^2+N_2^2+N_3^2) \\ &+ (1+\alpha+\beta)(2N_1N_2+2N_1N_3+2N_2N_3) \\ &- 3(\alpha+\beta)(N_1N_2+N_1N_3+N_2N_3).\end{aligned}$$

Hence,

$$\frac{dV}{dt} = -\frac{P}{S^4}\frac{\alpha+\beta-2}{2}[(N_1-N_2)^2+(N_1-N_3)^2+(N_2-N_3)^2] \le 0.$$

The LaSalle Invariance Principle (see Ch.X, Thm.1.3 of [**90**]) implies that ω is contained in the largest connected invariant set in $\{N : \frac{dV}{dt} = 0\}$. It must also belong to the set T defined in Remark 3.25. $\frac{dV}{dt} = 0$ precisely on the invariant line and on the boundary of the positive orthant. The only invariant set on the line that belongs to T is $\{E^*\}$. But $\omega \neq \{E^*\}$ has already been noted. Since ω is connected, $E^* \notin \omega$. Therefore, ω belongs to the intersection of T with the boundary of the positive orthant. As ω is compact and invariant, it must be contained in the heteroclinic cycle HC. Indeed, let $\breve{N}$ be in ω. Then there exists a total solution $N : \mathbb{R} \to \omega$ with $N(0) = \breve{N}$. As $t \to -\infty$, $N(t)$ converges towards one of the three boundary equilibria, let us say E_1. For large enough t, $N(t)$ lies on the one-dimensional unstable manifold of E_1 and thus on HC. Since HC is invariant, $\breve{N}$ is a point on HC as well.

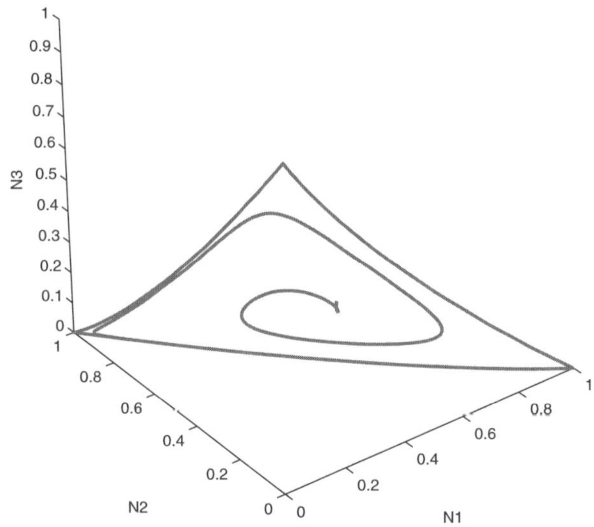

Figure 3.2. Phase portrait for May-Leonard equation with $\alpha = 0.8$ and $\beta = 1.5$

By the Butler-McGehee Theorem 8.8, applied to each E_i (recall Lemma 3.24), and by invariance of limit sets, ω must coincide with the heteroclinic cycle. □

An obvious consequence of Proposition 3.26 is that (3.24) is not uniformly weakly $\tilde{\rho}$-persistent for

$$\tilde{\rho}(N) = \min\{N_1, N_2, N_3\}.$$

Indeed, $\tilde{\rho}(N(t)) \to 0$ as $t \to \infty$,

$$\begin{aligned} 0 &= \liminf_{t\to\infty}(N_1(t) + N_2(t)) = \liminf_{t\to\infty}(N_1(t) + N_3(t)) \\ &= \liminf_{t\to\infty}(N_2(t) + N_3(t)), \end{aligned}$$

and

$$\limsup_{t\to\infty} N_i(t) = 1, \quad i = 1, 2, 3,$$

for each positive solution (i.e., when all three species are present) unless $N_1(0) = N_2(0) = N_3(0)$. See Figure 3.1 and 3.2. Notice that the last equation does not mean that the semiflow induced by (3.24) is uniformly weak ρ_i-persistent with $\rho_i(N_1, N_2, N_3) = N_i$ because it is essential that all three species are around. Recall, for example, that $N_1(t) \to 0$ as $t \to \infty$ if $N_1(0) > 0, N_2(0) = 0, N_3(0) > 0$ (Lemma 3.24).

The main reason for introducing the May-Leonard system is to point out that (uniform) $\tilde{\rho}$-persistence does not follow from the fact that every equilibrium point (or periodic orbit) on the boundary is unstable and even a uniform weak repeller for the interior. It is also a warning that cyclic connections among compact invariant sets on the boundary constitute potential ω-limit sets for positive solutions.

In Section 5.5.4, the tools of persistence theory will establish that the heteroclinic cycle HC is the compact attractor of neighborhoods of compact sets in $\mathbb{R}^3_+ \setminus L$ where L is the invariant line $\{N_1 = N_2 = N_3\}$.

An up-to-date reference on three-species Lotka-Volterra systems is Zeeman [**237**].

3.5. Exercises

Exercise 3.1. Show uniform weak parasite (disease) persistence for (3.1): there exists some $\epsilon > 0$ such that $\limsup_{t\to\infty} y(t) \geq \epsilon$ for all solutions of (3.2) and (3.3) with $N(0) > 0$ and $y(0) > 0$.

Exercise 3.2. Prove Remark 3.5. Hint: Use Perron-Frobenius Theorem or Gerschgorin's Circle Theorem [**136**].

Exercise 3.3. Give analogs of Theorem 3.4 and Proposition 3.6 if we replace Λ_i by $\mu_i^B N_i$ or explain why they cannot hold.

Exercise 3.4. Is there a (DFE) of (3.5) if we replace Λ_i by $\mu_i^B N_i$? What dynamical state is the analog of (DFE)?

Exercise 3.5. Prove Proposition 3.7 and verify the assertions made above concerning the spectrum of the Jacobian matrix J.

Exercise 3.6. Show that B, defined by (3.9), is irreducible if $\kappa_i, \gamma_i^E > 0$ for all i and either m^E or m^I is irreducible.

Exercise 3.7. Prove Theorem 3.8. Hint: Show that $z = (E, I)^*$ satisfies $z' \leq Bz$.

Exercise 3.8. Verify Remark 3.13. Justify the assertions made in the derivation of $\mathcal{R}_0$.

Exercise 3.9. Notice that in the LPA model (3.13) we have the additive decomposition $A(0) = B(0) + T(0)$ with

$$B(0) = \begin{pmatrix} 0 & 0 & d \\ 0 & 0 & 0 \\ 0 & 0 & 0 \end{pmatrix}, \qquad T(0) = \begin{pmatrix} 0 & 0 & 0 \\ p & 0 & 0 \\ 0 & q & r \end{pmatrix}.$$

Show that $\mathcal{R}_0$ equals the spectral radius of $B(0)(\mathbb{I} - T(0))^{-1}$.

Exercise 3.10. Explain why the conclusion of Corollary 3.18 is a substantially stronger assertion than instability of $x = 0$.

Exercise 3.11. Verify Remark 3.25. Also, in the case that $\alpha + \beta < 2$ holds, show that all the assertions remain valid but with inequalities reversed.

Exercise 3.12. Show that the largest invariant set of (3.24) contained in $T \cap \partial \mathbb{R}^3_+$ is the heteroclinic cycle. Hint: Show that the negative orbit of a point of $T \cap \partial \mathbb{R}^3_+$ that does not belong to the heteroclinic cycle must leave $T \cap \partial \mathbb{R}^3_+$.

Exercise 3.13. Show that the compact attractor A of bounded sets for (3.24) exists and describe precisely what it contains. Hint: Recall Exercise 2.18.

Exercise 3.14. Assume $0 < \alpha < 1 < \beta$ and $\alpha + \beta < 2$. Then every solution with $N_j(0) > 0$, $j = 1, 2, 3$, converges to E^*.

Chapter 4

Uniform Persistence

4.1. From uniform weak to uniform persistence

Let J be a closed time set, X a nonempty set, $\Phi : J \times X \to X$ a semiflow, and $\rho : X \to \mathbb{R}_+$. We set

$$\sigma(t, x) = \rho(\Phi(t, x)). \tag{4.1}$$

Notice that

$$\sigma(t, \Phi(r, x)) = \sigma(t + r, x). \tag{4.2}$$

Since the time-set J is closed, we can assume, without loss of generality, that $J = \mathbb{R}_+$ or $J = \mathbb{Z}_+$.

Proposition 4.1. *Let J be a closed time-set. Then either $J = \mathbb{R}_+$ or there exists $n \in \mathbb{N}$ such that $J = \frac{1}{n}\mathbb{Z}_+$.*

Proof. Let $\alpha = \inf\{t \in J : t > 0\}$. If $\alpha = 0$, then there is a sequence $\{t_k\}_{k \in \mathbb{N}} \subset J \cap (0, \infty)$ such that $t_k \to 0$ as $k \to \infty$. By the property of time-set, we have $\bigcup_{k \in \mathbb{N}}(t_k\mathbb{Z}_+) \subset J$. This means that J is dense in $\mathbb{R}_+$ and so $J = \mathbb{R}_+$ since J is closed. Now if $\alpha > 0$, then since J is closed, $\alpha \in J$ and so $\alpha\mathbb{Z}_+ \subset J$. If there is $t \in J \setminus (\alpha\mathbb{Z}_+)$, then there is $m \in \mathbb{Z}_+$ such that $m\alpha < t < (m+1)\alpha$. But then $0 < (m+1)\alpha - t < \alpha$ and $(m+1)\alpha - t \in J$, contradicting the minimality of α. Hence $J = \alpha\mathbb{Z}_+$. Now since $1 \in J$, there is $n \in \mathbb{Z}_+$ such that $1 = n\alpha$. Hence $\alpha = \frac{1}{n}$. □

4.1.1. A persistence result for general time-sets.

The endemic model (0.1) with $p = 0$ shows that uniform weak persistence does not imply uniform persistence in general. The basic idea consists in assuming the existence of a set B which is sort of attracting for the semiflow. We first assume that X is just a nonempty set. Recall that $J = \mathbb{R}_+$ or $J = \mathbb{Z}_+$.

We make the following assumptions ♣.

♣ There exist a subset B of X and a sequence (B_k) of subsets of X such that the following properties hold:

- ♣$_0$ For every $x \in B$, $\sigma(t, x)$ is a continuous function of $t \geq 0$.
- ♣$_1$ There are no $y \in B$, $s, t \in J$ such that $\rho(y) > 0$, $\sigma(s, y) = 0$, and $\sigma(s + t, y) > 0$.
- ♣$_2$ For every $k \in \mathbb{N}$ and every $x \in X$, $\rho(x) > 0$, there exists some $t_k \in J$ such that $\Phi(t, x) \in B_k$ for all $t \geq t_k$, $t \in J$.
- ♣$_3$ If (y_k) is a sequence in X with $y_k \in B_k$ for all $k \in \mathbb{N}$, then, after possibly choosing a subsequence, there exists some $y \in B$ such that $\sigma(s, y_k) \to \sigma(s, y)$ as $k \to \infty$, uniformly for s in any set $[0, t] \cap J$, $t \in (0, \infty)$.

A typical situation is $B \subset B_{k+1} \subset B_k$ and $B = \bigcap_{k=1}^{\infty} B_k$. But this is not necessary for the following result.

Theorem 4.2. *Let $J = \mathbb{R}_+$ or $J = \mathbb{Z}_+$. Under the assumptions ♣, the semiflow Φ is uniformly ρ-persistent, whenever it is uniformly weakly ρ-persistent.*

Proof. Suppose that Φ is uniformly weakly ρ-persistent, but not uniformly ρ-persistent. Choose $\epsilon > 0$ such that

$$\limsup_{t \to \infty} \sigma(t, x) > \epsilon \qquad \forall x \in X, \rho(x) > 0.$$

Let (ϵ_j) be a sequence in $(0, \epsilon)$ such that $\epsilon_j \to 0$ as $j \to \infty$. Since Φ is not uniformly ρ-persistent, we find a sequence (x_j) in X, $\rho(x_j) > 0$, such that

$$\liminf_{t \to \infty} \sigma(t, x_j) < \epsilon_j.$$

We find sequences (r_j), (s_j), $(t_j), (u_j), (v_j)$ in J such that $r_j \to \infty$ and $s_j, v_j \leq 1$,

$$\begin{aligned}
&\sigma(r_j, x_j) \geq \epsilon, \quad \Phi(r_j, x_j) \in B_j, \\
&\sigma(r_j + s_j + t_j, x_j) < \epsilon_j, \\
&\sigma(r_j + s_j + s, x_j) \leq \epsilon \qquad \forall s \in [0, t_j + u_j] \cap J, \\
&\sigma(r_j + s_j + t_j + u_j + v_j, x_j) \geq \epsilon.
\end{aligned}$$

We set $y_j = \Phi(r_j, x_j)$. By definition of σ,

$$\begin{aligned}&\sigma(0, y_j) \geq \epsilon, \quad y_j \in B_j,\\ &\sigma(s_j + t_j, y_j) < \epsilon_j,\\ &\sigma(s_j + s, y_j) \leq \epsilon \qquad \forall s \in [0, t_j + u_j] \cap J,\\ &\sigma(s_j + t_j + u_j + v_j, y_j) \geq \epsilon.\end{aligned}$$

By assumption ♣$_3$, possibly after choosing a subsequence, there exists some $y \in B$ such that

$$\begin{aligned}&\sigma(s, y_k) \to \sigma(s, y), \qquad k \to \infty,\\ &\text{uniformly for } s \text{ in any set } [0, t] \cap J, t \in (0, \infty).\end{aligned} \tag{4.3}$$

In particular, $\rho(y) = \sigma(0, y) = \lim_{j\to\infty} \sigma(0, y_j) \geq \epsilon > 0$. After choosing subsequences again, $s_j \to s^*$, $v_j \to v^*$ as $j \to \infty$, with appropriate $s^*, u^* \in [0, 1] \cap J$.

Claim: $t_j + u_j \to \infty$ as $j \to \infty$.

Suppose not. After choosing subsequences, we find some $t > 0$ such that $t_j + u_j \leq t$ for all $j \in \mathbb{N}$. After choosing subsequences again, $u_j \to u^*$, $t_j \to t^*$, for appropriate $t^*, u^* \in [0, t] \cap J$. By ♣$_0$ and (4.3),

$$\sigma(s^* + t^*, y) = \lim_{j\to\infty} \sigma(s_j + t_j, y) = \lim_{j\to\infty} \sigma(s_j + t_j, y_j) \leq \limsup_{j\to\infty} \epsilon_j = 0$$

and

$$\sigma(s^* + t^* + u^* + v^*, y) = \lim_{j\to\infty} \sigma(s_j + t_j + u_j + v_j, y_j) \geq \epsilon,$$

contradicting assumption ♣$_1$ (recall $\rho(y) > 0$).

Let $s \in J$. Since $u_j + t_j \to \infty$ as $j \to \infty$, as we have just shown, $u_j + t_j \geq s$ and $\sigma(s_j + s, y_j) \leq \epsilon$ for sufficiently large j. Then $\sigma(s^* + s, y) = \lim_{j\to\infty} \sigma(s_j + s, y_j) \leq \epsilon$. Since $s \in J$ has been arbitrary and $s^* \leq 1$, $\sigma(r, y) \leq \epsilon$ for all $r \in J, r \geq 1$, while $\rho(y) > 0$. This contradicts the choice of ϵ at the beginning of our proof. □

From the proof of Theorem 4.2, we record the following technical detail for later use.

Corollary 4.3. *Assume ♣$_0$, ♣$_1$, and ♣$_3$. Let $\epsilon > 0$ and $\limsup_{t\to\infty} \sigma(t, x) > \epsilon$ for all $x \in X$ with $\rho(x) > 0$. Then there exist no sequences (y_j) in X and (s_j), (t_j), (u_j), (v_j) in J such that $v_j \leq 1$ and*

$$\begin{aligned}&\sigma(0, y_j) \geq \epsilon, \quad y_j \in B_j,\\ &\sigma(s_j + t_j, y_j) \to 0, \quad j \to \infty,\\ &\sigma(s_j + s, y_j) \leq \epsilon \qquad \forall s \in [0, t_j + u_j] \cap J,\\ &\sigma(s_j + t_j + u_j + v_j, y_j) \geq \epsilon.\end{aligned}$$

There may be examples where we need to use Theorem 4.2 in this generality, for instance, in models of structured populations where the state space has a topology which is not metrizable. But typically we can work in a metric space and recast the theorem in a more user-friendly form with the following assumptions $\tilde{\clubsuit}$.

$\tilde{\clubsuit}_0$ X is a metric space, $J = \mathbb{R}_+$ or $J = \mathbb{Z}_+$, and $\sigma = \rho \circ \Phi : J \times X \to X$ is continuous.

Further, there exists a nonempty compact subset B of X such that the following properties hold:

$\tilde{\clubsuit}_1$ There are no $y \in B$, $s, t \in J$ such that $\rho(y) > 0$, $\sigma(s, y) = 0$, and $\sigma(s + t, y) > 0$.

$\tilde{\clubsuit}_2$ For every $x \in X$, $\rho(x) > 0$, $d(\Phi(t, x), B) \to 0$ as $t \to \infty$.

Remark 4.4. The continuity of σ in $\tilde{\clubsuit}_0$ can be relaxed by $\clubsuit_0$ and σ being state-continuous uniformly in finite time: For any $x \in X$, $t > 0$, $\epsilon > 0$, there exists some $\delta > 0$ such that $|\sigma(s, y) - \sigma(s, x)| < \epsilon$ whenever $s \in [0, t]$, $y \in X$, $d(y, x) < \delta$.

The proof that the continuity of σ implies that σ is state-continuous uniformly in finite time is similar to the proof of Lemma 1.34 (a).

Theorem 4.5. *Under the assumptions $\tilde{\clubsuit}$, the semiflow Φ is uniformly ρ-persistent, whenever Φ is uniformly weakly ρ-persistent.*

Proof. Obviously $\tilde{\clubsuit}_0$ implies $\clubsuit_0$. To check the remaining assumptions, we set

$$B_k = U_{1/k}(B) = \{x \in X; d(x, B) < 1/k\}.$$

By definition, $B_{k+1} \subset B_k$ and $\tilde{\clubsuit}_2$ implies $\clubsuit_2$. We only need to check $\clubsuit_3$. Let (y_k) be a sequence in B_k. By definition, we find elements $z_k \in B$ such that $d(y_k, z_k) < 1/k$. Since B is compact, after choosing a subsequence, $z_k \to y$ as $k \to \infty$ for some $y \in B$. After choosing a subsequence of (y_k), by the triangle inequality,

$$d(y_k, y) \le d(y_k, z_k) + d(z_k, y) < \frac{1}{k} + d(z_k, y) \to 0, \qquad k \to \infty.$$

$\clubsuit_3$ follows because σ is state-continuous, uniformly in finite time. □

4.1.2. Application to the SEIRS epidemic model in a patchy environment. We extend Proposition 3.9 for (3.5) to obtain uniform persistence of the disease. The disease is endemic in the sense that it is ultimately present in at least some patches. This result will be further extended in Theorem 4.23 to conclude that the disease is ultimately present in all patches. Recall the matrix B in (3.9).

Theorem 4.6. *If B is irreducible and $\mathbf{s}(B) > 0$, then there exists $\epsilon > 0$ such that*

$$\liminf_{t\to\infty}\left(\sum_i I_i(t)\right) > \epsilon$$

for all solutions of (3.5) satisfying $\sum_i (I_i(0) + E_i(0)) > 0$.

Proof. We use Proposition 3.9 and Theorem 4.5 with $X = \mathbb{R}_+^{4n}$, $J = \mathbb{R}_+$, and $\rho(x) = \sum_{i=1}^n I_i$, where $x = (S, E, I, R)$, each component being an n-vector. Choose the set B in Theorem 4.5 as $B = \{x \in X : \sum_i N_i \le c\}$, where $c > 0$ is as in Theorem 3.4. Then ($\tilde{\clubsuit}$) hold. □

4.2. From uniform weak to uniform persistence: Discrete case

Let J be a closed time set, X a nonempty set, $\Phi : J \times X \to X$ a semiflow, and $\rho : X \to \mathbb{R}_+$. We set

$$\sigma(t, x) = \rho(\Phi(t, x)). \tag{4.4}$$

Notice that

$$\sigma(t, \Phi(r, x)) = \sigma(t + r, x). \tag{4.5}$$

Here we consider discrete semiflows $J = \mathbb{Z}_+$. We make the following assumptions ♠.

(♠) There exist a nonempty subset B of X and a sequence (B_k) of subsets of X such that $B \subset B_{k+1} \subset B_k$ for all $k \in \mathbb{N}$ and the following properties hold:

- ♠$_1$ For every $k \in \mathbb{N}$ and every $x \in X$, $\rho(x) > 0$, there exists some $\tau_k \in \mathbb{N}$ such that $\Phi(t, x) \in B_k$ for all $t \ge \tau_k$.
- ♠$_2$ If (y_k) is a sequence in X such that $y_k \in B_k$ for all $k \in \mathbb{N}$ and $0 < \inf_{k=1}^\infty \rho(y_k) \le \sup_{k=1}^\infty \rho(y_k) < \infty$, then, after possibly choosing a subsequence, there exists some $y \in B$ such that, for all s in $\mathbb{Z}_+$,
$$\lim_{k\to\infty} \sigma(s, y_k) = \sigma(s, y).$$
- ♠$_3$ For each $\epsilon > 0$ there exists some $\delta > 0$ and $k \in \mathbb{N}$ such that $x \in B_k$, $\Phi_1(x) \in B_k$, $\rho(x) \ge \epsilon$ imply $\rho(\Phi_1(x)) \ge \delta$.

Theorem 4.7. *Let Φ be a discrete semiflow such that the assumptions (♠) are satisfied. Then Φ is uniformly ρ-persistent, whenever it is uniformly weakly ρ-persistent.*

Proof. Suppose that Φ is uniformly weakly ρ-persistent, but not uniformly ρ-persistent. Choose $\epsilon > 0$ such that

$$\limsup_{t\to\infty} \sigma(t,x) > \epsilon \qquad \forall x \in X, \rho(x) > 0.$$

Let (ϵ_j) be a sequence in $(0,\epsilon)$ such that $\epsilon_j \to 0$ as $j \to \infty$. Since Φ is not uniformly ρ-persistent, we find a sequence (x_j) in X, $\rho(x_j) > 0$, such that

$$\liminf_{t\to\infty} \sigma(t,x_j) < \epsilon_j.$$

Using this last inequality, $\limsup_{t\to\infty} \sigma(t,x_j) > \epsilon$, and $\spadesuit_1$, we find sequences (r_j), (t_j), (u_j) in $\mathbb{Z}_+$ such that $\tau_j < r_j \to \infty$ as $j \to \infty$,

$$\begin{aligned}
&\sigma(r_j, x_j) \ge \epsilon, \qquad \Phi(t,x_j) \in B_j,\ t \ge r_j,\\
&\sigma(s+1+r_j, x_j) \le \epsilon, \qquad 0 \le s \le t_j + u_j,\\
&\sigma(t_j+1+r_j, x_j) < \epsilon_j,\\
&\sigma(1+u_j+t_j+1+r_j, x_j) \ge \epsilon.
\end{aligned}$$

We set $y_j = \Phi(r_j+1, x_j)$. Then $y_j = \Phi_1(\Phi(r_j,x_j))$. Corresponding to ϵ above, we choose $\delta > 0$ and $k \in \mathbb{N}$ according to ($\spadesuit_3$). For $j \ge k$, $y_j = \Phi_1(\Phi(r_j,x_j)) \in B_k$ and $\Phi(r_j,x_j)) \in B_k$. Since $\rho(\Phi(r_j,x_j)) \ge \epsilon$, $\rho(y_j) \ge \delta > 0$ for $j \ge k$ by ($\spadesuit_3$). By definition of σ,

$$\begin{aligned}
&\rho(y_j) = \sigma(0,y_j) \in [\delta,\epsilon], \quad y_j \in B_j \quad \forall j \ge k,\\
&\sigma(t_j, y_j) < \epsilon_j,\\
&\sigma(s,y_j) \le \epsilon \qquad \forall s \in [0, t_j+u_j] \cap \mathbb{Z}\\
&\sigma(t_j+u_j+1, y_j) \ge \epsilon.
\end{aligned}$$

By assumption $\spadesuit_2$, after possibly choosing a subsequence, there exists some $y \in B$ such that

$$\lim_{j\to\infty} \sigma(s,y_j) = \sigma(s,y) \qquad \forall s \in \mathbb{Z}_+.$$

In particular, $\rho(y) = \sigma(0,y) = \lim_{j\to\infty} \sigma(0,y_j) \ge \delta > 0$.

Claim: $t_j + u_j \to \infty$ as $j \to \infty$.

Suppose not. After choosing subsequences, we find some $t > 0$ such that $t_j + u_j \le t$ for all $j \in \mathbb{N}$. After choosing subsequences again, $t_j = t^*$, $u_j = u^*$ for appropriate $t^*, u^* \in [0,t] \cap \mathbb{Z}_+$. Then

$$\sigma(t^*,y) = \lim_{j\to\infty} \sigma(t_j,y_j) \le \limsup_{j\to\infty} \epsilon_j = 0,$$

contradicting assumption $\spadesuit_3$. Indeed, since $\rho(y_j) \ge \delta$, $j \ge k$ and $\Phi(t,y_j) \in B_j$, $t \ge 0$, $\spadesuit_3$ implies there is $\eta > 0$ and $k_1 \in \mathbb{N}$ such that for $j \ge k_1$ we have $\rho(\Phi(1,y_j)) \ge \eta$. Iterating this procedure a finite number of times, we find there exists $\kappa > 0$ such that $\rho(\Phi(t^*,y_j)) \ge \kappa$ for all large j, contradicting that $\rho(\Phi(t^*,y_j)) = \sigma(t^*,y_j) < \epsilon_j \to 0$.

Let $s \in \mathbb{Z}_+$. Since $u_j + t_j \to \infty$ as $j \to \infty$, as we have just shown, $u_j + t_j \geq s$ and, using the second of the four conditions above, $\sigma(s, y_j) \leq \epsilon$ for sufficiently large j. Then $\sigma(s, y) = \lim_{j\to\infty} \sigma(s, y_j) \leq \epsilon$. Since $s \geq 0$ has been arbitrary, $\sigma(s, y) \leq \epsilon$ for all $s \geq 0$, while $\rho(y) > 0$. This contradicts the choice of ϵ at the beginning of our proof. $\square$

As before, we can formulate a more user-friendly version for metric spaces. We make the following assumptions.

($\tilde{\spadesuit}$) X is a metric space, $\rho : X \to \mathbb{R}_+$ is continuous and $\Phi_1 : X \to X$ is continuous. There exists a nonempty subset B of X such that the following properties hold:

$\tilde{\spadesuit}_1$ For every $x \in X$, $\rho(x) > 0$, we have $\Phi(t, x) \to B$ as $t \to \infty$.

$\tilde{\spadesuit}_2$ If $0 < c_1 < c_2$, then the set $\{c_1 \leq \rho \leq c_2\} \cap B$ is compact.

$\tilde{\spadesuit}_3$ For each $\epsilon > 0$ there exists some $\delta > 0$ such that $x \in B$, $\rho(x) \geq \epsilon$ imply $\rho(\Phi_1(x)) \geq \delta$.

We define B_k as the collection of all those $y \in X$ such that there exists some $z \in B$ with

$$d(y, z) < \frac{1}{k}, \qquad |\rho(y) - \rho(z)| < \frac{1}{k}, \qquad |\rho(\Phi_1(y)) - \rho(\Phi_1(z))| < \frac{1}{k}.$$

Since ρ and Φ_1 are continuous, B_k is an open subset of X which contains B, $B_{k+1} \subset B_k$ (Exercise 4.1). Indeed,

$$B_k = \bigcup_{z\in B} B_{1/k}(z) \cap \rho^{-1}(\rho(z) - \frac{1}{k}, \rho(z) + \frac{1}{k})$$
$$\cap (\rho \circ \Phi_1)^{-1}(\sigma(1, z) - \frac{1}{k}, \sigma(1, z) + \frac{1}{k}).$$

So $\tilde{\spadesuit}_1$ implies $\spadesuit_1$ by Definition 2.1.

To check $\spadesuit_2$, let (y_k) be a sequence in X such that $y_k \in B_k$ and $0 < \inf_{k=1}^\infty \rho(y_k) \leq \sup_{k=1}^\infty \rho(y_k) < \infty$. Then there exists a sequence (z_k) in B such that $d(y_k, z_k) \to 0$, $\rho(y_k) - \rho(z_k) \to 0$ and $\rho(\Phi_1(y_k)) - \rho(\Phi_1(z_k)) \to 0$ as $k \to \infty$. We conclude that $0 < \inf_{k=1}^\infty \rho(z_k) \leq \sup_{k=1}^\infty \rho(z_k)) < \infty$. By $\tilde{\spadesuit}_2$, after choosing a subsequence, $z_k \to y$ for some $y \in B$. So $y_k \to y$ as $k \to \infty$, after choosing a subsequence. Since ρ and Φ_1 are continuous,

$$\sigma(s, y_k) = \rho(\Phi_1^s(y_k)) \to \rho(\Phi_1^s(y)) = \sigma(s, y), \qquad k \to \infty.$$

To check $\spadesuit_3$, let $\epsilon > 0$. By $\tilde{\spadesuit}_3$ there exist some $\delta > 0$ such that $x \in B$, $\rho(x) \geq \epsilon/3$ implies $\rho(\Phi_1(x)) \geq 3\delta$. Choose $k \in \mathbb{N}$ such that $\frac{1}{k} < \min\{\epsilon/3, \delta\}$. Let $y \in B_k$, $\rho(y) \geq \epsilon$. Then there exists some $x \in B$ such that $|\rho(x) - \rho(y)| < \epsilon/3$ and $|\rho(\Phi_1(x)) - \rho(\Phi_1(y))| < \delta$. Then $\rho(x) \geq \epsilon/3$. So $\rho(\Phi_1(x)) \geq 3\delta$ and $\rho(\Phi_1(y)) > \delta$.

Corollary 4.8. *Let Φ be a discrete semiflow such that the assumptions $(\tilde{\spadesuit})$ are satisfied. Then Φ is uniformly ρ-persistent, whenever it is uniformly weakly ρ-persistent.*

Remark 4.9. If B of $\tilde{\spadesuit}$ is compact and $X \setminus \rho^{-1}(0)$ is forward invariant, then $\tilde{\spadesuit}_3$ holds.

4.3. Application to a metered endemic model of SIR type

We formulate a host-parasite model for a wildlife population where the host has a very short reproductive season while the disease spreads continuously over the year. Because of these different time scales, we will work with a difference equation where the right-hand side involves the time-map of a differential equation. Hybrid models of this type have been called *discrete time metered* models [**32**, Chap. 7], *sequential-continuous* models [**20**, Sec. 3.6], or *semi-discrete models* [**199**]. They can be considered special cases of *impulsive* or *pulsed* differential equations. Since Clarke (1976) [**32**, Chap. 7] has the priority in naming these models, we refer to them as *discrete time metered* models or simply *metered models.* Metered epidemic models have been used before to model helminthic infections in sheep (see [**182**] and the references therein). In [**51**], they are used to analyze the persistence of viral infections in salamander populations. Difference equations which involve time-maps of ordinary differential equations have also been employed in genetic models [**89**].

Let S_n, I_n, R_n be the number of susceptible, infective, and recovered individuals in year n immediately before the reproductive season and $S(t)$, $I(t)$, $R(t)$ the respective numbers between two reproductive seasons, $0 \le t \le 1$. Let N_n be population size in year n immediately before the reproductive season and $N(t)$ the population size during the year. Then

$$
\begin{aligned}
&N_n = S_n + I_n + R_n,\\
&S(0) = \beta N_n + S_n, \quad I(0) = I_n, \quad R(0) = R_n,\\
&\left.\begin{aligned}
S' &= -\tfrac{\kappa(N)}{N} SI - \mu S + \theta R\\
I' &= \tfrac{\kappa(N)}{N} SI - (\mu + \alpha + \gamma) I\\
R' &= \gamma I - (\mu + \theta) R\\
N &= S + I + R
\end{aligned}\right\} \quad 0 \le t \le 1,\\
&S_{n+1} = S(1), \quad I_{n+1} = I(1), \quad R_{n+1} = R(1).
\end{aligned} \tag{4.6}
$$

β is now the per capita number of offspring in a reproductive season. $\kappa(N)$ is the per capita rate of contacts at population size N which are effective if they occur between a susceptible and infective individual. All parameters are assumed to be nonnegative, $\mu > 0$, $\beta > 0$, $\gamma + \alpha > 0$. κ is an increasing locally Lipschitz continuous function, $\kappa(N) > 0$ for $N > 0$. We show that the host population is uniformly persistent if $(\beta+1)e^{-\mu} > 1$ and $\kappa(0) = 0$.

We introduce the fractions of susceptible, infective and recovered individuals, $u = S/N$, $y = I/N$, $z = R/N$ and reformulate the model,

$$
\begin{aligned}
N(0) =&(\beta+1)N_n, & u(0) &= \frac{\beta+u_n}{\beta+1},\\
y(0) =&\frac{y_n}{\beta+1}, & z(0) &= \frac{z_n}{\beta+1},
\end{aligned}
\tag{4.7}
$$

$$
\left.
\begin{aligned}
u' =&\ -\kappa(N)uy + \theta z + \alpha uy\\
y' =&\ \kappa(N)uy - (\alpha+\gamma)y + \alpha y^2\\
z' =&\ \gamma y - \theta z + \alpha yz\\
N' =&\ -N(\mu + \alpha y)
\end{aligned}
\right\}
\qquad 0 \le t \le 1,
\tag{4.8}
$$

$$
\begin{aligned}
N_{n+1} =&N(1), & u_{n+1} &= u(1),\\
y_{n+1} =&y(1), & z_{n+1} &= z(1).
\end{aligned}
\tag{4.9}
$$

We first notice that solutions are nonnegative if the initial data are nonnegative. Further $(u+y+z)' = \alpha y(u+y+z-1)$. This shows that the plane $u+y+z=1$ is invariant. We choose $X = \{(N,u,y,z) \in \mathbb{R}^4_+; u+y+z=1\}$ as the state space.

4.3.1. Uniform persistence of the host. We set $\rho(N,u,y,z) = N$. We first show that the discrete dynamical system induced by this metered model is uniformly weakly ρ-persistent. Let us assume that the system is not uniformly weakly ρ-persistent. By the semiflow property, we can assume that $N_0 > 0$ and $N_n \le \epsilon$ for all $n \in \mathbb{N}$. Then always $N(t) \le \epsilon$. From the differential equations for y in (4.8),

$$y' \le y(\kappa(\epsilon) - \alpha - \gamma + \alpha y).$$

Notice that $y(0) \le \frac{1}{\beta+1}$. Since we have assumed that $\alpha + \gamma > 0$, we can choose $\epsilon > 0$ so small that $\xi = \kappa(\epsilon) - \alpha - \gamma + \frac{\alpha}{\beta+1} < 0$. Then y is decreasing between reproductive seasons. This implies $y_{n+1} \le \frac{y_n}{\beta+1} \to 0$ as $n \to \infty$. Since $(\beta+1)e^{-\mu} > 1$, there exists some $\nu > \mu$ such that $(\beta+1)e^{-\nu} > 1$. There exists some $n \in \mathbb{N}$ such that $\mu + \alpha y < \nu$ and $N' \ge -\nu N$ during all years after year n. So $N(1) \ge N(0)e^{-\nu}$ and, for $j \ge n$, $N_{j+1} \ge (\beta+1)e^{-\nu}N_j \to \infty$, a contradiction.

We check the assumptions $\tilde{\spadesuit}$. Let Ψ be the semiflow induced by the differential equations in (4.8) and T the map

$$T(N,u,y,z) = \Big((\beta+1)N, \frac{\beta+u}{\beta+1}, \frac{y}{\beta+1}, \frac{z}{\beta+1}\Big).$$

Then $\Phi_1 = \Psi_1 \circ T$ is continuous and $\vec{x}_{n+1} = \Phi_1(\vec{x}_n)$, $\vec{x}_n = (N_n, u_n, y_n, z_n)$.

We choose $B = X$ and $\tilde{\spadesuit}_1$ is trivially satisfied. Further,

$$B \cap \{\rho(N,u,y,z) \le c\} = \big\{(N,u,y,z) \in \mathbb{R}^4_+; N \le c, u+y+z=1\big\}$$

is bounded and closed in $\mathbb{R}^4$ and so compact. This means that $\tilde{\spadesuit}_2$ holds. Since $N' \ge -(\mu+\alpha)N$, $\rho(\Phi_1(x)) \ge (\beta+1)e^{-(\mu+\alpha)}\rho(x)$. This implies $\tilde{\spadesuit}_3$. Corollary 4.8 implies the following result.

Theorem 4.10. *Assume* $(\beta+1)e^{-\mu} > 1$ *and* $\kappa(0) = 0$. *Then the host population in model (4.6) is uniformly persistent, i.e., there exists some* $\epsilon > 0$ *such that, for all nonnegative solutions of (4.6), we have* $\liminf_{n\to\infty} N_n \ge \epsilon$ *whenever* $N_0 > 0$.

4.3.2. Uniform weak persistence of the parasite. We turn to disease persistence. Let $\kappa_\infty = \lim_{N\to\infty}\kappa(N)$. κ_∞ can be finite or infinite.

Proposition 4.11. *Assume that* $\mathcal{R}_0^{\text{host}} = (\beta+1)e^{-\mu} > 1$ *and* $\mathcal{R}_0^{\text{par}} = \frac{e^{\kappa_\infty - \alpha - \gamma}}{\beta+1} > 1$. *Then the disease (i.e., the parasite) persists uniformly weakly in the following sense: There exists some* $\epsilon > 0$ *such that* $\limsup_{n\to\infty} y_n \ge \epsilon$ *for all nonnegative solutions of (4.7), (4.8), (4.9) with* $y_0 > 0$, $N_0 > 0$.

Proof. Assume that the disease does not persist uniformly weakly. Then there exists some $\epsilon > 0$, which we can choose as small as we want, and a solution with $y_0 > 0$, $N_0 > 0$, and $\limsup_{n\to\infty} y_n < \epsilon$. Notice that $y_n > 0$ and $N_n > 0$ for all $n \in \mathbb{Z}_+$. By a shift in time, we can assume that $y_n < \epsilon$ for all $n \in \mathbb{Z}_+$. From the differential equation for y in (4.8), $y' \ge -(\alpha+\gamma)y$. This implies that $y(1) \ge y(t)e^{-(\gamma+\alpha)(1-t)}$. Hence, $y(t) \le y_{n+1}e^{\gamma+\alpha} \le \epsilon e^{\gamma+\alpha}$. From the differential equation for N in (4.8), $N' \ge -N(\mu + \alpha\epsilon e^{\gamma+\alpha})$. Hence

$$N(t) \ge N(0)\exp\big(-(\mu+\alpha\epsilon e^{\gamma+\alpha})t\big).$$

By (4.7) and (4.9),

$$N_{n+1} \ge N_n(\beta+1)\exp\big(-(\mu+\alpha\epsilon e^{\gamma+\alpha})\big). \tag{4.10}$$

Since $(\beta+1)e^{-\mu} > 1$, we can choose $\epsilon > 0$ so small such that $N_n \to \infty$ as $n \to \infty$. This means, that after sufficiently many years, we can choose N as large as we want in the differential equation for y in (4.8). From the differential equation for z in (4.8),

$$z' \le (\gamma+\alpha)\epsilon e^{\gamma+\alpha} - \theta z.$$

We integrate this inequality,

$$z(t) \le z(0)e^{-\theta t} + \int_0^t (\gamma+\alpha)\epsilon e^{\gamma+\alpha}e^{-\theta s}ds.$$

By (4.7),

$$z(t) \le \frac{z_n}{\beta+1} + \epsilon\zeta(1-e^{-\theta}), \qquad \zeta = \frac{(\gamma+\alpha)}{\theta}e^{\gamma+\alpha}. \tag{4.11}$$

Since $z_n \le 1$, $z^\infty = \limsup_{n\to\infty} z_n$ exists and

$$z^\infty \le \frac{z^\infty}{\beta+1} + \epsilon\zeta(1-e^{-\theta}).$$

We solve for z^∞,

$$z^\infty < \frac{\epsilon\zeta}{1-\frac{1}{\beta+1}} = \frac{(\beta+1)\epsilon\zeta}{\beta}.$$

If n is sufficiently large, during year n in (4.8),

$$z \le \frac{\epsilon\zeta}{\beta}(1+\beta)$$

and

$$y + z \le \epsilon\Big(e^{\gamma+\alpha} + \frac{\zeta(1+\beta)}{\beta}\Big) =: \epsilon\xi. \tag{4.12}$$

From the differential equation for y in (4.8),

$$y'/y \ge \kappa(N)(1-\epsilon\xi) - (\alpha+\gamma).$$

Let $\tilde\kappa \in (0,\kappa_\infty)$. If we choose $\epsilon > 0$ small enough, there exists $m \in \mathbb{N}$ such that $\kappa(N(t))(1-\epsilon\xi) \ge \tilde\kappa$ in all years after the mth year. Hence,

$$\frac{y(1)}{y(0)} \ge \exp\Big(\tilde\kappa - (\alpha+\gamma)\Big).$$

This implies

$$\frac{y_{n+1}}{y_n} \ge \frac{\exp\big(\tilde\kappa-(\alpha+\gamma)\big)}{\beta+1}, \qquad n \ge m.$$

Choosing $\tilde\kappa$ sufficiently close to κ_∞, we have $y_{n+1} \ge \nu y_n$ for sufficiently large $n \in \mathbb{N}$, with $\nu > 1$ being independent of n. This implies $y_n \to \infty$ as $n \to \infty$ contradicting $y_n \le 1$. □

4.4. From uniform weak to uniform persistence for time-set $\mathbb{R}_+$

If the time-set is $\mathbb{R}_+$, we can exploit time-continuity properties of σ. We consider the following scenario ♡.

♡$_0$ For every $x \in X$, $\sigma(t,x)$ is a continuous function of $t \ge 0$.

There exist a nonempty subset B of X and a sequence (B_k) of subsets of B such that $B \subset B_{k+1} \subset B_k$ for all $k \in \mathbb{N}$ and the following properties hold:

♡$_1$ For every $k \in \mathbb{N}$ and every $x \in X$, $\rho(x) > 0$, there exists some $t_k \ge 0$ such that $\Phi(t,x) \in B_k$ for all $t \ge t_k$.

$\heartsuit_2$ If (y_k) is a sequence in X such that $y_k \in B_k$ for all $k \in \mathbb{N}$ and $\rho(y_k) = \rho(y_1) > 0$ for all $k \in \mathbb{N}$, then, after possibly choosing a subsequence, there exists some $y \in B$ such that $\sigma(s, y_k) \to \sigma(s, y)$ as $k \to \infty$, uniformly for s in every interval $[0, t]$, $t \in (0, \infty)$.

$\heartsuit_3$ There are no $x \in B$, $r, s > 0$ such that $\rho(x) > 0$, $\sigma(s, x) = 0$, and $\sigma(s + r, x) > 0$.

Theorem 4.12. *Let Φ be a semiflow with time-set $J = \mathbb{R}_+$ such that the assumptions ($\heartsuit$) are satisfied. Then Φ is uniformly ρ-persistent, whenever it is uniformly weakly ρ-persistent.*

Proof. We merely sketch the changes in the proof of Theorem 4.7. Using $\heartsuit_0$, we may choose r_j such that $\sigma(r_j, x_j) = \epsilon$, $\sigma(s, x_j) < \epsilon$, $r_j < s < t_j + u_j$, $\sigma(t_j, x_j) < \epsilon_j$, and $\sigma(t_j + u_j, x_j) = \epsilon$. Let $y_j = \Phi(r_j, x_j)$ so $\rho(y_j) = \epsilon$. Applying $\heartsuit_2$, we obtain $y \in B$ so that $\sigma(s, y_k) \to \sigma(s, y)$. The stronger uniform convergence of $\sigma(s, y_k) \to \sigma(s, y)$ is required in the proof of the claim that $t_j + u_j \to \infty$ since for continuous time we cannot assert that $t_j = t^*$ and $u_j = u^*$ but only that convergence (of a subsequence) holds. We need this stronger convergence to conclude that $\sigma(t_j, y_j) \to \sigma(t^*, y) = 0$ and $\sigma(t^* + u^*, y) = \epsilon$. But then $\heartsuit_3$ is used to obtain a contradiction. The remainder of the proof follows that of Theorem 4.7. □

The following set of assumptions is less general, but more user-friendly if it applies.

$\tilde{\heartsuit}_0$ X is a metric space, ρ is uniformly continuous, and $\sigma = \rho \circ \Phi$ is continuous on $\mathbb{R}_+ \times X$.

There exist a nonempty subset B of X such that the following properties hold:

$\tilde{\heartsuit}_1$ For every $x \in X$, $\rho(x) > 0$, $d(\Phi(t, x), B) \to 0$ as $t \to \infty$.

$\tilde{\heartsuit}_2$ If $0 < \epsilon_1 < \epsilon_2 < \infty$, then $B \cap \{\epsilon_1 \le \rho(x) \le \epsilon_2\}$ is compact.

$\tilde{\heartsuit}_3$ There are no $x \in B$, $r, s > 0$ such that $\rho(x) > 0$, $\sigma(s, x) = 0$, and $\sigma(s + r, x) > 0$.

Alternatively, we can make the following assumptions.

$\hat{\heartsuit}_0$ X is a metric space, ρ is continuous, and $\sigma = \rho \circ \Phi$ is continuous from $\mathbb{R}_+ \times X$ to R_+.

There exist a nonempty subset B of X such that the following properties hold:

$\hat{\heartsuit}_1$ For every $x \in X$, $\rho(x) > 0$, $\Phi(t, x) \to B$ as $t \to \infty$.

$\hat{\heartsuit}_2$ If $0 < \epsilon_1 < \epsilon_2 < \infty$, then $B \cap \{\epsilon_1 \le \rho(x) \le \epsilon_2\}$ is compact.

$\hat{\heartsuit}_3$ There are no $x \in B$, $r, s > 0$ such that $\rho(x) > 0$, $\sigma(s, x) = 0$, and $\sigma(s + r, x) > 0$.

Theorem 4.13. *Let Φ be a semiflow with time-set $J = \mathbb{R}_+$ such that the assumptions $(\tilde{\heartsuit})$ or $(\hat{\heartsuit})$ are satisfied. Then Φ is uniformly ρ-persistent, whenever it is uniformly weakly ρ-persistent.*

We apply these results to the endemic model (3.2) and (3.3) for a fertility-reducing disease. We choose the state space $X = \mathbb{R}_+ \times [0, 1] = \{(N, y); N \geq 0, 0 \leq y \leq 1\}$. We choose $B = X$ and $\rho(N, y) = N$. The assumptions $\tilde{\heartsuit}$ (or $\hat{\heartsuit}$) are readily checked, and Theorem 4.13 together with Theorem 3.3 implies uniform host persistence for $q \in (0, 1]$.

Theorem 4.14. *Let $\beta > \mu$ and $q \in (0, 1]$. Then there exists some $\epsilon > 0$ such that* $\liminf_{t\to\infty} N(t) \geq \epsilon$ *for all solutions of (3.2) and (3.3) with $N(0) > 0$.*

Recall that, for $q = 0$, this statement only holds for lim sup and the host is only uniformly weakly persistent. The reason for this is that uniform weak host persistence only holds on the state space $\tilde{X} = \{(N, y); N \geq 0, 0 \leq y < 1\}$ and $\tilde{X} \cap \{\epsilon_1 \leq N \leq \epsilon_2\}$ is not complete (in particular, not compact) if $0 < \epsilon_1 < \epsilon_2 < \infty$.

Remark 4.15. In Exercise 4.4, the reader is asked to prove that Theorem 4.13 holds under either Assumption $\tilde{\heartsuit}$ or under Assumption $\hat{\heartsuit}$.

4.5. Persistence à la Baron von Münchhausen

Here we assume from the beginning that X is a metric space and $\rho : X \to \mathbb{R}_+$ is continuous.

We first consider a discrete semiflow with $\Phi_1 : X \to X$ being continuous and make the following assumptions.

$\diamondsuit_0$ There exists a closed subset B of X such that, for all $x \in X$, $\rho(x) > 0$, there exists some $t_x \in \mathbb{Z}_+$ such that $\Phi(t, x) \in B$ for all $t \geq t_x, t \in \mathbb{Z}_+$.

Further, if $\epsilon > 0$ is sufficiently small, there are $D \subset X$ and $\delta \in (0, \epsilon)$ with the following properties:

$\diamondsuit_1$ $D \cap B$ is compact.

$\diamondsuit_2$ If $x \in B \cap D$ and $t \in \mathbb{N}$ such that $\Phi(\mathbb{N} \times \{x\}) \subset B$ and $\rho(x) \geq \epsilon$, $\sigma(t+1, x) \geq \epsilon$, and $\sigma(s, x) < \epsilon$ for $s = 1, \ldots, t$, then $\sigma(s, x) > 0$ for $s = 1, \ldots, t$.

Figure 4.1. Baron von Münchhausen lifts himself and his horse out of the swamp — Drawing by Theodor Hosemann (1807-1875)

$\diamondsuit_3$ If $x \in B \setminus D$ and $t \in \mathbb{N}$ such that $\Phi(\mathbb{N} \times \{x\}) \subset B$ and $\rho(x) \geq \epsilon$, $\sigma(t+1, x) \geq \epsilon$ and $\sigma(s, x) < \epsilon$ for $s = 1, \ldots, t$, then $\sigma(s, x) \geq \delta$ for $s = 1, \ldots, t$.

Theorem 4.16. *Under the assumptions $\diamondsuit$, the semiflow Φ is uniformly ρ-persistent, whenever Φ is uniformly weakly ρ-persistent.*

Proof. Suppose that Φ is uniformly weakly ρ-persistent, but not uniformly ρ-persistent. Then there exists some $\epsilon > 0$ such that

$$\limsup_{t \to \infty} \sigma(t, x) > \epsilon \qquad \forall x \in X, \rho(x) > 0.$$

We can assume that $\epsilon > 0$ is small enough for D, δ, t to exist such that $\diamondsuit_{1,2,3,4}$ hold. Let (ϵ_j) be a sequence in $(0, \epsilon)$ such that $\epsilon_j \to 0$ as $j \to \infty$. Since Φ is not uniformly persistent, there exists a sequence (x_j) in X, $\rho(x_j) > 0$, such that $\liminf_{t \to \infty} \sigma(t, x_j) < \epsilon_j$.

So there exist sequences (r_j), (u_j) in $\mathbb{Z}_+$ and (s_j) in $\mathbb{N}$ such that

$$\begin{aligned}
&\sigma(r_j, x_j) \geq \epsilon, \\
&\sigma(s + r_j, x_j) < \epsilon, \qquad s = 1, \ldots, s_j + u_j, \\
&\sigma(s_j + r_j, x_j) < \epsilon_j, \\
&\sigma(1 + u_j + s_j + r_j, x_j) \geq \epsilon.
\end{aligned}$$

For each j, r_j can be chosen arbitrarily large and so we can arrange that $r_j \to \infty$ as $j \to \infty$ and, by $\diamondsuit_0$, $\Phi(s + r_j, x_j) \in B$ for all $s \in \mathbb{Z}_+$. We set $y_j = \Phi(r_j, x_j)$. Then

$$\begin{aligned} &\sigma(0, y_j) \geq \epsilon, && \Phi(s, y_j) \in B \quad \forall s \in \mathbb{Z}_+, \\ &\sigma(s, y_j) < \epsilon, && s = 1, \dots, s_j + u_j, \\ &\sigma(s_j, y_j) < \epsilon_j, \\ &\sigma(1 + u_j + s_j, y_j) \geq \epsilon. \end{aligned}$$

Case 1: There are infinitely many $j \in \mathbb{N}$ such that $y_j \in D$.

After choosing a subsequence, $y_j \in B \cap D$ for all $j \in \mathbb{N}$. Since $B \cap D$ is compact by assumption $\diamondsuit_1$, $y_j \to y$ as $j \to \infty$, with some $y \in B \cap D$ after choosing another subsequence. Since Φ is continuous and B is closed, $\Phi(s, y) = \lim_{j \to \infty} \Phi(s, y_j) \in B$ for all $s \in \mathbb{Z}_+$. By $\diamondsuit_2$, $s_j + u_j \to \infty$. Since ρ and Φ are continuous, $\rho(y) \geq \epsilon$ and $\sigma(s, y) \leq \epsilon$ for all $s \in \mathbb{N}$, contradicting the choice of ϵ at the beginning.

Case 2: There are infinitely many $j \in \mathbb{N}$ such that $y_j \notin D$.

After choosing a subsequence, $y_j \in B \setminus D$ for all $j \in \mathbb{N}$. By $\diamondsuit_3$, with $t = u_j + s_j$, $\sigma(s_j, y_j) \geq \delta > 0$ for all $j \in \mathbb{N}$. Since $\delta > 0$ does not depend on j, this contradicts $\sigma(s_j, y_j) \to 0$ as $j \to \infty$. □

We formulate an analogous result for continuous time-set. The proof is left as an exercise.

Theorem 4.17. *Let X be a metric space. Let $\Phi : \mathbb{R}_+ \times X \to X$ be a state-continuous semiflow on time-set $\mathbb{R}_+$. Let $\rho : X \to \mathbb{R}_+$, and define $\sigma : \mathbb{R}_+ \times X \to \mathbb{R}_+$ by $\sigma(t, x) = \rho\big(\Phi(t, x)\big)$. Assume that σ is continuous.*

Assume there exists a closed set $B \subset X$ such that

- $\blacklozenge_0$ *for each $x \in X \cap \{\rho > 0\}$, there is $\tau \in \mathbb{R}_+$ such that $\Phi\big([\tau, \infty) \times \{x\}\big) \subset B$.*

Also, assume that there exists $\tilde{\varepsilon} \in (0, \infty]$ such that:

For each $\varepsilon \in (0, \tilde{\varepsilon})$, there is a set $D \subset X$ and there is $\delta > 0$ with the following properties:

- $\blacklozenge_1$ *$B \cap D \cap \{\rho = \varepsilon\}$ is compact.*
- $\blacklozenge_2$ *If $x \in X \cap D$ and $t \in \mathbb{R}_+$ such that $\Phi(\mathbb{R}_+ \times \{x\}) \subset B$, $\rho(x) = \epsilon = \sigma(t, x)$ and $\sigma(s, x) < \epsilon$ for all $s \in (0, t)$, then $\sigma(s, x) > 0$ for all $s \in [0, t]$.*
- $\blacklozenge_3$ *If $x \in X \setminus D$ and $t \in \mathbb{R}_+$ such that $\Phi(\mathbb{R}_+ \times \{x\}) \subset B$, $\rho(x) = \epsilon = \sigma(t, x)$ and $\sigma(s, x) < \epsilon$ for all $s \in (0, t)$, then $\sigma(s, x) \geq \delta$ for all $s \in [0, t]$.*

Then Φ is uniformly ρ-persistent if it is uniformly weakly ρ-persistent.

4.5.1. Uniform parasite persistence in the SI model with fertility reduction. As a short application, we prove uniform parasite persistence in the SI model for a fertility-reducing diseases,

$$N' = N\Big(\beta(1-y) - \mu + (q\beta - \alpha)y\Big),$$
$$y' = y\Big((\kappa N - \alpha - \beta)(1-y) - q\beta y\Big).$$

We assume $q \in (0,1]$. As mentioned several times, uniform persistence does not hold for $q = 0$. We already know that the host is uniformly persistent (Theorem 4.14) and the parasite is uniformly weakly persistent (Exercise 3.1). More precisely, there exists some $\epsilon_0 > 0$ such that

$$\liminf_{t\to\infty} N(t) \geq \epsilon_0$$

for all solutions with $N(0) > 0$ and

$$\limsup_{t\to\infty} y(t) \geq \epsilon_0$$

for all solutions with $N(0) > 0$ and $y(0) > 0$. This suggests the state space $X = \{(N,y); N > 0, 0 \leq y \leq 1\}$. Since we are interested in parasite persistence, we choose $\rho(N,y) = y$. Then the solution semiflow is uniformly weakly ρ-persistent.

Further, we choose $B = \{(N,y); N \geq \epsilon_0/2, 0 \leq y \leq 1\}$. By choice of ϵ_0, we have that $\blacklozenge_0$ holds.

For each $\epsilon > 0$, we choose $D_\epsilon = \{(N,y); N \leq N_\epsilon, 0 \leq y \leq 1\}$ with $N_\epsilon > 0$ still to be determined. Then $B \cap D_\epsilon = [\epsilon_0/2, N_\epsilon] \times [0,1]$ is compact.

Now, for $\epsilon \in (0,1)$, choose $N_\epsilon > 0$ so large that

$$(\kappa N_\epsilon - \alpha - \beta)(1-\epsilon) - q\beta\epsilon > 0.$$

So, if $x = (N_0, y_0) \in X \setminus D$ and $\epsilon = \rho(x)$, then $N_0 > N_\epsilon$ and $y_0 = \epsilon$ and the solution with $N(0) = N_0$ and $y(0) = y_0$ satisfies $y'(0) > 0$. This implies that the antecedent in $\blacklozenge_2$ does not occur, and $\blacklozenge_2$ is trivially satisfied.

$\blacklozenge_3$ is satisfied because every solution with $y(0) > 0$ satisfies $y(t) > 0$ for all $t \geq 0$.

By Theorem 4.17, Φ is uniformly ρ-persistence. By our choice of state space and ρ, this translates into the following result.

Theorem 4.18. *If $\beta > \mu$ and $q \in (0,1]$, the parasite is uniformly persistent (and the disease uniformly endemic):*

There exists some $\epsilon > 0$ such that $\liminf_{t\to\infty} y(t) \geq \epsilon$ for all solutions with $N(0) > 0$ and $y(0) \in (0,1]$.

If we add more disease stages, applying persistence à la Münchhausen requires more effort, but it will become clearer why the assumptions have been formulated as they are.

4.5.2. Uniform parasite persistence in the metered SIRS model. Recall Section 4.3. As we only have uniform weak parasite persistence if both $N_0 > 0$ and $y_0 > 0$, we choose the state space as

$$X = \{(N, u, y, z) \in \mathbb{R}^4_+; u + y + z = 1, N > 0\}.$$

Since we are interested in parasite persistence, $\rho(N, u, y, z) = y$. By Theorem 4.10, the host persists uniformly. We choose $\epsilon_0 > 0$ such that $\liminf_{n\to\infty} N_n > \epsilon_0$ if $N_0 > 0$. We choose

$$B = \{(N, u, y, z) \in X; N \geq \epsilon_0\},$$

and $\diamondsuit_0$ is satisfied. For $\epsilon > 0$, which still needs to be determined and can be chosen arbitrarily small, we choose $D = \{(N, u, y, z) \in X; N \leq N_\epsilon\}$ where $N_\epsilon \in (\epsilon_0, \infty)$ will still be determined. However it will be chosen, the set $B \cap D$ is compact and $\diamondsuit_1$ is satisfied. From the differential equation for y in (4.8), $y' \geq -(\alpha + \gamma)y$ and so $y(1) \geq e^{-(\alpha+\gamma)}y(0)$. By (4.7) and (4.9),

$$y_{n+1} \geq \frac{e^{-(\alpha+\gamma)}}{\beta + 1} y_n. \tag{4.13}$$

This implies $\diamondsuit_2$.

Before we continue, we need a variation of constants type result for difference inequalities

$$z_n \leq a z_{n-1} + b_n, \qquad n \in \mathbb{N}. \tag{4.14}$$

Here $a > 0$ is a real number and $(b_n)_{n=1}^\infty$ is a sequence of real numbers. We divide by a^n and set $\tilde{z}_n = z_n a^{-n}$, $\tilde{b}_n = b_n a^{-n}$. Then

$$\tilde{z}_n \leq \tilde{z}_{n-1} + \tilde{b}_n.$$

Obviously any solution satisfies

$$\tilde{z}_n \leq \tilde{z}_0 + \sum_{j=1}^n \tilde{b}_j.$$

We return to the original variables,

$$z_n \leq z_0 a^n + \sum_{j=1}^n a^{n-j} b_j.$$

Trivially, this formula also holds if $a = 0$. If $b_n \leq b$ for $k = 0, \ldots, n$, and $a \neq 1$, then

$$z_n \leq z_0 a^n + b \sum_{k=0}^{n-1} a^k = z_0 a^n + b \frac{1 - a^n}{1 - a}. \tag{4.15}$$

To choose N_ϵ and $\delta = \delta_\epsilon$, let $N_0 > N_\epsilon$, $y_0 \ge \epsilon$, and $y_s \le \epsilon$ for $s = 1, \dots, t$, $t \in \mathbb{N}$. Since $(1+\beta)e^{-\mu} > 1$, we can choose ϵ so small that $N_n \ge N_0$ for all $n = 0, \dots, t$ by (4.10) and $N \ge N_0$ in the differential equations (4.8) which describe the dynamics between reproductive seasons. For $n = 1, \dots, t$, by (4.11), we have the difference inequality

$$z_n \le a z_{n-1} + b$$

with $a = \frac{1}{\beta+1}$ and $b = \epsilon\zeta(1 - e^{-\theta})$, $\zeta = \frac{\gamma+\alpha}{\theta} e^{\gamma+\alpha}$. By (4.15), since $z_0 \le 1$,

$$z_n \le a^n + \frac{\beta+1}{\beta}\epsilon\zeta(1 - e^{-\theta}). \tag{4.16}$$

Since $a < 1$, we can fix an $n_\epsilon \in \mathbb{N}$ such that $a^{n_\epsilon} < \frac{\beta+1}{\beta}\epsilon\zeta e^{-\theta}$. Then the right-hand side of (4.16) is smaller than $\frac{\beta+1}{\beta}\epsilon\zeta$ for $n \ge n_\epsilon$.

Case 1: $t < n_\epsilon$.

Then, by (4.13),

$$y_n \ge \Big(\frac{e^{-(\alpha+\gamma)}}{\beta+1}\Big)^{n_\epsilon - 1} =: \delta, \qquad n = 1, \dots, t.$$

Case 2: $t \ge n_\epsilon$.

By the choice of n_ϵ,

$$z_n \le \frac{\beta+1}{\beta}\epsilon\zeta, \qquad n = n_\epsilon, \dots, t.$$

By the same consideration as in Section 4.3.2, during every year from n_ϵ to $t-1$, we have $y + z \le \epsilon\xi$ in (4.8), with an appropriate constant ξ which does not depend on ϵ (cf. (4.12)). Hence, during any year from n_ϵ to $t-1$,

$$\frac{y'}{y} \ge \kappa(N_0)(1 - \epsilon\xi) - (\alpha + \gamma) =: \theta_\epsilon$$

and

$$\frac{y_{n+1}}{y_n} \ge \frac{e^{\theta_\epsilon}}{\beta+1}.$$

By choosing N_ϵ sufficiently large and $\epsilon > 0$ sufficiently small, we achieve that $\frac{e^{\theta_\epsilon}}{\beta+1} > 1$ if $N_0 > N_\epsilon$. In combination with (4.13),

$$y_n \ge \begin{cases} \epsilon\Big(\frac{e^{-(\alpha+\gamma)}}{\beta+1}\Big)^n, & n = 1, \dots, n_\epsilon - 1, \\ \epsilon\Big(\frac{e^{-(\alpha+\gamma)}}{\beta+1}\Big)^{n_\epsilon - 1}\Big(\frac{e^{\theta_\epsilon}}{\beta+1}\Big)^{n - n_\epsilon + 1}, & n = n_\epsilon, \dots, t. \end{cases}$$

$\diamondsuit_3$ is satisfied with $\delta = \epsilon\big(\frac{e^{-(\alpha+\gamma)}}{\beta+1}\big)^{n_\epsilon - 1}$.

Since all its assumptions have been verified, Theorem 4.16 implies uniform persistence of the parasite.

Theorem 4.19. *Assume that* $\mathcal{R}_0^{\text{host}} = (\beta+1)e^{-\mu} > 1$, $\kappa(0) = 0$, *and* $\mathcal{R}_0^{\text{par}} = \frac{e^{\kappa_\infty - \alpha - \gamma}}{\beta+1} > 1$. *Then the disease (i.e., the parasite) persists uniformly in the following sense: There exists some* $\epsilon > 0$ *such that* $\liminf_{n\to\infty} y_n \geq \epsilon$ *for all nonnegative solutions of (4.7), (4.8), (4.9) with* $y_0 > 0$.

4.5.3. Incorporating an exposed class into the metered endemic model. We consider the total host population size, N, and the fractions of susceptible, exposed, infective and recovered individuals, $u = S/N$, $x = E/N$, $y = I/N$, $z = R/N$,

$$
\begin{aligned}
N(0) =&(\beta+1)N_n, \qquad u(0) = \frac{\beta+u_n}{\beta+1}, \\
x(0) =&\frac{x_n}{\beta+1}, \quad y(0) = \frac{y_n}{\beta+1}, \quad z(0) = \frac{z_n}{\beta+1},
\end{aligned}
\tag{4.17}
$$

$$
\left.
\begin{aligned}
u' =& \; -\kappa(N)uy + \theta z + \alpha uy \\
x' =& \; \kappa(N)uy - \eta x + \alpha xy \\
y' =& \; \eta x - (\alpha+\gamma)y + \alpha y^2 \\
z' =& \; \gamma y - \theta z + \alpha yz \\
N' =& \; -N(\mu + \alpha y)
\end{aligned}
\right\}
\qquad 0 \leq t \leq 1,
\tag{4.18}
$$

$$
\begin{aligned}
N_{n+1} =&N(1), \qquad u_{n+1} = u(1), \\
x_{n+1} =&x(1), \quad y_{n+1} = y(1), \quad z_{n+1} = z(1).
\end{aligned}
\tag{4.19}
$$

Host persistence can be shown in a very similar way as without the exposed class. Introduce $w = x + y$. Then $0 \leq w \leq 1 - z \leq 1$. We add the differential equations for x and y in (4.18),

$$
\begin{aligned}
w' &= y\big(\kappa(N)(1-w-z) - \alpha(1-w) - \gamma\big), \\
w(0) &\leq \frac{1}{\beta+1}.
\end{aligned}
$$

Starting from here we can argue almost exactly as before. If the total population size is eventually low, the exposed and infectious fractions die out and in turn the population size tends to infinity, a contradiction which prove uniform weak host persistence. Uniform host persistence follows from Corollary 4.8.

We turn to parasite persistence. We only sketch the proof of uniform weak parasite persistence. This time we let $\rho = w = x + y$. We have the differential inequalities

$$x' \geq -\eta x, \qquad y' \geq -(\alpha+\gamma)y.$$

We add,

$$w' \geq -\nu w, \qquad \nu = \max\{\eta, \alpha+\gamma\}.$$

So $w_{n+1} \geq w_n \frac{e^{-\nu}}{\beta+1}$ and we see that $w_n > 0$ for all $n \in \mathbb{N}$ if $w_0 > 0$.

We can assume that w_n and $w(t)$ are as small as we want them to be. Then N_n tends to infinity as $n \to \infty$, and we can assume that $N(t)$ is arbitrarily large and $\kappa(N(t))$ as close to κ_∞ as wanted. So, after sufficiently many years,

$$\begin{aligned} x' &\geq \tilde{\kappa} y - \eta x, \\ y' &\geq \eta x - (\alpha + \gamma) y, \end{aligned}$$

where we can have $\tilde{\kappa}$ as close to κ_∞ as wanted by making N large enough and choosing $\epsilon > 0$ small enough. Let

$$A = \begin{pmatrix} -\eta & \tilde{\kappa} \\ \eta & -(\alpha + \gamma) \end{pmatrix}.$$

Let $v(t)$ be the column vector with components $x(t), y(t)$ and v_n the column vector with components x_n, y_n. The differential inequality can now be rewritten in matrix form as

$$v' \geq Av.$$

A is the special case of an irreducible quasipositive matrix because its off-diagonal elements are strictly positive. By the Perron-Frobenius theory (see [**203**, Thm.A.5], [**217**, Sec.A.8], e.g.), the eigenvalue with the largest real part is real and is associated with a strictly positive eigenvector v^* of the transposed matrix A^*, $A^* v^* = \tilde{\lambda} v^*$. Let $\langle \cdot, \cdot \rangle$ denote the standard inner product in $\mathbb{R}^2$. Notice that $\langle v^*, v \rangle \geq \min\{v_1^*, v_2^*\}(x + y)$. So $\langle v^*, v(t) \rangle > 0$ and $\langle v^*, v_n \rangle > 0$ if $x_0 + y_0 > 0$. Further,

$$\frac{d}{dt} \langle v^*, v \rangle = \langle v^*, v' \rangle \geq \langle v^*, Av \rangle = \langle A^* v^*, v \rangle = \tilde{\lambda} \langle v^*, v \rangle.$$

We integrate this inequality,

$$\langle v^*, v(1)) \rangle \geq \langle v^*, v(0) \rangle e^{\tilde{\lambda}}.$$

Since $v(0) = \frac{1}{\beta + 1} v_n$,

$$\langle v^*, v_{n+1}) \rangle \geq \langle v^*, v_n \rangle \frac{e^{\tilde{\lambda}}}{\beta + 1}.$$

The eigenvalue of A with the larger real part is

$$\tilde{\lambda} = \frac{1}{2} \Big(-(\alpha + \gamma + \eta) + \sqrt{(\alpha + \gamma + \eta)^2 + 4\eta(\tilde{\kappa} - (\alpha + \gamma))} \Big).$$

More suggestively, it can be written as

$$\tilde{\lambda} = \frac{2\eta(\tilde{\kappa} - \alpha - \gamma)}{\alpha + \gamma + \eta + \sqrt{(\alpha + \gamma - \eta)^2 + 4\eta\tilde{\kappa}}}.$$

If $\frac{e^{\tilde{\lambda}}}{\beta+1} > 1$, we see that $\langle v^*, v_n \rangle \to \infty$ as $n \to \infty$, a contradiction, because $\langle v^*, v_n \rangle \le \max\{v_1^*, v_2^*\}(x_n + y_n) \le \max\{v_1^*, v_2^*\}$. Let

$$\lambda = \frac{2\eta(\kappa_\infty - \alpha - \gamma)}{\alpha + \gamma + \eta + \sqrt{(\alpha + \gamma - \eta)^2 + 4\eta\kappa_\infty}}.$$

If $\frac{e^{\lambda}}{\beta+1} > 1$, we can also make $\frac{e^{\tilde{\lambda}}}{\beta+1} > 1$ because $\tilde{\kappa}$ can be chosen as close to κ_∞ as wanted, and the contradiction results as just shown.

Theorem 4.20. *Let $\mathcal{R}_0^{\text{host}} = (\beta+1)e^{-\mu} > 1$ and $\frac{e^{\lambda}}{\beta+1} > 1$. Then the parasite persists uniformly weakly in the sense that there exists some $\epsilon > 0$ such that $\limsup_{n\to\infty}(x_n + y_n) > \epsilon$ for all solutions with $x_0 + y_0 > 0$, $N_0 > 0$.*

4.6. Navigating between alternative persistence functions

Different functions ρ provide persistence of different strength. For instance, if the state space is $X = C_+(K)$, the cone of nonnegative continuous functions on a compact set K, then persistence with respect to $\tilde{\rho}(f) = \min_K f$ is stronger than persistence with respect to $\breve{\rho}(f) = \int_K f$ which, in turn, is stronger than persistence with respect to $\tilde{\rho}(f) = \max_K f = \|f\|$. The concept of ρ-persistence makes it possible to navigate between alternative functions ρ.

Theorem 4.21. *Let J be a time-set, X a metric space, and $\Phi : J \times X \to X$ a state-continuous semiflow.*

Let $\rho : X \to \mathbb{R}_+$ be continuous, Φ be uniformly ρ-persistent, and $\tilde{\rho} : X \to \mathbb{R}_+$ be lower semi-continuous.

Further, assume that there exists a closed subset set C of X with the following properties:

(i) *For every $x \in X$ with $\rho(x) > 0$, $\Phi(t,x) \to C$ as $t \to \infty$.*
(ii) *For every $\epsilon > 0$, $C_\epsilon = C \cap \{\rho(x) \ge \epsilon\}$ is compact, and every total Φ-trajectory $\phi : J \cup (-J) \to C_\epsilon$ satisfies $\tilde{\rho}(\phi(0)) > 0$.*

Then there exists some $\epsilon_0 > 0$ such that

$$\liminf_{t\to\infty} \tilde{\rho}(\Phi(t,x)) \ge \epsilon_0$$

for all $x \in X$ with $\rho(x) > 0$.

Recall that $\tilde{\rho}$ is lower semi-continuous if $\liminf_{k\to\infty} \tilde{\rho}(x_k) \ge \tilde{\rho}(x)$ whenever $x_k \to x$ as $k \to \infty$. Condition (ii) can be equivalently phrased as follows:

For every $\epsilon > 0$, $C_\epsilon = C \cap \{\rho(x) \ge \epsilon\}$ is compact and $\tilde{\rho}(x) > 0$, $x \in K$, where K is the largest invariant subset of C_ϵ.

Proof. Since Φ is uniformly ρ-persistent, there exists some $\epsilon_1 > 0$ such that

$$\liminf_{t\to\infty} \rho(\Phi(t,x)) \geq \epsilon_1$$

for all $x \in X$ with $\rho(x) > 0$. Choose $\epsilon \in (0, \epsilon_1)$.

- Step 1: Φ is asymptotically compact on all singleton sets $\{x\}$ with $\rho(x) > 0$.

Let $x \in X$, $\rho(x) > 0$ and (t_k) be a sequence in J such that $t_k \to \infty$ as $k \to \infty$. Let

$$U_k = \{x \in X; \exists y \in C; d(x,y) < 1/k, |\rho(x) - \rho(y)| < 1/k\}.$$

Then U_k is open and contains C. By (i), after choosing a subsequence, $\Phi(t_k, x) \in U_k$ for all $k \in \mathbb{N}$. By definition of U_k, there exist $y_k \in C$ such that

$$d(\Phi(t_k,x), y_k) < 1/k \text{ and } |\rho(\Phi(t_k,x)) - \rho(y_k)| < 1/k, \qquad k \in \mathbb{N}.$$

Since $\liminf_{k\to\infty} \rho(\Phi(t_k,x)) \geq \epsilon_1$, also $\liminf_{k\to\infty} \rho(y_k) \geq \epsilon_1$. After choosing subsequences, $y_k \in C_\epsilon$ for all $k \in \mathbb{N}$. Since C_ϵ is compact, $y_k \to y$ as $k \to \infty$ after choosing subsequences again, where $y \in C_\epsilon$. Then $\Phi(t_k, x) \to y$ as $k \to \infty$. So Φ is asymptotically compact on $\{x\}$.

- Step 2: A compact attractor, K, of points x with $\rho(x) > 0$.

By Theorem 2.11 and Step 1, $\omega(x)$ is nonempty, compact, and invariant. A similar argument shows that $\omega(x) \subset C_\epsilon$. Let

$$\tilde{K} = \bigcup_{x\in X,\, \rho(x)>0} \omega(x).$$

Then $\tilde{K}$ is invariant, $\Phi(t,x) \to \tilde{K}$ as $t \to \infty$ for every $x \in X$ with $\rho(x) > 0$, and $\tilde{K}$ is contained in the compact set C_ϵ. Let K be the closure of $\tilde{K}$. Then K inherits all these properties and is compact (recall Exercise 1.12).

- Step 3: $0 < \inf_{x\in K} \tilde{\rho}(x) =: \epsilon_0$.

Let $x \in K$. Since K is invariant, by Theorem 1.40, there is a Φ-trajectory $\phi : J \cup (-J) \to K$ with $\phi(0) = x$. Since ϕ takes all its values in C_ϵ, by (ii), $\tilde{\rho}(x) > 0$. Since K is compact and $\tilde{\rho}$ is lower semi-continuous, $\epsilon_0 = \inf_{x\in K} \tilde{\rho}(x) > 0$.

- Step 4: Finale

Let $x \in X$, $\rho(x) > 0$. We claim that $\liminf_{t\to\infty} \tilde{\rho}(\Phi(t,x)) \geq \epsilon_0$. Suppose not. Since $\Phi(t,x) \to K$ as $t \to \infty$ and K is compact, there exists a sequence (t_k) in J with $t_k \to \infty$ as $k \to \infty$ and $\Phi(t_k, x) \to y$ with some $y \in K$ and $\lim_{k\to\infty} \tilde{\rho}(\Phi(t_k,x)) < \epsilon_0$. Since $\tilde{\rho}(y) \geq \epsilon_0$ by Step 3, this contradicts the lower semi-continuity of $\tilde{\rho}$. □

The following corollary is immediate from Theorem 2.28

Corollary 4.22. *Let J be a time-set, X a metric space, and $\Phi : J \times X \to X$ a state-continuous semiflow.*

Let $\rho : X \to \mathbb{R}_+$ be continuous and Φ be uniformly ρ-persistent and $\tilde{\rho} : X \to \mathbb{R}_+$ be lower semi-continuous.

Further, assume

(i) *Φ is point-dissipative and asymptotically smooth,*

(ii) *if $\phi : \hat{J} \to X$, $\hat{J} = J \cup (-J)$, is a total Φ-trajectory with pre-compact range and $\inf_{t \in \hat{J}} \rho(\phi(t)) > 0$, then $\tilde{\rho}(\phi(0)) > 0$.*

Then there exists some $\epsilon_0 > 0$ such that

$$\liminf_{t \to \infty} \tilde{\rho}(\Phi(t, x)) \geq \epsilon_0$$

for all $x \in X$ with $\rho(x) > 0$.

4.6.1. The SEIRS epidemic model for patchy host populations revisited. Recall that E_i and I_i are the numbers of exposed and infected individuals in patch i, respectively (Section 3.2). We show that the disease is endemic in each patch.

Theorem 4.23. *Let the matrix B in (3.9) be irreducible. Then there exists some $\epsilon > 0$ such that*

$$\liminf_{t \to \infty} \min_i I_i(t) > \epsilon \qquad \text{and} \qquad \liminf_{t \to \infty} \min_i E_i(t) > \epsilon$$

for all solutions of (3.5) with $\sum_i (E_i(0) + I_i(0)) > 0$.

Proof. Define For $x = (S, E, I, R)$ define

$$\rho(x) = \sum_{i=1}^{n} (E_i + I_i), \qquad \tilde{\rho}(x) = \min_{i=1}^{n} \min\{E_i, I_i\}.$$

By Theorem 4.6, the solution semiflow Φ is uniformly ρ-persistent. We apply Theorem 4.21 with $C = \{x \in X; \sum_i N_i \leq c\}$, where $c > 0$ is as in Theorem 3.4. To check (ii), let $\phi = (S, E, I, R)$ be a total Φ-trajectory with range in C and $\rho(\phi(t)) > 0$ for all $t \in \mathbb{R}$, so a solution of (3.5) that is defined and bounded on $\mathbb{R}$ and satisfies $\sum_i (E_i(t) + I_i(t)) > 0$ for all $t \in \mathbb{R}$. Then $E_i(t) > 0$ and $I_i(t) > 0$ for all $t \in \mathbb{R}$ by (3.10) and $\tilde{\rho}(\phi(0)) > 0$. So Theorem 4.21 applies and the statement of this theorem follows. □

If there is no attracting set C as in Theorem 4.21, sometimes the fluctuation method can be used (Theorem A.14); see the upcoming Theorem 4.29.

4.7. A fertility reducing endemic with two stages of infection

In the Introduction and some of the sections of this chapter we used an SI type endemic model with reduced fertility of infected individuals to illustrate the applicability and the limits of our persistence results. This has lead to the analysis of the model to be scattered all over the chapter. Further, this model was deliberately chosen to be simple to avoid distracting technicalities. Consequently, the model has simple dynamics which could be established by making exclusive use of the Poincaré-Bendixson theory. In this section, we therefore look at a more realistic version of the model where the infection has two stages (an incubation stage and an infectiousness stage, e.g.) and perform a comprehensive and uninterrupted persistence analysis. The increase from two to three dimensions of the system has the consequence that the endemic equilibrium can be unstable and the dynamics become more interesting. We keep the feature that the Lotka-Volterra predator prey model is a limiting case with failing uniform persistence and dissipativity; so all assumptions must be carefully checked before the general persistence results can be applied. Since the host population is assumed to grow exponentially in the absence of the disease (similar to the human population of a developing country), we cannot even assume the existence of a compact attractor of points (Definition 2.21). In some cases, after having established uniform persistence of the disease, it can be shown that the disease limits the growth of the host population which results in a compact attractor of points in the state space where both host and parasite are present. Differently from the SI model, we have not been able to establish a compact attractor of neighborhoods of compact sets.

4.7.1. The model. We consider an infectious disease that, at any time $t \geq 0$, divides the population, with density N, into a susceptible part with density $S(t)$ and an infected part with density $I(t)$,

$$N(t) = S(t) + I(t). \tag{4.20}$$

The infected stage has two infected substages,

$$I(t) = I_1(t) + I_2(t). \tag{4.21}$$

The model is a system of three ordinary differential equations,

$$\begin{aligned} S' &= (\beta - \mu)S + \beta(q_1 I_1 + q_2 I_2) - (\kappa_1 I_1 + \kappa_2 I_2)S, \\ I_1' &= (\kappa_1 I_1 + \kappa_2 I_2)S - (\mu + \alpha_1 + \gamma)I_1, \\ I_2' &= \gamma I_1 - (\mu + \alpha_2)I_2. \end{aligned} \tag{4.22}$$

Here β and μ are the natural per capita birth and death rates, respectively. The infection can reduce fertility: $q_j \in [0, 1]$ is the factor by which the birth rate of an infected individual in stage j is reduced compared with the birth rate of a susceptible individual.

The parameters α_j denote the extra per capita death rates due to the infection in stage j. The parameter γ is the per capita transition rate from the first into the second infection stage. The parameters κ_j are per capita infections rates: $1/\kappa_j$ is the average time it takes for an average susceptible individual to become infected when being in contact with one infected unit in stage j. If $\kappa_1 = 0$, stage 1 can be identified with the incubation stage; if $\kappa_2 = 0$, stage 2 can be identified with the removed stage. We assume that α_1, α_2 and κ_1, κ_2 are nonnegative,

$$\kappa_1 + \kappa_2 > 0, \qquad \gamma > 0;$$

if γ were zero, the model would reduce to the two-dimensional one we have analyzed before. We also assume that

$$\beta > \mu > 0$$

such that the population increases exponentially in the absence of the disease.

Notice that, for $\kappa_1 = \kappa_2$, $\alpha_1 = \alpha_2$, and $q_j = 0$, we can add the second and third equation to yield

$$S' = (\beta - \mu)S - \kappa_2 SI, \qquad I' = \kappa_2 SI - (\mu + \alpha_1)I. \tag{4.23}$$

This equation again has the form of the Lotka-Volterra predator-prey system which, in the positive first quadrant, is strongly persistent and uniformly weakly persistent, but not uniformly persistent.

4.7.2. Endemic equilibrium and its stability. In order to show that the model displays nontrivial dynamics, we look at the endemic equilibrium. Its equations result form (4.22) by setting the time-derivative equal to 0,

$$\begin{aligned} 0 =& (\beta - \mu)S^* + \beta(q_1 I_1^* + q_2 I_2^*) - (\kappa_1 I_1^* + \kappa_2 I_2^*)S^*, \\ 0 =& (\kappa_1 I_1^* + \kappa_2 I_2^*)S^* - (\mu + \alpha_1 + \gamma)I_1^*, \\ 0 =& \gamma I_1^* - (\mu + \alpha_2)I_2^*. \end{aligned} \tag{4.24}$$

We substitute the third equation into the second and divide by I_1^*,

$$(\kappa_1(\mu + \alpha_2) + \kappa_2\gamma)S^* = (\mu + \alpha_1 + \gamma)(\mu + \alpha_2). \tag{4.25}$$

We also substitute the third into the first equation,

$$0 = (\beta - \mu)(\mu + \alpha_2)S^* + \beta I_1^*(q_1(\mu + \alpha_2) + q_2\gamma) - I_1^*(\kappa_1(\mu + \alpha_2) + \kappa_2\gamma)S^*.$$

This shows that the endemic equilibrium exists if and only if

$$(\kappa_1(\mu + \alpha_2) + \kappa_2\gamma)S^* > \beta(q_1(\mu + \alpha_2) + q_2\gamma). \tag{4.26}$$

Equivalently, by (4.25), the endemic equilibrium exists if and only if

$$(\mu+\alpha_1+\gamma)(\mu+\alpha_2) > \beta(q_1(\mu+\alpha_2)+q_2\gamma). \tag{4.27}$$

The inequality can be rewritten in a more meaningful way as

$$1 > \frac{\beta q_1}{\mu+\alpha_1+\gamma} + \frac{\beta q_2}{\mu+\alpha_2}\frac{\gamma}{\mu+\alpha_1+\gamma} =: \mathcal{R}. \tag{4.28}$$

$\mathcal{R}$ can be interpreted as the reproduction number of infected hosts. $\frac{1}{\mu+\alpha_1+\gamma}$ is the average length of the first infected stage. So $\frac{\beta q_1}{\mu+\alpha_1+\gamma}$ is the expected offspring produced by one average infected individual during the first stage. $\frac{1}{\mu+\alpha_2}$ is the average length of the second infected stage, while $\frac{\gamma}{\mu+\alpha_1+\gamma}$ is the probability of surviving the first infected stage. So $\frac{\beta q_2}{\mu+\alpha_2}\frac{\gamma}{\mu+\alpha_1+\gamma}$ is the expected offspring produced by one average infected individual during the second infected stage.

The endemic equilibrium exists if q_1 and q_2 are sufficiently small and/or α_1 and α_2 are sufficiently large.

The Jacobian matrix at this equilibrium is

$$J = \begin{pmatrix} a & \beta q_1 - \kappa_1 S^* & \beta q_2 - \kappa_2 S^* \\ b & \kappa_1 S^* - (\mu+\alpha_1+\gamma) & \kappa_2 S^* \\ 0 & \gamma & -(\mu+\alpha_2) \end{pmatrix}, \tag{4.29}$$

with

$$\begin{aligned} a =& \beta - \mu - \kappa_1 I_1^* - \kappa_2 I_2^* \overset{(4.24)}{=} -\beta \frac{q_1 I_1^* + q_2 I_2^*}{S^*} = \beta - \mu - b, \\ b =& \kappa_1 I_1^* + \kappa_2 I_2^* > 0. \end{aligned} \tag{4.30}$$

Notice that $a < 0$. We have from the equilibrium relations that

$$\mathrm{tr} J = a + \kappa_1 S^* - (\mu+\alpha_1+\gamma) - (\mu+\alpha_2) < 0. \tag{4.31}$$

Further, when J_k denotes the determinant of the submatrix of J where the kth row and column have been canceled,

$$\begin{aligned} J_1 =& (\mu+\alpha_1+\gamma)(\mu+\alpha_2) - S^*(\kappa_1(\mu+\alpha_2)+\kappa_2\gamma) \overset{(4.25)}{=} 0, \\ J_2 =& -a(\mu+\alpha_2), \\ J_3 =& a(\kappa_1 S^* - (\mu+\alpha_1+\gamma)) + b(\kappa_1 S^* - \beta q_1). \end{aligned} \tag{4.32}$$

Since $J_1 = 0$,

$$\det J = -b\Big(S^*(\kappa_1(\mu+\alpha_2)+\kappa_2\gamma) - \beta\big(q_1(\mu+\alpha_2)+q_2\gamma\big)\Big). \tag{4.33}$$

We notice that $\det J < 0$ if and only if the endemic equilibrium exists. We apply [**53**, Thm.A.2]. The Routh-Hurwitz coefficient satisfies

$$\begin{aligned}\hbar J =& \det J - (\text{tr } J)(J_1 + J_2 + J_3)\\ =& - b\Big(S^*(\kappa_1(\mu+\alpha_2) + \kappa_2\gamma) - \beta\big(q_1(\mu+\alpha_2) + q_2\gamma\big)\Big)\\ & - \Big(a + \kappa_1 S^* - (\mu+\alpha_1+\gamma) - (\mu+\alpha_2)\Big)\\ & \Big(-a(\mu+\alpha_2) + a(\kappa_1 S^* - (\mu+\alpha_1+\gamma)) + b(\kappa_1 S^* - \beta q_1)\Big).\end{aligned}$$

We factor out common terms and simplify,

$$\begin{aligned}\hbar J =& - bS^*\Big(\kappa_2\gamma + \big(a + \kappa_1 S^* - (\mu+\alpha_1+\gamma)\big)\kappa_1\Big)\\ & + b\beta\Big(q_1(\mu+\alpha_2) + q_2\gamma + q_1\big(a + \kappa_1 S^* - (\mu+\alpha_1+\gamma) - (\mu+\alpha_2)\big)\Big)\\ & + a^2(\mu+\alpha_1+\mu+\alpha_2+\gamma-\kappa_1 S^*)\\ & - a\Big(\mu+\alpha_1+\mu+\alpha_2+\gamma-\kappa_1 S^*)^2.\end{aligned}$$

We rewrite $\hbar J$ in a condensed form,

$$\begin{aligned}\hbar J =& - bS^* a\kappa_1 - bS^* c\\ & + b\beta q_1 a + b\beta\Big(q_2\gamma + q_1\big(\kappa_1 S^* - (\mu+\alpha_1+\gamma)\big)\Big)\\ & + a^2(\mu+\alpha_1+\mu+\alpha_2+\gamma-\kappa_1 S^*)\\ & - a\Big(\mu+\alpha_1+\mu+\alpha_2+\gamma-\kappa_1 S^*)^2\end{aligned}$$

with

$$c = \kappa_2\gamma + \big(\kappa_1 S^* - (\mu+\alpha_1+\gamma)\big)\kappa_1. \tag{4.34}$$

We substitute the equilibrium relations,

$$\begin{aligned}c =& \frac{1}{(\kappa_1(\mu+\alpha_2) + \kappa_2\gamma))}\\ & \Big(\kappa_2\gamma\big(\kappa_1(\mu+\alpha_2)+\kappa_2\gamma\big) + \kappa_1^2(\mu+\alpha_1+\gamma)(\mu+\alpha_2)\\ & - (\mu+\alpha_1+\gamma)\kappa_1\big(\kappa_1(\mu+\alpha_2)+\kappa_2\gamma\big)\Big).\end{aligned}$$

We simplify

$$c = \frac{\kappa_2\gamma}{\kappa_1(\mu+\alpha_2)+\kappa_2\gamma}\Big(\kappa_1\alpha_2 + \kappa_2\gamma - (\alpha_1+\gamma)\kappa_1\Big).$$

We regroup,

$$c = \frac{\kappa_2\gamma}{\kappa_1(\mu+\alpha_2)+\kappa_2\gamma}\Big(\kappa_1(\alpha_2-\alpha_1) + (\kappa_2-\kappa_1)\gamma\Big). \tag{4.35}$$

Notice that, as $q_1, q_2 \to 0$, $a \to 0$ and $b \to \beta - \mu$. So $\hbar J \to -(\beta - \mu)S^* c$ and $J_1 + J_2 + J_3 \to (\beta - \mu)\kappa_1 S^* > 0$. By the Routh-Hurwitz criterion [**53**, Thm.A.2], the endemic equilibrium is unstable (with one negative eigenvalue and two complex conjugate eigenvalues with positive real part) if

$$0 < \kappa_1(\alpha_2 - \alpha_1) + (\kappa_2 - \kappa_1)\gamma \tag{4.36}$$

and q_1 and q_2 are sufficiently small. Condition (4.36) is satisfied if the first stage of infection is an initial stage that has lower infection and disease death rates than the second stage.

Computational evidence shows that the endemic equilibrium can also be unstable if $q_1 = 1 = q_2$ and $\kappa_1 = 0 < \kappa_2$ [**77**, Sec.2].

4.7.3. Reformulation of the model. We first rewrite the model (4.22) in terms of N, I_1, and I_2,

$$\begin{aligned} N' =&(\beta - \mu)N - (\beta[1 - q_1] + \alpha_1)I_1 - (\beta[1 - q_2] + \alpha_2)I_2, \\ I_1' =&(\kappa_1 I_1 + \kappa_2 I_2)(N - I_1 - I_2) - (\mu + \alpha_1 + \gamma)I_1, \\ I_2' =&\gamma I_1 - (\mu + \alpha_2)I_2. \end{aligned} \tag{4.37}$$

We introduce the fractions of infected individuals in the first and second stage as

$$f_j = \frac{I_j}{N}, \tag{4.38}$$

and rewrite the model in terms of the total population density N and the fractions of infected individuals f_1 and f_2,

$$\begin{aligned} N' =&N\Big(\beta - \mu - (\beta[1 - q_1] + \alpha_1)f_1 - (\beta[1 - q_2] + \alpha_2)f_2\Big), \\ f_1' =&N(\kappa_1 f_1 + \kappa_2 f_2)(1 - f_1 - f_2) - \gamma f_1 \\ &- (\beta + \alpha_1)f_1(1 - f_1) + (\beta + \alpha_2)f_1 f_2 - q_1\beta f_1^2 - q_2\beta f_1 f_2, \\ f_2' =&\gamma f_1 - (\beta + \alpha_2)f_2 + (\beta + \alpha_1)f_1 f_2 + (\beta + \alpha_2)f_2^2 \\ &- q_1\beta f_1 f_2 - q_2\beta f_2^2. \end{aligned} \tag{4.39}$$

Define the fraction of infected individuals

$$\bar{f} = f_1 + f_2. \tag{4.40}$$

Then

$$\begin{aligned} \bar{f}' =&N(\kappa_1 f_1 + \kappa_2 f_2)(1 - \bar{f}) - (\beta + \alpha_1)f_1(1 - \bar{f}) \\ &- (\beta + \alpha_2)f_2(1 - \bar{f}) - q_1\beta f_1\bar{f} - q_2\beta f_2\bar{f}. \end{aligned} \tag{4.41}$$

It follows that the state space $\{N \geq 0, f_1, f_2 \geq 0, f_1 + f_2 \leq 1\}$ is forward invariant, consistent with the interpretation of $\bar{f}$ as the fraction of infected individuals, and solutions exists for all forward times.

This formulation has the advantage that there is only one potentially unbounded component of the solution, N, as compared to three potentially unbounded components in the first two model formulations.

4.7.4. Persistence of the host. We define

$$\kappa^\diamond = \max\{\kappa_1, \kappa_2\}. \tag{4.42}$$

By assumption, $\kappa^\diamond > 0$. From (4.41),

$$\bar{f}' \leq N\kappa^\diamond \bar{f}(1-\bar{f}) - \beta \bar{f}(1-\bar{f}). \tag{4.43}$$

Proposition 4.24. *If $q_2 > 0$, the host population is uniformly weakly persistent,*

$$\limsup_{t\to\infty} N(t) \geq \frac{\beta}{\kappa^\diamond}$$

for all solutions with $N(0) > 0$, $f_1(0) \in [0,1]$, and $f_2 \in [0,1]$.

Proof. Assume that $N^\infty < \beta/\kappa^\diamond$. Then $\bar{f}'(t) < 0$ for all sufficiently large t and $\bar{f}$ is eventually decreasing and has a limit $\bar{f}(\infty)$. Since N, f_1, f_2 and $\bar{f}$ are bounded, we see from (4.39) and (4.41) that their second derivatives are bounded and thus their first derivatives are uniformly continuous on $\mathbb{R}_+$. By Barbalat's lemma (Lemma A.16), $\bar{f}'(t) \to 0$ as $t \to \infty$. By (4.43),

$$0 \leq (N^\infty \kappa^\diamond - \beta)\bar{f}(\infty)(1-\bar{f}(\infty)). \tag{4.44}$$

This implies $\bar{f}(\infty) = 0$ or $\bar{f}(\infty) = 1$. Suppose $\bar{f}(\infty) = 1$. By (4.41), since $q_2 > 0$, $f_2(t) \to 0$ as $t \to \infty$ and so $f_1(t) \to 1$ as $t \to \infty$. By Barbalat's lemma, $f_2'(t) \to 0$ as $t \to \infty$. From the equation for f_2 in (4.39) we obtain the contradiction $0 = -\gamma$.

This implies $f_1(t) + f_2(t) \to 0$ as $t \to \infty$, and the host population increases exponentially as seen from the N-equation in (4.39), a contradiction. □

Remark 4.25. Let $q_2 = 0$ and $q_1 > 0$. Then uniform weak host persistence holds in a restricted sense:

$$\limsup_{t\to\infty} N(t) \geq \frac{\beta}{\kappa^\diamond}$$

for all solutions with $N(0) > 0$ and with $f_1(0) > 0$ or $f_1(0) + f_2(0) < 1$.

If $q_2 = 0$, the set $\{f_1 = 0, f_2 = 1\}$ is invariant and, on this set, $N(t) \to 0$ as $t \to \infty$.

If $q_2 = 0 = q_1$, uniform weak host persistence holds in an even more restricted sense:

$$\limsup_{t\to\infty} N(t) \geq \frac{\beta}{\kappa^\diamond}$$

for all solutions with $N(0) > 0$ and $f_1(0) + f_2(0) < 1$.

If $q_2 = 0 = q_1$, the set $\{f_1 + f_2 = 1\}$ is invariant and on this set $N(t) \to 0$ as $t \to \infty$.

The condition $f_1(0) + f_2(0) < 1$ is equivalent to $S(0) > 0$, i.e., some hosts are susceptible at the beginning.

Proof. We notice from (4.43) that $\bar{f}(t) = f_1(t) + f_2(t)$ is decreasing. So $\bar{f}$ remains bounded away from 1, if $\bar{f}(0) < 1$. In particular, $\bar{f}(\infty) < 1$.

If $\bar{f}(0) = 1$ and $q_1 > 0$ and $f_1(0) > 0$, then $\bar{f}'(0) < 0$ and $\bar{f}(t) < 1$ for $t > 0$. Since $\bar{f}$ is decreasing, $\bar{f}(\infty) < 1$ again.

If $N^\infty \kappa^\diamond < \beta$, in both cases, $\bar{f}(\infty) = 0$, and $N(t)$ increases exponentially if $N(0) > 0$.

The statements concerning the invariant sets follow from the appropriate differential equations. By the N-equation in (4.39), in either case $N' \le -\mu N$ and $N(t) \to 0$ as $t \to \infty$. □

By Theorem 4.13, the host population is uniformly persistent.

Theorem 4.26. *If $q_2 > 0$, then the host population is uniformly persistent: There exists some $\epsilon > 0$ such that*

$$\liminf_{t\to\infty} N(t) > \epsilon$$

for all solutions with $N(0) > 0$, $f_1(0) \in [0,1]$, $f_2(0) \in [0,1]$.

Proof. We define the state space as

$$X = \{(N, f_1, f_2);\ N > 0,\ f_1, f_2 \ge 0,\ f_1 + f_2 \le 1\}$$

and the persistence function as $\rho(N, f_1, f_2) = N$. By Proposition 4.24, the solution semiflow Φ induced by (4.39) is uniformly weakly ρ-persistent. To check the assumptions of Theorem 4.13, we set $B = X$. Then $\tilde{\heartsuit}_1$ is trivially satisfied. As for $\tilde{\heartsuit}_2$, if $0 < \epsilon_1 < \epsilon_2 < \infty$,

$$\begin{aligned} &B \cap \{\epsilon_1 \le \rho(x) \le \epsilon_2\} \\ =&\{(N, f_1, f_2);\ \epsilon_1 \le N \le \epsilon_2,\ f_1, f_2 \ge 0,\ f_1 + f_2 \le 1\} \end{aligned}$$

is compact. As for $\tilde{\heartsuit}_3$, if $N(0) > 0$, then $N(t) > 0$ for all $t > 0$ by the form of the N-equation in (4.39). By Theorem 4.13, Φ is uniformly ρ-persistent, and the statement of this theorem follows from the choice of ρ. □

Remark 4.27. The previous proof fails if $q_2 = 0 = q_1$. By Remark 4.25, we need to work with the state space

$$X = \{(N, f_1, f_2);\ N > 0,\ f_1, f_2 \ge 0,\ f_1 + f_2 < 1\}$$

and

$$\begin{aligned}&B \cap \{\epsilon_1 \leq \rho(x) \leq \epsilon_2\}\\=&\{(N, f_1, f_2);\ \epsilon_1 \leq N \leq \epsilon_2,\ f_1, f_2 \geq 0,\ f_1 + f_2 < 1\}\end{aligned}$$

is not compact. By a similar token, the proof fails if $q_2 = 0 < q_1$.

4.7.5. Persistence of the disease. We turn to disease persistence. From now on we assume

$$\kappa_2 > 0.$$

This is the case where we have seen that the endemic equilibrium can be unstable (if it exists). The case $\kappa_2 = 0 < \kappa_1$ is left as an exercise (Exercise 4.10). Recall the standing assumptions $\gamma > 0$ and $\beta > \mu > 0$. Let

$$\alpha^\diamond = \max\{\alpha_1, \alpha_2\}.$$

By (4.39),

$$N' \geq N\Big(\beta - \mu - (\beta + \alpha^\diamond)\bar{f}\Big).$$

This implies that $N(t) \to \infty$ if $N(0) > 0$ and $\bar{f}^\infty$ is small enough.

Proposition 4.28. *The disease is uniformly weakly persistent in the following sense:*

$$\limsup_{t\to\infty}(f_1(t) + f_2(t)) \geq \frac{\beta - \mu}{\beta + \alpha^\diamond}$$

for all solutions with $N(0) > 0$ and $f_1(0) + f_2(0) > 0$.

Proof. By (4.41),

$$\bar{f}' \geq -(2\beta + \alpha^\diamond)\bar{f}.$$

This implies that $\bar{f}(t) = f_1(t) + f_2(t) > 0$ for all $t \geq 0$. Suppose that $\bar{f}^\infty < \frac{\beta-\mu}{\beta+\alpha^\diamond}$. It follows from our preceding considerations that $N(t) \to \infty$ as $t \to \infty$. Let $N^\sharp > 0$ be a number that will be chosen later. After a shift in time, we can assume that $N(t) \geq N^\sharp$ for all $t \geq 0$ and $\bar{f}(0) > 0$. Then, for some $\delta > 0$ independent of the solution,

$$\begin{aligned}f_1' \geq& N^\sharp(\kappa_1 f_1 + \kappa_2 f_2)\delta - \gamma f_1 - (2\beta + \alpha_1) f_1,\\ f_2' \geq& \gamma f_1 - (2\beta + \alpha_2) f_2.\end{aligned}$$

Then $f' \geq Af$, where $f(t)$ is the column vector with coordinates $f_1(t)$ and $f_2(t)$ and

$$A = \begin{pmatrix} N^\sharp\kappa_1\delta - \gamma - (2\beta + \alpha_1) & N^\sharp\kappa_2\delta \\ \gamma & -(2\beta + \alpha_2)\end{pmatrix}.$$

The matrix A is quasipositive and, since $\kappa_2 > 0$ and $\gamma > 0$, irreducible. If we choose $N^\sharp$ large enough, the trace of A is positive and so is the spectral bound of A. By Corollary A.27, $f_1(t) + f_2(t) \to \infty$ as $t \to \infty$, a contradiction. □

The persistence approach à la Münchhausen implies uniform persistence.

Theorem 4.29. *If $q_2 > 0$, the disease is uniformly persistent in the following sense: There exists some $\epsilon > 0$ such that*

$$\liminf_{t\to\infty} f_2(t) > \epsilon$$

for all solutions with $N(0) > 0$ and $f_1(0) + f_2(0) > 0$.

Proof. We apply Theorem 4.17. We choose the state space

$$X = \{(N, f_1, f_2); N > 0, f_j \in [0, 1], j = 1, 2\}$$

and the persistence function

$$\rho(N, f_1, f_2) = f_1 + f_2 = \bar{f}.$$

By Proposition 4.28, the solution semiflow induced by (4.39) is uniformly weakly ρ-persistent. In the language of Theorem 4.17,

$$\sigma(t, x) = \bar{f}(t), \qquad x = (N(0), f_1(0), f_2(0)).$$

By Theorem 4.26, there exists some $\epsilon_0 > 0$ such that $\liminf_{t\to\infty} N(t) > \epsilon_0$ for all solutions starting in X. We define a closed subset B of X by

$$B = \{(N, f_1, f_2);\ N \ge \epsilon_0,\ f_j \in [0, 1], j = 1, 2\}.$$

Then Assumption $\blacklozenge_0$ is satisfied. By (4.41), $\bar{f}' \ge -(2\beta + \alpha^\diamond)\bar{f}$. So Assumption $\blacklozenge_2$ holds. Let $N^\sharp > 0$ be a number to be chosen later. We define

$$D = \{(N, f_1, f_2);\ 0 < N \le N^\sharp,\ f_j \in [0, 1], j = 1, 2\}.$$

Then $B \cap D$ is compact and Assumption $\blacklozenge_1$ is satisfied. By (4.39),

$$\begin{aligned} N' \ge& N\Big(\beta - \mu - (\beta + \alpha^\diamond)\bar{f}\Big), \\ f_1' \ge& N(\kappa_1 f_1 + \kappa_2 f_2)(1 - \bar{f}) - \gamma f_1 - (2\beta + \alpha_1) f_1, \\ f_2' \ge& \gamma f_1 - (2\beta + \alpha_2) f_2. \end{aligned} \tag{4.45}$$

Choose

$$\tilde{\epsilon} = \frac{\beta - \mu}{\beta + \alpha^\diamond}, \qquad \tilde{\epsilon} \in (0, 1).$$

To check $\blacklozenge_3$, let $\epsilon \in (0, \tilde{\epsilon})$, $t \in \mathbb{R}_+$, and $x \in X \setminus D$ such that $\Phi(\mathbb{R}_+ \times \{x\}) \subset B$, $\rho(x) = \epsilon = \sigma(t, x)$ and $\sigma(s, x) < \epsilon$ for all $s \in (0, t)$. This means for our model that $N(0) \ge N^\sharp$, $\bar{f}(0) = \epsilon = \bar{f}(t)$ and $\bar{f}(t) < \epsilon$ for $s \in (0, t)$. The differential inequality for N in (4.45) implies that $N \ge N^\sharp$ on $[0, t]$. Then, on $[0, t]$,

$$\begin{aligned} f_1' \ge& N^\sharp(\kappa_1 f_1 + \kappa_2 f_2)(1 - \tilde{\epsilon}) - \gamma f_1 - (2\beta + \alpha_1) f_1, \\ f_2' \ge& \gamma f_1 - (2\beta + \alpha_2) f_2. \end{aligned}$$

Let $f(s)$ be the column vector with $f_1(s)$ and $f_2(s)$ with coordinates. Then, on $[0, t]$,

$$f' \ge Af, \qquad A = \begin{pmatrix} N^\sharp\kappa_1(1 - \tilde{\epsilon}) - \gamma - (2\beta + \alpha_1) & N^\sharp\kappa_2(1 - \tilde{\epsilon}) \\ \gamma & -(2\beta + \alpha_2) \end{pmatrix}.$$

The matrix A is quasipositive and, since $\kappa_2 > 0$ and $\gamma_2 > 0$, irreducible. We choose $N^\sharp$ so large that the trace of A is positive and so is the spectral bound of A. By Corollary A.27, there exists some $\xi > 0$ such that $\bar{f}(s) \geq \xi \bar{f}(0)$. So $\bar{f}(s) \geq \delta$ with $\delta = \xi\epsilon$. This finishes the check of $\blacklozenge_2$.

By Theorem 4.17, the solution semiflow is uniformly ρ-persistent: There exists some $\hat{\epsilon} > 0$ such that

$$\liminf_{t\to\infty} \bar{f}(t) > \hat{\epsilon}$$

for all solutions with $N(0) > 0$ and $\bar{f}(0) > 0$. By (4.45),

$$f_2' \geq \gamma \bar{f} - (2\beta + \alpha_2 + \gamma) f_2.$$

This implies that f_2 is bounded away from 0 and, by the fluctuation method

$$\liminf_{t\to\infty} f_2(t) \geq \frac{\gamma}{2\beta + \alpha_2 + \gamma} \liminf_{t\to\infty} \bar{f}(t) \geq \frac{\gamma}{2\beta + \alpha_2 + \gamma} \hat{\epsilon}. \qquad \square$$

4.7.6. Uniform eventual boundedness of the host. We derive conditions under which the disease limits the growth of the host population. Assume that $f_1(0) + f_2(0) > 0$, $N(0) > 0$, and

$$\liminf_{t\to\infty} N(t) > N_\sharp > 0. \tag{4.46}$$

Here $N_\sharp > 0$ will be chosen later. Since the disease persists uniformly (Theorem 4.29), there exists some $\epsilon > 0$ (independent of the initial data) such that

$$\liminf_{t\to\infty} f_2(t) > \epsilon.$$

After a shift in time we can assume that $N(t) > N_\sharp$ and $f_2(t) > \epsilon$ for all $t \geq 0$. From (4.21),

$$\begin{aligned} f_1' \geq& N_\sharp \kappa_2 \epsilon (1 - f_1 - f_2) - \gamma f_1 \\ &- (\beta + \alpha_1) f_1 (1 - f_1) + (\beta + \alpha_2) f_1 f_2 - q_1 \beta f_1^2 - q_2 \beta f_1 f_2, \\ f_2' =& \gamma f_1 - (\beta + \alpha_2) f_2 + (\beta + \alpha_1) f_1 f_2 + (\beta + \alpha_2) f_2^2 - q_1 \beta f_1 f_2 - q_2 \beta f_2^2. \end{aligned}$$

We derive a system in terms of $\bar{f} = f_1 + f_2$ and f_2,

$$\begin{aligned} \bar{f}' &\geq N_\sharp \kappa_2 \epsilon (1 - \bar{f}) - (2\beta + \alpha^\diamond) \bar{f}, \\ f_2' &= \gamma \bar{f} - (\beta + \alpha_2 + \gamma) f_2 + (\beta[1 - q_1] + \alpha_1)(\bar{f} - f_2) f_2 \\ &\quad + (\beta[1 - q_2] + \alpha_2) f_2^2. \end{aligned}$$

Let $\delta \in (0, 1)$, which will be chosen later. By choosing $N_\sharp$ large enough, the fluctuation method (Theorem A.14) implies that $\bar{f}_\infty > 1 - \delta$. By a shift in time, we can assume that $\bar{f}(t) \geq 1 - \delta$ for all $t \geq 0$. Then

$$\begin{aligned} f_2' \geq& \gamma(1 - \delta) - (\beta + \alpha_2 + \gamma) f_2 \\ &+ (\beta[1 - q_1] + \alpha_1)(1 - \delta - f_2) f_2 + (\beta[1 - q_2] + \alpha_2) f_2^2. \end{aligned} \tag{4.47}$$

By the fluctuation method,

$$\begin{aligned}0 \geq& \gamma(1-\delta) - (\beta + \alpha_2 + \gamma) f_{2\infty} \\ &+ (\beta[1-q_1] + \alpha_1)(1-\delta - f_{2\infty}) f_{2\infty} + (\beta[1-q_2] + \alpha_2) f_{2\infty}^2.\end{aligned}$$

This leads to a quadratic inequality for $f_{2\infty}$,

$$\begin{aligned}0 \geq& \gamma(1-\delta) - \big(q_1\beta + \delta\beta[1-q_1] + \alpha_2 - \alpha_1(1-\delta) + \gamma\big) f_{2\infty} \\ &+ \big(\beta[q_1 - q_2] + \alpha_2 - \alpha_1\big) f_{2\infty}^2.\end{aligned} \tag{4.48}$$

We rewrite the equation for N in (4.39) in terms of $\bar{f} = f_1 + f_2$ and f_2,

$$N' = N\Big(\beta - \mu - (\beta[1-q_1] + \alpha_1)\bar{f} - (\beta[q_1 - q_2] + \alpha_1 - \alpha_2) f_2\Big).$$

Since $\bar{f} \geq 1 - \delta$,

$$N' \leq N\Big(q_1\beta - \mu + \delta\beta[1-q_1] - \alpha_1(1-\delta) - (\beta[q_1 - q_2] + \alpha_2 - \alpha_1) f_2\Big).$$

We restrict our analysis to the case that the first stage of infection is the initial stage where the parasite has not yet developed its full virulence,

$$\beta[q_1 - q_2] + \alpha_2 - \alpha_1 > 0. \tag{4.49}$$

This inequality is in particular satisfied if the disease death rate (denoted by α_j) and the fertility reduction (related to $1 - q_j$) are larger in the second than in the first stage. By (4.49),

$$\limsup_{t\to\infty} \frac{N'(t)}{N(t)} \leq q_1\beta - \mu + \delta\beta[1-q_1] - \alpha_1(1-\delta) - (\beta[q_1 - q_2] + \alpha_2 - \alpha_1) f_{2\infty}. \tag{4.50}$$

Assume we can show that

$$f_{2\infty} > \frac{q_1\beta - \mu - \alpha_1}{\beta[q_1 - q_2] + \alpha_2 - \alpha_1} =: f_2^\sharp. \tag{4.51}$$

We can then choose $\delta > 0$ so small that the right-hand side of (4.50) is negative and $N(t) \to 0$ as $t \to \infty$, a contradiction to (4.46).

To show that $f_{2\infty} > f_2^\sharp$, we can assume that $f_2^\sharp > 0$. By (4.48), $f_{2\infty}$ satisfies a quadratic inequality

$$0 \geq a_0 - a_1 x + a_2 x^2 \tag{4.52}$$

with $a_j > 0$ and $a_0 - a_1 + a_2 < 0$,

$$\begin{aligned}a_0 =& \gamma(1-\delta), \\ a_1 =& q_1\beta + \delta\beta[1-q_1] + \alpha_2 - \alpha_1(1-\delta) + \gamma, \\ a_2 =& \beta[q_1 - q_2] + \alpha_2 - \alpha_1.\end{aligned}$$

Since the right-hand side of (4.52) describes a parabola, if $x \in [0,1]$ is a solution of this inequality and $a_0 - a_1 x^\sharp + a_2 (x^\sharp)^2 > 0$, then $x \ge x^\sharp$. So it is sufficient to show ($x^\sharp = f_2^\sharp$) that

$$\begin{aligned} 0 <& \gamma(1-\delta) - (q_1\beta + \delta\beta[1-q_1] + \alpha_2 - \alpha_1(1-\delta) + \gamma) f_2^\sharp \\ &+ (\beta[q_1 - q_2] + \alpha_2 - \alpha_1)(f_2^\sharp)^2. \end{aligned} \tag{4.53}$$

Here $\delta > 0$ can be chosen arbitrarily small. By continuity of the right-hand side in δ, it is sufficient to show that

$$0 < \gamma - (q_1\beta + \alpha_2 - \alpha_1 + \gamma) f_2^\sharp + (\beta[q_1 - q_2] + \alpha_2 - \alpha_1)(f_2^\sharp)^2. \tag{4.54}$$

We substitute $f_2^\sharp$ from (4.51) into this equation and find that (4.54) is equivalent to

$$\begin{aligned} 0 < &\gamma(\beta[q_1 - q_2] + \alpha_2 - \alpha_1) - (q_1\beta + \alpha_2 - \alpha_1 + \gamma)(q_1\beta - \mu - \alpha_1) \\ &+ (q_1\beta - \mu - \alpha_1)^2. \end{aligned}$$

This inequality equivalently simplifies to

$$0 < \gamma(\mu - \beta q_2 + \alpha_2) - (\mu + \alpha_2)(q_1\beta - \mu - \alpha_1). \tag{4.55}$$

This is equivalent to (4.28), the sufficient and necessary condition for the existence of an endemic equilibrium, namely that the expected amount of offspring of an average infected individual is strictly less than 1.

Proposition 4.30. *Let $q_2 > 0$, $\beta[q_1 - q_2] + \alpha_2 - \alpha_1 > 0$, and (4.28). Then the host population is uniformly weakly eventually bounded: There exists some $N^\sharp > 0$ such that*

$$\liminf_{t\to\infty} N(t) < N_\sharp$$

for all solutions with $f_1(0) + f_2(0) > 0$ and $N(0) > 0$.

Persistence theory allows us to replace the limit inferior by the limit superior.

Theorem 4.31. *Let $q_2 > 0$, $\beta[q_1 - q_2] + \alpha_2 - \alpha_1 > 0$, and (4.28). Then the host population is uniformly eventually bounded: There exists some $N^\sharp > 0$ such that*

$$\limsup_{t\to\infty} N(t) < N^\sharp$$

for all solutions with $f_1(0) + f_2(0) > 0$ and $N(0) \ge 0$.

By Proposition 4.24, $N^\sharp > \beta/\kappa^\diamond$.

Proof. Only solutions with $N(0) > 0$ are of interest; otherwise $N(t) = 0$ for all $t \ge 0$. We choose the state space

$$X = \{(N, f_1, f_2); N > 0, 0 < f_1 + f_2 \le 1\}$$

and the persistence function $\rho(N, f_1, f_2) = \frac{1}{1+N}$. Let Φ be the solution semiflow induced by (4.39). By Proposition 4.30, Φ is uniformly weakly ρ-persistent. We apply Theorem 4.13. By Theorem 4.26 and Theorem 4.29, both host and disease are uniformly persistent: There exists some $\epsilon > 0$ such that

$$\liminf_{t\to\infty} N(t) > \epsilon, \qquad \liminf_{t\to\infty} f_2(t) > \epsilon$$

for all solutions starting in X. To satisfy the assumptions of Theorem 4.13, we define

$$B = \{(N, f_1, f_2); N \geq \epsilon, \epsilon \leq f_1 + f_2 \leq 1\}.$$

Then $\Phi(t,x) \to B$ as $t \to \infty$ for all $x \in X$. Condition $\hat{\heartsuit}_3$ is trivially satisfied. For $0 < \epsilon_1 < \epsilon_2 < \infty$, the set

$$\begin{aligned} &B \cap \{\epsilon_1 \leq \rho(x) \leq \epsilon_2\} \\ =&\big\{(N, f_1, f_2); N \geq \epsilon, \epsilon_2^{-1} - 1 \leq N \leq \epsilon_1^{-1} - 1, \epsilon \leq f_1 + f_2 \leq 1\big\} \end{aligned}$$

is compact. All assumptions of Theorem 4.13 are satisfied and Φ is uniformly ρ-persistent. By choice of ρ, the host is uniformly eventually bounded. □

4.7.7. Persistence of the susceptible and first-stage infected part of the host population. The fraction of susceptibles is given by $1 - \bar{f} = 1 - f_1 - f_2$.

Theorem 4.32. *Let the assumptions of Theorem 4.31 hold. Then the fractions of susceptible and of infected individuals in both stages are uniformly persistent in the following sense: There exists some $\epsilon > 0$ such that*

$$\liminf_{t\to\infty}(1 - f_1(t) - f_2(t)) > \epsilon, \qquad \liminf_{t\to\infty} f_j(t) > \epsilon, \quad j = 1, 2,$$

for all solutions with $N(0) > 0$, $f_1(0) + f_2(0) > 0$.

Proof. Uniform persistence of the fraction of stage-two infected hosts follows from Theorem 4.29. We apply the fluctuation method (Theorem A.14) to (4.41). By Theorem 4.31, $\bar{f}^\infty = 0$ or

$$0 \leq N^\sharp \kappa^\diamond (1 - \bar{f}^\infty) - \beta(1 - \bar{f}^\infty) - q_2 \beta f_{2\infty} \bar{f}^\infty.$$

This implies

$$\bar{f}^\infty \leq \frac{N^\sharp \kappa^\diamond - \beta}{N^\sharp \kappa^\diamond - \beta[1 - q_2 f_{2\infty}]} < 1,$$

Since $f_{2\infty} \geq \epsilon > 0$ with ϵ independent of the initial data, the susceptible fraction of the host population is uniformly persistent. Since the host is uniformly persistent as well (Theorems 4.26), there exists some $\epsilon > 0$ such that the inferior limits of $1 - \bar{f}$, N and f_2 exceed ϵ for all solutions with $N(0) > 0$ and $f_1(0) + f_2(0) > 0$. After a shift in time we can assume that $1 - \bar{f}(t)$, $N(t) \geq \epsilon$ and $f_2(t) \geq \epsilon$ for all $t \geq 0$. By (4.45),

$$f_1' \geq \epsilon^3 \kappa_2 - \gamma f_1 - (2\beta + \alpha_1) f_1,$$

for all $t \geq 0$. The fluctuation method (Theorem A.14) implies that

$$f_{1\infty} \geq \frac{\epsilon^3 \kappa_2}{\gamma + 2\beta + \alpha_1}. \qquad \square$$

4.7.8. A compact attractor of points. We can summarize the results of the last two subsections in the language of compact attractors (Definition 2.21).

Theorem 4.33. *Let the assumptions of Theorem 4.32 hold. Then the solution semiflow of (4.39) has a compact attractor of points for the state space*

$$\{(N, f_1, f_2);\ N > 0,\ f_1, f_2 \geq 0,\ 0 < f_1 + f_2 \leq 1\}.$$

The compact attractor is contained in the set

$$\{(N, f_1, f_2);\ N > 0,\ f_1, f_2 > 0,\ f_1 + f_2 < 1\}.$$

Differently from the analogous two-dimensional SI model, where we could use phase-plane analysis, we could not establish the existence of a compact attractor of neighborhoods of compact sets (Definition 2.21).

4.8. Exercises

Exercise 4.1. Let X be a metric space with metric d, B a nonempty subset of X, and let $f_j : X \to \mathbb{R}$ be continuous, $j = 1, \dots, n$, and $\epsilon > 0$. Then

$$\big\{x \in X : \exists y \in B : d(x, y) < \epsilon, |f_j(x) - f_j(y)| < \epsilon, j = 1, \dots, n\big\}$$

is an open subset of X.

Exercise 4.2. Verify Remark 4.9.

Exercise 4.3. Prove Theorem 4.12.

Exercise 4.4. Verify Remark 4.15.

Exercise 4.5. Find conditions under which the host uniformly persists in the following endemic model:

$$\begin{aligned}
S' &= \beta N - \frac{\kappa(N)}{N} SI - \mu S + \theta R, \\
E' &= \frac{\kappa(N)}{N} SI - (\mu + \eta) E, \\
I' &= \eta E - (\mu + \alpha + \gamma) I, \\
R' &= \gamma I - (\mu + \theta) R, \\
N &= S + E + I + R.
\end{aligned} \tag{4.56}$$

Prove your result.

Exercise 4.6. Show that the parasite goes extinct in model (4.7), (4.8), (4.9), i.e., $y_n \to 0$ as $n \to \infty$, if $\mathcal{R}_0^{\mathrm{par}} = \frac{\exp(\kappa_\infty - \alpha - \gamma)}{\beta + 1} < 1$. If you want to impress us, show that it even holds if $\mathcal{R}_0^{\mathrm{par}} \le 1$.

Exercise 4.7. Prove Theorem 4.17.

Exercise 4.8. Find conditions for uniform parasite persistence in model (4.56) in Exercise 4.5. Prove your result.

Hint: Use Theorem 4.17.

Exercise 4.9. Consider the following SI endemic model with fertility reducing disease:

$$\begin{aligned} N' &= N\Big(\beta(1-y) - \mu + (q\beta - \alpha)y\Big), \\ y' &= y\Big((\kappa N - \alpha - \beta)(1-y) - q\beta y\Big). \end{aligned}$$

Here $\beta > \mu > 0$, $\kappa > 0$ and $\alpha \ge 0$ and $q \in (0,1]$. Assume that $0 < q\beta < \mu + \alpha$. Show that the disease imposes bounds on the host population size that are eventually uniform: There exists some $c > 0$ such that $\limsup_{t\to\infty} N(t) < c$ for all nonnegative solutions N and y with $N(0) > 0$ and $y(0) \in (0,1]$.

Hint: Use $\rho(N) = \frac{1}{1+N}$ and the state space $X = \{(N, y); N > 0, 0 < y \le 1\}$. Recall Theorems 3.3, 4.14, and 4.18.

Exercise 4.10. If $\kappa_2 = 0 < \kappa_1$ and $q_2 > 0$, uniform disease persistence holds for the model (4.39) in the following form: There exists some $\epsilon > 0$ such that $\liminf_{t\to\infty} f_1(t) \ge \epsilon$ for every solution with $N(0) > 0$ and $f_1(0) > 0$.

Chapter 5

The Interplay of Attractors, Repellers, and Persistence

5.1. An attractor of points facilitates persistence

In this section, we examine the consequences of the existence of a compact attractor of points (as guaranteed in Theorem 2.28) for a semiflow Φ that is uniformly weakly ρ-persistent. It is then possible to shift the positivity requirements in Section 4.2 from the semiflow to its total trajectories [**216**, Thm.2.6]. This is important for structured population models, in particular, for infectious disease models with infection-age structure (Chapter 9) or with explicit stages that are not exponentially distributed [**69**] .

Let X be a metric space, and $J = \mathbb{R}_+$ or $J = \mathbb{Z}_+ = \{0\} \cup \mathbb{N}$. Assume that the semiflow $\Phi : J \times X \to X$ is state-continuous, $\rho : X \to \mathbb{R}_+$ is continuous and not identically zero.

Throughout this section, we assume the following two additional hypotheses.

(H0) Φ has a compact attractor A which attracts all points in X.

(H1) There exists no total trajectory $\phi : \hat{J} \to A$ such that $\rho(\phi(0)) = 0$ and $\rho(\phi(-r)) > 0$ and $\rho(\phi(t)) > 0$ for some $r, t \in J$.

Observe that (H1) is equivalent to $\tilde{\clubsuit}_1$ of Theorem 4.5 with $B = A$ (see Theorem 1.40):

(H1′) There exists no $y \in A$, $s, t \in J$, such that $\rho(y) > 0$, $\rho(\Phi(s, y)) = 0$, and $\rho(\Phi(s + t, y)) > 0$.

The following set is closed and forward invariant (and possibly empty),

$$X_0 = \{x \in X;\ \forall t \in J : \rho(\Phi(t, x)) = 0\}. \tag{5.1}$$

Recall that Φ is uniformly weakly ρ-persistent, if there exists $\eta > 0$ such that

$$\limsup_{t\to\infty} \rho(\Phi_t(x)) > \eta, \qquad \text{whenever } \rho(x) > 0,$$

and is uniformly ρ-persistent if we can replace lim sup by lim inf above. The following result says roughly that uniform weak persistence plus a compact attractor of points implies uniform strong ρ-persistence.

Theorem 5.1. *If $X_0 = \emptyset$, then $\rho(x) > 0$ for all $x \in A$ and there exists some $\eta > 0$ such that $\liminf_{t\to\infty} \rho(\Phi(t, x)) \geq \eta$ for all $x \in X$.*

Proof. Suppose $\rho(x) = 0$ for some $x \in A$. Since A is invariant, there exists a total trajectory $\phi : \hat{J} \to A$ with $\phi(0) = x$. Since $X_0 = \emptyset$, $\rho(\phi(t)) > 0$ for some $t \in J$. By (H1), $\rho(\phi(-r)) = 0$ for all $r \in J$. If $z \in \alpha(\phi)$, there exist $t_j \in J$, $t_j \to \infty$ such that $\phi(-t_j) \to z$. For all $t \in J$, since ρ and Φ_t are continuous,

$$\rho(\Phi_t(z)) = \lim_j \rho(\Phi_t(\phi(-t_j))) = \lim_j \rho(\phi(t - t_j)) = 0,$$

implying that $z \in X_0 = \emptyset$, a contradiction.

We conclude that $\rho(x) > 0$ for all $x \in A$. Since ρ is continuous, there exists $\delta > 0$ such that $\rho(x) \geq \delta$ for all $x \in A$. As A attracts all points of X and $V = \{x \in X;\ \rho(x) > \delta/2\}$ is a neighborhood of A, for each $x \in X$ there exists $T > 0$ such that $\Phi(t, x) \in V$ for all $t \in J_T$. So

$$\liminf_{t\to\infty} \rho(\Phi_t(x)) \geq \delta/2, \ \ x \in X. \qquad \square$$

The following are immediate consequences of Theorem 4.5 and Remark 4.4.

Theorem 5.2. *If $X_0 \neq \emptyset$, $\rho \circ \Phi$ is continuous, and Φ is uniformly weakly ρ-persistent, then Φ is uniformly ρ-persistent.*

Remark 5.3. In Theorem 5.2, the continuity of $\rho \circ \Phi$ can be replaced by the following weaker hypothesis which will be used in Chapter 9.10.

(H2) $\sigma := \rho \circ \Phi$ is state-continuous, uniformly in finite time, and $\rho \circ \phi$ is continuous for every total Φ-trajectory ϕ with range in A.

The second part of (H2) is equivalent to assuming that $\sigma(t, x)$ is a continuous function of $t \in J$ for every $x \in A$; cf. Theorem 1.40.

The following argument will be used repeatedly.

Lemma 5.4. *Assume (H2). Let $\epsilon > 0$ and $\limsup_{t\to\infty} \sigma(t,x) > \epsilon$ for all $x \in X$ with $\rho(x) > 0$. Then there exist no sequences (y_j) in X and (s_j), (t_j), (u_j), (v_j) in J such that $v_j \le 1$ and $y_j \to A$ as $j \to \infty$ and*

$$\begin{aligned}
&\sigma(0, y_j) \ge \epsilon, \\
&\sigma(s_j + t_j, y_j) \to 0, \quad j \to \infty, \\
&\sigma(s_j + s, y_j) \le \epsilon \qquad \forall s \in [0, t_j + u_j] \cap J, \\
&\sigma(s_j + t_j + u_j + v_j, y_j) \ge \epsilon.
\end{aligned}$$

Proof. We use Corollary 4.3. We can find sets B_j such that $d(B_j, A) \to 0$ as $j \to \infty$ and $y_j \in B_j$ for all $j \in \mathbb{N}$. Then $\clubsuit_3$ follows from (H2) as does $\clubsuit_0$. $\clubsuit_1$ is a consequence of (H1). □

5.2. Partition of the global attractor under uniform persistence

Throughout this section we assume that $J = \mathbb{Z}_+$ or $J = \mathbb{R}_+$ and that the semiflow $\Phi : J \times X \to X$ is state-continuous, uniformly in finite time. Further, let $\rho : X \to \mathbb{R}_+$ be continuous.

Recall the forward invariant set

$$X_0 = \{x \in X;\ \rho(\Phi_t(x)) = 0 \quad \forall t \in J\}.$$

Finally, we assume that Φ has a compact attractor, A, of neighborhoods of compact sets in X. Recall the Definition 2.21. Sufficient conditions for such an attractor to exist are given in Theorems 2.30 and 2.31.

Definition 5.5. A subset D of X is *uniformly ρ-positive* if there exists some $\delta > 0$ such that $\rho(x) \ge \delta$ for all $x \in D$. The semiflow Φ is *eventually uniformly ρ-positive on $D \subset X$* if $\Phi(J_r \times D)$ is uniformly ρ-positive for some $r \in J$ where $J_r = \{t \in J : t \ge r\}$.

Every uniformly ρ-positive subset is contained in $X \setminus X_0$. In order to put our main result into perspective, we first consider the case $X_0 = \emptyset$.

Theorem 5.6. *Assume that $X_0 = \emptyset$ and*

(H1) *there exists no total Φ-trajectory with range in A such that $\rho(\phi(-r)) > 0$, $\rho(\phi(0)) = 0$, and $\rho(\phi(s)) > 0$ with $r, s \in J$.*

Then the compact attractor A is uniformly ρ-positive, and ρ-persistence holds even more uniformly: There exists some $\epsilon > 0$ such that for every compact K in X there exist some neighborhood U of K and some $r \in J$ such that

$$\rho(\Phi(t,x)) \ge \epsilon, \qquad x \in U, t \in J_r;$$

further, if B is a subset of X that is attracted by A, then there exists some $r \in J$ such that

$$\rho(\Phi(t,x)) \geq \epsilon, \qquad x \in B, t \in J_r.$$

Proof. Since ρ is continuous and strictly positive on A by Theorem 5.2, the compact set A is uniformly ρ-positive. Let $\epsilon = \inf \rho(A)/2$. Then $\tilde{U} = \{x \in X; \rho(x) > \epsilon\}$ is an open neighborhood of A. If K is a compact subset of X, it has a neighborhood U that is attracted by A (Theorem 2.30), and there exists some $r \in J$ such that $\Phi(J_r \times U) \in \tilde{U}$. The same argument works for any subset B of X which is also attracted by A. □

We will show that the attractor A divides into three parts if X_0 is nonempty and Φ is uniformly persistent. Our result is inspired by Theorem 3.2 in [**95**].

5.2.1. Persistence à la Caesar.

"Attractor est omnis divisus in partes tres."

Theorem 5.7. *Assume that $X_0 \neq \emptyset$, Φ is uniformly weakly ρ-persistent, $\rho \circ \Phi$ is continuous, and*

(H1) *there exists no total Φ-trajectory with range in A such that $\rho(\phi(-r)) > 0$, $\rho(\phi(0)) = 0$, and $\rho(\phi(s)) > 0$ with $r, s \in J$.*

Then the attractor A is the disjoint union

$$A = A_0 \cup C \cup A_1$$

of three invariant sets A_0, C, and A_1. A_0 and A_1 are compact and the following hold.

(a) *$A_0 = A \cap X_0$ is the compact attractor of compact subsets of X_0; more strongly, every compact subset K of X_0 has a neighborhood in X_0 that is attracted by A_0. Further, A_0 attracts every subset of X_0 that is attracted by A.*

(b) *A_1 is uniformly ρ-positive and is the compact attractor of neighborhoods of compact sets in $X \setminus X_0$. It also attracts all sets that are attracted by A and on which Φ is eventually uniformly ρ-positive. In particular, A_1 is stable.*

(c) *If $x \in X \setminus A_1$ and ϕ is a total Φ-trajectory through x with pre-compact range, then $\phi(t) \to A_0$ for $t \to -\infty$.*

If $x \in X \setminus A_0$ and ϕ is total Φ-trajectory through x with pre-compact range, then $\phi(t) \to A_1$ for $t \to \infty$.

In particular, the set C consists of those points $x \in A$ through which there exists a total trajectory ϕ with $\phi(-t) \to A_0$ and $\phi(t) \to A_1$ as $t \to \infty$.

We call A_1 the (ρ-)*persistence attractor* of Φ and A_0 the (ρ-)*extinction attractor* of Φ.

Magal and Zhao [**158**] construct examples (with time-set $\mathbb{Z}_+$ and time-set $\mathbb{R}_+$) in which A_1 does not attract all bounded uniformly ρ-positive sets, though A attracts all bounded sets.

Remark 5.8. (a) If $\bar{\Omega}(X \setminus X_0)$, the closure of $\bigcup_{x\in X\setminus X_0} \omega(x)$, is stable, then it coincides with the persistence attractor A_1.

(b) In Theorem 5.7, the continuity of $\rho \circ \Phi$ can be replaced by the following weaker hypothesis.

(H2) $\rho \circ \phi$ is continuous for every total Φ-trajectory ϕ with range in A. $\rho \circ \Phi$ is state-continuous, uniformly in finite time.

Recall the overall assumption that Φ is state-continuous, uniformly in finite time. This implies that $\rho \circ \Phi$ is state-continuous, uniformly in finite time, provided that ρ is uniformly continuous.

(c) As in Theorem 5.6, ρ-persistence is uniform in the following stronger sense: There exists some $\epsilon > 0$ such that for every compact subset K of $X \setminus X_0$ there exist some neighborhood U of K and some $r \in J$ with

$$\rho(\Phi(t,x)) > \epsilon, \qquad x \in U, t \in J_r.$$

Further, for every B which is attracted by A and on which Φ is eventually uniformly ρ-positive, there exists some $r \in J$ such that

$$\rho(\Phi(t,x)) > \epsilon, \qquad x \in B, t \in J_r.$$

(d) Neither (H1) nor (H2) are needed for the existence and the properties of A_0 in part (a) of Theorem 5.7. Later (Theorem 5.19), we will see that (H1) or (H2) are not needed for the properties of A_0 in part (c) either.

Proof of Theorem 5.7. (a) To show that $A_0 = A \cap X_0$ is invariant, let $x \in A \cap X_0$. Since A is invariant, there exists a total trajectory ϕ with values in A and $\phi(0) = x$. Since $\phi(0) \in X_0$ and X_0 is forward invariant, $\phi(t) \in X_0$ for all $t \in J$. Suppose that $\phi(-r) \notin X_0$ for some $r \in J$. Then $0 < \rho(\Phi(s, \phi(-r)) = \rho(\phi(s - r))$ for some $s \in J$. Since Φ is uniformly weakly ρ-persistent, $\limsup_{t\to\infty} \rho(\phi(t)) \geq \eta$, a contradiction. So ϕ takes all its values in $A \cap X_0$. By Theorem 1.40, $A \cap X_0$ is invariant.

Now assume that M is a subset of X_0 that is attracted by A. Then $\omega(M)$ is a nonempty compact subset of A and attracts M. Since X_0 is forward invariant and closed, $\omega(M) \subset X_0$ by Definition 2.7. So $\omega(M) \subset A_0$, and A_0 attracts M by Theorem 2.11. Now let K be a compact subset of X_0. Then K has a neighborhood U that is attracted by A. $U \cap X_0$ is a neighborhood of K in X_0 and is attracted by A_0.

(b) We now assume (H1). By Theorem 5.2, Φ is uniformly ρ-persistent.

Define $A_2 = \bar{\Omega}(X \setminus X_0)$ as the closure of $\bigcup_{x \in X \setminus X_0} \omega(x)$, $A_2 \subset A$. Since Φ is uniformly ρ-persistent, A_2 is uniformly ρ-positive, compact and invariant and attracts all points in $X \setminus X_0$.

Step 1: For sufficiently small $\delta > 0$, Φ is eventually uniformly ρ-positive on $U = U_\delta(A_2)$ and A attracts U.

Since A is the compact attractor of neighborhoods of compact sets in X, A attracts an open neighborhood V of the compact set A_2. Choose $\delta > 0$ so small that $U = U_\delta(A_2) \subset V$. Then A attracts U.

Suppose that, for any $\delta > 0$, Φ is not eventually uniformly ρ-positive on $U_\delta(A_2)$. By negation of the second part of Definition 5.5, there exists a sequence (x_n) in X and a sequence (j_n) in J such that $j_n \to \infty$, $x_n \to A_2$, and $\rho(\Phi(j_n, x_n)) \to 0$ as $n \to \infty$. Choose $\epsilon > 0$ such that $\inf \rho(A_2) > \epsilon$ and $\liminf_{t \to \infty} \rho(\Phi(t, x)) > \epsilon$ for all $x \in X \setminus X_0$. After choosing subsequences, $x_n \to x$ for some $x \in A_2$, $\rho(x_n) > \epsilon$, $\rho(\Phi(j_n, x_n)) < \epsilon$ and $\liminf_{t \to \infty} \rho(\Phi(t, x_n)) > \epsilon$. We find sequences (r_n), (s_n), (u_n) and (v_n) in J such that $r_n + s_n \le j_n$, $s_n \le 1$, $v_n \le 1$ and

$$\begin{aligned}
&\rho(\Phi(r_n, x_n)) \ge \epsilon, \qquad \rho(\Phi(j_n + u_n + v_n, x_n)) \ge \epsilon, \\
&\rho(\Phi(s, x_n)) \le \epsilon, \quad s \in [r_n + s_n, j_n + u_n], \\
&\epsilon > \rho(\Phi(j_n, x_n)) \to 0, \quad n \to \infty.
\end{aligned}$$

After choosing subsequences, (r_n) is bounded or $r_n \to \infty$. In the first case, since Φ is state-continuous uniformly in finite time, $d(\Phi(r_n, x_n), \Phi(r_n, x)) \to 0$ and $\Phi(r_n, x_n) \to A_2$ because $\Phi(r_n, x)$ is an element of the invariant set A_2. In the second case, $\Phi(r_n, x_n) \to A$. In either case, after choosing subsequences, $\Phi(r_n, x_n) \to y$ for some $y \in A$. We set $y_n = \Phi(r_n, x_n)$, $t_n = j_n - r_n - s_n$. Since Φ is a semiflow, with $\sigma = \rho \circ \Phi$,

$$y_n \to y \in A, \qquad \sigma(s_n + t_n, y_n) \to 0, \qquad n \to \infty,$$

$$\left. \begin{aligned} &\rho(y_n) \ge \epsilon, \qquad \sigma(s_n + t_n + u_n + v_n, y_n) \ge \epsilon \\ &\sigma(s_n + s, y_n) < \epsilon, \qquad s \in [0, t_n + u_n] \cap J \end{aligned} \right\} n \in \mathbb{N}.$$

This contradicts Lemma 5.4.

Step 2: Φ is asymptotically compact on U. If K is a compact subset of $X \setminus X_0$, then $A_1 = \omega(U)$ attracts a neighborhood of K. A_1 is uniformly ρ-positive.

By Step 1, there exists some $r \in J$ such that $B = \Phi(J_r \times U)$ is uniformly ρ-positive. Since A attracts U, $A_1 = \omega(U)$ is a nonempty compact subset of $\bar{B}$ that attracts U.

Since ρ is continuous, A_1 is uniformly ρ-positive. By Proposition 2.29, $\omega(U)$ attracts a neighborhood of any compact set in $\bigcup \Phi^{-1}(U) = X \setminus X_0$. The last equality follows because U is a neighborhood of A_2 and A_2 attracts all points in $X \setminus X_0$.

Step 3: $A_1 = \omega(U)$ attracts all subsets of X which are attracted by A and on which Φ is eventually uniformly ρ-positive.

Let M be a subset of $X \setminus X_0$ that is attracted by A and on which Φ is eventually uniformly ρ-positive. As before, $\omega(M)$ is uniformly ρ-positive and a compact subset of $X \setminus X_0$. By Step 2, A_1 attracts a neighborhood of $\omega(M)$. Then A_1 also attracts M.

The stability of A_1 follows from Theorem 2.39.

Step 4: Let $x \in X \setminus A_0$.

Let ϕ be a total Φ-trajectory through x with pre-compact range. Then the closure of the range is compact and invariant (Exercise 1.12) and thus a subset of A. So $x \in A \setminus A_0$. Since $A_0 = A \cap X_0$, $x \in X \setminus X_0$ and $\phi(t) \to A_1$ as $t \to \infty$ because A_1 attracts all points in $X \setminus X_0$.

Step 5: Let $x \in X \setminus A_1$.

Let ϕ be a total Φ-trajectory through x with pre-compact range.

Then the closure of the range is compact and invariant and thus a subset of A. So $x \in A \setminus A_1$. We can assume $x = \phi(0)$.

Then $\phi(\mathbb{R})$ is not uniformly ρ-positive. Otherwise the closure of $\phi(\mathbb{R})$ would be uniformly ρ-positive which is a compact invariant set; by Step 3, the closure of $\phi(\mathbb{R})$ would be attracted by A_1 and thus be a subset of A_1 contradicting $\phi(0) \notin A_1$.

Since $\phi(\mathbb{R})$ is not uniformly ρ-positive, there exists a sequence (s_n) in $J \cup (-J)$ such that $\rho(\phi(s_n)) \to 0$. Suppose that $\rho(y) > 0$ for some $y \in \alpha(\phi)$. Then there exists some $\epsilon > 0$ such that $\epsilon < \limsup_{r\to-\infty} \rho(\phi(r))$ and $\epsilon < \liminf_{t\to\infty} \rho(\Phi(t,x))$ for all $x \in X$ with $\rho(x) > 0$. In particular, $\liminf_{t\to\infty} \rho(\phi(t)) > \epsilon$. Since $\rho \circ \phi$ is continuous by (H2), we find sequences (r_n) and (t_n) in $J \cup (-J)$ such that $r_n < s_n < t_n$ and

$$\rho(\phi(r_n)) \geq \epsilon, \qquad \rho(\phi(t_n)) \geq \epsilon,$$
$$\rho(\phi(s)) \leq \epsilon, \quad s \in (r_n, t_n).$$

Set $y_n = \phi(r_n)$. After choosing subsequences, $y_n \to y$ for some $y \in A$. We set $\tilde{s}_n = s_n - r_n$ and $\tilde{t}_n = t_n - r_n$. Let $\sigma = \rho \circ \Phi$. Since ϕ is a Φ-trajectory,

$\rho(\phi(r+t)) = \sigma(t, \phi(r))$. Then $0 < \tilde{s}_n < \tilde{t}_n$ and

$$\begin{aligned} \rho(y_n) \geq \epsilon, &\quad \sigma(\tilde{t}_n, y_n) \geq \epsilon, \\ \sigma(s, y_n) \leq \epsilon, &\quad s \in (0, \tilde{t}_n), \\ \sigma(\tilde{s}_n, y_n) \to 0, &\quad n \to \infty. \end{aligned}$$

This contradicts Lemma 5.4. So ρ is zero on the invariant set $\alpha(\phi)$ and $\alpha(\phi) \in A_0$.

Step 6: Proof of Remark 5.8 (a).

Assume that A_2 is stable. Then, for any neighborhood V of A_2, there exists some neighborhood U such that $\Phi(J \times U) \subset V$. We can assume U to coincide with the set U in Steps 1 to 3. Then $A_1 = \omega(U) \subset \bar{V}$. Since V has been an arbitrary neighborhood of A_2, $A_1 \subset A_2$. Since A_2 is compact and uniformly ρ-positive, A_1 attracts A_2 by step 2. Since A_2 is also invariant, $A_2 \subset A_1$. □

The persistence attractor is connected if X is the closed convex subset of a Banach space and ρ is concave: $\rho(\xi x + (1-\xi)\tilde{x}) \geq \xi \rho(x) + (1-\xi)\rho(\tilde{x})$.

Proposition 5.9. *Let X be the closed convex subset of a Banach space and d the metric induced by the norm. Let the assumptions of Theorem 5.7 be satisfied and the persistence function ρ be concave. Then the persistence attractor, A_1, is connected.*

Proof. The convex hull of A_1, $\operatorname{co}(A_1)$, is defined as the set of convex combinations $\sum_{j=1}^n \xi_j x_j$ of elements $x_1, \ldots, x_n \in A_1$ where $n \in \mathbb{N}$, $\xi_j \geq 0$ and $\sum_{j=1}^n \xi_j = 1$. Since ρ is concave,

$$\rho\Big(\sum_{j=1}^n \xi_j x_j\Big) \geq \sum_{j=1}^n \xi_j \rho(x_j) \geq \inf \rho(A_1).$$

Since A_1 is uniformly ρ-positive, so is its convex hull and the closure of its convex hull which equals the closed convex hull [**62**, Sec.V.2]. By a theorem of Mazur [**62**, Theorem V.2.6], the closed convex hull of a compact set is also compact. By Theorem 5.7, A_1 attracts its closed convex hull (a connected set) and is connected by Theorem 2.24 (a). □

Remark 5.10. It is more difficult to find a satisfactory condition for the connectedness of the extinction attractor A_0. Of course, one could flatly assume that X_0 is convex and apply Mazur's theorem as above. The convexity of X_0 follows from the following three assumptions: X is a convex subset of a Banach space, ρ is convex and $\rho(\Phi(t,x)) = 0$ for all $t > 0$ whenever $\rho(x) = 0$. Then $X_0 = X \cap \{\rho = 0\}$ which is a convex set.

The set C is nonempty, e.g., if the compact global attractor is flow-connected (Definition 2.14), in particular if X is the closed convex subset of a Banach space (Theorem 2.24). A more general condition for justifying our Caesar citation about the three parts is the following.

Proposition 5.11. *Let the assumptions of Theorem 5.7 be satisfied. Then the set C is nonempty if and only if $\partial X_0 \neq \emptyset$.*

Proof. If $\partial X_0 = \emptyset$, A_0 and the closure of $C \cup A_1 \subset X \setminus X_0$ are disjoint. Assume that C is nonempty and consider some $x \in C$. By Theorem 5.7, there exists a total Φ-trajectory ϕ with values in A such that $\phi(0) = x$, $\phi(-t) \to A_0$ and $\phi(t) \to A_1$ as $t \to \infty$. The last implies that the range of ϕ is contained in $C \cup A_1$, so $\alpha(\phi)$ is contained in the closure of $C \cup A_1$ which is disjoint from A_0, contradicting $\phi(-t) \to A_0$ as $t \to \infty$.

Now assume that ∂X_0 is nonempty. Since X_0 is closed, there exist some $x \in X_0$ and a sequence (x_n) in $X \setminus X_0$ such that $x_n \to x$ as $n \to \infty$. Since ρ is continuous, $\rho(x_n) \to \rho(x) = 0$. Further, $\{x_n; n \in \mathbb{N}\} \cup \{x\}$ is compact and thus attracted by A. Since A_1 is uniformly ρ-positive and attracts each x_n,

$$\epsilon = \inf \rho(A_1) > 0 \quad \text{and} \quad \liminf_{t\to\infty} \rho(\Phi(t, x_n)) \geq \epsilon.$$

For each n there exists some $s_n \in J$ such that

$$\rho(\Phi(s_n, x_n)) \geq \epsilon/2, \qquad \rho(\Phi(s, x_n)) < \epsilon/2 \quad \text{for all } s \in [0, s_n). \tag{5.2}$$

The sequence (s_n) is unbounded. Otherwise, since $\rho \circ \Phi$ is state-continuous, uniformly in finite time, $\rho(\Phi(s_n, x_n)) - \rho(\Phi(s_n, x)) \to 0$, a contradiction because $\rho(\Phi(s_n, x)) = 0$ by definition of X_0. After choosing subsequences, we can assume that $s_n \to \infty$. Then $\Phi(s_n, x_n) \to A$. After choosing subsequences again, $\Phi(s_n, x_n) \to y \in A$. We define

$$\phi_n(s) = \begin{cases} \Phi(s_n + s, x_n), & s > -s_n, \\ x_n, & s \leq -s_n. \end{cases}$$

Since $s_n \to \infty$, for each $s \in J$, $\phi_n(s) \to A$ as $n \to \infty$ and the sequence $(\phi_n(s))_{n\in\mathbb{N}}$ has a convergence subsequence because A is compact. Using a diagonalization procedure, we can assume that, after choosing subsequences,

$$\lim_{n\to\infty} \phi_n(j) = \lim_{n\to\infty} \Phi(s_n + j, x_n)$$

exists for each $j \in \mathbb{Z}$. If $J = \mathbb{R}_+$, let $r \in [0, 1)$. For sufficiently large $n \in \mathbb{N}$,

$$\phi_n(j + r) = \Phi_r(\Phi(s_n + j, x_n)) \to \Phi_r\big(\lim_{n\to\infty} \phi_n(j)\big) = \lim_{n\to\infty} \Phi(s_n + j + r, x_n).$$

So,

$$\phi(s) := \lim_{n\to\infty} \phi_n(s) = \lim_{n\to\infty} \Phi(s_n + s, x_n)$$

exists for all $s \in \hat{J} = J \cup (-J)$. Since Φ is state-continuous, ϕ is a total Φ-trajectory with values in A. By (5.2), for all $s \in \mathbb{N}$, $\rho(\phi(-s)) \leq \epsilon/2$ and

$\phi(-s) \notin A_1$. Moreover, $\rho(\phi(0)) \geq \epsilon$ and $\phi(0) \notin A_0$. Since A_0 is invariant, $\phi(-s) \notin A_0$ for all $s \in \mathbb{N}$. Thus $\phi(-s) \in C$ for all $s \in \mathbb{N}$ and C is not empty. □

In Section 8.7 and Section 9.9 we will illustrate techniques for showing that the persistence attractor is a singleton set formed by an equilibrium. These techniques involve Lyapunov type functions or involve the Fourier transform (Section B.3). Under the assumptions of Theorem 5.7, such an equilibrium automatically is locally asymptotically stable and is globally asymptotically stable on $X \setminus X_0$.

5.2.2. An elementary example: scalar difference equations. Let $f : \mathbb{R}_+ \to \mathbb{R}_+$ be continuous, $f(0) = 0$, and $f(x) > 0$ for $x > 0$. Assume

$$\liminf_{x \to 0} \frac{f(x)}{x} > 1 > \limsup_{x \to \infty} \frac{f(x)}{x}.$$

We consider the discrete time dynamical system Φ with time-set $\mathbb{Z}_+$ and state space $X = \mathbb{R}_+$ given by the iterates of f, $\Phi_n = f^n$, which is induced by the difference equation $x_n = f(x_{n-1})$, $n \in \mathbb{N}$, with given x_0. By the right-hand side of our inequality, there exists some $c > 0$ such that $f(x) < x$ for all $x \geq c$.

Let $\tilde{c}$ be the larger of c and $1 + \max f([0, c])$ and $U = [0, \tilde{c})$. Then U is open in X and, for any $x \in X$, $\Phi(t, x) \in [0, c) \subset U$ for some t. We claim that $f(U) \subset U$.

Indeed, if $x \leq c$, then $f(x) < 1 + \max f([0, c])$. If $x \in [c, \tilde{c}]$, then $f(x) \leq x \leq \tilde{c}$.

Since U is bounded and $\Phi_1(U) \subset U$, Φ is asymptotically compact on U. So Φ has a compact attractor, A, of compact sets in X by Theorem 2.31.

By the left-hand side of the inequality above, there exists some $\epsilon > 0$ such that $f(x) > x$ for all $x \in (0, \epsilon]$. An easy contradiction argument (notice $\Phi(t, x) > 0$ for all $t \in \mathbb{N}_+, x > 0$) implies that $\limsup_{t \to \infty} \Phi(t, x) \geq \epsilon$ for all $x > 0$. This means that Φ is uniformly weakly ρ-persistent for $\rho(x) = x$, $X_0 = \{0\}$. By Theorem 5.7, we have a partition $A = A_0 \cup C \cup A_1$. The extinction attractor is $A_0 = \{0\}$. The persistence attractor, A_1, is the compact attractor of compact sets in $(0, \infty)$. By Proposition 5.9, A_1 is connected. So A_1 is an interval $[a, b]$ with $0 < a \leq b < \infty$. Further A_1 is invariant,

$$A_1 = [a, b] = f([a, b]).$$

The endpoints of the persistence attractor interval satisfy

$$a = g(a, b), \qquad b = g(b, a),$$

where g is the following monotone representation of f,

$$g(y,z) = \begin{cases} \inf f([y,z]); & y \le z, \\ \sup f([z,y]); & y \ge z. \end{cases}$$

The two-argument function g is increasing in $y \ge 0$ and decreasing in $z \ge 0$ and satisfies $g(x,x) = f(x)$. This is why it is called a monotone representation [**212**] or decomposition [**201**] of f. This specific monotone composition has been used in [**211, 212**]; others have been suggested in [**195**].

We now assume that f is unimodal, i.e., there exists some $x^\diamond$ such that f is strictly increasing on $[0, x^\diamond]$ and strictly decreasing on $[x^\diamond, \infty)$. In particular, f takes a local and global maximum $f^{\max}$ at $x^\diamond$. Since f is unimodal, it is easy to evaluate the monotone representation g.

There are three possible cases:

Case 1: $0 < a \le b \le x^\diamond$, $a < x^\diamond$.

Then f is increasing on $[a,b]$ and $a = f(a)$ and $b = f(b)$ and the endpoints of the persistence attractor interval are positive fixed points of f in $(0, x^\diamond]$.

Case 2: $a \le x^\diamond \le b$.

Then $b = f(x^\diamond) = f^{\max}$ and

$$a = \min\{f(a), f(b)\} = \min\{f(a); f(f^{\max})\}.$$

If there is no fixed point of f in $[0, x^\diamond]$, then

$$a = f(f^{\max}).$$

Case 3: $x^\diamond \le a \le b$.

Then f is decreasing on $[a,b]$ and $a = f(b)$ and $b = f(a)$; in particular, a and b are period-two points of f. This does not exclude that they may be fixed points of f, but the latter only happens if $a = b$ and A_1 is the singleton set formed by a fixed point of f in $[x^\diamond, \infty)$.

Since $b = f(a)$, we have the estimates $b \le f^{\max}$ and, since f is decreasing on $[x^\diamond, \infty)$, $a = f(b) \ge f(f^{\max})$.

If f has no fixed points in $(0, x^\diamond)$, case 1 cannot occur, and we can conclude the following from cases two and three.

Proposition 5.12. *Assume that f is unimodal and has no fixed points in $(0, x^\diamond)$. Then the persistence attractor A_1 is a compact interval satisfying*

$$A_1 \subset [f(f^{\max}), f^{\max}].$$

5.3. Repellers and dual attractors

Our terminology follows [**215**].

Definition 5.13. Let $\emptyset \neq M \subset X$.

a) M is called a *weak repeller* if $\limsup_{t\to\infty} d(\Phi_t(x), M) > 0$ for all $x \in X \setminus M$.

b) M is called a *uniform weak repeller* if there exists some $\epsilon > 0$ such that $\limsup_{t\to\infty} d(\Phi_t(x), M) \geq \epsilon$ for all $x \in X \setminus M$.

c) M is called a *uniform repeller* if there exists some $\epsilon > 0$ such that $\liminf_{t\to\infty} d(\Phi_t(x), M) \geq \epsilon$ for all $x \in X \setminus M$.

Definition 5.14. Let $\emptyset \neq M \subset X$.

(a) A neighborhood V of M is called an *isolating neighborhood* of M in X if every compact invariant $K \subset V$ is a subset of M.

(b) M is called *isolated* if it has an isolating neighborhood.

Remark 5.15. (a) If M is closed, M is a uniform repeller if and only if Φ is uniformly ρ-persistent with $\rho(x) = d(x, M)$.

(b) If M is compact and M has an isolating neighborhood V, V can be assumed to be a closed isolating neighborhood.

(c) If M is compact, M is a uniform repeller if and only if it has a neighborhood V such that for each $x \notin M$ there is some $r \in J$ such that $\Phi_t(x) \notin V$ for all $t \in J_r$.

V is called a *repeller neighborhood* of M.

Proof. (a) Recall that $d(x, M) = 0$ if and only if $x \in M$ (Exercise 1.3).

(b) Let M be compact and $\tilde{V}$ an isolating neighborhood of M. We can assume that $\tilde{V}$ is open. Since M is compact, by Exercise 1.8, there exists some $\epsilon > 0$ such that $U_{2\epsilon}(M) \subset \tilde{V}$. Then $V = \bar{U}_\epsilon(M) := \{x \in X; d(x, M) \leq \epsilon\} \subset U_{2\epsilon}(M) \subset \tilde{V}$ is a closed isolating neighborhood of M. □

The following result is patterned after Theorem 2.1 in [**107**] where it is remarked that it is a special case of the famous Ura-Kimura Theorem. See also Theorem 2 in [**104**]. The more recent paper of Garay and Hofbauer [**79**] contains interesting historical observations and notes the role of Zubov. See also Theorem VI 1.1 in [**16**].

Theorem 5.16. *Assume that Φ is continuous and has a compact attractor A that attracts all points in X. Let M be a nonempty compact forward invariant set. Then M is a uniform repeller if and only if it is a weak repeller and isolated.*

Proof. One direction is immediate by Remark 5.15 (c).

Now assume that M is a weak repeller and isolated. Define $\rho(x) = d(x, M)$. Since M is forward invariant, $M = X_0$ and hypotheses (H0) and

(H1) of Theorem 5.2 are satisfied. Let U be an isolating open neighborhood of M. Then $\epsilon = \inf \rho(X \setminus U) > 0$. Let $x \in X \setminus M$. Since M is a weak repeller, $\omega(x) \cap (X \setminus M) \neq \emptyset$. Since U isolates M, $\omega(x) \cap (X \setminus U) \neq \emptyset$. So there exists some $y \in \omega(x)$ such that $\rho(y) \geq \epsilon$ and Φ is uniformly weakly ρ-persistent. By Theorem 5.2, Φ is uniformly ρ-persistent and M is a uniform repeller. □

Theorem 5.17. *Let Φ be continuous and let A be the compact attractor of neighborhoods of compact sets in X. Let M be a forward invariant closed proper subset of X. Assume that M is a uniform weak repeller.*

Then M is a uniform repeller and the attractor A is the disjoint union

$$A = A_0 \cup C \cup A_1$$

of three invariant sets A_0, C, and A_1. A_0 and A_1 are compact.

$A_0 = A \cap M$ is the compact attractor of compact sets in M; more strongly, every compact subset K of M has a neighborhood in M that is attracted by A_0. Further, A_0 attracts every subset of M that is attracted by A.

A_1 is disjoint from M and attracts a neighborhood of each compact set in $X \setminus M$. In particular, A_1 is stable.

Further, if $x \in X \setminus A_1$ and ϕ is a total trajectory through x with pre-compact range, then $\phi(t) \to A_0$ for $t \to -\infty$.

If $x \in X \setminus A_0$ and ϕ is a total trajectory through x with pre-compact range, then $\phi(t) \to A_1$ for $t \to \infty$.

The set C consists of those points $x \in A$ through which there exists a bounded total trajectory ϕ with $\phi(-t) \to A_0$ and $\phi(t) \to A_1$ as $t \to \infty$. C is nonempty if $\partial M \neq \emptyset$.

We call A_1 the compact attractor dual to M.

Proof. Let $\rho(x) = d(x, M)$. Since M is a uniform weak repeller, Φ is uniformly weakly ρ-persistent. Since M is forward invariant, $M = X_0$ in the language of Theorem 5.7. Since M is also closed, (H1) is satisfied as well (recall Exercise 1.3). Then we have the partition $A = A_0 \cup C \cup A_1$ with pairwise disjoint sets as provided by Theorem 5.7. Since $A_0 = A \cap M$, $A_1 \cup C \subset X \setminus M$. All except the final statement follows from Theorem 5.7; Proposition 5.11 implies the last assertion. □

Proposition 5.18. *Let X be a subset of a Banach space and d the metric induced by the norm. Let the assumptions of Theorem 5.17 be satisfied and let M be convex. Then A_0 is connected.*

Proof. By a theorem by Mazur [**62**, Theorem V.2.6], the closed convex hull of A_0 is a convex compact subset of M and is attracted by A_0 by Theorem 5.17. So A_0 is connected by Theorem 2.24. □

With these results we return to the case that Φ is uniformly weakly ρ-persistent but that (H1) does not necessarily hold.

Theorem 5.19. *Assume that $X_0 \neq \emptyset$ and that ρ and Φ are continuous and Φ is uniformly weakly ρ-persistent and has a compact attractor A of neighborhoods of compact sets.*

Then the attractor A is the disjoint union

$$A = A_0 \cup C \cup A_1$$

of three invariant sets A_0, C, and A_1. A_0 and A_1 are compact.

$A_0 = A \cap X_0$ attracts all compact subsets of X_0; more strongly, every compact subset K of $X \setminus X_0$ has a neighborhood in X_0 that is attracted by A_0. Further, A_0 attracts every subset of X_0 that is attracted by A.

A_1 attracts all points in $X \setminus X_0$ and all compact sets in $A \setminus A_0$.

Further, if $x \in X \setminus A_1$ and ϕ is a total Φ-trajectory through x with pre-compact range, then $\phi(t) \to A_0$ for $t \to -\infty$.

If $x \in X \setminus A_0$ and ϕ is a total Φ-trajectory through x with pre-compact range, then $\phi(t) \to A_1$ for $t \to \infty$.

In particular, the set C consists of those points $x \in A$ through which there exists a total trajectory ϕ with $\phi(-t) \to A_0$ and $\phi(t) \to A_1$ as $t \to \infty$.

If ρ is uniformly continuous, then A_1 attracts a neighborhood of each compact subset in $X \setminus X_0$ and is stable in X; it also attracts all sets in $X \setminus X_0$ which are attracted by A and on which Φ is eventually uniformly ρ-positive.

We mention that A_1 is uniquely determined as a compact invariant subset of $A \setminus A_0$ that attracts all compact subsets of $A \setminus A_0$. The main difference to Theorem 5.7 is that A_1 is not necessarily uniformly ρ-positive. If it were, A_1 would be the same as in Theorem 5.7.

Proof. The statements concerning A_0 follow exactly as in the proof of Theorem 5.7. If ρ is merely continuous, we apply Theorem 5.17 to the state space A with $M = A_0$. If $x \in A \setminus A_0$, then $x \notin X_0$ and so $\limsup_{t\to\infty} \rho(\Phi(t,x)) > \epsilon$ with some ϵ that does not depend on x. The closed set $\{x \in X; \rho(x) \geq \epsilon\}$ has a distance $\delta > 0$ from the compact set A_0. Hence, $\limsup_{t\to\infty} d(\Phi(t,x), A_0) \geq \delta$ for all $x \in A \setminus A_0$. A_1 is now chosen as the attractor dual to A_0.

Let $x \in X \setminus X_0$. By our overall assumptions, A attracts $\{x\}$ and $\omega(x)$ is a nonempty compact subset of A and attracts x. Since Φ is uniformly weakly ρ-persistent, $\rho(\Phi(s,x)) \geq \epsilon$ for some $s > 0$. This implies that $d(\Phi(s,x), A_0) \geq \delta$. Since A_0 is a uniform repeller, $\inf_{y \in \omega(x)} d(y, A_0) > 0$. Since A_1 attracts every compact subset of $A \setminus A_0$ and $\omega(x)$ is invariant, $\omega(x) \subset A_1$ and A_1 attracts $\{x\}$.

The statements concerning the total Φ-trajectories follow because every total Φ-trajectory with pre-compact range takes its values in A.

Now let ρ be uniformly continuous. Since Φ is uniformly weakly ρ-persistent, we can choose $\epsilon > 0$ such that $\limsup_{t\to\infty} \rho(\Phi(t,x)) > \epsilon$ for all $x \in X \setminus X_0$. Now choose $\delta > 0$ such that $|\rho(y) - \rho(z)| < \epsilon$ whenever $d(y,z) \leq \delta$. Since $\rho(y) = 0$ all $y \in X_0$, $\limsup_{t\to\infty} d(\Phi(t,x), X_0) \geq \delta$ for all $x \in X \setminus X_0$. So $M = X_0$ is a uniform weak repeller and a forward invariant closed proper subset of X. We now choose A_1 as the attractor dual to X_0 which exists according to Theorem 5.17. This provides the same set as before because $A_1 \subset A \setminus A_0$. But now, all compact subsets of $X \setminus X_0$ have a neighborhood that is attracted by A_1. By Theorem 2.39, A_1 is stable.

Now let B be a set which is attracted by A and on which Φ is eventually uniformly ρ-positive. Since A is compact, $\omega(B)$ is nonempty and attracts B, and $\omega(B) \subset A$. Further, $\omega(B)$ is uniformly ρ-positive and thus a subset of $X \setminus X_0$. So $\omega(B)$ is attracted by A_1, and B itself is attracted by A_1 (Theorem 2.11). □

5.4. The cyclic competition model of May and Leonard revisited

For an application of repellers and their dual attractors, we revisit the three species competition model in Ch. 3.3.4,

$$
\begin{aligned}
N_1' &= N_1[1 - N_1 - \alpha N_2 - \beta N_3],\\
N_2' &= N_2[1 - \beta N_1 - N_2 - \alpha N_3],\\
N_3' &= N_3[1 - \alpha N_1 - \beta N_2 - N_3],
\end{aligned}
\tag{5.3}
$$

and the assumptions

$$
0 < \alpha < 1 < \beta, \qquad \alpha + \beta > 2. \tag{5.4}
$$

Besides some obvious invariant sets, there are the invariant line $L = \{N_1 = N_2 = N_2\}$ and the invariant heteroclinic cycle HC that connects the boundary equilibria E_j, $j = 1, 2, 3$, with total orbits situated in the boundary of the positive orthant.

It is the purpose of this section to demonstrate the versatility of persistence theory by using it to complete the analysis of the case in which none of the three species persists uniformly.

Since the heteroclinic orbit HC attracts all solutions starting outside of L by Proposition 3.26, L is a uniform repeller. We will show that HC is the compact attractor dual to L.

Theorem 5.20. *The heteroclinic orbit HC is the compact attractor of neighborhoods of compact sets in $\mathbb{R}^3_+ \setminus L$. In particular, HC is stable.*

We first recall (Remark 3.25) that all nonzero solutions are attracted to the compact set

$$T = \{N \in \mathbb{R}^3_+; 2(\alpha+\beta)^{-1} \leq N_1 + N_2 + N_3 \leq 1\}.$$

Further, the semiflow Φ that is induced by (5.3) is eventually bounded on bounded sets. By Theorem 2.33, Φ has a compact attractor of bounded sets, A, that is a subset of T.

Let A_1 be the attractor of Φ dual to the repeller L which exists according to Theorem 5.17. We will show that A_1 coincides with the heteroclinic cycle HC. Obviously, HC is a subset of A_1. Consider the Lyapunov function V in Proposition 3.26. We apply Theorem 2.53. By Proposition 3.26, $V \circ \phi$ is continuously differentiable for any total trajectory ϕ with values in A_1 and has nonpositive derivative. Assume that $\frac{d}{dt}(V \circ \phi) \equiv 0$, then, by the proof of Proposition 3.26, ϕ takes its values in the boundary of the positive orthant (and in T). As $t \to -\infty$, $\phi(t)$ converges to one of the boundary equilibria. So, for t close to $-\infty$, $\phi(t)$ lies on the one-dimensional unstable manifold of one of the boundary equilibria and is an element of HC. Since HC is invariant, ϕ takes its values in HC. Theorem 2.53 implies that A_1 coincides with HC.

5.5. Attractors at the brink of extinction

In this section we assume that the time set J is $\mathbb{Z}_+$ or $\mathbb{R}_+$ and that Φ is state-continuous and has a compact attractor, A, of compact sets in X.

Proposition 5.21. *Assume that $X_0 \neq \emptyset$ and that there is no total trajectory ϕ with range in A and $\rho(\phi(0)) > 0$ and $\rho(\phi(J_r)) = \{0\}$ for some $r \in J$. Then $A_0 = A \cap X_0$ is a compact attractor of compact sets in X_0 and attracts all subsets of X_0 that are attracted by A.*

Proof. To show that $A_0 = A \cap X_0$ is invariant, let $x \in A \cap X_0$. Since A is invariant, by Theorem 1.40 there exists a total trajectory ϕ with values in A and $\phi(0) = x$. Since $\phi(0) \in X_0$ and X_0 is forward invariant, $\phi(t) \in X_0$ for all $t \in J$. Suppose that $\phi(-r) \notin X_0$ for some $r \in J$. Then $0 < \rho(\Phi(t, \phi(-r)) = \rho(\phi(t-r))$ for some $t \in J$. After an appropriate shift of the trajectory, this contradicts our assumption. So ϕ takes all its values in $A \cap X_0$. This implies that $A \cap X_0$ is invariant.

Now let M be a subset of X_0 that is attracted by A. By Proposition 2.10, $\omega(M)$ is a nonempty compact subset of A and attracts M. Since X_0 is forward invariant and closed, $\omega(M) \subset X_0$ by Definition 2.7. Thus $\omega(M) \subset A_0$, and A_0 attracts M by Theorem 2.11. In particular, A_0 attracts every compact subset of X_0. □

Proposition 5.22. *Let $X_0 \neq \emptyset$ and A be a compact attractor of compact sets in X. Let $\tilde{A}$ be a compact invariant subset of X_0 that attracts all compact subsets of X_0. (Such a set exists, namely $A \cap X_0$.) Further, assume that $\rho(\Phi(t, x)) \to 0$ uniformly for all $x \in A$. Then $\tilde{A} = A$.*

Proof. Since A is invariant, $A \subset X_0$ and $\tilde{A}$ attracts A. Since A also attracts $\tilde{A}$ and both sets are invariant, they are equal. □

5.6. An attractor under two persistence functions

In this section, we assume that the time-set J is $\mathbb{Z}_+$ or $\mathbb{R}_+$ and that we have two continuous functions $\rho, \tilde{\rho} : X \to \mathbb{R}_+$. Let X_0 and $\tilde{X}_0$ be the forward invariant sets

$$X_0 = \{x \in X;\ \rho(\Phi_t(x)) = 0 \quad \forall t \in J\},$$
$$\tilde{X}_0 = \{x \in X;\ \tilde{\rho}(\Phi_t(x)) = 0 \quad \forall t \in J\}.$$

Theorem 5.23. *Let Φ be state-continuous, uniformly in finite time. Assume that there is a compact attractor A of neighborhoods of compact sets in X. Assume that X_0 is not empty and (H1) in Theorem 5.7 holds for ρ.*

Assume that Φ is uniformly ρ-persistent, but

$$\tilde{\rho}(\Phi(t, x)) \to 0, \qquad t \to \infty, \tag{5.5}$$

uniformly for x in all compact invariant subsets of A on which ρ is positive.

Let $\tilde{A}$ be a uniformly ρ-positive compact invariant subset of $\tilde{X}_0$ which attracts all compact subsets of $\tilde{X}_0 \setminus X_0$. (Such a set exists and is uniquely determined.)

Then $\tilde{A}$ attracts neighborhoods of all compact sets in $X \setminus X_0$, and it attracts all subsets of X which are attracted by A and on which ρ is eventually uniformly ρ-positive. $\tilde{A}$ is stable.

Remark 5.24. The assumption (5.5) can be replaced by

> $\rho(\phi(t)) \to 0$ as $t \to \infty$ uniformly for all total Φ-trajectories ϕ whose range is contained in A and uniformly ρ-positive.

Recall Theorem 1.40.

Proof. By Theorem 5.7, there exists a uniformly ρ-positive compact attractor A_1 of neighborhoods of compact sets in $X \setminus X_0$. A_1 also attracts all subsets of X which are attracted by A and on which Φ is eventually uniformly ρ-positive. A_1 is stable. By assumption, since $A_1 \subset A$, $\tilde{\rho}(\Phi(t,x)) \to 0$ as $t \to \infty$ uniformly for $x \in A_1$. Since A_1 is invariant, $A_1 \subset \tilde{X}_0$. A_1 has the properties attributed to $\tilde{A}$ in the statement of this theorem. Now assume that $\tilde{A}$ is given as stated. Then, A_1 attracts $\tilde{A}$. By assumption $\tilde{A}$ also attracts A_1. Since they are both invariant, $\tilde{A} = A_1$. □

5.7. Persistence of bacteria and phages in a chemostat

Phages are viruses that attack bacteria. Lenski and Levin [**140**] model phage growth on a bacterial host which consumes a limiting nutrient in a chemostat by the system

$$\begin{aligned} R'(t) =& D(R^\diamond - R(t)) - f(R(t))S(t), \\ S'(t) =& (f(R(t)) - D)\, S(t) - kS(t)P(t), \\ I'(t) =& kS(t)P(t) - DI(t) - e^{-D\tau}kS(t-\tau)P(t-\tau), \\ P'(t) =& -DP(t) - kS(t)P(t) + be^{-D\tau}kS(t-\tau)P(t-\tau). \end{aligned} \tag{5.6}$$

R is the resource supporting bacterial growth, S are uninfected bacteria, I are phage-infected bacteria, and P are phages. $R^\diamond$ is input nutrient concentration supplied to bacteria, D is the dilution rate of the chemostat, and $f(R)$ is the specific growth rate of bacteria at resource level R. The specific growth rate f is typically taken to be of Monod type:

$$f(R) = \frac{mR}{a+R}$$

where $m, a > 0$. However, we need only assume that $f : \mathbb{R}_+ \to \mathbb{R}_+$ is C^1 and

$$f(0) = 0, \;\; f'(R) > 0, \;\; f(\infty) < \infty. \tag{5.7}$$

A recent mathematical analysis of the model was carried out by Beretta et al. [**13**].

Phages attach to the cell surface of a bacterium and inject their DNA into it. This causes the bacterium to begin to synthesize viral DNA and viral proteins in order to make new virus. After a time τ, called the latent period, synthesis is complete, and the bacterium lyses, releasing the new viruses. Latent periods vary by bacterial type but are usually in the half-hour to hour range. Let b denote the average number of progeny released when an infected bacterial cell lyses. The factor $e^{-D\tau}$ in the equations accounts for the fraction of infected bacteria that survive being washed out of the chemostat during the latent period.

Two important assumptions are made in formulating the model: (1) nutrient uptake by infected cells is negligible, and (2) infected cells do not grow and divide. Phage binding to infected cells has also been neglected. See [**205**] for a treatment of a more general model that includes some of these features. We have scaled out the yield constant, a positive number multiplying $f(R)$ in the equation for R.

Observe that the I equation is decoupled for the remaining equations. So we can consider only the R, S, P subsystem and use I as an auxiliary variable:

$$
\begin{aligned}
R'(t) &= D(R^\diamond - R(t)) - f(R(t))S(t), \\
S'(t) &= (f(R(t)) - D)\, S(t) - kS(t)P(t), \\
P'(t) &= -DP(t) - kS(t)P(t) + be^{-D\tau}kS(t-\tau)P(t-\tau).
\end{aligned} \tag{5.8}
$$

Nonnegative initial data for S and P must be prescribed on $[-\tau, 0]$ but only $R(0)$ need be prescribed (though it may be sometimes convenient to prescribe R on $[-\tau, 0]$ as well):

$$
\begin{aligned}
R(0) &= R_0(0), \\
S(s) &= S_0(s), \quad s \in [-\tau, 0], \\
P(s) &= P_0(s), \quad s \in [-\tau, 0],
\end{aligned} \tag{5.9}
$$

with given continuous functions $R_0, S_0, P_0 : [-\tau, 0] \to \mathbb{R}_+$.

We assume that susceptible bacteria are viable in the absence of phage. By this we mean that the phage-free system

$$
\begin{aligned}
R'(t) &= D(R^\diamond - R(t)) - f(R(t))S(t), \\
S'(t) &= (f(R(t)) - D)S(t),
\end{aligned} \tag{5.10}
$$

has a unique positive equilibrium $(\bar{R}, \bar{S})$,

$$
\bar{R} = f^{-1}(D), \qquad \bar{S} = R^\diamond - \bar{R}. \tag{5.11}
$$

Since f strictly increases, this is the case if and only if the following threshold condition holds, which we assume from now on,

$$
f(R^\diamond) > D, \tag{5.12}
$$

and which implies

$$
\bar{R} < R^\diamond. \tag{5.13}
$$

It is well known that this equilibrium attracts all solutions of (5.10) with $S(0) > 0$. Of course, (5.10) contains no delay terms so appropriate initial data are nonnegative values for $R(0), S(0)$.

System (5.8) has two boundary equilibria, namely

$$
E_R = (R^\diamond, 0, 0) \text{ and } E_S = (\bar{R}, \bar{S}, 0).
$$

In addition, it may have a (unique) positive equilibrium $E_C = (R^*, S^*, P^*)$ corresponding to the coexistence of phages and bacteria. It is determined by the relations:

$$S^* = \frac{D}{k(be^{-D\tau} - 1)}, \quad S^* = \frac{D(R^\diamond - R^*)}{f(R^*)}, \quad kP^* = f(R^*) - D. \tag{5.14}$$

Since f is strictly increasing and $f(\bar{R}) - D = 0$ and $R^\diamond = \bar{R} + \bar{S}$,

$$R^* > \bar{R} \quad \text{and } S^* < \bar{S}. \tag{5.15}$$

Existence of the coexistence equilibrium depends on the phage reproduction number

$$\bar{\mathcal{R}} := \frac{be^{-D\tau} k\bar{S}}{D + k\bar{S}}. \tag{5.16}$$

Lemma 5.25. *E_C exists with all positive components if and only if (5.12) and $\bar{\mathcal{R}} > 1$ hold.*

Proof. Suppose first that E_C exists. From (5.14) we have $\bar{R} < R^* < R^\diamond$. Since f is strictly decreasing,

$$S^* = \frac{D(R - R^*)}{f(R^*)} < \frac{D(R^\diamond - \bar{R})}{f(\bar{R})} = R^\diamond - \bar{R} = \bar{S}.$$

Therefore, $\bar{S} > S^* = \frac{D}{k(be^{-D\tau}-1)}$, implying that $\bar{\mathcal{R}} > 1$.

Conversely, if (5.12) and $\bar{\mathcal{R}} > 1$ hold, then $k\bar{S}(be^{-D\tau} - 1) > D$ and there is a unique value of $S^* \in (0, \bar{S})$ satisfying $kS^*(be^{-D\tau} - 1) = D$. But then $R^* \in (\bar{R}, R^\diamond)$ is uniquely defined by the relation in (5.14) due to monotone decrease of $R \to \frac{D(R^\diamond - R)}{f(R)}$ and similarly for P^*. □

$\bar{\mathcal{R}} = b \cdot e^{-D\tau} \cdot \frac{k\bar{S}}{D+k\bar{S}}$ has a simple biological interpretation. Imagine adding a single hypothetical phage to the chemostat at the phage-free equilibrium E_S. Two possibilities can occur: it can bind to and infect a bacterium, or it can be washed out before this happens. The probability of the first event is $\frac{k\bar{S}}{D+k\bar{S}}$. Similarly, there are two possibilities for the resulting infected cell, namely that it is washed out before releasing the phages synthesized inside it or that it releases these phages before being washed out. The probability of the latter is $e^{-D\tau}$. Therefore, the expected value of the number of progeny resulting from adding a single phage to the chemostat at equilibrium E_S is $\bar{\mathcal{R}}$. It may be called the *Critical Phage Reproduction Number or Ratio.*

More generally, we introduce the *Phage Reproduction Number at constant level S* by

$$\mathcal{R}(S) := \frac{be^{-D\tau} kS}{D + kS}. \tag{5.17}$$

The same consideration as above provides the interpretation that this is the expected value of the number of progeny resulting from adding a single phage to the chemostat if the susceptible bacteria are held at constant level S. $\mathcal{R}$ is a strictly increasing function of S, $\bar{\mathcal{R}} = \mathcal{R}(\bar{S})$, and the S-value of the coexistence equilibrium E_C in (5.14) (if it exists) is characterized by

$$\mathcal{R}(S^*) = 1. \tag{5.18}$$

The following result can be shown by a linearized stability analysis. We omit the proof because we will prove stronger results in Theorem 5.39 and Theorem 5.36.

Proposition 5.26. *Let (5.12) hold. Then E_S is locally asymptotically stable if the critical phage reproduction number satisfies $\bar{\mathcal{R}} < 1$, and unstable if $\bar{\mathcal{R}} > 1$.*

The central question is whether or not the phages and bacteria can coexist. Our aim here is to show that the phages P and bacteria S persist if (5.12) and $\bar{\mathcal{R}} > 1$ hold.

Existence of a unique nonnegative solution of the initial value problem (5.8) with nonnegative initial data (5.9) can be argued simply by the method of steps. It is simpler to explain in the abstract setting of a system

$$x'(t) = F(x(t), x(t-\tau)), \tag{5.19}$$

where $F : \mathbb{R}^6_+ \to \mathbb{R}$ is continuously differentiable and $x_3(s) = R(0)$ for $s \in [-\tau, 0]$. (5.8) has the form (5.19) where

$$F(x,y) = \begin{pmatrix} D(R^\diamond - R) - f(R)S \\ (f(R) - D)S - kSP \\ -DP - kSP + be^{-D\tau}\tilde{S}\tilde{P} \end{pmatrix} \qquad \begin{cases} x = (R, S, P), \\ y = (\tilde{R}, \tilde{S}, \tilde{P}). \end{cases}$$

Theorem 5.27. *The initial value problem (5.8) with nonnegative initial data (5.9) has a unique nonnegative solution $(R(t), S(t), P(t))$ defined for all $t \geq 0$.*

Proof. Assume that an initial condition $x(s) = \xi(s)$, $-\tau \leq s \leq 0$, is given. Then for $t \in [0, \tau]$, $x(t)$ must satisfy the nonautonomous ODE $x'(t) = F(x(t), \xi(t-\tau))$ with initial condition $x(0) = \xi(0)$. Assuming this ODE has a solution defined for all $t \in [0, \tau]$, we can repeat the process on the interval $[\tau, 2\tau]$, etc. Positivity of solutions, continuous dependence on initial data, and continuation can all be treated using standard ODE theory. Continuation of our nonnegative solution to $t \geq 0$ requires an a priori bound. To this end we define

$$I(t) = \int_{t-\tau}^{t} kS(r)P(r)e^{-D(t-r)}dr, \qquad t \geq 0. \tag{5.20}$$

Then $I(t) \geq 0$ and I satisfies the differential equation in (5.6). Let $U(t) = R(t) + S(t) + I(t) + P(t)/b$. We add the equations in (5.6),

$$U' \leq D[R^\diamond - U],$$

which implies that

$$U(t) = U(0)e^{-Dt} + R^\diamond(1 - e^{-Dt})$$

and hence

$$\limsup_{t\to\infty} \left(R(t) + S(t) + I(t) + \frac{P(t)}{b}\right) \leq R^\diamond \tag{5.21}$$

and

$$R(t) + S(t) + I(t) + \frac{P(t)}{b} \leq \max\{U(0), R^\diamond\}, \ t \geq 0. \tag{5.22}$$

Since solutions are nonnegative, they exist for all $t \geq 0$ and are bounded. □

We choose the following state space for (5.8),

$$X = C([-\tau, 0], \mathbb{R}^3_+)$$

Given $x : [-\tau, c) \to \mathbb{R}^3_+$ for $c > 0$ and $0 \leq t < c$, define $x_t \in C([-\tau, 0], \mathbb{R}^3_+)$ by $x_t(s) = x(t + s)$, $-\tau \leq s \leq 0$. This notation is traditional in the theory of delay differential equations. See [**94**]. This means that we also prescribe prehistories for R. This is not necessary, but allows a unified treatment. We now show that (5.8) induces a semiflow on X. This is most effectively done by introducing some more abstraction, a functional-differential equation,

$$x'(t) = G(x_t) \tag{5.23}$$

with $G : X = C([-\tau, 0], \mathbb{R}^3_+) \to \mathbb{R}^3$, defined by

$$G(\xi) = F(\xi(0), \xi(-\tau)), \ \xi \in X, \tag{5.24}$$

with F from (5.19).

Proposition 5.28. *Assume that the FDE $x' = G(x_t), x_0 = \xi$, has a unique solution x on $[0, \infty)$ for all $\xi \in X$. Then the definition*

$$\Phi(t, \xi) = x_t$$

defines a semiflow on X.

Proof. Let $\xi \in X$ and x the solution of $x' = G(x_t)$, $x_0 = \xi$. Let $r \geq 0$. Set $y(t) = x(t + r)$. Further, let z be the solution of $z' = G(z_t), z_0 = x_r$. We want to show that $y \equiv z$. Now $y'(t) = x'(t + r) = G(x_{t+r})$. Further, for $s \in [-\tau, 0]$, $x_{t+r}(s) = x(t + r + s) = y(t + s) = y_t(s)$. So $y'(t) = G(y_t)$ and $y_0 = x_r$. Since we have assumed that solutions are unique, $y \equiv z$ and $y_t = z_t$. By the definition of Φ, $x_r = \Phi(r, \xi)$, $z_t = \Phi(t, x_r)$, $y_t = x_{t+r} = \Phi(t + r, \xi)$. Hence, $\Phi(t + r, \xi) = \Phi(t, \Phi(r, \xi))$. □

Theorem 5.29. *Equations (5.8) and (5.9) induce a continuous semiflow* $\Phi : \mathbb{R}_+ \times X \to X$ *defined by*

$$\Phi(t, (R_0, S_0, P_0)) = (R_t, S_t, P_t).$$

The semiflow Φ *has a compact attractor of bounded sets.*

Proof. Knowing that solutions are bounded uniformly for initial data in bounded sets facilitates the proof that (5.8) generates a continuous semiflow on X. Here again the abstract setting (5.19) makes the ideas more clear. Let ϕ and ψ be two initial data satisfying

$$R(0) + S(0) + I(0) + \frac{1}{b}P(0) \leq L,$$
$$I(0) = \int_{-\tau}^{0} kS(r)P(r)e^{Dr}\,dr,$$

for some L. Let $x(t,\phi)$ and $x(t,\psi)$ be the respective solutions evaluated at time t. Since these solutions are bounded on $[0,T]$ by a constant M and since F is Lipschitz on this bounded set, after writing the integral equation satisfied by each solution, subtracting, and taking norms, we may use Gronwall's inequality to show the existence of $K = K(L,M,T) > 0$ such that

$$\sup_{0\leq t\leq T} |x(t,\phi) - x(t,\psi)| \leq K \sup_{-\tau\leq s\leq 0} |\phi(s) - \psi(s)|.$$

The semiflow Φ generated by (5.8) is point dissipative by (5.21). To show it is asymptotically smooth we use the abstract setting (5.19). If B is a forward invariant bounded closed set in X, then it follows that $|x'(t)|$ is uniformly bounded by a constant K on $t \geq 0$ independent of the solution starting in B. It follows immediately from the Ascoli-Arzela Theorem [**164**, Ch.8.3] that Φ is asymptotically compact on B because $\{x_t(\phi) : \phi \in B, t \geq \tau\}$ is an equicontinuous and equibounded subset of $C([-\tau,0],\mathbb{R}^3)$. Since B was arbitrary with the indicated properties, Φ is asymptotically smooth. Finally, Φ is eventually bounded on bounded sets. Indeed, if B is bounded in X, then its elements satisfy a uniform bound $R(0)+S(0)+I(0)+\frac{1}{b}P(0) \leq M$ for some $M > 0$ dependent only on B. Then (5.22) implies the boundedness of $\Phi([0,\infty)\times B)$. Theorem 2.52 shows that the semiflow induced by (5.8) has a compact attractor of bounded sets. □

It is intuitively obvious that the presence of phages should keep bacterial density below the level that it could achieve without phages. To see this, we consider bounded total Φ-trajectories. By Theorem 5.29, there exists a compact attractor of bounded sets which equals the union of the ranges of the total trajectories (Theorem 2.20). Bounded total Φ-trajectories correspond to nonnegative solutions of (5.6) that exist and are bounded on all of $\mathbb{R}$.

Proposition 5.30. *Let $R, S, I, P : \mathbb{R} \to \mathbb{R}_+$ be bounded and solve (5.6) for all $t \in \mathbb{R}$. Then*

$$R(t) + S(t) + I(t) + \frac{P(t)}{b} \le R^\diamond, \quad S(t) + I(t) \le \bar{S}, \quad R(t) \ge \bar{R}, \qquad t \in \mathbb{R}.$$

Corollary 5.31. *Let $c > 0$, $\epsilon > 0$. Then there exists some $r > 0$ such that*

$$R(t)+S(t)+I(t)+\frac{P(t)}{b} \le R^\diamond+\epsilon, \quad S(t)+I(t) \le \bar{S}+\epsilon, \quad R(t) \ge \bar{R}-\epsilon,$$

for all $t \ge r$ and all nonnegative solutions of (5.6) on $\mathbb{R}_+$ with $R(0)+I(0) \le c$ and $S(s) + P(s) \le c$, $s \in [-\tau, 0]$.

Proof of Proposition 5.30. Let $U(t) = R(t) + S(t) + I(t) + \frac{P(t)}{b}$. Then U is bounded on $\mathbb{R}$ and, as we noted before, $U' \le D[R^\diamond - U]$, but this time for all $t \in \mathbb{R}$. Notice that, for any $t \in \mathbb{R}$, if $U(t) > R^\diamond$, then $U'(t) < 0$.

Suppose that there is some $t \in \mathbb{R}$ with $U(t) > R^\diamond$. Then $U'(t) < 0$ and U is strictly decreasing in a neighborhood of t.

We claim that $U(r) \ge U(t)$ for all $r \le t$. If not, then there exists some $r < t$ such that $U(r) < U(t)$. Then U takes its maximum on $[r, t]$ at some point $s \in [r, t]$. Since U is strictly decreasing in a neighborhood of t, $U(s) > U(t) > U(r)$. So U also takes a local maximum at s and $U'(s) = 0$. But $U(s) > U(t) > R^\diamond$ and so $U'(s) < 0$, a contradiction.

Since $U(r) \ge U(t)$ for all $r \le t$,

$$U'(r) \le D[R^\diamond - U(r)] \le D[R^\diamond - U(t)] < 0$$

and U' is negative and bounded away from 0 on $(-\infty, t]$. This contradicts the boundedness of U on $\mathbb{R}$. So $U(t) \le R^\diamond$ for all $t \in \mathbb{R}$.

We turn to $S + I =: M$. We add the differential equations,

$$M' \le f(R)S - DM \le f(U - M)M - DM \le [f(R^\diamond - M) - D]M.$$

Here we have used twice that f is increasing and our previous result that $U \le R^\diamond$. Notice that, if $t \in \mathbb{R}$ and $M(t) > \bar{S}$, then

$$M'(t) < [f(R^\diamond - \bar{S}) - D]M = [f(\bar{R}) - D]M = 0.$$

Recall (5.11). Since f is strictly increasing, the same arguments as before show that this cannot occur because it implies that M is unbounded, a contradiction. So $S + I = M \le \bar{S}$ on $\mathbb{R}$.

We turn to R. From the differential equation for R, using that $S \le \bar{S}$,

$$R' \ge D(R^\diamond - R) - f(R)\bar{S}.$$

Notice that, for all $t \in \mathbb{R}$, $R(t) < \bar{R}$ implies

$$R'(t) > D(R^\diamond - \bar{R}) - f(\bar{R})\bar{S} = 0.$$

Recall again (5.11). Assume that $R(t) < \bar{R}$ for some $t \in \mathbb{R}$. Then $R'(t) > 0$ and R is strictly increasing in a neighborhood of t.

We claim that $R(r) \le R(t)$ for all $r \le t$. If not, then there exists some $r < t$ such that $R(r) > R(t)$. R takes its minimum on $[r,t]$ at some point $s \in [r,t]$. Since R is strictly increasing in a neighborhood of t, $R(s) \le R(t) < R(r)$. So R also takes a local minimum at s and $R'(s) = 0$. But $R(s) \le R(t) < \bar{R}$, and so $R'(s) > 0$, a contradiction.

Since $R(r) \le R(t)$ for all $r \le t$,

$$R'(r) \ge D(R^\diamond - R(t)) - f(R(t))\bar{S} = R'(t) > 0, \qquad r < t.$$

So R' is positive and bounded away from 0 on $(-\infty, t]$. This contradicts the nonnegativity of R on $\mathbb{R}$. □

It is apparent that, if phages go extinct, then solutions with $S(0) > 0$ converge to E_S. The fluctuation method makes this rigorous.

Lemma 5.32. *If $P(t) \to 0$ as $t \to \infty$, then $R(t) \to \bar{R}$ and $S(t) \to \bar{S}$ as $t \to \infty$, for all solutions of (5.6) with $S(0) > 0$.*

Proof. Define $N = R + S$. By (5.6),

$$\begin{aligned} N' &= D(R^\diamond - N) - kPS, \\ S' &= (f(N-S) - D - kP)S. \end{aligned}$$

We apply the fluctuation method to the differential equation for N and obtain $0 = D(R^\diamond - N_\infty) = D(R^\diamond - N^\infty)$. So $N(t) \to R^\diamond$ as $t \to \infty$. Since $f(R^\diamond) > D$ and $S(t) > 0$ for all $t \ge 0$, $S'(t) > 0$ if $t > 0$ is sufficiently large and $S(t)$ is sufficiently small. Thus $S_\infty > 0$. The fluctuation method applied to the differential equation for S provides $0 = (f(R^\diamond - S_\infty) - D)S_\infty$. Thus $f(R^\diamond - S_\infty) = D = f(\bar{R})$. Since f is strictly increasing, $R^\diamond - S_\infty = \bar{R} = R^\diamond - \bar{S}$. This implies $S_\infty = \bar{S}$. Since $S^\infty \le \bar{S}$ by Corollary 5.31, $S(t) \to \bar{S}$ as $t \to \infty$. □

Next we show that the susceptible bacteria persist uniformly.

Theorem 5.33. *If $S(0) > 0$, then*

$$S^\infty \ge \min\Big\{\frac{De^{D\tau}}{bk}, \bar{S}\Big\}.$$

Further, the susceptible bacteria persist uniformly: There exists some $\epsilon_S > 0$ such that $S_\infty \ge \epsilon_s$ for all solutions with $S(0) > 0$.

Proof. Suppose that S^∞ is smaller than this minimum. We apply the fluctuation method to the differential equation for P,

$$0 \le -DP^\infty + be^{-D\tau}kS^\infty P^\infty.$$

Since $be^{-D\tau}kS^{\infty} < D$, $P^{\infty} = 0$. By Lemma 5.32, $S(t) \to \bar{S}$ as $t \to \infty$, a contradiction.

For $x_0 = (R_0, S_0, P_0) \in C([-\tau, 0], \mathbb{R}^3_+)$, define $\rho(x_0) = S_0(0)$. Then $\rho(\Phi(t, x_0)) = S(t)$. Notice that $\rho(x_0) > 0$ implies $\rho(\Phi(t, x_0)) > 0$ for all $t \geq 0$. By choice of ρ and our previous result, the semiflow Φ is uniformly weakly ρ-persistent. Since Φ has a compact attractor of bounded sets, Φ is uniformly ρ-persistent by Theorem 5.2. □

The analysis of the differential equation for P in (5.6) faces the difficulty that the SP-terms come with different signs. For this reason we apply the Laplace transform,

$$\hat{P}(\lambda) = \int_0^\infty e^{-\lambda t} P(t)dt,$$

which exists for all $\lambda > 0$ because P is bounded,

$$\lambda \hat{P}(\lambda) - P(0) = -D\hat{P}(\lambda) - k\widehat{SP}(\lambda) + kbe^{-D\tau} \int_0^\infty e^{-\lambda t} S(t-\tau)P(t-\tau)dt.$$

Notice that

$$\begin{aligned}\int_0^\infty e^{-\lambda t} S(t-\tau)P(t-\tau)dt &= \int_{-\tau}^\infty e^{-\lambda(t+\tau)} S(t)P(t)dt \\ &= e^{-\lambda\tau}\Big(\widehat{SP}(\lambda) + \int_{-\tau}^0 e^{-\lambda t} S(t)P(t)dt\Big).\end{aligned}$$

We substitute this equation into the previous one,

$$\begin{aligned}(\lambda + D)\hat{P}(\lambda) =& k\big(be^{-(D+\lambda)\tau} - 1\big)\widehat{SP}(\lambda) + P(0) \\ &+ kbe^{-(D+\lambda)\tau} \int_{-\tau}^0 e^{-\lambda t} S(t)P(t)dt.\end{aligned} \tag{5.25}$$

In the following, we will make frequent use of the phage reproduction function $\mathcal{R}$ in (5.17), sometimes without further reference.

Proposition 5.34. *If $P(0) > 0$, then $\mathcal{R}(S_\infty) \leq 1$.*

Proof. Let $\delta > 0$. After a shift in time we can assume that $S(t) \geq S_\infty - \delta$ for all $t \geq 0$ and so $\widehat{SP}(\lambda) \geq (S_\infty - \delta)\hat{P}(\lambda)$. We substitute this inequality into (5.25),

$$(\lambda + D)\hat{P}(\lambda) \geq k[be^{-(D+\lambda)\tau} - 1]_+(S_\infty - \delta)\hat{P}(\lambda).$$

Here $[r]_+ = \max\{r, 0\}$ denotes the positive part of a number r. Since $P(0) > 0$, $P(t) > 0$ for all $t \geq 0$. We divide by $\hat{P}(\lambda) > 0$,

$$\lambda + D \geq k[be^{-(D+\lambda)\tau} - 1]_+(S_\infty - \delta).$$

Since this inequality holds for all $\lambda > 0$ and $\delta > 0$, we can take the limit as $\lambda, \delta \to 0$,

$$D \geq k[be^{-D\tau} - 1]_+ S_\infty.$$

Recall (5.17) and conclude that $\mathcal{R}(S_\infty) \leq 1$. □

Proposition 5.35. *If $\mathcal{R}(S^\infty) < 1$, then $\int_0^\infty P(t)dt < \infty$ and $P(t) \to 0$ as $t \to \infty$.*

Proof. By (5.25), since $\lambda > 0$,

$$D\hat{P}(\lambda) \leq k\big[be^{-D\tau} - 1\big]_+ \widehat{SP}(\lambda) + kbe^{-D\tau} \int_{-\tau}^0 S(t)P(t)dt + P(0).$$

Here $[r]_+$ denotes the positive part of a real number r, $[r]_+ = \max\{r, 0\}$. Since $\mathcal{R}(S^\infty) < 1$, we find some $S^\sharp > S^\infty$ such that $\mathcal{R}(S^\sharp) < 1$. Pick some $P^\sharp > bR^\diamond$. After a shift in time we can assume that $S(t) \leq S^\sharp$ and $P(t) \leq P^\sharp$ for all $t \geq -\tau$ (Corollary 5.31). In particular, $\widehat{SP}(\lambda) \leq S^\sharp \hat{P}(\lambda)$. We substitute these inequalities into the last displayed equation,

$$\Big(D - kS^\sharp\big[be^{-D\tau} - 1\big]_+\Big)\hat{P}(\lambda) \leq kbe^{-D\tau}\tau S^\sharp P^\sharp + P(0).$$

Since $\mathcal{R}(S^\sharp) < 1$ by assumption, the expression on the left-hand side is positive and

$$\hat{P}(\lambda) \leq \frac{kbe^{-D\tau}\tau S^\sharp P^\sharp + P(0)}{D - kS^\sharp\big[be^{-D\tau} - 1\big]_+}.$$

By Beppo Levi's theorem of monotone convergence we can take the limit as $\lambda \to 0$ and obtain that $\hat{P}(0) = \int_0^\infty P(t)dt < \infty$. Since the derivative of P is bounded and thus P is uniformly continuous on $\mathbb{R}_+$, this implies $P(t) \to 0$ as $t \to \infty$ (Exercise 5.2). □

We are now in a position to address uniform persistence of phage.

Theorem 5.36. *Let $\bar{\mathcal{R}} > 1$. Then the susceptible bacteria and the phages persist uniformly: There exists some $\epsilon > 0$ such that $S_\infty \geq \epsilon$ and $P_\infty \geq \epsilon$ for all solutions with $S(0) > 0$ and $P(0) > 0$.*

Proof. We define $\rho : X \to \mathbb{R}_+$ by

$$\rho(x_0) = \min\{S_0(0), P_0(0)\}, \qquad x_0 = (R_0, S_0, P_0).$$

Then $\rho(\Phi(t, x_0)) = \min\{S(t), P(t)\}$. It is sufficient to show that Φ is uniformly ρ-persistent.

Notice that $\rho(x_0) > 0$ implies $\rho(\Phi(t, x_0)) > 0$ for all $t \geq 0$. Since Φ has a compact attractor of bounded sets by Theorem 5.29, it is sufficient by Theorem 5.2 to show that Φ is uniformly weakly ρ-persistent.

Suppose not. Let $\epsilon > 0$, to be determined later. Then there exists a solution with $S(0) > 0$ and $P(0) > 0$ such that $\min\{S, P\}^\infty < \epsilon$. Then

$S(t) > 0$ and $P(t) > 0$ for all $t \geq 0$. Recall that the susceptible bacteria persist uniformly (Theorem 5.33): There exists some $\epsilon_S > 0$ such that $S_\infty > \epsilon_S$. Let $\epsilon \in (0, \epsilon_S)$. Then $P^\infty < \epsilon$.

We apply the fluctuation method to the system for $M = R + S$ and S,

$$\begin{aligned} M' &= D(R^\diamond - M) - kSP, \\ S' &= (f(M - S) - D)S - kSP, \end{aligned}$$

and use $S \leq M$, to obtain

$$\begin{aligned} 0 \geq& D(R^\diamond - M_\infty) - kM_\infty\epsilon, \\ 0 \geq& (f(M_\infty - S_\infty) - D)S_\infty - kS_\infty\epsilon. \end{aligned}$$

Let $\delta > 0$, to be determined. Choosing $\epsilon > 0$ small enough, we obtain from the inequalities above that

$$M_\infty > R^\diamond - \delta, \qquad S_\infty > \bar{S} - \delta.$$

Since $\bar{\mathcal{R}} = \mathcal{R}(\bar{S}) > 1$ and $\mathcal{R}$ in (5.17) is strictly increasing, we can choose $\delta > 0$ so small that $\mathcal{R}(S_\infty) > \mathcal{R}(\bar{S} - \delta) > 1$. This contradicts Proposition 5.34 which provides $\mathcal{R}(S_\infty) \leq 1$. □

Corollary 5.37. *If* $\bar{\mathcal{R}} > 1$, *then* $0 < S_\infty \leq S^* \leq S^\infty \leq \bar{S}$ *for all nonnegative solutions of (5.6) with* $S(0) > 0$ *and* $P(0) > 0$.

Proof. Consider a solution with $S(0) > 0$ and $P(0) > 0$. Then $\mathcal{R}(S_\infty) \leq 1$ by Proposition 5.34. By Theorem 5.36, $S_\infty > 0$ and $P_\infty > 0$. So $\mathcal{R}(S^\infty) \geq 1$ by the contraposition of Proposition 5.35. Since $\mathcal{R}$ is strictly increasing and $\mathcal{R}(S^*) = 1$, $S_\infty \leq S^* \leq S^\infty$. The estimate $S^\infty \leq \bar{S}$ follows from Corollary 5.31. □

The persistence condition $\bar{\mathcal{R}} > 1$ is sharp because the following holds.

Theorem 5.38. *Let* $\bar{\mathcal{R}} < 1$. *Then the phage goes extinct:* $\int_0^\infty P(t)\,dt < \infty$ *and* $P(t) \to 0$ *as* $t \to \infty$. *Further, all solutions with* $S(0) > 0$ *converge to the boundary equilibrium* $E_S = (\bar{R}, \bar{S}, 0)$.

Proof. By Corollary 5.31, $S^\infty \leq \bar{S}$. Since $\mathcal{R}$ is strictly increasing, $1 > \bar{\mathcal{R}} = \mathcal{R}(\bar{S}) > \mathcal{R}(S^\infty)$. The first part of the assertion now follows from Proposition 5.35. If $S(0) > 0$, by Lemma 5.32, $R(t) \to \bar{R}$ and $S(t) \to \bar{S}$. □

The attraction properties of the phage-free equilibrium can be improved by using the results from Section 5.6.

Theorem 5.39. *Let* $\bar{\mathcal{R}} < 1$. *Then the phage-free equilibrium* $E_S = (\bar{R}, \bar{S}, 0)$ *is stable. Further, if* $c > \epsilon > 0$, $(R(t), S(t), P(t)) \to E_S$ *as* $t \to \infty$ *uniformly for all nonnegative solutions of (5.8) with* $R(0) \leq c$, $S(0) \geq \epsilon$, *and* $S(s) + P(s) \leq c$, $-\tau \leq s \leq c$.

The rest of this section is devoted to the proof of this result. The uniformity of convergence requires that we redo the Laplace transform analysis with truncated Laplace transforms.

Lemma 5.40. *Let $\bar{\mathcal{R}} < 1$. Then $P(t) \to 0$ as $t \to \infty$ uniformly for all nonnegative solutions of (5.8) that exist and are bounded on all of $\mathbb{R}$.*

Proof. Consider a solution (R, S, I, P) of (5.6) that is defined and bounded on $\mathbb{R}$. Then, for $\eta > 0$,

$$\begin{aligned}\int_0^T e^{\eta t} P'(t)dt =& - D \int_0^T e^{\eta t} P(t)dt - k \int_0^T e^{\eta t} S(t)P(t)dt \\ &+ be^{-D\tau} k \int_0^T e^{\eta t} S(t-\tau)P(t-\tau)dt.\end{aligned}$$

We integrate by parts and do appropriate substitutions,

$$\begin{aligned}& e^{\eta T} P(T) - P(0) - \eta \int_0^T e^{\eta t} P(t)dt \\ =& - D \int_0^T e^{\eta t} P(t)dt - k \int_0^T e^{\eta t} S(t)P(t)dt \\ &+ be^{-D\tau} k \int_{-\tau}^{T-\tau} e^{\eta(t+\tau)} S(t)P()dt.\end{aligned}$$

Since P and S are nonnegative, we obtain the following estimate for $T > \tau$ after some reorganization,

$$\begin{aligned}(D-\eta) \int_0^T e^{\eta t} P(t)dt \le & P(0) + k\big(be^{-(D-\eta)\tau} - 1\big) \int_0^T e^{\eta t} S(t)P(t)dt \\ &+ kbe^{-(D-\eta)\tau} \int_{-\tau}^0 e^{\eta t} S(t)P(t)dt.\end{aligned}$$

By Lemma 5.30, $S(t) \le \bar{S}$ and $P(t) \le bR^\diamond$.

$$\begin{aligned}(D-\eta) \int_0^T e^{\eta t} P(t)dt \le & P(0) + k\big[be^{-(D-\eta)\tau} - 1\big]_+ \int_0^T e^{\eta t} \bar{S} P(t)dt \\ &+ kbe^{-(D-\eta)\tau} \int_{-\tau}^0 e^{\eta t} dt\, b(R^\diamond)^2.\end{aligned}$$

We reorganize,

$$\begin{aligned}& \big(D - \eta - k\big[be^{-(D-\eta)\tau} - 1\big]_+ \bar{S}\big) \int_0^T e^{\eta t} P(t)dt \\ \le & P(0) + kbe^{-(D-\eta)\tau} \int_{-\tau}^0 e^{\eta t} dt\, b(R^\diamond)^2.\end{aligned}$$

Since $\bar{\mathcal{R}} < 1$,

$$D - \eta - k\big[be^{-(D-\eta)\tau} - 1\big]_+ \bar{S} > 0$$

for $\eta = 0$ and so also for small $\eta \in (0, D)$. We divide by this positive term and obtain

$$\int_0^T e^{\eta t} P(t)dt \leq c_\eta$$

where c_η is a uniform bound for all $T > \tau$ and all solutions of (5.6) that are defined and bounded on all of $\mathbb{R}$. By Beppo Levi's lemma of monotone convergence,

$$\int_0^\infty e^{\eta t} P(t)dt \leq c_\eta.$$

For such solutions,

$$P'(t) \leq -DP(t) - ke^{-D\tau} bS(t-\tau)P(t-\tau).$$

We recall $S(t) \leq \bar{S}$ and integrate

$$P(t) \leq P(0)e^{-Dt} + ke^{-D\tau} b\bar{S} \int_0^t e^{-D(t-s)} P(s-\tau)ds.$$

We recall $P(s) \leq bR^\diamond$ for all $s \in R$,

$$P(t) \leq bR^\diamond e^{-Dt} + ke^{-D\tau} b^2 \bar{S} R^\diamond \tau + kb\bar{S} \int_0^t e^{-D(t-s)} P(s)ds.$$

Let $0 < T < t$. Then

$$\begin{aligned}\int_0^t e^{-D(t-s)} P(s)ds \leq & \int_0^T e^{-D(t-s)} P(s)ds + \int_T^t e^{-\eta s} P(s) e^{\eta s} ds \\ \leq & e^{-D(t-T)} \int_0^\infty P(s)ds + e^{-\eta T} \int_0^\infty P(s) e^{\eta s} ds \\ \leq & (e^{-D(t-T)} + e^{-\eta T}) c_\eta.\end{aligned}$$

We combine the last two estimates and obtain that $P(t) \to 0$ as $t \to \infty$ uniformly for all nonnegative solutions of (5.8) that exist and are bounded on all of $\mathbb{R}$. □

Lemma 5.41. *If $R, S : \mathbb{R} \to \mathbb{R}_+$ are bounded and solve the phage-free system (5.10) and $\inf S(\mathbb{R}) > 0$, then $R \equiv \bar{R}$ and $S \equiv \bar{S}$.*

Proof. We already know from Lemma 5.30 that $R \geq \bar{R}$, $S \leq \bar{S}$ and $S + R \leq R^\diamond$. Let $Q = R + S$. We add the differential equation in (5.10),

$$Q' = D(R^\diamond - Q), \qquad S' = (f(Q - S) - D)S. \tag{5.26}$$

If Q is not identically equal to $R^\diamond$, either $Q(t) > R^\diamond$ for all $t \in \mathbb{R}$ or $Q(t) < R^\diamond$ for all $t \in R$ because of uniqueness of solutions. In the first case, Q is strictly decreasing and not bounded in backward time; in the second case Q is strictly increasing and not bounded in forward time. So $Q \equiv R^\diamond$ and

$$S' = f(R^\diamond - S) - D)S.$$

We already know that $S(t) \leq \bar{S}$ for all $t \in \mathbb{R}$ and have assumed that S is bounded away from 0. Notice that the right-hand side is 0 for $S = \bar{S}$. If S is not identically equal to $\bar{S}$ or to 0, $0 < S(t) < \bar{S}$ for all $t \in \mathbb{R}$ and S is strictly increasing. This implies that S'/S is bounded away from 0 on $(-\infty, 0]$ and $S(t) \to 0$ as $t \to -\infty$, a contradiction. □

Proof of Theorem 5.39. We apply Theorem 5.23 and the remark thereafter. Recall the semiflow Φ established in Theorem 5.28. The total trajectories have the form $\phi(t) = (R_t, S_t, P_t)$ where R, S, P are solutions of (5.8) that are defined on all of $\mathbb{R}$. For $x = (R, S, P) \in X = C([-\tau, 0], \mathbb{R}^3_+)$ we define

$$\rho(x) = S(0), \qquad \tilde{\rho}(x) = P(0).$$

Identify $X_0 = \{S \equiv 0\}$ and $\tilde{X}_0 = \{P \equiv 0\}$. By Theorem 5.33, Φ is uniformly ρ-persistent. Further, $\rho(\phi(t)) = S(t)$ and $\tilde{\rho}(\phi(t)) = P(t)$, $t \in \mathbb{R}$, if ϕ is a total Φ-trajectory. By Lemma 5.40, $\tilde{\rho}(\phi(t)) \to 0$ uniformly for all total Φ-trajectories with range in the compact attractor A.

By Theorem 5.23 and the subsequent remark, there exists a unique uniformly ρ-positive compact invariant set $\tilde{A}$ in $\tilde{X}_0$ which attracts all compact subsets of $\tilde{X}_0 \setminus X_0$. Further, $\tilde{A}$ is stable. By Lemma 5.41, $\tilde{A}$ is the singleton set containing the phage-free equilibrium $E_S = (\bar{R}, \bar{S}, 0)$.

This means that $\tilde{A} = \{E_S\}$ attracts all compact sets in $\tilde{X}_0 \setminus X_0$. By Theorem 5.23, E_S is stable and $\{E_S\}$ attracts all compact sets in $X \setminus X_0$.

Let B be a bounded uniformly ρ-positive set with bound c. Since

$$S' \geq -(D + kP)S \geq -(D + \tilde{c})S, \qquad \tilde{c} = k \max\{c, bR^\diamond\},$$

by (5.22), we have $S(t) \geq S(0)e^{-(D+\tilde{c})t}$. So the closure of $\Phi_\tau(B)$ is contained in $X \setminus X_0$. It is compact by the proof of Theorem 5.29 and thus attracted by $\{E_S\}$. Hence, $\{E_S\}$ attracts B. This finishes the proof of Theorem 5.39. □

5.8. Exercises

Exercise 5.1. Formulate a resource-bacterium-phage model where the time between infection and lysis is not fixed but infected bacteria burst at a constant rate.

Exercise 5.2. Let $u : \mathbb{R}_+ \to \mathbb{R}_+$ be uniformly continuous and $\int_0^\infty u(t)dt < \infty$. Show: $u(t) \to 0$ as $t \to \infty$.

Exercise 5.3. Redo the analysis in this chapter for the ODE model you formulated in Exercise 5.1.

Exercise 5.4. Show that if $f(R^\diamond) < D$ holds, then $S(t) \to 0$ and $P(t) \to 0$ for (5.8). Use $\limsup_{t\to\infty} R(t) \leq R^\diamond$.

Chapter 6

Existence of Nontrivial Fixed Points via Persistence

If a uniformly ρ-persistent semiflow has a compact attractor of compact sets and some more conditions are satisfied, the semiflow has a ρ-positive fixed point. The additional assumptions given in [**158, 238**] entail that the set $X \cap \{\rho > 0\}$ is forward invariant and convex and that Φ_t is compact (or condensing) for every $t \in J$, $t > 0$. If $J = \mathbb{Z}_+$, it is enough that Φ_1 is compact or condensing. Below we will drop the forward invariance of $X \cap \{\rho > 0\}$, but assume that ρ is concave. A possible application of such a modification may be the endemic model with infection age (Chapter 9); but there we will be able to prove the existence of an endemic equilibrium directly, while proving the additional compactness type conditions for Φ would be a challenge.

The equilibria of autonomous semiflows with time-set $\mathbb{R}_+$ often obey equations which can be much more easily analyzed than the equations inducing the semiflow. This even holds for general physiologically structured population models [**57**]. Using persistence to show existence of nontrivial equilibria would not only be redundant, but sometimes counterproductive because one needs to extract information from the equilibrium equations to study the stability of the equilibria. See Section 4.7. Most of the models we have presented so far or will present belong to those which allow a relatively easy analysis of equilibria. One exception is the SEIRS endemic model in a patchy host population (Chapter 3.2) where we have used uniform weak persistence to derive the existence of endemic equilibria (alternatively we could have waited until we proved uniform persistence and then used the result in [**158, 238**]). Similarly it may be hardly effective to show the existence of nontrivial fixed points for maps using persistence theory, the more so as a host of fixed point results is available in the literature including Krasnosel'skii's classic work [**128**, Chap.4.2]. This holds in particular for nonlinear matrix models (Chapter 7.2). The phenomenon that uniform persistence often implies the existence of a nontrivial fixed point can still be useful, however, as it shows that necessary assumptions for the existence of the nontrivial fixed point are also necessary for uniform persistence.

A more interesting class of semiflows for using persistence to establish the existence of nontrivial equilibria are the discrete semiflows that result as period-maps of periodic differential equations or as time-maps in metered models. Examples for the first kind can be found in [**158**, Sec.5.4] and [**238**, Chap.3.1, Chap.5, 8].

6.1. Nontrivial fixed points in the global compact attractor

Definition 6.1. Let X be the closed convex subset of a Banach space and $T : X \to X$ continuous. Let β be the ball or the diameter measure of noncompactness. Then T is called *condensing* if $\beta(T(K)) < \beta(K)$ for every bounded subset K of X with $\beta(K) > 0$.

Theorem 6.2. *We assume that $J = \mathbb{Z}_+$ or $J = \mathbb{R}_+$ and that the semiflow $\Phi : J \times X \to X$ is continuous. Further, let X be the closed convex subset of a Banach space and let $\rho : X \to \mathbb{R}_+$ be continuous and concave.*

We assume that Φ is uniformly weakly ρ-persistent, has a compact attractor A of neighborhoods of compact sets in X and that Φ_t is condensing for $t = 1$ if $J = \mathbb{Z}_+$ and for all $t \in (0, 1]$ if $J = \mathbb{R}_+$.

Finally, we assume that

(H1) *there exists no total Φ-trajectory with range in A such that $\rho(\phi(-r)) > 0$, $\rho(\phi(0)) = 0$, and $\rho(\phi(s)) > 0$ with $r, s \in J$.*

Then there exists some $x^\diamond \in X$ with $\rho(x^\diamond) > 0$ and $\Phi(t, x^\diamond) = x^\diamond$ for all $j \in J$.

Proof. By Theorem 5.6 and Theorem 5.7, there exists a compact uniformly ρ-positive set A_1 which is the compact attractor of neighborhoods of compact sets in $X \setminus X_0$. X_0 is possibly empty. By Proposition 5.9 and its proof, the convex closed hull of A_1, denoted by K, is uniformly ρ-positive and compact. We choose an open neighborhood B of K close enough to K that it is attracted to A_1 and uniformly ρ-positive. Since Φ is continuous, we can also arrange that $\Phi(J \times B)$ is bounded. Let S be the closed convex hull of

$\Phi(J \times B)$. Then S is closed, convex, and bounded and $\Phi_t(B) \subset S$ for all $t \in J$.

Let us first assume that $J = \mathbb{Z}_+$. Then Φ_1 is condensing and has a fixed point $x^\diamond$ in B by the Hale-Lopes fixed point theorem [**91**, L.2.6.5, L.2.6.6] [**238**, L.1.3.4]. Since B is uniformly ρ-positive, $\rho(x^\diamond) > 0$.

Now let $J = \mathbb{R}_+$ and let Φ_t be condensing for all $t \in (0,1]$. Choose a sequence (s_n) in $(0,1]$ with $s_n \to 0$ as $n \to \infty$. By the Hale-Lopes fixed point theorem, for each $n \in \mathbb{N}$ there exists some $x_n \in X$ with $\Phi(s_n, x_n) = x_n$, $\rho(x_n) > 0$. Since A_1 attracts x_n, $x_n \in A_1$. Since A_1 is compact, after choosing a subsequence, $x_n \to x^\diamond$ as $n \to \infty$ with some $x^\diamond \in A_1$. Let $t > 0$. For every $n \in \mathbb{N}$ there exist some $k_n \in \mathbb{Z}_+$ such that $k_n s_n \le t \le (k_n + 1)s_n$. Since $s_n \to 0$, $k_n s_n \to t$. Since Φ is continuous,

$$\Phi(t, x^\diamond) = \lim_{n\to\infty} \Phi(k_n s_n, x_n) = \lim_{n\to\infty} x_n = x^\diamond. \qquad \square$$

We relax the assumption that ρ is concave by working with a second persistence function. Recall $\hat{J} = J \cup (-J)$.

Theorem 6.3. *We assume that $J = \mathbb{Z}_+$ or $J = \mathbb{R}_+$ and that the semiflow $\Phi : J \times X \to X$ is continuous. Further, let X be the closed convex subset of a Banach space and $\rho, \tilde{\rho} : X \to \mathbb{R}_+$ be continuous and $\tilde{\rho}$ concave.*

We assume that Φ is uniformly weakly ρ-persistent, has a compact attractor of neighborhoods of compact sets in X and that Φ_t is condensing for $t = 1$ if $J = \mathbb{Z}_+$ and for all $t \in (0,1]$ if $J = \mathbb{R}_+$.

Finally, we assume that

- **(H1)** *there exists no total Φ-trajectory with pre-compact range such that $\rho(\phi(-r)) > 0$, $\rho(\phi(0)) = 0$, and $\rho(\phi(s)) > 0$ with $r, s \in J$,*
- ($\tilde{\mathbf{H}}$1) *if ϕ is a total Φ-trajectory with pre-compact range and* $\inf \rho(\phi(\hat{J}))$ > 0, *then $\tilde{\rho}(\phi(0)) > 0$,*
- ($\tilde{\mathbf{H}}$2) *every compact convex set that is uniformly $\tilde{\rho}$-positive is also uniformly ρ-positive.*

Then there exists some $x^\diamond \in X$ with $\rho(x^\diamond) > 0$, $\tilde{\rho}(x^\diamond) > 0$, and $\Phi(t, x^\diamond) = x^\diamond$ for all $j \in J$.

Proof. We sketch how the previous proof needs to be modified. We first show that the persistence attractor A_1 is uniformly $\tilde{\rho}$-positive. Let $x \in A_1$. Then there exists a total Φ-trajectory ϕ with range in A_1 and $\phi(0) = x$. Since A_1 is uniformly ρ-positive, $\inf \rho \circ \phi > 0$. By ($\tilde{\mathbf{H}}_1$), $\tilde{\rho}(x) > 0$. Since A_1 is positive and $\tilde{\rho}$ is continuous, A_1 is uniformly $\tilde{\rho}$-positive.

Since $\tilde{\rho}$ is concave, by Proposition 5.9 and its proof, the closed convex hull K of A_1 is uniformly $\tilde{\rho}$-positive and compact. By ($\tilde{\mathbf{H}}$2), K is uniformly

ρ-positive. Now the proof proceeds as for Theorem 6.2 and Φ has a fixed point $x^\diamond$ with $\rho(x^\diamond) > 0$. Since $x^\diamond$ is attracted to A_1, $x_1 \in A_1$ and $\tilde{\rho}(x^\diamond) > 0$. □

6.2. Periodic solutions of the Lotka-Volterra predator-prey model

We present an example for which one cannot use the results in [**158, 238**] or Theorem 6.2 because there is not even a compact attractor of points. Our procedure should work in many other situations, but it is quite cumbersome to formulate it for a general scenario. So we choose an easy-looking example and hope to convey the main idea clearly enough that it can be used in other contexts.

We consider a periodic Lotka-Volterra predator-prey system,

$$\begin{aligned} P' &= \gamma(t)P - PQ, \\ Q' &= PQ - \mu(t)Q. \end{aligned} \tag{6.1}$$

The per capita predation rate and biomass conversion rate are assumed to be constant and have been scaled out. The function γ gives the time-dependent per unit growth rate of prey biomass and $\mu(t)$ the per unit loss rate of predator biomass. The functions γ and μ are assumed to be bounded and continuous and μ strictly positive. We further assume that γ and μ are periodic with the same period τ and $\int_0^\tau \gamma(t)dt > 0$. Since the system with constant coefficients is not uniformly persistent, there is little hope that the system with periodic coefficients is uniformly persistent. So we cannot use the results in [**158**] or [**238**].

We will combine techniques for establishing uniform weak persistence with suitable perturbations of the system,

$$\begin{aligned} P' &= \gamma(t)P - PQ + \delta - \delta P^2, \\ Q' &= PQ - \mu(t)Q + \delta - \delta Q^2, \end{aligned} \tag{6.2}$$

where $\delta \in (0,1)$. We have used this approach before in Theorem 3.11, but there we can alternatively prove uniform persistence and the general results that uniform persistence implies the existence of a nontrivial fixed point. The point of the next proposition is that the estimates are independent of the perturbation parameter δ.

Proposition 6.4. *There exist* $0 < \epsilon < 1 < c < \infty$ *such that*

$$\liminf_{t\to\infty} P(t) \le c, \qquad \limsup_{t\to\infty} P(t) \ge \epsilon$$

and

$$\liminf_{t\to\infty} Q(t) \le c$$

for all solutions P and Q and all $\delta \in (0,1)$.

Proof. Let $c > 1$, to be determined later and $\liminf_{t\to\infty} P(t) > c$. After a shift in time, we can assume that $P(t) \geq c$ for all $t \geq 0$. Then

$$Q' \geq (c - \mu^\diamond)Q(t) - \delta Q^2, \qquad \mu^\diamond = \sup \mu.$$

Q is bounded away from 0 because $\delta > 0$. Assume $c > \mu^\diamond$. By the fluctuation method, $Q_\infty \geq (c - \mu^\diamond)/\delta > c - \mu^\diamond$.

Since $c > 1$ and $P(t) \geq c$,

$$P' \leq \gamma^\diamond P - PQ, \qquad \gamma^\diamond = \sup \gamma.$$

Choose $c > \mu^\diamond + \gamma^\diamond$. Then, for large t, $P' \leq (\gamma^\diamond + \mu^\diamond - c)P$ and $P(t) \to 0$ as $t \to 0$, a contradiction.

Let $\epsilon > 0$, to be determined later, and $\limsup_{t\to\infty} P(t) < \epsilon$. Without loss of generality, we can assume that $\epsilon < 1$. After a shift in time, we can assume that $P(t) < \epsilon$ for all $t \geq 0$. Then $Q' \leq (\epsilon - \mu_\diamond)Q + \delta$ with $\mu_\diamond = \inf \mu$. Since μ is positive, continuous, and periodic, $\mu_\diamond > 0$. Choose $\epsilon < \mu_\diamond/2$. Then

$$Q(t) \leq Q(0)e^{-t\mu_\diamond/2} + \frac{2\delta}{\mu_\diamond}.$$

After shifting forward in time by a suitable multiple of periods, we can assume that $Q(t) \leq \frac{3\delta}{\mu_\diamond}$ for all $t \geq 0$ and

$$P' \geq \gamma(t)P - \frac{3\delta}{\mu_\diamond}P + \delta - \delta P^2, \qquad t \geq 0.$$

By choosing $\epsilon < \mu_\diamond/3$, since $P(t) \leq \epsilon \leq 1$ and $\delta < 1$,

$$P' \geq \gamma(t)P - \epsilon P.$$

We integrate this differential inequality,

$$P(t) \geq P(0)\exp(t(\bar{\gamma}(t) - \epsilon)), \qquad \bar{\gamma}(t) = (1/t)\int_0^t \gamma(s)ds.$$

Notice that $\bar{\gamma}(t) \to \bar{\gamma}(\tau)$ [**217**, Prop.3.3]. This implies $P(t) \to \infty$ as $t \to \infty$ if ϵ is chosen small enough compared to $\bar{\gamma}(\tau)$, a contradiction. The upper estimate for Q follows similarly. □

Theorem 6.5. *There exists a nontrivial periodic solution of (6.1).*

Proof. Fix $\delta \in [0,1)$. All solutions satisfy the inequality

$$P' \leq \gamma^\diamond P + \delta(1 - P^2), \qquad \gamma^\diamond = \sup \gamma.$$

Choose $P^\diamond_\delta = P^\diamond$ such that $\gamma^\diamond P^\diamond + \delta(1 - (P^\diamond)^2) \leq 0$. Then any solution with $P(0) \leq P^\diamond$ satisfies $P(t) \leq P^\diamond$ for all $t \geq 0$. Consider such a solution, then

$$Q' \leq P^\diamond Q + \delta(1 - Q^2).$$

Choose $Q^\diamond = Q_\delta^\diamond > 0$ such that

$$0 \geq P_\delta^\diamond Q_\delta^\diamond + \delta(1 - (Q_\delta^\diamond)^2).$$

Then the rectangle $C_\delta = [0, P_\delta^\diamond] \times [0, Q_\delta^\diamond]$ is forward invariant. Let K_δ, $0 \leq \delta < 1$, be the period map associated with (6.2). For $\delta \in (0,1)$, the map K_δ maps the closed convex set C_δ in $\mathbb{R}^2$ into itself and thus has a fixed point $(\breve{P}_\delta, \breve{Q}_\delta)$ in C_δ by Brouwer's fixed point theorem. Let (P_δ, Q_δ) be the associated periodic solution with period τ, $(P_\delta(0), Q_\delta(0)) = (\breve{P}_\delta, \breve{Q}_\delta)$. By Proposition 6.4, $0 < \epsilon \leq \sup P_\delta$ and $\inf P_\delta \leq c < \infty$ for all $\delta \in (0,1)$. It follows from the differential equations and the τ-periodicity that there exist $0 < \tilde{\epsilon} < \tilde{c} < \infty$ such that $\tilde{c} \geq \breve{P}_\delta = P_\delta(\tau) \geq \tilde{\epsilon}$ and $\tilde{c} \geq \breve{Q}_\delta$ for all $\delta \in (0,1)$. By the Bolzano-Weierstraß theorem, there exists a sequence $\delta_n \to 0$ such that $\breve{P}_{\delta_n} \to P_0$ and $\breve{Q}_{\delta_n} \to Q_0$ as $n \to \infty$. Since the solutions of ODE continuously depend on their initial data and on their vector field,

$$(P_0, Q_0) = \lim_{n\to\infty} (\breve{P}_{\delta_n}, \breve{Q}_{\delta_n}) = \lim_{n\to\infty} K_{\delta_n}(\breve{P}_{\delta_n}, \breve{Q}_{\delta_n}) = K_0(P_0, Q_0).$$

So the solution of (6.1) with initial data (P_0, Q_0) is periodic and $\tilde{\epsilon} \leq P_0 \leq \tilde{c}$ and $0 \leq Q_0 \leq \tilde{c}$. Then $Q_0 \neq 0$ as well by the Q-equation in (6.1). □

6.3. Exercises

Exercise 6.1. Use Theorem 6.2 to establish Theorem 3.11.

Chapter 7

Nonlinear Matrix Models: Main Act

The year-to-year development of populations is often modeled by systems

$$x(n+1) = F(x(n)), \quad n \in \mathbb{Z}_+, \tag{7.1}$$

where $F : \mathbb{R}^m_+ \to \mathbb{R}^m_+$. Of particular interest are nonlinear matrix models where $F(x) = A(x)x$ with nonnegative square matrices $A(x)$ the entries of which continuously depend on x. See Section 3.3 where we have presented sufficient conditions for uniform weak persistence. Important examples are the LPA model (for flour beetles) and the Leslie model in demographics which have been introduced there. We now turn to uniform persistence.

7.1. Forward invariant balls and compact attractors of bounded sets

The following results, similar in character to Theorem 1.2.2 in [**38**], gives sufficient conditions for the existence of a compact attractor of bounded sets.

Lemma 7.1. *Assume that $F : \mathbb{R}^m_+ \to \mathbb{R}^m_+$ is continuous and that there exist some $R > 0$ and such that $|F(x)| \le |x|$ whenever $|x| \ge R$.*

(a) *Then there exists some $\tilde{R} > 0$ such that*

$$F(\mathbb{R}^m_+ \cap \bar{B}_s) \subset \bar{B}_s, \qquad s \ge \tilde{R},$$

for all closed balls $\bar{B}_s$ with radius s and the origin as center.

(b) *If $|F(x)| < |x|$ whenever $|x| \ge R$, then*

$$\liminf_{n\to\infty} |F^n(x)| \le R, \qquad \limsup_{n\to\infty} |F^n(x)| \le \max\{\tilde{R}, R\}.$$

Proof. (a) Suppose that such an $\tilde{R} > 0$ does not exist. Then there exist sequences (s_n) in $\mathbb{R}_+$ and (x_n) in $\mathbb{R}^m_+$ such that $s_n \to \infty$ and $|x_n| \le s_n < |F(x_n)|$. Since $s_n > R$ for large n, this is a contradiction.

(b) Suppose that $x \in \mathbb{R}^m_+$ and $\liminf_{n\to\infty} |F^n(x)| > R$. Then there exists some $N \in \mathbb{N}$ such that $|F^n(x)| > R$ for all $n \ge N$. By assumption,

$$|F(F^n(x))| \le |F^n(x)|, \qquad n \ge N,$$

and $(|F^n(x)|)_{n=N}^\infty$ is a decreasing sequence which is bounded below by R. Thus the sequence $(F^n(x))_{n=N}^\infty$ is bounded and has compact closure C, $C \subseteq \mathbb{R}^m_+ \setminus B_R$. Since F is continuous, there exists some $\xi \in (0,1)$ such that $|F(y)| \le \xi |y|$ for all $y \in C$. So

$$|F(F^n(x))| \le \xi |F^n(x)|, \qquad n \ge N.$$

By induction, $|F^{N+n}(x)| \le \xi^n |F(x(N))|$ for all $n \in \mathbb{N}$ and $F^n(x) \to 0$ as $n \to \infty$, a contradiction.

In particular, for each $\epsilon > 0$, there exists some $n \in \mathbb{N}$ such that $|F^n(x)| \le R + \epsilon$. Let $\check{R} \ge R + \epsilon$ and $\check{R} \ge \tilde{R}$. By part (a), $|F^j(x)| \le \check{R}$ for all $j \ge n$. Since $\epsilon > 0$ was arbitrary, the last assertion follows. □

Proposition 7.2. *Let $F : \mathbb{R}^m_+ \to \mathbb{R}^m_+$ be continuous. Assume that there is some $R > 0$, some vector $y \in \mathbb{R}^m_+$ and some nonnegative matrix D such that $\mathbf{r}(D) < 1$ and*

$$F(x) \le y + Dx, \qquad x \in \mathbb{R}^m_+, \quad |x| \ge R.$$

If $R = 0$, then

$$x(n) \le \sum_{i=0}^{n-1} D^i v + D^n x(0) \to (I - D)^{-1} y, \ n \to \infty. \tag{7.2}$$

In the general case $R \ge 0$, after introducing an equivalent norm, there exists some $\tilde{R} > 0$ such that

$$F(\mathbb{R}^m_+ \cap \bar{B}_s) \subset \bar{B}_s, \qquad s \ge \tilde{R},$$

for all closed balls $\bar{B}_s$ with radius $s \ge \tilde{R}$ and the origin as center. Further, there exists some $\check{R} > 0$ such that

$$\limsup_{n\to\infty} |F^n(x)| \le \check{R}, \qquad x \in \mathbb{R}^m_+.$$

Proof. First assume $R = 0$. We have $x(n+1) \le v + Dx(n)$, $n \ge 0$ and the estimate (7.2) follows by iteration because $D \ge 0$. The limit results from $(\mathbb{I} - D)^{-1} = \sum_{i=0}^\infty D^i \ge 0$ and $D^n \to 0$ because $\mathbf{r}(D) < 1$.

Now, consider the general case $R \ge 0$. By Proposition A.24, there exists some $\zeta \in (\mathbf{r}(D), 1)$ and a norm $|\cdot|'$ on $\mathbb{R}^m$ such that

$$|Dx|' \le \zeta |x|', \qquad x \in \mathbb{R}^m.$$

Further, $|\cdot|'$ is monotone, i.e., $|x|' \le |y|'$ whenever $0 \le x \le y$. Since the new norm is equivalent to the original one, there exists $R' > 0$ such that

$$F(x) \le y + Dx, \qquad |x|' \ge R'.$$

If $|x|' \ge R'$,

$$|F(x)|' \le |y|' + |Dx|' \le |y|' + \zeta |x|'.$$

Choose some $\xi \in (\zeta, 1)$. Then, if $|x|' \ge R'$,

$$|F(x)|' \le |y|' + (\zeta - \xi)R' + \xi |x|'.$$

Increasing R' as needed, the assumptions of Lemma 7.1 are satisfied and the assertions follow. □

Corollary 7.3. *Under the assumptions of Proposition 7.2, the semiflow induced by F has a compact attractor of bounded sets.*

Proof. By Proposition 7.2, the semiflow induced by F is point-dissipative and eventually bounded on bounded sets. It is asymptotically smooth because every bounded set has compact closure in finite dimensions. So the semiflow has a compact attractor of bounded sets by Theorem 2.33. □

Example 7.4. Assume for the Leslie model (3.14), (3.15) that there exists $M > 0$ such that $\sum_i x_i \beta_i(x) \le M$, $x \ge 0$, and that the functions τ_j are all bounded. Then the hypotheses of Proposition 7.2 are met with

$$y = \begin{pmatrix} M \\ 0 \\ \vdots \\ 0 \end{pmatrix}, \qquad D = \begin{pmatrix} 0 & 0 & \dots & 0 \\ \sup \tau_1 & 0 & \dots & 0 \\ \vdots & \vdots & & \vdots \\ 0 & \dots & \sup \tau_{m-1} & 0 \end{pmatrix}$$

because $\mathbf{r}(D) = 0$.

The LPA model satisfies the hypotheses of Proposition 7.2 with $y = (d/a, 0, 0)^*$ and

$$D = \begin{pmatrix} 0 & 0 & 0 \\ p & 0 & 0 \\ 0 & q & r \end{pmatrix}$$

provided that $r \in [0, 1)$.

7.2. Existence of nontrivial fixed points

For nonlinear matrix models, the assumptions that guarantee uniform weak persistence and the existence of large forward invariant ball also imply the existence of nonzero fixed points. They are typically less stringent than the conditions which provide nontrivial fixed points via uniform persistence (see Chapter 6).

Theorem 7.5. *Let $F : \mathbb{R}_+^m \to \mathbb{R}_+^m$ be continuous. Assume that there is some $R > 0$, some vector $y \in \mathbb{R}_+^m$ and some nonnegative matrix D such that $\mathbf{r}(D) < 1$ and*

$$F(x) \le y + Dx, \qquad x \in \mathbb{R}_+^m, \quad |x| \ge R.$$

Further, let F be differentiable at 0, $F'(0)$ be nonnegative, and $\mathbf{r}(F'(0)) > 1$. Assume that $F'(0)$ is irreducible or $F(x) = A(x)x$ with nonnegative matrices $A(x)$ depending continuously on x.

Then there exists some $x \in \mathbb{R}_+^m$, $x \ne 0$, such that $F(x) = x$.

Proof. We consider the maps $F_n : \mathbb{R}_+^m \to \mathbb{R}_+^m$ defined by

$$F_n(x) = F(x) + (1/n)\mathbf{1}, \qquad x \in \mathbb{R}_+^m.$$

If $|x| \ge R$,

$$F_n(x) \le y + \mathbf{1} + Dx.$$

By the proofs of Proposition 7.2 and Lemma 7.1, there exists some $s > R$ such that $F_n(\mathbb{R}_+^m \cap \bar{B}_s) \subset \bar{B}_s$ for all $n \in \mathbb{N}$. By Brouwer's fixed point theorem [**49**, Ch.1.Thm.3.2], there exist $x_n \in \mathbb{R}_+^m \cap \bar{B}_s$ such that $F_n(x_n) = x_n$. By construction, $x_n \ge (1/n)\mathbf{1}$ and so $\rho(x_n) > 0$.

We claim that there exists some $\delta > 0$ such that $|x_n| \ge \delta$ for all $n \in \mathbb{N}$.

Let $F'(0) = A_0$ and v be a nonnegative eigenvector of A_0^* associated with the spectral radius r_0 of $F'(0)$ (Theorem A.25) and $\rho(x) = \langle v, x\rangle$.

If $A_0 = F'(0)$ is irreducible, the eigenvector v is strictly positive and we can arrange that ρ is the restriction of $|\cdot|$ to $\mathbb{R}_+^m$. The same proof as for Theorem 3.17 provides $\delta > 0$ and $\gamma > 1$ such that

$$\rho(F_n(x)) \ge \rho(F(x)) \ge \gamma\rho(x)$$

whenever $x \ge 0$ and $|x| \le \delta$. Suppose $\rho(x_n) < \delta$ for some n. Then $\rho(x_n) = \rho(F_n(x_n)) \ge \gamma\rho(x_n)$, a contradiction.

Let $F(x) = A(x)x$ and $A(x)$ depend continuously on x. Let $|\cdot|$ be any norm. By Lemma 3.20, for any $\eta \in (0,1)$, there is some $\delta > 0$ such that

$$|x| \le \delta \implies F(x) \ge \eta A(0)x.$$

Let $r_0 > 1$ denote the spectral radius of $A(0) = A_0$. If $|x| \le \delta$,

$$\rho(F_n(x)) \ge \eta\langle v, A_0x\rangle = \eta\langle A_0^*v, x\rangle = \eta r_0\rho(x).$$

We choose $\eta \in (0,1)$ so that $\eta r_0 > 1$. If x_n is the fixed point of F_n and $|x_n| \le \delta$, since $\rho(x_n) > 0$,

$$\rho(x_n) = \rho(F_n(x_n)) > \rho(x_n),$$

a contradiction.

In either case, there exists some $\delta > 0$ such that $|x_n| \geq \delta$ for all $n \in \mathbb{N}$. Since $x_n \in \bar{B}_s$, $x_n \to x$ after choosing a subsequence. We take the limit as $n \to \infty$ and obtain $|x| \geq \delta$ and $F(x) = x$. □

Once it is known that a fixed point is not zero, stronger positivity properties can easily be derived.

Theorem 7.6. *Assume that F is irreducible: For any proper nonempty subset S of $\{1, \dots, m\}$ there exist $i \in \{1, \dots, m\} \setminus S$ and $j \in S$ such that $F_i(x) > 0$ whenever $x \in \mathbb{R}_+^m$, $x_j > 0$.*

Then any nonzero fixed point $x = F(x)$ in $\mathbb{R}_+^m$ is strictly positive.

Proof. For any $x \in \mathbb{R}_+^m$, since F is irreducible, $\mathrm{supp}(x)$ is a proper subset of $\mathrm{supp}(x) \cup \mathrm{supp}(F(x))$ if it is neither empty nor all of $\{1, \dots, m\}$. This implies that $\mathrm{supp}(x) = \{1, \dots, n\}$ if $x = F(x)$, $x \neq 0$. □

The following conclusion follows from one of the equivalent characterizations of an irreducible matrix (Section A.5).

Corollary 7.7. *Assume that there exists an irreducible matrix Q such that $supp(Q) \subset suppA(x))$ for all $x \in \mathbb{R}_+^m$. Then any nonzero fixed point $x = A(x)x$ in $\mathbb{R}_+^m$ is strictly positive.*

Example 7.8. Consider the Leslie model (3.14) and (3.15), and assume that $\mathcal{R}_0 > 1$, $\sup_{x \in \mathbb{R}_+^m} \sum_{j=1}^m \beta_j(x) x_j < \infty$, and all the functions τ_j are bounded. Then there exists a nonzero fixed point $x = A(x)x$. This fixed point is strictly positive if $\beta_m(x) > 0$ and $\tau_j(x) > 0$ for all $x \in \mathbb{R}_+^m$, $j = 1, \dots, m-1$.

There also exists a strictly positive fixed point in the LPA-model if $\mathcal{R}_0 > 1$ (even if $r = 0$).

Proof. For the existence part, combine Theorem 7.5 and Example 7.4. For the strict positivity choose the following irreducible matrix

$$\begin{pmatrix} 0 & \cdots & 0 & 1 \\ 1 & 0 & \cdots & 0 \\ \vdots & \ddots & \ddots & \vdots \\ 0 & \cdots & 1 & 0 \end{pmatrix}.$$

□

7.3. Uniform persistence and persistence attractors

Let $F : \mathbb{R}_+^m \to \mathbb{R}_+^m$ be continuous and differentiable at 0 and let $F'(0)$ be represented by a nonnegative Jacobian matrix A_0.

Theorem 7.9. *Suppose:*

(a) $\mathbb{R}^m_+ \setminus \{0\}$ *is forward invariant under* F.

(b) *There exist* $r_0 > 1$ *and* $v \gg 0$ *such that* $A_0^* v \geq r_0 v$, $A_0 = F'(0)$.

(c) *The discrete semiflow induced by* F *is point-dissipative.*

Then (7.1) is uniformly ρ*-persistent for* $\rho(x) = |x|$ *where* $|\cdot|$ *is the chosen norm on* $\mathbb{R}^m$.

If the induced semiflow has a compact attractor of bounded sets, F *has a fixed point different from* 0 *and there exists a compact connected set* A_1 *in* $\mathbb{R}^m \setminus \{0\}$, $F(A_1) = A_1$, *which attracts every compact set in* $\mathbb{R}^m \setminus \{0\}$.

A sufficient condition for the existence of a compact attractor of bounded sets is presented in Proposition 7.2 and Corollary 7.3.

Proof. We apply Theorem 2.23. By Theorem 3.17, the semiflow Φ induced by F is uniformly weakly ρ-persistent. Recall that $|\cdot|$ is equivalent to $|\langle \cdot, v\rangle|$. Since Φ is point-dissipative, there exists a compact set B such that, for every $x \in \mathbb{R}^m_+$, $\Phi(n, x) \in B$ for all sufficiently large n and $\tilde{\clubsuit}_2$ is satisfied. $\tilde{\clubsuit}_1$ holds by (a). $\tilde{\clubsuit}_0$ follows from the continuity of ρ and F. Existence of a nonzero fixed point follows from Theorem 6.2.

Now assume that there is a compact attractor of bounded sets. Notice that $X_0 = \{x \in \mathbb{R}^m_+, F^n(x) = 0 \ \forall n \in \mathbb{Z}_+\} \subset \{0\}$. Since F maps $X \setminus \{0\}$ into itself, hypothesis (H1) of Theorem 5.6 and Theorem 5.7 is satisfied, and most of the statements concerning A_1 follow from these theorems. The connectedness of A_1 follows from Proposition 2.24 and Proposition 5.9. Notice that we can choose $\rho(x) = \sum_{j=1}^m x_j$ which is concave. □

A very similar proof derives the following result from Theorem 3.21.

Theorem 7.10. *Let* F *be of nonlinear matrix form* $F(x) = A(x)x$, *and* $A_0 = A(0)$. *Assume:*

(a) *There exist some vector* $v > 0$ *and* $r_0 > 1$ *such that* $A_0^* v \geq r_0 v$.

(b) *If* $\langle x, v\rangle > 0$, *then* $\langle F(x), v\rangle > 0$.

(c) *There exists some* $n \in \mathbb{N}$ *and* $c > 0$ *such that* $|F^n(x)| \leq c\langle x, v\rangle$ *for all* $x \in \mathbb{R}^m_+$.

(d) *The semiflow induced by* F *is point-dissipative.*

Then the semiflow induced by F *is uniformly weakly* ρ*-persistent for* $\rho = \langle \cdot, v\rangle$.

If the induced semiflow has a compact attractor of bounded sets, there exists a compact set A_1, $F(A_1) = A_1$, *such that* $\inf_{x \in A_1} \langle x, v\rangle > 0$ *and* A_1 *attracts any compact set* K *in* $\{x \in \mathbb{R}^m_+; \exists n \in \mathbb{Z}_+ : \langle F^n x, v\rangle > 0\}$.

Example 7.11. For the Leslie model assume that

$$\sup_{x\in\mathbb{R}^m_+} x_i\beta_i(x) < \infty, \qquad i = 1,\dots,m,$$

and that all transition functions τ_i, $i = 1,\dots,m-1$, are bounded, $\mathcal{R}_0 > 1$. Further, assume that $\tau_i(x) > 0$ for $i = 1,\dots,m-1$, $x \in \mathbb{R}^m_+$. Finally, let $j \in \{1,\dots,m\}$ such that $\beta_j(x) > 0$ for all $x \in \mathbb{R}^m_+$ and that, if $j < m$, then $\beta_k(x) = 0$ for all $x \in \mathbb{R}_+$, $k \in \{j+1,\dots,m\}$.

Then the semiflow induced by the Leslie model is uniformly ρ_j-persistent for $\rho_j(x) = \sum_{i=1}^j x_i$. Further, there exists a compact set A_1, $F(A_1) = A_1$ such that $\inf_{x\in A_1}\sum_{i=1}^j x_i > 0$ and A_1 attracts every compact set K with $\inf_{x\in K}\sum_{i=1}^j x_i > 0$.

Proof. Combine Example 3.22, Corollary 7.3, Example 7.4, and Theorem 7.10. □

The next uniform persistence result is based on another result allowing transition from uniform weak to uniform persistence.

Theorem 7.12. *Suppose:*

(a) $\mathbb{R}^m_+ \setminus \{0\}$ *is forward invariant under* F.

(b) *There exist* $r_0 > 1$ *and* $v \gg 0$ *such that* $A_0^* v \ge r_0 v$, $A_0 = F'(0)$.

(c) *For every* $\epsilon > 0$ *there is some* $\delta > 0$ *such that* $|F(x)| \ge \delta$ *whenever* $x \in \mathbb{R}^n$, $|x| \ge \epsilon$.

Then (7.1) is uniformly ρ*-persistent for* $\rho(x) = |x|$ *where* $|\cdot|$ *is the chosen norm on* $\mathbb{R}^n$.

Proof. We apply Theorem 2.26. By Theorem 3.17, the semiflow Φ induced by F is uniformly weakly ρ-persistent. Recall that $|\cdot|$ is equivalent to $|\langle\cdot, v^*\rangle|$. We choose $B = \mathbb{R}^m_+$. Then all assumptions are satisfied. □

Example 7.13. A simple age-structured model of a species that reproduces only in its second year of life and does not live to a third year (a biennial plant, e.g.) is given by

$$\begin{aligned} x_1(n+1) &= \beta x_2(n), \\ x_2(n+1) &= \frac{x_1(n)}{a + b x_2(n)}, \end{aligned} \tag{7.3}$$

with $\beta, a, b > 0$. Then

$$A_0 = F'(0) = \begin{pmatrix} 0 & \beta \\ 1/a & 0 \end{pmatrix}.$$

and

$$A_0^2 = \begin{pmatrix} \beta/a & 0 \\ 0 & \beta/a \end{pmatrix}.$$

This implies that $\mathbf{r}(A_0^2) = \beta/a$ and $r_0 := \mathbf{r}(A_0) = \sqrt{\beta/a}$. Since A_0 is irreducible, by the Perron-Frobenius theorem there exists some $v \in \mathbb{R}_+^2$ with both coordinates positive and $A^*v = r_0 v$. We choose the norm $|x| = \max\{x_1, x_2\}$. To check (c) of Theorem 7.12, let $\epsilon > 0$ and $|x| \geq \epsilon$, $x \in \mathbb{R}^2$.

Case 1: $x_2 \geq \epsilon$.

Then $F_1(x) \geq \beta\epsilon$.

Case 2: $x_2 < \epsilon$.

Then $x_1 \geq \epsilon$ and $F_2(x) \geq \frac{\epsilon}{a+b\epsilon}$.

In either case,

$$|F(x)| \geq \delta := \min\left\{\beta\epsilon, \frac{\epsilon}{a+b\epsilon}\right\}.$$

By Theorem 7.12, the total population persists uniformly if $\beta/a > 1$.

Another simple age-structured model of a species that reproduces only in its second year of life and does not live to a third year is given by

$$x_1(n+1) = \frac{fx_2(n)}{1+ax_1(n)+bx_2(n)}, \tag{7.4}$$

$$x_2(n+1) = px_1(n), \quad n = 0, 1, 2, \ldots, \tag{7.5}$$

where $a, b, f > 0$ and $0 < p < 1$.

Remark 7.14. In Exercise 7.3, we ask the reader to show:

(a) Notice that $x_1(n) \leq f/b$ for $n \geq 2$ and $x_2(n) \leq pf/b$ for $n \geq 3$ and that Φ is dissipative and eventually bounded on bounded sets.

(b) If $pf < 1$, $x = 0$ is the compact attractor of bounded sets.

(c) Φ is ρ-persistent for $\rho(x) = x_1 + x_2$ when $fp > 1$.

(d) Φ has a period-two orbit

$$(0, d) \to (c, 0) \to (0, d) \to \cdots$$

when $pf > 1$. This shows that while ρ-persistence holds, age-class persistence may fail.

(e) every nontrivial orbit starting on the boundary of $\mathbb{R}_+^2$ converges to this period two orbit or its phase shift if $pf > 1$.

7.4. Stage persistence

In the previous example, when $fp > 1$, the semiflow induced by (7.4) is uniformly ρ-persistent for $\rho(x_1, x_2) = x_1 + x_2$, meaning that the population persists. However, it also has a period-two orbit

$$(0, d) \to (c, 0) \to (0, d) \to \cdots$$

which attracts every nontrivial orbit starting on the boundary of $\mathbb{R}^2_+$. A period-two orbit of this form is the special case of a *synchronous orbit* which is defined to be a nonzero orbit for which $x(n) \in \partial\mathbb{R}^m_+$ for all n [**40, 45**].

In this section we are interested in $\tilde{\rho}$-persistence, where

$$\tilde{\rho}(x) = \min_i x_i$$

(some use $\tilde{\rho}(x) = \prod_i x_i$). In the context of our example, $\tilde{\rho}$-persistence would represent persistence of every age-class. For the LPA model, it means persistence of every developmental stage.

While the existence of nontrivial synchronous orbits does not preclude $\tilde{\rho}$-persistence, because the premise of $\tilde{\rho}$-persistence is that initial data satisfy $\tilde{\rho}(x) > 0$ (see Section 8.6), it does preclude the stronger (and desirable) conclusion that there exists $\epsilon > 0$ such that

$$\liminf_{n\to\infty} \tilde{\rho}(x(n)) \geq \epsilon, \quad x(0) \neq 0.$$

Kon [**121**] has given sufficient conditions to rule out synchronous orbits which we extend below.

F is called *strongly positive* if for all $c > 0$ there exists some $n \in \mathbb{N}$ such that $F^n(x) \gg 0$ for each $x \in \mathbb{R}^m_+$, $0 < |x| \leq c$.

Let

$$X_0 = \{x \in \mathbb{R}^m_+ : \tilde{\rho}(\Phi(n, x)) = 0, \ n = 0, 1, 2, \ldots\}.$$

X_0 contains all synchronous orbits. If F is strongly positive, then $X_0 \subset \{0\}$ with equality holding if also $F(0) = 0$. The latter is equivalent to the nonexistence of nontrivial asynchronous orbits.

Theorem 7.15. *Suppose:*

(a) *The discrete semiflow induced by F is point-dissipative.*

(b) $\mathbb{R}^m_+ \setminus \{0\}$ *is forward invariant under F.*

(c) *F is strongly positive.*

(d) *There exist $r_0 > 1$ and $v \gg 0$ such that $A_0^* v \geq r_0 v$, $A_0 = F'(0)$.*

Then there exists some $\epsilon > 0$ such that $\liminf_{n\to\infty} \tilde{\rho}(x(n))$ for any solution of (7.1) with $x(0) \in \mathbb{R}^m_+ \setminus X_0$.

Proof. By Theorem 7.12, the semiflow Φ induced by F is uniformly ρ-persistent. We apply Corollary 4.22. Now let $(x(n))_{n\in\mathbb{Z}}$ be a bounded solution of $x(n+1) = F(x(n))$, $n \in \mathbb{Z}$, with $\inf_{n\in\mathbb{Z}} \rho(x(n)) > 0$. In particular, $x(n) \neq 0$, $n \in \mathbb{Z}$. Since F is strongly positive, $x(0) = F^n(x(-n)) \gg 0$ for some $n \in \mathbb{N}$ and $\tilde{\rho}(x(0)) > 0$. By Theorem 7.12, there exists some $\epsilon > 0$ such that $\liminf_{n\to\infty} \tilde{\rho}(x(n)) \geq \epsilon$ for every solution $(x(n))$ of (7.1) with $\rho(x(0)) > 0$, i.e., $x(0) \neq 0$. If $x(0) \in \mathbb{R}^m_+ \setminus X_0$, then $x(k) \neq 0$ for some $k \in \mathbb{N}$ and so $\liminf_{n\to\infty} \tilde{\rho}(x(n)) \geq \epsilon$ as well. $\square$

Recall that a nonnegative matrix Q is *primitive* [**15**] if there exists $k \geq 1$ such that $Q^k \gg 0$. Primitivity is a stronger property than irreducibility.

Notice that a matrix Q is primitive if and only if the map F_0, $F_0(x) = Qx$, is strongly positive. More generally, we have the following relation. Remember the notion of support of a matrix or vector, Definition 3.19.

Lemma 7.16. *Let $F : \mathbb{R}^m_+ \to \mathbb{R}^m_+$ and let Q a primitive square matrix of size m such that, if $(i,j) \in \operatorname{supp}(Q)$, then $F_i(x) > 0$ for all $x \in \mathbb{R}^m_+$ with $x_j > 0$. Then F is strongly positive.*

Proof. We first show that

$$\operatorname{supp}(F^n(x)) \supseteq \operatorname{supp}(Q^n x), \qquad x \in \mathbb{R}^m_+, x \neq 0, n \in \mathbb{N}. \tag{7.6}$$

Indeed, let $n = 1$ and $i \in \operatorname{supp}\,(Qx)$. Then there exists some j such that $(i,j) \in \operatorname{supp}(Q)$ and $x_j > 0$. By assumption, $i \in \operatorname{supp}(F(x))$.

Assume that (7.6) holds for $k = 1, \ldots, n$. Then, for all $x \in \mathbb{R}^n_+$,

$$\operatorname{supp}(F^{n+1}(x)) \supseteq \operatorname{supp}(QF^n(x)) \supseteq \operatorname{supp}(Q(Q^n x)) = \operatorname{supp}(Q^{n+1}x).$$

By strong induction, $\operatorname{supp}(F^n(x)) \subset \operatorname{supp}(Q^n x)$ for all $n \in \mathbb{N}$. Since Q is primitive, there exists some $n \in \mathbb{N}$ such that all entries of Q^n are positive. Then, if $x \neq 0$, $Q^n x$ is a strictly positive vector, i.e., $\operatorname{supp}(Q^n x) = m$, and so $F^n(x)$ is a strictly positive vector. $\square$

We turn to the special case of nonlinear matrix models. The entries of the matrices $A(x)$ below depend continuously on x.

Theorem 7.17. *Let F be of the form $F(x) = A(x)x$. Assume:*

(1) *The induced semiflow has a compact attractor of bounded sets.*
(2) *There exists a primitive matrix Q such that*
$$\operatorname{supp}(Q) \subset \operatorname{supp}(A(x)), \quad x \in \mathbb{R}^m_+.$$
(3) $\mathbf{r}(A(0)) > 1$.

Then Φ is uniformly persistent in the optimal sense that there exists $\epsilon > 0$ such that

$$\liminf_{n\to\infty} \min_{j=1}^{m} x_j(n) \geq \epsilon$$

for any solution of (7.1) with $x(0) > 0$. Further there exist $c > 1$ and a compact set A_1 in $[c^{-1}, c]^m$ with $F(A_1) = A_1$ that attracts every compact set in $\mathbb{R}^m \setminus \{0\}$.

Proof. Assumption (2) and Lemma 7.16 imply that the maps F and F_0, $F(x) = A(x)x$, $F_0(x) = A(0)x$, are strongly positive. Since $F(0) = 0$, $\mathbb{R}^m \setminus \{0\}$ is forward invariant under F. $A(0)$ is primitive and thus irreducible. By the Perron-Frobenius Theorem A.25, the hypotheses of Theorem 7.15 are satisfied for the minimum function $\tilde{\rho}(x) = \min_1^m x_j$ and the induced semiflow is uniformly persistent in the optimal sense above. As noted before, since $F(0) = 0$, $X_0 = \mathbb{R}^m \setminus \{0\}$. We check assumption (H1) of Theorem 5.7. Let $\tilde{\rho}(x) > 0$. Suppose $\tilde{\rho}(F(x)) = 0$. Then $x_j > 0$ for all $j = 1, \ldots, m$, and, by (2), $(Qx)_i = 0$ for some $i \in \{1, \ldots, m\}$. This implies that the ith row of Q consists of zeroes only, and Q is not primitive. Since Q is primitive by assumption, (H1) in Theorem 5.7 holds and the existence of the compact attractor A_1 follows. A_1 is uniformly $\tilde{\rho}$-positive. The connectedness of A_1 follows from Theorem 5.9 because $\tilde{\rho}$ is concave. □

Sufficient conditions for point-dissipativity and for the existence of a compact attractor of bounded sets were given in Proposition 7.2.

The LPA model of Costantino et al. [**36**], described in (3.13) of Chapter 3, is defined by

$$A(x) = \begin{pmatrix} 0 & 0 & d\exp(-ax_1 - bx_3) \\ p & 0 & 0 \\ 0 & q\exp(-cx_3) & r \end{pmatrix} \tag{7.7}$$

where $p, q, r \in (0, 1]$ are survival probabilities, a, b, c are coefficients related to cannibalism and $d > 0$ to fecundity. There, we noted that

$$\mathbf{r}(A(0)) > 1 \Leftrightarrow \mathcal{R}_0 = \frac{pqd}{1-r} > 1$$

where $\mathcal{R}_0$ is the basic reproduction number [**44**]. The same equivalence holds for "< 1". See Remark 3.13. We already know that the LPA has a strictly positive fixed point if $\mathcal{R} > 1$ (Example 7.8).

We summarize our results for the LPA model.

Proposition 7.18. *If $\mathcal{R}_0 < 1$, then $\{0\}$ attracts bounded subsets of $\mathbb{R}^3_+$.*

If $\mathcal{R}_0 > 1$, then the LPA model is uniformly $\tilde{\rho}$-persistent in the sense of Theorem 7.17.

Proof. The first assertion was noted in Chapter 3, using Proposition 3.12 and (3.16). The persistence assertion follows from Theorem 7.17 with

$$Q = \begin{pmatrix} 0 & 0 & 1 \\ 1 & 0 & 0 \\ 0 & 1 & 1 \end{pmatrix}.$$

Q^4 is a matrix with all entries positive. Proposition 7.2 implies the existence of a compact attractor of bounded sets. □

We mention that the optimal persistence result in Proposition 7.18 only holds if $r > 0$, i.e., that some adults live longer than one unit of time. For $r = 0$ and $\mathcal{R}_0 > 1$, there is only uniform persistence of the total population, but not of the stages. A bifurcation analysis of the case $r = 0$ has been performed in [**40**].

We turn to Leslie models.

Theorem 7.19. *Assume that all transition functions τ_j are bounded and that $\sup_{x \in \mathbb{R}_+^m} x_j \beta_j(x) < \infty$, $j = 1, \ldots, m$.*

Further, assume that, for all $x \in \mathbb{R}_+^m$, $\beta_m(x) > 0$, $\beta_{m-1}(x) > 0$ and $\tau_j(x) > 0$, $j = 1, \ldots, m-1$.

If $\mathcal{R}_0 > 1$, the Leslie model is uniformly $\tilde{\rho}$-persistent in the stronger sense of Theorem 7.17, and there exists a fixed point with positive $\tilde{\rho}$-value.

Proof. We apply Theorem 7.17. Its hypothesis (1) is satisfied by the first assumption. $\mathbf{r}(A(0)) > 1$ because $\mathcal{R}_0 > 1$. To fulfill hypothesis (2), choose Q as the $m \times m$ matrix

$$Q = \begin{pmatrix} 0 & 0 & \cdots & 1 & 1 \\ 1 & 0 & \cdots & 0 & 0 \\ 0 & 1 & \cdots & 0 & 0 \\ \vdots & \vdots & \ddots & \vdots & \vdots \\ 0 & 0 & \cdots & 1 & 0 \end{pmatrix}. \tag{7.8}$$

Then supp $(Q) \subset$ supp $(A(x))$ for all $x \in \mathbb{R}_+^m$. Notice that Q is primitive if and only if there exists some $n \in \mathbb{N}$ such that $Q^n x \gg 0$ for all $x > 0$. Further, notice

$$Qx = \begin{pmatrix} x_{m-1} + x_m \\ x_1 \\ \vdots \\ x_{m-1} \end{pmatrix}.$$

This implies that $\sharp$ supp$(Q(x)) \geq \sharp$ supp(x) with $\sharp$ denoting the cardinality of a set. Further, $\sharp$ supp$(Q(x)) < \sharp$ supp(x) if $x_m = 0 < x_{m-1}$. Since Q is also cyclically shifting the vectors, $\sharp$ supp $(Q^m(x)) > \sharp$ supp (x) if $x \in \mathbb{R}_+^m$

is not the zero vector but still has zero coordinates. So $Q^{m^2}(x) >> 0$ if $x > 0$. □

For comparison, consider a Leslie model where reproduction only occurs every other year, i.e., m is even and $\beta_j(x) = 0$ if j is odd and $\beta_j(x) > 0$ if j is even. Then the matrices $A(x)$ map vectors of the form $(0, *, \cdots, 0, *)$ onto vectors of the form $(*, 0, \cdots, *, 0)$ and vice versa, and $\tilde{\rho}$-persistence does not hold in the stronger sense of Theorem 7.17.

7.5. Exercises

Exercise 7.1. Let $F : \mathbb{R}_+^m \to \mathbb{R}_+^m$ be continuously differentiable. Assume that there is some $R > 0$, some vector $v \in \mathbb{R}_+^m$ and some nonnegative matrix D such that $\mathbf{r}(D) < 1$ and

$$F(x) \leq y + Dx, \qquad x \in \mathbb{R}_+^m, \quad |x| \geq R.$$

Further, let $F'(0)$ be nonnegative, and $\mathbf{r}(F'(0)) > 1$.

Show that there exists some $x \in \mathbb{R}_+^m$, $x \neq 0$, such that $F(x) = x$.

Hint: $F(x) \geq \displaystyle\int_0^1 F'(sx)x ds.$

Exercise 7.2. Consider the Leslie model with

$$A(x) = \begin{pmatrix} 0 & \beta_2(x) & \beta_3(x) & 0 \\ \tau_1(x) & 0 & 0 & 0 \\ 0 & \tau_2(x) & 0 & 0 \\ 0 & 0 & \tau_3(x) & 0 \end{pmatrix}$$

with $\beta_j(x) > 0$ for $j = 2, 3$, $\tau_j(x) > 0$, $j = 1, 2, 3$, $x \in \mathbb{R}_+^4$. Find the right conditions such that there exists some $\epsilon > 0$ such that

$$\liminf_{n\to\infty} \min_{j=1}^{4} x_j(n) \geq \epsilon$$

for all solutions with $x_1(0) + x_2(0) + x_3(0) > 0$.

Exercise 7.3. Verify Remark 7.14.

Chapter 8

Topological Approaches to Persistence

8.1. Attractors and repellers

Let X be a metric space with metric d, J a time-set and $\Phi : J \times X \to X$ a continuous semiflow.

Definition 8.1. A nonempty subset of X is called *locally attracting* if it attracts a neighborhood of itself.

An characterization of locally attracting sets is given by the following result. Although it (this result) is not used in the sequel, its proof is a nice application of Theorem 5.7.

Theorem 8.2. *Let X be compact and invariant and the semiflow Φ continuous. If M is a proper subset of X that is compact and forward invariant, then M is locally attracting if and only if*

(1) *M is isolated,*
(2) *there is no full trajectory with $\phi(0) \notin M$ and $\phi(-t) \to M$ as $t \to \infty$.*

The proof of this result provides the following additional information.

Remark 8.3. (a) If M is locally attracting, there exists a uniform repeller R with the following properties.

(i) $R \cap M = \emptyset$, R is compact and forward invariant, and $X \setminus R$ is forward invariant.

(ii) M attracts a neighborhood of every compact set in $X \setminus R$ and contains every compact invariant set in $X \setminus R$.

(iii) If $x \in X \setminus M$ and ϕ is a total trajectory through x, $\phi(-t) \to R$ as $t \to \infty$.

(b) If M attracts an open neighborhood U of itself, then $R_U = \bigcap \Phi^{-1}(X \setminus U)$ is a nonempty, compact, forward invariant uniform repeller, $R_U \subset X \setminus U \subset X \setminus M$, and M attracts a neighborhood of all compact subsets of $X \setminus R_U$.

Recall that $\bigcap \Phi^{-1}(X \setminus U) = \{x \in X; \forall t \in J : \Phi(t, x) \in X \setminus U\}$. The subsequent proof will freely (and without further reference) use the notation and the results of Exercises 1.16 and 1.17.

Proof of Theorem 8.2 and Remark 8.3. "$(a) \Rightarrow$". Let M attract a neighborhood U. To show that U is an isolating neighborhood of M, let $K \subset U$ and K compact and invariant. By Theorem 2.11, $K \subset \omega(U) \subset M$ and (1) holds.

To show (2), let ϕ be a total trajectory with $\phi(0) \in X \setminus M$ and $\phi(-t) \to M$ as $t \to \infty$. Since X is compact, $\alpha(\phi)$ is nonempty, compact, invariant, and a subset of M. Since U is a neighborhood of M, there exists some $r \in J$ such that $\phi(-t) \in U$ for all $t \in J_r$. For sufficiently large $n \in \mathbb{N}$, $\phi(0) = \Phi(n, \phi(-n)) \in \omega(U) \subset M$, a contradiction.

"(a) $\Leftarrow$". Suppose that (1) and (2) hold. By Remark 5.15, M has a compact isolating neighborhood U. Hypotheses (2) implies that $\alpha(\phi) \cap (X \setminus U) \neq \emptyset$ for every total trajectory ϕ such that $\phi(0) \notin M$. As $X \setminus U$ is open we may conclude that $\phi(-t) \in X \setminus U$ for some arbitrarily large $t \in J$, if $\phi(0) \notin M$.

Step 1: Let K be a compact set contained in $X \setminus M$. Then there exists some $T \geq 1$ such that, if ϕ is a total trajectory with $\phi(0) \in K$, $\phi(-t) \in X \setminus U$ for some $t \in [1, T] \cap J$.

To show step 1, suppose that such a $T \geq 1$ does not exist. Then there exist total trajectories ϕ_n, $n \in \mathbb{N}$, $\phi_n(0) \in K$, such that $\phi_n(-t) \in U$ for all $t \in [1, n] \cap J$. Since K is compact, we may assume that $\phi_n(0) \to \phi(0) \in K$, where ϕ is a total trajectory in X. By Proposition 1.42, $\phi_n \to \phi$ uniformly on compact subsets of $-J \cup J$. By the considerations preceding step 1, $\phi(-t) \in X \setminus U$ for some arbitrarily large $t \in J$. Fix one of these, say $t_0 \geq 1$. It follows that $\phi_n(-t_0) \in X \setminus U$ for all large n which implies that $t_0 > n$ for such n, a contradiction. This proves step 1.

Step 2: Set $B = \bigcup \Phi^{-1}(\overline{X \setminus U})$. Then $B \subset X \setminus M$, B is compact, $X \setminus B = \bigcap \Phi^{-1}(\breve{U})$, and $\Phi_t^{-1}(B) \subset B$ for all $t \in J$.

Here $\breve{U}$ denotes the interior of U. Since $\overline{X \setminus U} \subset X \setminus M$,

$$B \subset \bigcup \Phi^{-1}(X \setminus M) = X \setminus \bigcap \Phi^{-1}(M).$$

Since M is forward invariant, $M \subset \bigcap \Phi^{-1}(M)$ and $B \subset X \setminus M$. Further,

$$X \setminus B = \bigcap \Phi^{-1}(X \setminus \overline{X \setminus U}) = \bigcap \Phi^{-1}(\breve{U}).$$

To prove the compactness of B which is a subset of the compact set X, it is sufficient to show that B is closed. We first show that

$$B = \bigcup_{t \in J, 0 \leq t \leq T} \Phi_t^{-1}(\overline{X \setminus U}), \tag{8.1}$$

with $T > 1$ from step 1 with $K = \overline{X \setminus U}$. Let $t \in J$, $t > T$, and $x \in \Phi_t^{-1}(\overline{X \setminus U})$. Then $\Phi(t, x) \in \overline{X \setminus U}$. Since X is invariant, there exists a total Φ-trajectory ϕ such that $\phi(-t) = x$. Then $\phi(0) = \Phi(t, \phi(-t)) = \Phi(t, x) \in \overline{X \setminus U}$. By step 1, there exists some $s \in J$, $1 \leq s \leq T$, such that $\phi(-s) \in X \setminus U$. Then

$$\Phi(t - s, x) = \Phi(t - s, \phi(-t)) = \phi(-s) \in X \setminus U,$$

and $x \in \Phi_{t-s}^{-1}(X \setminus U)$. We have shown

$$\left. \begin{array}{r} t > T, \\ x \in \Phi_t^{-1}(\overline{X \setminus U}) \end{array} \right\} \Longrightarrow x \in \Phi_r^{-1}(X \setminus U) \text{ for some } r \in [t - T, t - 1].$$

We can repeat this conclusion as long as $r > T$ and finally find some $r \in [0, T]$ such that $x \in \Phi_r^{-1}(X \setminus U)$. This proves (8.1).

To show that B is closed, let (x_n) be a sequence in B and $x \in X$ such that $x_n \to x$ as $n \to \infty$. By (8.1), there exist a sequence (t_n) in $J \cap [0, T]$ such that $\Phi(t_n, x_n) \in \overline{X \setminus U}$. After choosing subsequences, $t_n \to t$ for some $t \in J$. Since Φ is continuous, $\Phi(t, x) = \lim_{n \to \infty} \Phi(t_n, x_n) \in \overline{X \setminus U}$. So $x \in B$.

> **Step 3:** The set $R = \bigcap \Phi^{-1}(B) = \bigcap_{t \in J} \Phi_t^{-1}(B)$ is nonempty, compact, forward invariant, and a uniform repeller, $R \subset B \subset X \setminus M$. M attracts all points in $X \setminus R$.

Since B is a nonempty subset of X, it contains an element $x \in B$. Since X is invariant, for each $t \in J$, $x = \Phi_t(y)$ for some $y \in X$ and $y \in \Phi_t^{-1}(B)$ which is a nonempty set. As $\Phi_{t+s}^{-1}(B) = \Phi_s^{-1}(\Phi_t^{-1}(B)) \subset \Phi_s^{-1}(B)$ by step 2, the family of compact nonempty sets $\{\Phi_t^{-1}(B)\}_{t \in J}$ is decreasing; so R is nonempty and compact. It is also forward invariant.

If $y \notin R$, then $y \notin \Phi_r^{-1}(B)$ for some $r \in J$, so $\Phi_r(y) \in X \setminus B = \bigcap \Phi^{-1}(\breve{U})$ by step 2, and consequently $\Phi_t(y) \in \breve{U}$ for all $t \in J_r$. Since U is compact, $\omega(y) \subset U$ and $\omega(y) \subset M$ because U is an isolating neighborhood of M. So

$\Phi_t(y) \to M$ as $t \to \infty$ for all $y \notin R$. Since $M \cap R = \emptyset$, R is a uniform repeller. Further, $X \setminus R$ is forward invariant.

Step 4: M is locally attracting and (i), (ii), (iii) in (b) hold.

We apply Theorem 5.7 with $\rho(x) = d(x, B)$. Notice that $X_0 = R$. Since, by step 3, M attracts all points in $X \setminus X_0$ and $B \subset X \setminus M$, Φ is uniformly ρ-persistent. To check (H1), let ϕ be a total Φ-trajectory with $\rho(\phi(0)) = 0$. Since B is closed, $\phi(0) \in B$ by Exercise 1.3. By step 2, for each $t \in J$, $\Phi_t^{-1}(\{\phi(0)\}) \subset B$, i.e., $\phi(-t) \in B$. This implies $\rho(\phi(-t)) = 0$, and thus (H1) holds.

Let A_1 be the persistence attractor according to Theorem 5.7. Then A_1 is uniformly ρ-positive. By choice of ρ, $A_1 \subset X \setminus B$. By step 2, for all $t \in J$,

$$\Phi_t(A_1) = A_1 \subset \bigcap \Phi^{-1}(\breve{U}).$$

This implies $A_1 \subset \breve{U}$. Since $\breve{U}$ is an isolating neighborhood of M and A_1 is compact and invariant, $A_1 \subset M$. Since $M \subset X \setminus X_0$, a neighborhood of M is attracted by A_1 and so by M.

Proof of Remark 8.3(b).

Consider an open neighborhood U of $M \neq X$ which is attracted by M. Then

$$R_U = \{x \in X;\ \forall t \in J : \Phi(t, x) \in X \setminus U\} = \bigcap_{t \in J} \Phi_t^{-1}(X \setminus U)$$

is closed, $R_U \subset X \setminus U \subset X \setminus M$. Then $X \setminus R_U \supseteq U$. Further, R_U is forward invariant. Since $X \setminus R_U = \bigcup \Phi^{-1}(U)$, by Proposition 2.29, $\omega(U) \subset M$ attracts a neighborhood of all compact subsets of $X \setminus R_U$.

If R_U were empty, $\omega(U)$ would contain all compact invariant subsets of X, in particular X itself, $X \subset \omega(U) \subset M$, a contradiction. Since M attracts all compact subsets of $X \setminus R_U$ and $M \cap R_U = \emptyset$, R_U is a uniform repeller. $\square$

The reader is asked to prove the following remark in Exercise 8.1.

Remark 8.4. Let X be compact and invariant and suppose that M is a proper subset that is closed, invariant and isolated. Then either M is locally attracting, or it is a repeller, or there exists $x \notin M$ such that $\omega(x) \subset M$ and there exists a full orbit ϕ with $\phi(0) \notin M$ such that $\alpha(\phi) \subset M$.

8.2. Chain transitivity and the Butler-McGehee lemma

In this section X is a metric space and $\Phi : J \times X \to X$ is a continuous semiflow. For the definition below, we assume that $J = [0, \infty)$ or $J = \{0\} \cup \mathbb{N} = \mathbb{Z}_+$. The following definitions are due to Conley [**33**].

Definition 8.5. A forward invariant set $A \subset X$ is said to be *chain transitive* if for any $a, b \in A$ and for any $\epsilon > 0$ and $s \in J_1$ there is a finite sequence of points $x_1 = a, x_2, \ldots, x_m, x_{m+1} = b$ in A and times $t_1, t_2, \ldots, t_m$ in J such that

$$t_i \geq s \quad \text{and} \quad d(\Phi_{t_i}(x_i), x_{i+1}) \leq \epsilon, \ \ i = 1, \ldots, m.$$

The sequence $\{x_1, x_2, \ldots, x_{m+1};\ t_1, t_2, \ldots, t_m\}$ is called an (ϵ, s)-chain in A connecting a to b. A is said to be chain recurrent if for every $a \in A$, $\epsilon > 0$ and $s \geq 1$ there is an (ϵ, s)-chain in A connecting a to itself.

Sometimes the term "internally chain transitive" is used to stress that the points x_i (and therefore their forward orbits) of the chain are required to belong to A.

Obviously, chain transitivity implies chain recurrence for A; it can be shown that the converse is true if A is connected. See [**33, 170**]. One of the key properties of limit sets is that they are chain transitive. The proof below follows [**101**].

It is easy to imagine an (ϵ, s)-chain as the time series one might observe of a real system subjected to small disturbances. Jacobs, Metz and Gyllenberg [**113, 88**] use the notion of chain transitivity to define a special notion of attractors for biological population models. Schreiber [**191**] uses it to give postulates for the meaning of persistence. A persistent population model should have a state x supporting all populations for which there is no (ϵ, s)-chain connecting x to any extinction state for some $\epsilon > 0$.

Proposition 8.6. *If $x \in X$ is such that $\overline{\Phi(J \times \{x\})}$ is compact, then $\omega(x)$ is chain transitive. If there is a total trajectory ϕ through x with $\overline{\phi(-J)}$ compact, then $\alpha(\phi)$ is chain transitive.*

Proof. Let $\omega = \omega(x)$ be the omega limit set of a forward orbit $\Phi(J \times \{x\})$ with compact closure. Fix $a, b \in \omega$ and let $\epsilon > 0$ and $s \in J_1$ be given. Since the set $(J \cap [s, 2s]) \times \omega$ is compact, there is a $\delta \in (0, \epsilon/3)$ such that for any $t \in J \cap [s, 2s]$ and $u, v \in U \equiv N_\delta(\omega)$ with $d(u, v) < \delta$, we have $d(\Phi(t, u), \Phi(t, v)) < \epsilon/3$. As $\Phi(t, x) \to \omega$, there exists $T > 0$ such that $\Phi(t, x) \in U$ for all $t \in J_T$.

Since $a, b \in \omega$, there exist $T_1 > T$ and $T_2 \geq T_1 + 2s$, $T_1, T_2 \in J$, such that $d(\Phi(T_1, x), a) < \delta$ and $d(\Phi(T_2, x), b) < \delta$. Let m be the unique integer satisfying

$$t_{m+1} := T_2 - T_1 - sm \in [s, 2s).$$

Since $T_2 - T_1 > s$, $m \geq 1$. We first construct a $(0, s)$-chain in X from $\Phi(T_1, x)$ to $\Phi(T_2, x)$ by defining

$$y_i = \Phi(T_1 + (i-1)s, x), \ i = 1, 2, \ldots, m+1, \qquad y_{m+2} = \Phi(T_2, x),$$

and

$$t_i = s, \quad i = 1, \ldots, m.$$

Then $\Phi(t_i, y_i) = y_{i+1}$, $i = 1, \ldots, m+1$. To find a chain in A, we notice that $y_i \in U$, $i = 2, \cdots, m$. So we can choose $z_2, \ldots, z_{m+1} \in \omega$ such that $d(z_i, y_i) < \delta$. Let $z_1 = a$ and $z_{m+2} = b$. It follows that $d(z_i, y_i) < \delta$ for $i = 1, \ldots, m+2$ and

$$\begin{aligned} d(\Phi(t_i, z_i), z_{i+1}) &\leq d(\Phi(t_i, z_i), \Phi(t_i, y_i)) + d(y_{i+1}, z_{i+1}) \\ &< \epsilon/3 + \epsilon/3, \; i = 1, 2, \ldots, m+1. \end{aligned}$$

This proves that

$$\{a = z_1, z_2, \ldots, z_{m+1}, z_{m+2} = b; \; t_1, t_2, \ldots, t_{m+1}\}$$

is an (ϵ, s)-chain in ω connecting a to b. Hence, ω is chain transitive. The case of the alpha limit set is proved in a similar fashion. □

Lemma 8.7. *Let Φ be a semiflow on X such that Φ is continuous and X is compact and invariant and chain transitive under Φ. Then X neither contains a proper subset that is a uniform weak repeller nor a locally attracting forward invariant proper subset.*

Proof. Suppose that X contains a proper subset that is a uniform weak repeller or is locally attracting and forward invariant. In either case, X contains a locally attracting forward invariant proper subset A (Theorem 5.17). Let $b \in X \setminus A$ and U be a compact neighborhood of A that is attracted by A, $b \notin U$, and choose $\epsilon > 0$ so small that $\bar{N}_{2\epsilon}(A) = \{x \in X : d(x, A) \leq 2\epsilon\}$ is contained in the interior of U. Let $s \in J_1$ be so large that $\Phi(t, U) \subset N_\epsilon(A)$ for all $t \in J_s$. If $a \in A$, we will show that there is no (ϵ, s)-chain $\{x_1, x_2, \ldots, x_{m+1}; t_1, t_2, \ldots, t_m\}$ in X from a to b. Indeed, since A is forward invariant, $\Phi(t_1, x_1) \in A$ and so $x_2 \in U$. As $t_2 \geq s$, $\Phi(t_2, x_2) \in N_\epsilon(A)$. Again, since $d(x_3, \Phi(t_2, x_2)) < \epsilon$ we have $x_3 \in U$, so $\Phi(t_3, x_3) \in N_\epsilon(A)$ because $t_3 \geq s$. It should now be obvious that all $x_i \in U$, in particular, $x_{m+1} \neq b$. □

The Butler-McGehee lemma, here up-graded to a theorem, is one of the results to come from persistence theory which are of their own interest. It is usually stated for the case that Y (see below) is a limit set.

Theorem 8.8 (Butler-McGehee). *Let M be a closed isolated forward invariant set and let Y be a compact, invariant and chain transitive set in X. Suppose that $Y \cap M \neq \emptyset$ but $Y \not\subseteq M$.*

- *Then there exists an $x \in Y \setminus M$ such that $\Phi(t, x) \to M$ as $t \to \infty$.*
- *Further, there exists a full trajectory ϕ in Y with $\phi(0) \in Y \setminus M$ and $\phi(-t) \to M$ as $t \to \infty$.*

Proof. $M_1 = Y \cap M$ is a compact (M is closed!) isolated forward invariant set for the restriction of Φ to Y. Notice that $Y \setminus M = Y \setminus M_1$. Since $Y \not\subset M$, M_1 is a proper subset of Y.

If the first result is false, M_1 is a repeller in Y by Theorem 5.16; but this contradicts Lemma 8.7.

If the second result is false, M_1 is a locally attracting forward invariant proper subset in Y by Theorem 8.2; but this again contradicts Lemma 8.7. □

In the first incarnation of Theorem 8.8 by Freedman and Waltman [**76**], M was taken to be a saddle point for a system of ordinary differential equations (isolated by the Hartman-Grobman Theorem) and Y an omega or alpha limit set assumed to contain M but not to coincide with M. As a consequence, the limit set must contain a point of the stable (unstable) manifold of the saddle point distinct from M. This is one of the more useful applications of Theorem 8.8; see Proposition 3.26. Later, the result was extended [**23, 73, 235**]. The present version is patterned after the one in [**101**].

Let A and B be two compact subsets of X. Recall that the Hausdorff distance between A and B is defined by

$$d_H(A, B) := \max\{d(A, B), d(B, A)\}, \quad d(B, A) = \sup_{x \in B} d(x, A).$$

The argument below is adapted from [**101, 206**].

Lemma 8.9. *Let Φ and $\{\Phi^n\}_{n\geq 1}$ be continuous semiflows on X. Let $(D_n)_{n\geq 1}$ be a sequence of nonempty compact subsets of X with $\lim_{n\to\infty} d_H(D_n, D) = 0$ for some nonempty compact subset D of X. Assume that for each $n \geq 1$, D_n is invariant for Φ^n and, for each $t \in J$, $\Phi^n(t, x) \to \Phi(t, x)$ uniformly for $x \in [D \cup (\bigcup_{n\geq 1} D_n)]$, then D is invariant for Φ. If, in addition, D_n is chain transitive for Φ_n and if for each $T > 0$, $\Phi^n \to \Phi$ uniformly on $([0, T] \cap J) \times [D \cup (\bigcup_{n\geq 1} D_n)]$, then it is chain transitive for Φ.*

Proof. Notice that $x \in D$ if and only if there exist $x_n \in D_n$ such that $x_n \to x$. Further, $K \equiv D \cup (\bigcup_n D_n)$ is compact.

Invariance of D: Let $x \in D$ and $x_n \in D_n$ such that $x_n \to x$ as $n \to \infty$. As $\Phi_t^n(x_n) \in D_n$ and $\Phi_t^n(x_n) \to \Phi_t(x)$, we conclude that $\Phi_t(x) \in D$ for $t \in J$ whenever $x \in D$. There exists $y_n \in D_n$ such that $\Phi_t^n(y_n) = x_n$ and we may assume, after renumbering a convergent subsequence, that $y_n \to y \in D$. Taking the limit in $\Phi_t^n(y_n) = x_n$ results in $\Phi_t(y) = x$.

Chain transitivity of D: Let $a, b \in D$, $\epsilon > 0$ and $s \geq 1$. By uniform continuity of Φ and uniform convergence of Φ^n on $(J \cap [0, 2s]) \times K$ there exists $\delta \in (0, \epsilon/3)$ and natural number N such that

$$d(\Phi^n(t,u), \Phi(t,v)) \leq d(\Phi^n(t,u), \Phi(t,u)) + d(\Phi(t,u), \Phi(t,v)) < \frac{\epsilon}{3}$$

for $u, v \in K$, $d(u,v) < \delta$, $n \geq N$, $t \in J \cap [0, 2s]$. Fix $n > N$ such that $d_H(D_n, D) < \delta$ and choose $x, y \in D_n$ with $d(x,a) < \delta$ and $d(y,b) < \delta$. As D_n is chain transitive for Φ^n, there is a (δ, s)-chain $\{z_1 = x, z_2, \ldots, z_{m+1} = y; t_1, \ldots, t_m\}$ from x to y in D_n for Φ^n. By adding new points to this chain, if necessary (but no new "jumps"), we may assume that $t_i \in [s, 2s], 1 \leq i \leq m$. For each $i = 2, \ldots, m$, we find $w_i \in D$ with $d(w_i, z_i) < \delta$. Set $w_1 = a$ and $w_{m+1} = b$. Then $d(w_i, z_i) < \delta$ for $i = 1, \ldots, m+1$ and

$$\begin{aligned} d(\Phi(t_i, w_i), w_{i+1}) &\leq d(\Phi(t_i, w_i), \Phi^n(t_i, z_i)) + d(\Phi^n(t_i, z_i), z_{i+1}) \\ &\quad + d(z_{i+1}, w_{i+1}) \\ &< \epsilon/3 + \delta + \delta < \epsilon, \qquad i = 1, \ldots, m. \end{aligned}$$

So $\{w_1 = a, w_2, \ldots, w_{m+1} = b; t_1, \ldots, t_m\}$ is an (ϵ, s) chain in D from a to b for Φ. □

Lemma 8.9 leads to a simple upper semicontinuity result for attractors (cf. [**154**] and the references therein).

Theorem 8.10. *Let $\Phi_\lambda : J \times X \to X$, $\lambda \in \Lambda$, be a family of semiflows on the metric space (X, d), where the index set Λ is also a metric space. Suppose that for each $t \in J$, the map $X \times \Lambda \to X$ given by*

$$(x, \lambda) \to \Phi_\lambda(t, x)$$

is continuous.

Further, let (A_λ) be a family of compact subsets of X, indexed by the same index set Λ, such that the following hold:

(1) *for each $\lambda \in \Lambda$, A_λ is invariant under Φ_λ ,*
(2) *$B = \overline{\bigcup_{\lambda \in \Lambda} A_\lambda}$ is compact.*

Then, if $\lambda_0 \in \Lambda$ and A_{λ_0} is a compact attractor of compact sets for Φ_{λ_0}, $d(A(\lambda), A(\lambda_0)) \to 0$ as $\lambda \to \lambda_0$.

Proof. If false, there exists $\epsilon > 0$ and sequences $\{\lambda_n\}$ and $\{x_n\}$ such that $\lambda_n \to \lambda_0$, $x_n \in A_{\lambda_n}$ and $x_n \notin U_\epsilon(A_0)$.

To simplify notation, let $\Phi_n = \Phi_{\lambda_n}$ and $A_n = A_{\lambda_n}$. By Proposition 1.24 and Exercise 1.9, we can assume, on passing to a subsequence if necessary, that $d_H(A_n, A) \to 0$ where A is a nonempty compact subset of X and that A contains a point $x = \lim_n x_n$ which is not in A_0.

By our continuity assumptions above, it follows that the hypotheses of Lemma 8.9 are satisfied:

$$\Phi_{\lambda_n}(t,\cdot) \to \Phi_{\lambda_0}(t,\cdot)$$

uniformly on B, for each $t \in J$. Lemma 8.9 implies that A is invariant for Φ_0. But then $A \subset A_0$ because A_0 attracts compact sets, and so $x \in A_0$. This contradiction proves the result. □

Theorem 8.10 can be particularly useful when the reference system has a globally asymptotically stable equilibrium. Under mild additional assumptions, the perturbed system should also have a globally asymptotically stable equilibrium. See [**154, 204**] for details.

8.3. Acyclicity implies uniform weak persistence

We have seen a variety of results which assert that uniform weak persistence implies uniform strong persistence. In this section we show that if, roughly speaking, there are no cycles of invariant sets on $\rho^{-1}\{0\}$ and if the set of orbits attracted to each such invariant set belongs to $\rho^{-1}\{0\}$, then we can conclude uniform weak persistence. This method requires a fairly complete understanding of the dynamics on $\rho^{-1}\{0\}$.

Let $J = \mathbb{R}_+$ or $J = \mathbb{Z}_+$. Assume that the semiflow $\Phi : J \times X \to X$ is continuous, and $\rho : X \to \mathbb{R}_+$ is continuous and not identically zero. Let

$$X_0 = \{x \in X;\ \rho(\Phi(t,x)) = 0\ \forall t \geq 0\},$$

which is closed and forward invariant. As our interest is in showing uniform weak ρ-persistence, we may assume that $X_0 \neq \emptyset$ by Theorem 5.2.

As in the previous section, we assume the following additional hypotheses hold without further mention.

(H) There exist a set $B \subset X$ and some $c > 0$ such that $\Phi_t(x) \to B$ as $t \to \infty$ for all $x \in X$ and $B \cap \{\rho \leq c\}$ has compact closure in X.

Remark 8.11. If ρ is uniformly continuous on X, we can replace $\Phi_t(x) \to B$ by $d(\Phi_t(x), B) \to 0$.

By (H), $B \cap X_0$ has compact closure and attracts points in the forward invariant set X_0. It will be useful to also consider the following nonempty invariant subset of $\overline{B \cap X_0}$:

$$\Omega = \bigcup_{x \in X_0} \omega(x). \tag{8.2}$$

We aim to give sufficient conditions for uniform weak ρ-persistence: There exists $\eta > 0$ such that

$$\limsup_{t\to\infty} \rho(\Phi_t(x)) > \eta \qquad \text{if } \rho(x) > 0.$$

Throughout the following consideration, we suppose that this is false. Then there exist $x_n \in X$ such that $\rho(x_n) > 0$ and

$$\limsup_{t\to\infty} \rho(\Phi_t(x_n)) \to 0, \ n \to \infty. \tag{8.3}$$

Lemma 8.12. *After choosing a subsequence, the positive orbit of each x_n has compact closure and $\omega(x_n)$ is a subset of the compact set $A = \overline{B \cap \{\rho \le c\}}$.*

Proof. We can assume that

$$\limsup_{t\to\infty} \rho(\Phi_t(x_n)) < c/2 \qquad \forall n \in \mathbb{N},$$

with $c > 0$ from (H). We claim that the positive orbit of x_n has compact closure. We fix $n \in \mathbb{N}$ and set $x = x_n$. It is sufficient to show that any sequence $(\Phi(s_j, x))$ with $s_j \in J$, $s_j \to \infty$ as $j \to \infty$, has a convergent subsequence. For $\epsilon \in (0, c/2)$, set

$$U_\epsilon = \{y \in X; \quad \exists z \in B : d(z, y) < \epsilon, |\rho(z) - \rho(y)| < \epsilon\}.$$

U_ϵ is an open neighborhood of B. By (H), there exists some $t_\epsilon \in J$ such that $\Phi_t(x) \in U_\epsilon$ for $t \ge t_\epsilon$; further, $\rho(\Phi_t(x)) \le c/2$ for $t \ge t_\epsilon$. So, whenever $s_j > t_\epsilon$, $\rho(\Phi(s_j, x)) \le c/2$ and we can find some $z_j \in B$ such that $d(z_j, \Phi(s_j, x)) < \epsilon$ and $|\rho(z_j) - \rho(\Phi(s_j, x))| < \epsilon$. Since this holds for every $\epsilon > 0$, we can find a sequence (z_j) in B such that $d(z_j, \Phi(s_j, x)) \to 0$ and $\rho(z_j) - \rho(\Phi(s_j, x)) \to 0$ as $j \to \infty$. For large j, $z_j \in B \cap \{\rho \le c\}$ which has a compact closure by (H). So, after choosing a subsequence, $z_j \to z$ for some $z \in \overline{B \cap \{\rho \le c\}}$. After choosing the analogous subsequence of (s_j), $\Phi(s_j, x) \to z$ as $j \to \infty$. This shows that the positive orbit of $x = x_n$ has compact closure. In particular, the ω-limit set $\omega(x_n)$ is nonempty. A similar proof as before shows that $\omega(x_n) \subset \overline{B \cap \{\rho \le c\}}$. □

Lemma 8.13. *The sequence $\{\omega(x_n)\}_{n\ge1}$, belonging to the compact set $A = \overline{B \cap \{\rho \le c\}}$ has a convergent subsequence which converges to a nonempty, compact, invariant, chain transitive subset $\omega \subset A$ in the compact metric space $(K(A), d_H)$, where $K(A)$ denotes the family of compact subsets of A. After renumbering the convergent subsequence, we may assume that*

$$d_H(\omega(x_n), \omega) \to 0, \quad n \to \infty.$$

A point $x \in X$ belongs to ω if and only if there is a sequence $y_n \in \omega(x_n)$ such that $y_n \to x$.

Proof. The metric space $(K(A), d_H)$ is compact [**10, 65**]. As $\omega(x_n) \subset A$, the convergence of some subsequence follows from the compactness of the space $(K(A), d_H)$. The subsequential limit is nonempty, compact, invariant and chain transitive by Lemma 8.9. The final assertion is well known. See e.g. [**65**]. □

If $x \in \omega$, then $y_n \to x$ for some $y_n \in \omega(x_n)$, so

$$\rho(x) = \lim_n \rho(y_n) = 0$$

by (8.3). Thus $\omega \subset \rho^{-1}(0)$ and, by its invariance, $\omega \subset X_0$.

Definition 8.14. Let $C, B \subset X_0$. C is said to be *chained* to B in X_0, written $C \mapsto B$, if there exists a total trajectory ϕ in X_0 with $\phi(0) \notin C \cup B$ and $\phi(-t) \to C$ and $\phi(t) \to B$ as $t \to \infty$. A finite collection $\{M_1, \ldots, M_k\}$ of subsets of X_0 is called *cyclic* if, after possibly renumbering, $M_1 \mapsto M_1$ in X_0 or $M_1 \mapsto M_2 \mapsto \cdots \mapsto M_j \mapsto M_1$ in X_0 for some $j \in \{2, \ldots, k\}$. Otherwise it is called *acyclic*.

An equivalent graph theoretic formulation of acyclicity may be useful. Let G be the directed graph with vertices $\{M_1, \ldots, M_k\}$ and with a directed edge from M_i to M_j whenever $M_i \mapsto M_j$. Then $\{M_1, \ldots, M_k\}$ is acyclic precisely when G has no cycles (or loops).

Definition 8.15. A set M in X is called *weakly ρ-repelling* if there is no $x \in X$ such that $\rho(x) > 0$ and $\Phi(t, x) \to M$ as $t \to \infty$.

M is called *uniformly weakly ρ-repelling* if there exists some $\epsilon > 0$ such that

$$\limsup_{t\to\infty} d(\Phi_t(x), M) \geq \epsilon \qquad \text{whenever } x \in X, \rho(x) > 0.$$

Proposition 8.16. *Let $\Omega \subset \bigcup_{i=1}^k M_i$ where each $M_i \subset X_0$ is isolated (in X), compact, forward invariant, and weakly ρ-repelling, $M_i \cap M_j = \emptyset$ if $i \neq j$. If Φ is not uniformly weakly ρ-persistent, then $\{M_1, \ldots, M_k\}$ is cyclic.*

Proof. Suppose $\omega \subset M_i$ for some i. Then, as $\omega(x_n) \to \omega$, $\omega(x_n) \subset U$ for all large n, where U is an isolating neighborhood of M_i in X. For such n, $\omega(x_n) \subset M_i$ and $\Phi(t, x_n) \to M_i$, a contradiction because M_i is weakly ρ-repelling. We conclude that $\omega \not\subset M_i$ for each i. However, $\omega \cap M_i \neq \emptyset$ for some i because the omega limit set of a point of ω belongs to Ω and hence to some M_i. By renumbering, we assume that $\omega \cap M_1 \neq \emptyset$. By the Butler-McGehee Theorem 8.8, there exists a total trajectory ϕ in ω with $\phi(0) \in \omega \setminus M_1$ and $\phi(-t) \to M_1$ as $t \to \infty$. Since $\omega \subset X_0$, ϕ takes all of its values in X_0. Thus $\omega(\phi) = \omega(\phi(0)) \subset \Omega$ and $\omega(\phi) \subset \bigcup_{i=1}^k M_i$.

Since $\omega(\phi)$ is chain transitive, we conclude that $\omega(\phi) \subset M_j$ for some j. Indeed, if $u \in \omega(\phi) \cap M_j$ is the initial point of some (ϵ, s)-chain in $\omega(\phi)$ where $\epsilon > 0$ is smaller than the distance between M_j and any M_i and $s \geq 1$, then the forward invariance of M_j implies that the chain cannot leave M_j. Thus $\omega(\phi) \subset M_j$ as asserted.

Therefore, $M_1 \mapsto M_j$ in X_0. If $j = 1$ we are done; otherwise after renumbering, we have $M_1 \mapsto M_2$. Since $\omega \cap M_2 \neq \emptyset$ and $\omega \not\subset M_2$, Theorem 8.8 again gives a total trajectory ψ with $\psi(0) \in \omega \setminus M_2$ and $\psi(-t) \to M_2$ as $t \to \infty$. An argument as above for ϕ gives that $\omega(\psi) \subset M_j$ for some j. Thus $M_1 \mapsto M_2 \mapsto M_j$ in X_0 and we are done if $j = 1$ or $j = 2$. Since the M_i are finite in number, iteration of this argument implies that $\{M_1, \ldots, M_k\}$ is cyclic. □

By contraposition, we obtain the following sufficient conditions for uniform ρ-persistence.

Theorem 8.17. *Let $\Omega \subset \bigcup_{i=1}^k M_i$ where each $M_i \subset X_0$ is isolated (in X), compact, invariant, and weakly ρ-repelling, $M_i \cap M_j = \emptyset$ if $i \neq j$. If $\{M_1, \ldots, M_k\}$ is acyclic, then Φ is uniformly weakly ρ-persistent.*

Butler, Freedman, and Waltman [**22**] and Butler and Waltman [**23**] seem to be the first authors to make use of the acyclicity hypothesis to obtain persistence results. Since then many authors have used it. See [**95, 238**]. A cycle of invariant sets can be an attractor, as shown for the May-Leonard system (3.24) in Proposition 3.26, leading to the failure of persistence to hold. However, acyclicity is not a necessary condition for persistence since a cycle may just as well be repelling.

We present a sufficient condition for a subset of X_0 to be isolated in X (adapted from [**52**, A]).

Lemma 8.18. *Let $M \subset X_0$, and M compact. Then M is isolated in X if M is uniformly weakly ρ-repelling and isolated in X_0.*

Proof. Let $M \subset X_0$ be compact, uniformly weakly ρ-repelling, and isolated in X_0. Then there exists some $\epsilon > 0$ such that

$$\limsup_{t \to \infty} d(\Phi_t(x), M) > \epsilon \qquad \text{whenever } x \in X, \rho(x) > 0.$$

Further, there exists some neighborhood W of M such that W isolates M in X_0: If N is compact, invariant and contained in $X_0 \cap W$, then $N \subset M$. We show that the neighborhood $W \cap U_\epsilon(M)$ of M isolates M in X. Let N be a compact invariant subset of $W \cap U_\epsilon(M)$. Suppose that N contains a point in $x \in X \setminus X_0$. Then $\rho(\Phi_r(x)) > 0$ for some $r \in J$ and $\limsup_{t\to\infty} d(\Phi_t(x), M) > \epsilon$. This implies that $\Phi_t(x) \notin N$ for some $t \in J$, contradicting the invariance of N. Hence, $N \subset X_0$. Since $W \supseteq N$ isolates N in X_0, $N \subset M$. □

The next result shows that each M_i in Theorem 8.17 needs to be uniformly weakly ρ-repelling for Φ to be uniformly weakly ρ-persistent.

Proposition 8.19. *Let $M \subset X_0$ be compact. If Φ is uniformly weakly ρ-persistent, then M is uniformly weakly ρ-repelling.*

Proof. Since Φ is uniformly weakly ρ-persistent, there exists some $\epsilon > 0$ such that

$$\limsup_{t\to\infty} \rho(\Phi_t(x)) > 2\epsilon \quad \text{whenever } \rho(x) > 0.$$

Since M is compact and ρ is continuous, there exists some $\delta > 0$ such that

$$|\rho(y) - \rho(x)| < \epsilon \quad \text{whenever } x, y \in U_\delta(M), d(x,y) < \delta.$$

Suppose that M is not uniformly weakly ρ-repelling. Then there exists some $x \in X$ with $\rho(x) > 0$ and

$$\limsup_{t\to\infty} d(\Phi(t,x), M) < \delta.$$

So, for sufficiently large $t \in J$, there exist $y_t \in M$ such that $d(\Phi(t,x), y_t) < \delta$. Since $M \subset X_0$, $\rho(y_t) = 0$ and

$$\rho(\Phi(t,x)) = \rho(\Phi(t,x)) - \rho(y_t) < \epsilon,$$

a contradiction. □

Theorem 8.20. *Let $\Omega \subset \bigcup_{i=1}^k M_i$ where each $M_i \subset X_0$ is compact, invariant, and isolated in X_0, $M_i \cap M_j = \emptyset$ if $i \neq j$, and $\{M_1, \dots, M_k\}$ is acyclic. Then Φ is uniformly weakly ρ-persistent if and only if each M_i is uniformly weakly ρ-repelling.*

Proof. One direction of our claim follows from Proposition 8.19 and the other from Theorem 8.17 and Lemma 8.18. □

Finally, we present a condition for a compact set in X_0 to be isolated in X_0.

Definition 8.21. Let $Y \subset X$ be forward invariant. A set M in Y is called *stable in Y* if for any neighborhood U of M and any $y \in M$ there exists some neighborhood V of y such that $\Phi(J \times (V \cap Y)) \subset U$.

M is called *locally asymptotically stable in Y* if it is stable in Y and there exists some neighborhood W of M such that $\Phi_t(x) \to M$ as $t \to \infty$ for every $x \in W \cap Y$.

Lemma 8.22. *Let $M \subset X_0$, and M compact. Then M is isolated in X_0 if M is locally asymptotically stable in X_0.*

Proof. Let $M \subset X_0$ be compact and locally asymptotically stable in X_0. Then there exists some neighborhood W of M such that $\Phi_t(x) \to M$ as $t \to \infty$ for all $x \in W \cap X_0$. We show that W isolates M in X_0. Let N be a compact invariant subset of $W \cap X_0$. Suppose that there exists a point $x \in N$ which is not in M. Since N is invariant, there exists a total trajectory ϕ, $\phi(0) = x$, which takes all of its values in N. Since N is compact, $\emptyset \neq \alpha(\phi) \in N$. Since $\alpha(\phi)$ is a nonempty subset of W and invariant, it contains a point $y \in M$. Choose an open neighborhood U of M such that $x \notin U$. Since M is stable, there exist some neighborhood V of y such that $\Phi_t(z) \in U$ for all $t \in J$ and $z \in V \cap X_0$. Since $y \in \alpha(\phi)$ there exists some $r \in J$ such that $\phi(-r) \in V$. Since also $\phi(-r) \in N \subset X_0$, $\phi(t-r) \in U$ for all $t \in J$, contradicting $\phi(0) = x \notin U$. This shows $N \subset M$. □

Rather than use the notion of an acyclic covering of Ω as in Theorem 8.17, some authors use the alternative notion of a Morse decomposition. Here we show that they are closely related.

Definition 8.23. Let Φ be a continuous semiflow on compact invariant set S. An ordered collection $\{M_1, \ldots, M_k\}$ of pairwise disjoint, compact and invariant subsets of S is called a *Morse decomposition* of S if for each $x \in S \setminus \bigcup_{i=1}^{k} M_i$ and each total trajectory ϕ with $\phi(0) = x$ there is an i with $\omega(\phi) \subset M_i$ and $j > i$ with $\alpha(\phi) \subset M_j$.

The equivalence between acyclic coverings and Morse decompositions was observed by Garay [**78**] for (two-sided) continuous flow on the boundary and by Freedman and So [**74**]. Here we follow [**101**]. In the context of persistence theory, the compact invariant set S might be viewed as the global attractor of $\Phi|_{X_0}$ if such exists.

Lemma 8.24. *A finite set $\{M_1, \ldots, M_k\}$ of pairwise disjoint, compact invariant sets in S is an acyclic covering of $\Omega(S) = \bigcup_{x \in S} \omega(x)$ if and only if, on reordering, it is a Morse decomposition of S.*

Proof. *Necessity.* We first claim that for any subcollection $\mathcal{M}$ of M_i's, there exists an element $D \in \mathcal{M}$ such that D cannot be chained to any element in $\mathcal{M}$. Otherwise, some subset of M_i's from $\mathcal{M}$ forms a cycle. Using the claim, we can reorder $\mathcal{M}_0 := \{M_1, \ldots, M_k\}$ by induction. First we choose an element, denoted by D_1, from the collection $\mathcal{M}_0$ such that D_1 cannot be chained to any element in $\mathcal{M}_0$. Having chosen $D_1, \ldots, D_m$, choose an element, denoted by D_{m+1}, from the collection $\mathcal{M}_m := \mathcal{M}_0 \setminus \{D_1, \ldots, D_m\}$ such that D_{m+1} cannot be chained to any element in $\mathcal{M}_m$. After k steps, we have a reordered collection $\mathcal{D} := \{D_1, \ldots, D_k\}$. For any $1 \leq i < j \leq k$, we have $D_i, D_j \in \mathcal{M}_{i-1}$. Therefore, by the choice of D_i, it cannot be chained to any element in $\mathcal{M}_{i-1}$, and hence D_i cannot be chained to D_j.

By assumption, for any $x \in S \setminus \bigcup_{i=1}^k D_i$ and any total trajectory ϕ through x we have $\omega(x) \subset \bigcup_{i=1}^k D_i$, and hence the invariant connectedness of $\omega(x)$ implies that $\omega(x) \subset D_i$ for some i. By Proposition 8.6, $\alpha = \alpha(\phi)$ is internally chain transitive. We claim that $\alpha \subset D_j$ for some j. Suppose not. Since $\alpha \subset S$ is compact and invariant, $\alpha \cap (\bigcup_{i=1}^k M_i) \neq \emptyset$, and hence there exists some M_{i_1} $(1 \leq i_1 \leq k)$ such that $\alpha \cap D_{i_1} \neq \emptyset$. By the Butler-McGehee Theorem 8.8, there exist $w_1 \in \alpha \setminus D_{i_1}$ and a total trajectory $\gamma(w_1) \subset \alpha$ such that $\alpha(\gamma(w_1)) \subset D_{i_1}$. Since $w_1 \in \alpha \subset S$, $\omega(w_1) \subset \bigcup_{i=1}^k D_i$, and hence, by the invariant connectedness of $\omega(w_1)$, there exists some D_{i_2} $(1 \leq i_2 \leq k)$ such that $\omega(w_1) \subset D_{i_2}$. Therefore, D_{i_1} is chained to D_{i_2}, i.e., $D_{i_1} \mapsto D_{i_2}$. Clearly, $\omega(w_1) \subset \alpha$. Then $\alpha \cap D_{i_2} \neq \emptyset$. Again, by Butler-McGehee, there exists $w_2 \in \alpha \setminus D_{i_2}$ and a total trajectory $\gamma(w_2) \subset \alpha$ and $\alpha(\gamma(w_2)) \subset D_{i_2}$. We can repeat the above argument to get an i_3 $(1 \leq i_3 \leq k)$ such that $D_{i_2} \mapsto D_{i_3}$. Since there are only finitely many D_m's, we will eventually arrive at a cyclic chain of some D_m, which contradicts the acyclicity condition. We have proved the claim so $\alpha \subset D_j$ for some j and it follows that $D_j \mapsto D_i$, and hence, by the ordering of $\{D_1, \ldots, D_k\}$, we must have $j > i$. Therefore, $\{D_1, \ldots, D_k\}$ is a Morse decomposition of S.

Sufficiency. The definition of a Morse decomposition implies that $\Omega(S) \subset \bigcup_{i=1}^k M_i$. We further claim that if $M_{i_1} \mapsto M_{i_2}$, then $i_1 > i_2$. Indeed, let $\gamma(x)$ be a total trajectory through some $x \notin M_{i_1} \cup M_{i_2}$ such that $\omega(x) \subset M_{i_2}$ and $\alpha(\gamma) \subset M_{i_1}$. If $x \in M_l$ for some l, we have $\omega(x) \subset M_l \cap M_{i_2}$ and $\alpha(\gamma) \subset M_l \cap M_{i_1}$, and hence $i_1 = l = i_2$, contradicting that $x \notin M_{i_1} \cup M_{i_2}$. It follows that $x \in S \setminus \bigcup_{i=1}^k M_i$. Since $\{M_1, \ldots, M_k\}$ is a Morse decomposition of S, there exist $j > i$ such that $\omega(x) \in M_i$ and $\alpha(\gamma) \in M_j$. Then we have $i_1 = j > i = i_2$. By this claim, it is easy to see that no subset of M_i's forms a cycle. Therefore, $\{M_1, \ldots, M_k\}$ is an acyclic covering of $\Omega(S)$. □

Remark 8.25. If $\{M_1, \ldots, M_k\}$ is a Morse decomposition of S, then each M_i is isolated in S. See [**101**].

8.4. Uniform persistence in a food chain

Consider a food chain were z eats y and y eats x. P. Turchin [**230**] refers to the model below as the Oksanen model, citing the paper of Oksanen et al. [**173**], although it was studied earlier by Freedman and Waltman [**75**]:

$$\begin{aligned} x' &= rx(1 - x/K) - \frac{1}{Y_{xy}} y g(x), \\ y' &= y g(x) - k_y y - \frac{1}{Y_{yz}} z h(y), \\ z' &= z h(y) - k_z z. \end{aligned} \tag{8.4}$$

For definiteness, we take the Holling type-II functional response for the predators:

$$g(x) = \frac{mx}{k+x}, \; h(y) = \frac{My}{L+y}.$$

Constants $r, K, m, k, M, L, Y_{xy}, Y_{yz}, k_y, k_z$ are all positive. By a simple scaling $y \to y/Y_{xy}$, and $z \to z/Y_{yz}$, we can scale out the unsightly parameters Y_{xy}, Y_{yz}, so we will hereafter assume

$$Y_{xy} = 1, \; Y_{yz} = 1.$$

There is a large body of literature on this system because of the chaotic nature of its positive solutions. Hastings and Powell [**97**] used numerical simulations to study the chaotic "tea cup" attractor but did not prove chaos. Muratori and Rinaldi [**171**] recognized that the time scale of growth is faster at lower trophic levels and they employ singular perturbation theory to study its periodic orbits. Chiu and Hsu [**30**] give conditions for top predator extinction. Klebanoff and Hastings [**119**] use codimension-two bifurcation theory in a neighborhood of the equilibrium where only x and y are present to give a rigorous proof of "small chaos". Further work using bifurcation theory may be found in [**131**]. See also the many references in these papers.

We answer some simple questions: can species x and y persist; can all three species persist? Our approach follows Freedman and Waltman [**75**].

The equations (8.4) generate a continuous flow on $X = \mathbb{R}^3_+$. If

$$g(K) > k_y, \tag{8.5}$$

then there are three equilibria on the boundary of $\mathbb{R}^3_+$:

$$E_0 = (0,0,0), \; E_x = (K,0,0), \; E_{xy} = (x^*, y^*, 0),$$

where $x^* > 0$ and $y^* > 0$ satisfy:

$$g(x^*) = k_y, \; y^* = (r/k_y)x^*(1 - x^*/K).$$

Setting $z = 0$ in (8.4) gives the classical Rosenzweig-MacArthur predator-prey model. In a remarkable paper with an interesting history, Liou and Cheng [**148**] consider this planar predator-prey system, among others. They prove that, if E_{xy} is unstable in the linear approximation, i.e., if

$$\frac{k}{K} < \frac{m - k_y}{m + k_y}, \tag{8.6}$$

then there is a unique periodic orbit

$$P = \{(\bar{x}(t), \bar{y}(t), 0) : 0 \le t \le T\}$$

surrounding E_{xy} in the $x - y$ plane. We let $T > 0$ denote its minimal period. Moreover, they show that P is orbitally asymptotically stable in the linear approximation (one Floquet exponent is negative, the other is one)

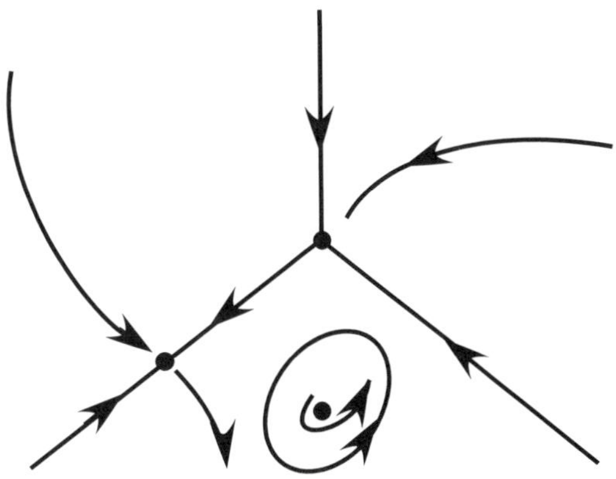

Figure 8.1. Dynamics on the boundary of the octant.

and that it attracts all points in the open quadrant except E_{xy} for the $z = 0$ subsystem. A shorter proof of these results uses the method in Kuang and Freedman [**133**]. See [**132**] for conditions implying the global stability of E_{xy} for more general planar systems.

We introduce the persistence function:

$$\rho(x, y, z) = \min\{x, y, z\}.$$

Then $X_0 = \rho^{-1}(0) = \partial\mathbb{R}^3_+$.

Define the four compact invariant sets:

$$M_0 = \{E_0\},\ M_x = \{E_x\},\ M_{xy} = \{E_{xy}\},\ M_P = P.$$

Every orbit on X_0 converges to one of these sets. Indeed, all orbits in the $y-z$ plane converge to E_0; all orbits on the open $x-z$ plane converge to E_x; as noted above, except for E_{xy}, all orbits in the open $x-y$ plane converge to the periodic orbit P.

Among the four invariant sets, we have the following chains:

$$M_0 \mapsto M_x \mapsto M_P,\ M_{xy} \mapsto M_P,\ \text{in } X_0.$$

Since no cycles are formed, $M = \{M_x, M_y, M_{xy}, M_P\}$ is an acyclic covering of the boundary limit sets. Figure 8.1 depicts the invariant sets.

In order to ensure persistence, we need to verify that M_v is weakly ρ-repelling for $v = 0, x, xy, P$. This is equivalent to showing that $W^s(M_v) \subset X_0$ where $W^s(M_v)$ denotes the stable manifold of M_v. This is automatically satisfied for M_0 and M_x. For it to be satisfied for M_{xy} we must assume that it is unstable to invasion by z. It may be easily verified that

$$h(y^*) > k_z \tag{8.7}$$

implies that E_{xy} is a repeller (three eigenvalues with positive real part), so

$$W^s(M_{xy}) = M_{xy} \subset X_0.$$

As for P, the corresponding variational equation is the T-periodic system given by

$$w' = \begin{pmatrix} a(t) & b(t) & 0 \\ c(t) & d(t) & f(t) \\ 0 & 0 & e(t) \end{pmatrix} w$$

where

$$e(t) = h(\bar{y}(t)) - k_z.$$

If $\Psi(t)$, $\Psi(0) = I$, denotes the fundamental matrix for this system, then the stability of P is determined by the eigenvalues of $\Psi(T)$: orbitally asymptotically stable if, except for the trivial eigenvalue one corresponding to the eigenvector $(\bar{x}'(0), \bar{y}'(0), 0)^T$, they are inside the unit circle of the complex plane, unstable if an eigenvalue lies outside of it. It is easy to see that Ψ has the same structure (pattern of zeros) as the coefficient matrix so the stability of P is determined by the eigenvalues of the upper left two-by-two and one eigenvalue is given by

$$\exp\left(\int_0^T e(s)ds\right).$$

This eigenvalue determines whether z can invade the periodic regime. As noted above, P is orbitally asymptotically stable in the linear approximation as a solution of the planar $x - y$ system so the nontrivial eigenvalue of the upper left two-by-two is inside the unit circle. Therefore, the key Floquet exponent is given by

$$\lambda := \frac{1}{T}\int_0^T h(\bar{y}(t))dt - k_z.$$

If $\lambda > 0$, then z can invade the periodic regime P. In that case, P is a saddle periodic orbit with two-dimensional stable set

$$W^s(P) = \{p = (x, y, 0) \neq E_{xy} : x > 0, y > 0\} \subset X_0.$$

We can now state our main result.

Proposition 8.26. *If (8.5) and (8.6) hold, then* (8.4) *is uniformly $\tilde{\rho}$-persistent with $\tilde{\rho} = \min\{x, y\}$: $\exists \epsilon > 0$ such that*

$$\liminf_{t\to\infty} x(t) > \epsilon,\ \liminf_{t\to\infty} y(t) > \epsilon,\ x(0), y(0) > 0.$$

If, in addition, (8.7) *holds and* $\lambda > 0$*, then* (8.4) *is uniformly* ρ*-persistent:* $\exists \epsilon > 0$ *such that*

$$\liminf_{t\to\infty} x(t) > \epsilon, \ \liminf_{t\to\infty} y(t) > \epsilon, \ \liminf_{t\to\infty} z(t) > \epsilon, \ x(0), y(0), z(0) > 0.$$

Notice that ρ-persistence fails if $h(y^*) < k_z$ or $\lambda < 0$ since the stable manifold of either E_{xy} or of P intersects the interior of $\mathbb{R}^3_+$. (8.7) and $\lambda > 0$ hold if parameter M is sufficiently large.

Proof. We prove the second assertion, leaving the first to the reader. See Exercise 8.6.

We note that M_0, M_x, M_{xy} are isolated compact invariant sets. Moreover, they are hyperbolic equilibria, so they are isolated in $\mathbb{R}^3_+$ by the Hartman-Grobman theorem (see [**177**]) or the saddle point property (Theorem III.6.1 in [**90**]). As their stable manifolds are contained in the boundary of the octant, each is weakly ρ-repelling.

Assume that $\lambda > 0$ and that P is not uniformly weakly ρ-repelling. Since h is continuous and P has a compact neighborhood, we can choose $\epsilon > 0$ such that

$$|h(\bar{y}(t) + u) - h(\bar{y}(t))| \leq \lambda/2, \qquad t \in \mathbb{R}, u \in (-\epsilon, \epsilon).$$

Let $\delta > 0$, to be determined later. Since P is not uniformly weakly ρ-repelling, there exists a solution of (8.4) with

$$\limsup_{t\to\infty} d((x(t), y(t), z(t)), P) < \delta, \qquad x(0) > 0, y(0) > 0, z(0) > 0.$$

By the form of the equations, $x(t) > 0$, $y(t) > 0$, $z(t) > 0$ for all $t \geq 0$. After a time shift, we can assume that $d((x(t), y(t), z(t)), P) < \delta$ for all $t \geq 0$. For each $n \in \mathbb{N}$, we can choose $r_n \in [0, T]$ such that

$$|x(nT) - \bar{x}(r_n)| < \delta, \quad |y(nT) - \bar{y}(r_n)| < \delta, \quad |z(nT)| < \delta.$$

Because solutions of (8.4) continuously depend on initial data and P has a compact neighborhood, we can choose $\delta > 0$ so small that

$$|y(nT + s) - \bar{y}(r_n + s)| < \epsilon, \qquad n \in \mathbb{N}, s \in [0, T].$$

We integrate the z-equation in (8.4),

$$\begin{aligned} z((n+1)T) =& z(nT) \exp\Big(\int_0^T h(y(nT+s))ds - k_z T \Big) \\ \geq & z(nT) \exp\Big(\int_0^T h(\bar{y}(r_n+s))ds - \frac{T\lambda}{2} - k_z T \Big). \end{aligned}$$

Since $\bar{y}$ is T-periodic,

$$z((n+1)T) \geq z(nT) \exp\Big(\int_0^T h(\bar{y}(s))ds - \frac{T\lambda}{2} - k_z T \Big) \geq z(nT) e^{T\lambda/2}.$$

Since $z(nT) > 0$ and $\lambda > 0$, $z(nT) \to \infty$, a contradiction.

So P is uniformly weakly ρ-repelling in $X = \mathbb{R}^3_+$. P is a locally asymptotically stable set in the $z = 0$ plane which is a relatively open subset of X_0. Thus P is isolated in X_0. By Lemma 8.22 and Lemma 8.18, P is isolated in X.

By Theorem 5.2 and Theorem 8.17, it suffices to show that there exists a compact attractor of bounded sets since we have already noted that the acyclicity hypothesis is satisfied.

The differential inequality $x' \le rx(1-x/K)$ leads immediately to lim sup $x(t) \le K$. In fact, it is easy to see that unless $(x(t), y(t), z(t)) \to (K, 0, 0)$ as $t \to \infty$, which holds precisely when $y(0) = 0$ and $z(0) > 0$, it follows that $x(t) \le K$ for all large t, say, $t \ge T$. Let $k_m = \min\{k_y, k_z\}$ and choose $R > 0$ such that $k_m R > K(r/4 + k_m) + 1$. Suppose that $x + y + z \ge R$ and $x \le K$. Then $y + z \ge R - K$, so

$$\begin{aligned}(x+y+z)' &= rx(1-x/K) - k_y y - k_z z\\ &\le rK/4 - k_m(R-K)\\ &\le -1, \quad t \ge T.\end{aligned}$$

We conclude that there exists $\tau > T$ such that $x(t) + y(t) + z(t) \le R$ and $x(t) \le K$ for all $t \ge \tau$. This proves the point dissipativity property of the flow. Now we show that bounded sets have bounded orbits. It is sufficient to find a family of large forward invariant sets such that every bounded set is contained in one. Consider the family of sets

$$Q(L) = \{(x, y, z) \in \mathbb{R}^3_+ : x \le L \text{ and } x + y + z \le R(L)\}, \ L > K$$

where $k_m R(L) = rK/4 + k_m L + 1$. Every bounded set is contained in $Q(L)$ for all large L. Forward invariance of $Q(L)$ holds since the inner product of the vector field with the outer normal to $Q(L)$ is negative at each boundary point of $Q(L)$. This is clear along $x = L$ since $L > K$. On $x + y + z = R(L)$ we have, as above,

$$\begin{aligned}(x+y+z)' &= rx(1-x/K) - k_y y - k_z z\\ &\le rK/4 - k_m(R(L) - L) = -1.\end{aligned}$$

By Remark 2.26, the semiflow is asymptotically smooth so Theorem 2.33 implies the existence of a compact attractor of bounded sets. □

8.5. The metered endemic model revisited

In Section 4.3, we showed uniform host persistence in a metered endemic model if $\kappa(0) = 0$ for the per capita infection function κ. Using the acyclicity approach, we want to derive sharp conditions for uniform host persistence if $\kappa(0) > 0$ in a reduced model of $S \to I \to S$ type,

$$
\begin{aligned}
&N_n = S_n + I_n, \\
&S(0) = \beta N_n + S_n, \quad I(0) = I_n, \\
&\left.\begin{aligned} S' &= -\tfrac{\kappa(N)}{N} SI - \mu S + \gamma I, \\ I' &= \tfrac{\kappa(N)}{N} SI - (\mu + \alpha + \gamma) I, \\ N &= S + I, \end{aligned}\right\} \quad 0 \le t \le 1, \\
&S_{n+1} = S(1), \qquad I_{n+1} = I(1).
\end{aligned} \tag{8.8}
$$

We introduce the fraction of infective individuals, $y = I/N$, and rewrite the system in terms of N and y,

$$
\begin{aligned}
&N(0) = (\beta + 1) N_n, \qquad y(0) = \frac{y_n}{\beta + 1}, \\
&\left.\begin{aligned} N' &= -N(\mu + \alpha y), \\ y' &= y\big((\kappa(N) - \alpha)(1 - y) - \gamma\big), \end{aligned}\right\} \quad 0 \le t \le 1, \\
&N_{n+1} = N(1), \qquad y_{n+1} = y(1).
\end{aligned} \tag{8.9}
$$

We assume $(\beta + 1)e^{-\mu} > 1$ such that the host population grows geometrically in the absence of the disease. In order to derive conditions for host persistence in the presence of the disease, we use the acyclicity approach of Section 8.3. We choose $X = \mathbb{R}_+ \times [0, 1]$ and $\rho(N, y) = N$. To satisfy (H), we choose $B = X$. Then $B \cap \{\rho \le c\} = [0, c] \times [0, 1]$ is compact. Further, $X_0 = \{0\} \times [0, 1]$. To figure out Ω, we study the dynamics on X_0. To this end we consider the equation

$$y' = y\big((\kappa_0 - \alpha)(1 - y) - \gamma\big) \tag{8.10}$$

with $\kappa_0 = \kappa(0)$.

If $\kappa_0 \le \alpha$, then $y' \le 0$ and $y_{n+1} \le \frac{y_n}{\beta+1}$ and $y_n \to 0$. In this case, $\Omega = \{(0, 0)\}$.

So we assume $\kappa_0 > \alpha$. Then

$$(y^{-1})' = -y^{-2} y' = -\xi y^{-1} + (\kappa_0 - \alpha), \tag{8.11}$$

with $\xi = \kappa_0 - \alpha - \gamma \le \kappa_0 - \alpha$. If $\xi \le 0$, $y^{-1}(1) \ge y^{-1}(0) + \kappa_0 - \alpha$ and $y_{n+1}^{-1} \ge y_n^{-1}(\beta + 1)$ which implies $y_n \to 0$ as $n \to \infty$ and again $\Omega = \{(0, 0)\}$.

So we assume that $\xi > 0$. We integrate (8.11),

$$y^{-1}(t) = y^{-1}(0) e^{-\xi t} + \frac{\kappa_0 - \alpha}{\xi}(1 - e^{-\xi t}). \tag{8.12}$$

We recall that $y_{n+1} = y(0)$ and obtain

$$y_{n+1}^{-1} = y_n^{-1}(\beta+1)e^{-\xi} + \frac{\kappa_0 - \alpha}{\xi}(1 - e^{-\xi}). \tag{8.13}$$

We still have $y_n \to 0$ as $n \to \infty$ if $(\beta+1)e^{-\xi} \geq 1$. So we assume $(\beta+1)e^{-\xi} < 1$. Then y_n has a limit y_∞ and y_∞^{-1} is a positive fixed point of (8.13). So

$$y_\infty = \frac{1-(\beta+1)e^{-\xi}}{1-e^{-\xi}} \frac{\xi}{\kappa_0 - \alpha}. \tag{8.14}$$

It also follows that y_∞^{-1} is a locally asymptotically stable fixed point of (8.13). The preceding discussion shows that $\Omega = \{(0,0)\}$ in the following cases:

(i) $\kappa_0 \leq \alpha + \gamma$.

(ii) $\kappa_0 > \alpha + \gamma$, but $(\beta+1)e^{-\xi} \geq 1$ for $\xi = \kappa_0 - \alpha - \gamma$.

If $\xi > 0$ and $(\beta+1)e^{-\xi} < 1$, then $\Omega = \{(0,0), (0, y_\infty)\}$. In either case Ω has an acyclic covering.

We illustrate the application of Theorem 8.20 for the most interesting case, $\Omega = \{(0,0), (0, y_\infty)\}$. To achieve that $(0, y_\infty)$ is uniformly weakly repelling, we substitute y_∞ back into (8.12),

$$\tilde{y}^{-1}(t) = y_\infty^{-1}(\beta+1)e^{-\xi t} + \frac{\kappa_0 - \alpha}{\xi}(1 - e^{-\xi t}) \tag{8.15}$$

and we assume that

$$(\beta+1)\exp\Big(-\mu - \alpha\int_0^1 \tilde{y}(t)dt\Big) > 1. \tag{8.16}$$

In order to show that $(0, y_\infty)$ is uniformly weakly ρ-repelling we assume the contrary. Then, for any $\epsilon > 0$ there exists a solution (N_n, y_n) such that $\limsup_{n\to\infty} N_n < \epsilon$ and $\limsup_{n\to\infty} |y_n - y_\infty| < \epsilon$. So, for sufficiently large n, $N_n < \epsilon$ and $|y_n - y_\infty| < \epsilon$. Let $\eta > 0$ be arbitrary. Notice that $\tilde{y}$ is the solution of (8.10) with $y(0) = (\beta+1)y_\infty$. It follows from the continuous dependence of solutions on their initial data, that, for sufficiently large n, the solution of (8.10) satisfies $|y(t) - \tilde{y}(t)| < \eta$ for all $t \in [0,1]$ if $\epsilon > 0$ is chosen small enough. If η is chosen small enough, we have $(\beta+1)\exp(-\mu - \alpha\int_0^1 y(t)dt) \geq 1 + \delta$ for some $\delta > 0$ which does not depend on n which are sufficiently large. This implies $N_{n+1} \geq N_n(1+\delta)$ for all sufficiently large n and $N_n \to \infty$ as $n \to \infty$, a contradiction.

It is even easier to prove that $(0,0)$ is uniformly weakly ρ-repelling. Further, $(0,0)$ is isolated in $X_0 = \{N_0 = 0\}$ because every solution in X_0 with $y_0 \neq 0$ converges to $(0, y_\infty)$. To formulate a theorem which summarizes our considerations, we evaluate (8.16). From (8.15),

$$\tilde{y}(t) = \frac{e^{\xi t}}{y_\infty^{-1}(\beta+1) + \frac{\kappa_0-\alpha}{\xi}(e^{\xi t} - 1)}.$$

We notice that, up to a constant factor, the numerator is the derivative of the denominator,

$$\begin{aligned}\int_0^1 \tilde{y}(t)dt =& \frac{1}{\kappa_0-\alpha}\ln\frac{y_\infty^{-1}(\beta+1)+\frac{\kappa_0-\alpha}{\xi}(e^\xi-1)}{y_\infty^{-1}(\beta+1)}\\ =& \frac{1}{\kappa_0-\alpha}\ln\frac{\beta+1+y_\infty\frac{\kappa_0-\alpha}{\xi}(e^\xi-1)}{\beta+1}.\end{aligned}$$

We substitute (8.14),

$$\int_0^1 \tilde{y}(t)dt = \frac{1}{\kappa_0-\alpha}\ln\frac{e^\xi}{\beta+1}.$$

So condition (8.16) becomes

$$\begin{aligned}1 <& (\beta+1)e^{-\mu}\Big(\frac{e^\xi}{\beta+1}\Big)^{\frac{\alpha}{\alpha-\kappa_0}} = e^{-\mu}e^{\frac{\xi\alpha}{\alpha-\kappa_0}}(\beta+1)^{1-\frac{\alpha}{\alpha-\kappa_0}}\\ =& e^{-\mu}e^{\frac{\xi\alpha}{\alpha-\kappa_0}}(\beta+1)^{\frac{\kappa_0}{\kappa_0-\alpha}}.\end{aligned}$$

Theorem 8.27. *Assume $\mathcal{R}_0^{\text{host}} = (\beta+1)e^{-\mu} > 1$ (such that the host grows geometrically in the absence of the disease). Then the host persists uniformly even in the presence of the disease if either (i) or (ii) are satisfied, or*

$$\begin{cases}(\beta+1)e^{-\xi} < 1 \ \ \textit{for} \ \ \xi = \kappa_0-\alpha-\gamma > 0, \ \kappa_0 = \kappa(0)\\ \textit{and}\\ e^{-\mu}e^{\frac{\xi\alpha}{\alpha-\kappa_0}}(\beta+1)^{\frac{\kappa_0}{\kappa_0-\alpha}} > 1.\end{cases}$$

Proof. By Theorem 8.17, the host persists uniformly weakly. Uniform persistence now follows in exactly the same way as in Section 4.3.1. □

8.6. Nonlinear matrix models (epilog): Biennials

The individuals of a biennial species only live for two years and only reproduce in the second year. Examples are biennial plants like parsley, Lunaria, silverbeet, Sweet William, colic weed, and carrot.

They can be modeled by a two-dimensional discrete time system

$$\left.\begin{aligned}J(n+1) &= M(n)g_1(J(n),M(n)),\\ M(n+1) &= J(n)g_2(J(n),M(n)),\end{aligned}\right\} n\in\mathbb{Z}_+. \tag{8.17}$$

Here $J(n)$ are the juvenile and $M(n)$ the mature (or adult) individuals in year n, where the juveniles are the individuals in their first year and the mature individuals are in their second year; the two stages coincide with two age classes. This is a special case of a discrete time system

$$x(n+1) = F(x(n)), \qquad n\in\mathbb{Z}_+, \tag{8.18}$$

where $x(n) = (J(n), M(n)) \in \mathbb{R}^2_+$ and $F : \mathbb{R}^2_+ \to \mathbb{R}^2_+$,

$$F(x) = \begin{pmatrix} f_1(x_1, x_2) \\ f_2(x_1, x_2) \end{pmatrix}, \qquad x = (x_1, x_2), \tag{8.19}$$

with

$$f_1(y, 0) = 0, \quad f_2(0, y) = 0, \qquad y \geq 0. \tag{8.20}$$

In (8.17),

$$f_1(x_1, x_2) = x_2 g_1(x_1, x_2), \qquad f_2(x_1, x_2) = x_1 g_2(x_1, x_2). \tag{8.21}$$

Systems of this form have been studied in [**45, 48, 63, 64**], mainly for interesting bifurcation and stability phenomena. We will investigate them from the point of view of population persistence.

Biennial species are a special case of semelparous species where individuals only reproduce once in a lifetime. Respective models have been studied in [**41, 42, 59, 122, 143**] where also more references can be found.

By (8.20),

$$F(\{0\} \times \mathbb{R}_+) \subset \mathbb{R}_+ \times \{0\}, \qquad F(\mathbb{R}_+ \times \{0\}) \subset \{0\} \times \mathbb{R}_+, \tag{8.22}$$

and $\partial\mathbb{R}^2_+$ is forward invariant under F. Moreover,

$$F^2(y, 0) = \begin{pmatrix} f_1(0, f_2(y, 0)) \\ f_2(0, f_2(y, 0)) \end{pmatrix} = \begin{pmatrix} f_1(0, f_2(y, 0)) \\ 0 \end{pmatrix} \tag{8.23}$$

and

$$F^2(0, y) = \begin{pmatrix} 0 \\ f_2(f_1(0, y), 0) \end{pmatrix}. \tag{8.24}$$

In particular, $\{0\} \times \mathbb{R}_+$ and $\mathbb{R}_+ \times \{0\}$ are forward invariant under F^2.

Assume that F is continuous and differentiable at (0,0), that the discrete semiflow induced by F is point-dissipative, and that $\mathbb{R}^2_+ \setminus \{(0, 0)\}$ and $(0, \infty)^2$ are forward invariant.

So solutions of (8.17) with $J(0) > 0$ and $M(0) = 0$ have the property that $J(2n) > 0$, $M(2n) = 0$ and $J(2n + 1) = 0$, $M(2n + 1) > 0$, i.e., in even years there are only juveniles and in odd years only adults. By switching the initial conditions, it will be the other way around. Solutions of this form have been called *synchronous* in [**40, 45**] as the two stages are synchronized as to appear and vanish alternately in one time unit. We will first establish conditions under which a biennial species persists uniformly, but possibly only in a synchronized fashion. This can be done more effectively for systems in the form (8.18) than for systems in the more special form (8.17). We will then look for conditions for uniform persistence of both stages for systems in the form (8.17). Naturally, both stages would need to be present initially.

Remark 8.28. For (8.17), i.e., (8.21), F is continuous and differentiable at (0,0) if g_1 and g_1 are continuous, and

$$F'(0) = \begin{pmatrix} 0 & g_1(0,0) \\ g_2(0,0) & 0 \end{pmatrix}.$$

$\mathbb{R}^2_+ \setminus \{0\}$, $(0,\infty)^2$, and $\mathbb{R}^2_+ \setminus (0,\infty)^2$ are forward invariant under F if $g_1(x) > 0$ and $g_2(x) > 0$ for all $x \in \mathbb{R}^2_+ \setminus \{0\}$.

The semiflow induced by F has a compact attractor if one of the following conditions hold:

(a) There is some $R > 0$, some vector $\breve{x} \in \mathbb{R}^2_+$ and numbers $a_1, a_2 \geq 0$ with $a_1 a_2 < 1$ such that

$$\left.\begin{array}{l} f_1(x) \leq \breve{x}_1 + a_1 x_2 \\ f_2(x) \leq \breve{x}_2 + a_2 x_1 \end{array}\right\} \text{ for } \left\{ \begin{array}{c} x = (x_1, x_2) \in \mathbb{R}^2_+, \\ x_1 + x_2 < R. \end{array}\right.$$

(b) (8.21) holds and f_1 and g_2 are bounded.

(c) (8.21) holds and f_2 and g_1 are bounded.

Proof. We prove the statements involving (a) and (b).

(a) Notice that

$$F(x) \leq \breve{x} + Dx, \quad |x| \geq R,$$

where

$$D = \begin{pmatrix} 0 & a_1 \\ a_2 & 0 \end{pmatrix}$$

has the spectral radius $\sqrt{a_1 a_2} < 1$. So the statement follows from Proposition 7.2 and Corollary 7.3.

(b) As before, but with

$$\breve{x} = \begin{pmatrix} \sup f_1 \\ 0 \end{pmatrix}, \qquad D = \begin{pmatrix} 0 & 0 \\ \sup g_2 & 0 \end{pmatrix}.$$

Notice that $\mathbf{r}(D) = 0$. □

By (8.20), the following partial derivatives exist and satisfy

$$\partial_1 f_1(y,0) = 0, \qquad \partial_2 f_2(0,y) = 0, \qquad y \geq 0. \tag{8.25}$$

So

$$F'(0) = \begin{pmatrix} 0 & \partial_2 f_1(0,0) \\ \partial_1 f_2(0,0) & 0 \end{pmatrix}. \tag{8.26}$$

F^2 is also differentiable at 0 and

$$(F^2)'(0) = (F'(0))^2 = \begin{pmatrix} \partial_2 f_1(0,0)\partial_1 f_2(0,0) & 0 \\ 0 & \partial_2 f_1(0,0)\partial_1 f_2(0,0) \end{pmatrix}. \tag{8.27}$$

Theorem 8.29. *Assume that the semiflow induced by F has a compact attractor of bounded sets and that $\partial_2 f_1(0,0)\partial_1 f_2(0,0) > 1$. Then the semiflow induced by F is uniformly persistent with respect to $\rho(x) = x_1 + x_2$ and there exists a fixed point of F in $(0,\infty)^2$.*

Further, the restrictions of the semiflow induced by F^2 on either $\mathbb{R}_+ \times \{0\}$ or $\{0\} \times \mathbb{R}_+$ are uniformly ρ-persistent with respect to $\rho_1(y,0) = y$ or $\rho_2(0,y) = y$, respectively:

there exists some $\epsilon > 0$ such that

$$\liminf_{n\to\infty} x_1(2n) \ge \epsilon \text{ and } \liminf_{n\to\infty} x_2(2n+1) \ge \epsilon$$

for all solutions of (8.18) with $x_1(0) > 0$ and $x_2(0) = 0$; further,

$$\liminf_{n\to\infty} x_1(2n+1) \ge \epsilon \text{ and } \liminf_{n\to\infty} x_2(2n) \ge \epsilon$$

for all solutions of (8.18) with $x_1(0) = 0$ and $x_2(0) > 0$.

In particular, there exist some $y^\diamond, z^\diamond > 0$ such that

$$F(y^\diamond, 0) = (0, z^\diamond) \quad \text{and} \quad F(0, z^\diamond) = (y^\diamond, 0).$$

Finally, there exists compact intervals I_1 and I_2 in $(0,\infty)$ with the following properties:

(1) $F(I_1 \times \{0\}) = \{0\} \times I_2, \quad F(\{0\} \times I_2) = I_1 \times \{0\}$,
(2) $I_1 \times \{0\}$ *is the compact attractor of all compact sets in* $(0,\infty) \times \{0\}$ *for the semiflow induced by* F^2,
(3) $\{0\} \times I_2$ *is the compact attractor of all compact sets in* $\{0\} \times (0,\infty)$ *for the semiflow induced by* F^2.
(4) $I_1 \times \{0\}$ *and* $\{0\} \times I_2$ *are locally asymptotically stable for the restriction on* $\partial\mathbb{R}^2_+$ *of the semiflow induced by* F^2.

The proof of this theorem also shows that

Remark 8.30. I_1 and I_2 are the compact attractors of compact sets in $(0,\infty)$ for the maps $\phi(y) = f_1(0, f_2(y,0))$ and $\psi(z) = f_2(f_1(0,z),0)$, respectively. Further, $y^\diamond = \phi(y^\diamond)$ and $z^\diamond = \psi(z^\diamond)$.

Proof. The population persistence result follows from Theorem 7.9 applied to F as does the existence of a nonzero fixed of F. By the form of F, the fixed point lies in $(0,\infty)^2$.

The persistence results for synchronous solutions follow from the same theorem applied to the restrictions of F^2 to $\mathbb{R}_+ \times \{0\}$ and $\{0\} \times \mathbb{R}_+$, respectively, or, more easily, applied to the functions ϕ and ψ in Remark 8.30. Recall (8.23) and (8.24). Theorem 7.9 also implies that there is some $y^\diamond > 0$ such that $F^2(y^\diamond, 0) = (y^\diamond, 0)$. By (8.22), for some $z^\diamond > 0$,

$F(y^\diamond, 0) = (0, z^\diamond)$ and $F(0, z^\diamond) = (y^\diamond, 0)$ and $F^2(0, z^\diamond) = (0, z^\diamond)$. By (8.23) and (8.23), $y^\diamond = \phi(y^\diamond)$ and $z^\diamond = \psi(z^\diamond)$ in Remark 8.30.

Let A_1 be the persistence attractor for the semiflow induced by F^2 on $\mathbb{R}_+ \times \{0\}$ which is the compact attractor of all compact sets in $(0, \infty) \times \{0\}$ (Theorem 5.7). By (8.23), $A_1 = I_1 \times \{0\}$ where I_1 is the compact attractor of all compact sets in $(0, \infty)$ for the semiflow induced by the map $\phi = f_1(0, f_2(\cdot, 0))$. Since I_1 is connected by Theorem 5.9, it is a compact interval in $(0, \infty)$. Similarly, the persistence attractor A_2 for the semiflow induced by F^2 on $\{0\} \times \mathbb{R}_+$ is of the form $A_2 = \{0\} \times I_2$ with a compact interval I_1 in $(0, \infty)$. A_2 is the compact attractor of all compact sets in $\{0\} \times (0, \infty)$, and I_2 is the compact attractor of all compact subsets of $(0, \infty)$ for the semiflow induced by $\psi = f_2(f_1(0, \cdot), 0)$. Since $F^2(A_1) = A_1$, also $F^2(F(A_1)) = F(A_1)$. Since $F(A_1) = F(I_1 \times \{0\}) \subset \{0\} \times (0, \infty)$ and $A_2 = \{0\} \times I_2$ contains all compact F^2-invariant subsets in $\{0\} \times (0, \infty)$, we have $F(A_1) \subset A_2$. By the same token, $F(A_2) \subset A_1$. So $A_2 = F^2(A_2) \subset F(A_1)$ and $A_1 = F^2(A_1) \subset F(A_2)$. □

We try to find handy conditions under which I_1 and I_2 are singleton sets.

This amounts to coming up with a property for $f : (0, \infty) \to (0, \infty)$ such that arbitrary compositions g of functions with these properties have a unique positive fixed point x^* and $g^n(x) \to x^*$ as $n \to \infty$, $x > 0$. The set $(0, \infty)$ is a complete metric space under the metric $d(x, y) = |\ln x - \ln y|$. This metric is a special case of the *part metric* (also called *Thompson metric*). Every interval $[a, b]$ with $0 < a < b < \infty$ is compact with respect to the metric d. The following is proved in Example 2.58.

Proposition 8.31. *If $f(x)/x$ decreases and $xf(x)$ increases in $x > 0$, then f is a contraction: $d(f(x), f(y)) \le d(x, y)$ for all $x, y > 0$. If both the decrease and the increase are strict, f is a strict contraction: $d(f(x), f(y)) < d(x, y)$ if $x \ne y$.*

Lemma 8.32. *The composition of a contraction and of a strict contraction is a strict contraction.*

Proposition 8.33. *Assume that*

(i) *$f_1(0, z)/z$ decreases and $zf_1(0, z)$ increases in $z > 0$,*

(ii) *$f_2(y, 0)/y$ decreases and $yf_2(y, 0)$ increases in $y > 0$,*

and that the decrease and increase is strict in (i) or in (ii).

Then, with the assumptions and the notation of Theorem 8.29, $I_1 = \{y^\diamond\}$ and $I_2 = \{z^\diamond\}$ and $y^\diamond, z^\diamond > 0$ are the unique solutions of

$$y^\diamond = f_1(0, z^\diamond) \quad \text{and} \quad z^\diamond = f_2(y^\diamond, 0). \tag{8.28}$$

Proof. The functions $f_1(0,\cdot)$ and $f_2(\cdot,0)$ are weak contractions with respect to the part metric on $(0,\infty)$ with one being a strict contraction. Then the compositions $f_1(0,f_2(\cdot,0))$ and $f_2(f_1(0,\cdot),0)$ are strict contractions with respect to the part metric. The assertion now follows from Theorem 2.56 and (8.23) and (8.24). $\square$

Remark 8.34. Assume that

(i) $\frac{d}{dz}(f_1(0,z)/z) \le 0$ and $\frac{d}{dz}(zf_1(0,z)) \ge 0$ for $z>0$,

(ii) $\frac{d}{dy}(f_2(y,0)/y) \le 0$ and $\frac{d}{dy}(yf_2(y,0)) \ge 0$ for $y>0$,

and that the inequalities are strict in (i) or in (ii).

Then, with the assumptions and the notation of Theorem 8.29, the points $(y^\diamond,0)$ and $(0,z^\diamond)$ are locally asymptotically stable fixed points of F^2 if

$$\partial_2 f_1(y^\diamond,0)\partial_1 f_2(0,z^\diamond) < 1,$$

and unstable if this product is greater than 1.

Proof. By the chain rule and by (8.25),

$$\begin{aligned}(F^2)'(y^\diamond,0) =& F'(0,z^\diamond)F'(y^\diamond,0)\\ =& \begin{pmatrix}\partial_1 f_1(0,z^\diamond) & \partial_2 f_1(0,z^\diamond)\\ \partial_1 f_2(0,z^\diamond) & 0\end{pmatrix}\begin{pmatrix}0 & \partial_2 f_1(y^\diamond,0)\\ \partial_1 f_2(y^\diamond,0) & \partial_2 f_2(y^\diamond,0)\end{pmatrix}\\ =& \begin{pmatrix}\partial_2 f_1(0,z^\diamond)\partial_1 f_2(y^\diamond,0) & ?\\ 0 & \partial_2 f_1(y^\diamond,0)\partial_1 f_2(0,z^\diamond)\end{pmatrix}.\end{aligned}$$

So the stability is determined by the entries in the diagonal. (i) implies that

$$|\partial_2 f_1(0,z)| \le \frac{f_1(0,z)}{z},$$

and (ii) implies that

$$|\partial_1 f_2(y,0)| \le \frac{f_2(y,0)}{y}.$$

One of the two inequalities is strict. So

$$|\partial_2 f_1(0,z^\diamond)\partial_1 f_2(y^\diamond,0)| < \frac{f_1(0,z^\diamond)}{z^\diamond}\frac{f_2(y^\diamond,0)}{y^\diamond} = 1.$$

So the spectral radius of $(F^2)'(y^\diamond,0)$ is greater (smaller) than 1 if the second diagonal entry $\partial_2 f_1(y^\diamond,0)\partial_1 f_2(0,z^\diamond)$ is greater (smaller) than 1. (Notice that this product is nonnegative.) This implies the assertion for $(y^\diamond,0)$. The proof of the one for $(0,z^\diamond)$ is similar. $\square$

In order to derive conditions for uniform persistence in both stages it is convenient to look at the special system (8.17), i.e., the vector field has the nonlinear matrix form,

$$f_1(y,z) = zg_1(y,z), \quad f_2(y,z) = yg_2(y,z). \tag{8.29}$$

Theorem 8.35. *Assume that g_1 and g_2 are continuous and $g_1(x) > 0$ and $g_2(x) > 0$ for $x \in \mathbb{R}^2_+ \setminus \{0\}$. Assume that the induced semiflow has a compact attractor of bounded sets.*

Further, assume that $g_1(0,0)g_2(0,0) > 1$ and

$$\inf_{z \in I_2} g_2(0,z) \inf_{y \in I_1} g_1(y,0) > 1, \tag{8.30}$$

where I_1 and I_2 are the intervals in Theorem 8.29.

Then there exists some $\epsilon > 0$ such that

$$\liminf_{n\to\infty} J(n) \geq \epsilon, \qquad \liminf_{n\to\infty} M(n) \geq \epsilon$$

for all solutions of (8.17) with $J(0) > 0$ and $M(0) > 0$.

Further, there exists some $c > 1$ and a compact connected set $A_1 \subset [c^{-1}, c]^2$, $F(A_1) = A_1$, that attracts every compact set in $(0,\infty)^2$.

Proof. For the persistence proof, we use Theorem 8.20 for the semiflow induced by F^2 and the persistence function $\tilde{\rho}(x_1, x_2) = \min\{x_1, x_2\}$.

We check the assumptions.

Both $\partial\mathbb{R}^2_+ = \{\tilde{\rho} = 0\}$ and $\mathbb{R}^m \setminus \partial\mathbb{R}^2_+ = (0,\infty)^2 = \{\tilde{\rho} > 0\}$ are forward invariant under F and thus under F^2. This implies that $X_0 = \partial\mathbb{R}^2_+$ where $X_0 = \{x \in \mathbb{R}^2_+; \forall n \in \mathbb{Z}_+ : \tilde{\rho}(F^{2n}(x)) = 0\}$. By Theorem 8.29, all solutions starting in $\partial\mathbb{R}^2_+$ are attracted by $(0,0)$, $A_1 = I_1 \times \{0\}$, or $A_2 = \{0\} \times I_2$, i.e., $\Omega \subset \{(0,0)\} \cup A_1 \cup A_2$. Since the semiflow induced by F is uniformly ρ-persistent for $\rho(x_1,x_2) = x_1 + x_2$ by the same theorem, $(0,0)$ is isolated in $\mathbb{R}^2_+$ and uniformly weakly $\tilde{\rho}$-repelling for the semiflow induced by F^2.

We show that A_1 and A_2 are uniformly weakly $\tilde{\rho}$-repelling for the semiflow induced by F^2. Consider a solution $(u(n))$ of $u(n+1) = F^2(u(n))$ with $u_1(0) > 0$ and $u_2(0) > 0$. This is equivalent to considering $(x(2n))$ where $x(n+1) = F(x(n))$ and $x_1(0) > 0$ and $x_2(0) > 0$. Then $x_1(n) > 0$ and $x_2(n) > 0$ for all $n \in \mathbb{N}$. Let $\delta > 0$, to be chosen later, and $\limsup_{n\to\infty} d(x(2n), A_1) < \delta$. Shifting forward in time we can assume that $d(x(2n), A_1) < \delta$ for all $n \in \mathbb{N}$. By (8.29),

$$x_2(2n+2) = x_1(2n+1)g_2(F(x(2n)) = x_2(2n)\, g_1(x(2n))\, g_2(F(x(2n))).$$

Let $\eta > 0$. Since F is continuous and $F(A_1) = A_2$, by choosing $\delta > 0$ small enough, we can achieve that $d(F(x(2n)), A_2) < \eta$. By (8.30), by choosing η and δ small enough, we can achieve that

$$\inf_{n\in\mathbb{N}} g_1(x(2n))\, g_2(F(x(2n))) > 1.$$

This implies that $\limsup_{n\to\infty} x_2(2n) = \infty$, a contradiction. The proof that $\{0\} \times I_2$ is uniformly weakly $\tilde{\rho}$-repelling is similar.

By Theorem 8.29, A_1 and A_2 are locally asymptotically stable under the restrictions to $\partial\mathbb{R}^2_+$ of the semiflow induced by F^2 and thus isolated in $\partial\mathbb{R}^2_+ = X_0$. Further, A_1, A_2, and $(0,0)$ are acyclic under F^2. By Theorem 8.20, the semiflow induced by F^2 is uniformly weakly $\tilde{\rho}$-persistent and so, a fortiori, is the semiflow induced by F.

As mentioned before, $\partial\mathbb{R}^2_+ = \{\tilde{\rho} = 0\}$ and $\mathbb{R}^m \backslash \partial\mathbb{R}^2_+ = (0,\infty)^2 = \{\tilde{\rho} > 0\}$ are forward invariant. This implies that $X_0 = \partial\mathbb{R}^2_+$ and hypothesis (H1) in Theorems 5.2 and 5.7 is satisfied. So the semiflow induced by F^2 is uniformly $\tilde{\rho}$-persistent, and there exists a compact attractor A_1 of compact sets in $(0,\infty)^2$ that is uniformly $\tilde{\rho}$-positive. The connectedness of A_1 follows from Theorem 5.9 because $\tilde{\rho}$ is concave. □

We combine Theorem 8.35 and Proposition 8.33.

Theorem 8.36. *Assume that g_1 and g_2 are continuous and $g_1(x) > 0$ and $g_2(x) > 0$ for $x \in \mathbb{R}^2_+ \setminus \{0\}$.*

Assume that $g_1(0,z)$ is decreasing and $z^2 g_1(0,z)$ is increasing in $z > 0$ and that $g_2(y,0)$ is decreasing and $y^2 g_2(y,0)$ is increasing in $y > 0$ and that these monotonicity properties are strict for g_1 or g_2. Further, assume that $g_1(0,0)g_2(0,0) > 1$ and that the induced semiflow has a compact attractor of bounded sets. Finally, assume that the unique solution $y^\diamond, z^\diamond > 0$ of

$$y^\diamond = z^\diamond g_1(0,z^\diamond) \quad \text{and} \quad z^\diamond = y^\diamond g_2(y^\diamond,0),$$

satisfies $g_2(0,z^\diamond)g_1(y^\diamond,0) > 1$. Then there exists some $\epsilon > 0$ such that

$$\liminf_{n\to\infty} J(n) \geq \epsilon, \qquad \liminf_{n\to\infty} M(n) \geq \epsilon,$$

for all solutions of (8.17) with $J(0) > 0$ and $M(0) > 0$. Further, there exists a compact connected attractor A_1 of compact sets in $(0,\infty)^2$ that is contained in some set $[c^{-1}, c]$, $c > 0$.

Remark 8.34 shows that the condition $g_2(0,z^\diamond)g_1(y^\diamond,0) > 1$ is close to being necessary for the persistence of both stages.

8.6.1. A generalized Beverton-Holt model. Let $x_1(n)$ and $x_2(n)$ denote the respective densities of juvenile and adult individuals in year n,

$$x_j(n+1) = x_k(n)\frac{\psi_j(x_j(n))}{\phi_j(x_j(n)) + a_j x_k(n)}, \qquad j,k \in \{1,2\}, j \neq k. \tag{8.31}$$

Here $\psi_j, \phi_j : \mathbb{R}_+ \to (0,\infty)$ are continuous, ψ_j bounded and ϕ_j bounded away from zero. Further, $a_j \geq 0$, $a_1 + a_2 > 0$.

Notice that there is intrastage competition of Beverton-Holt type the strength of which is measured by the coefficients a_1 and a_2. There is also interstage competition if ψ_j is decreasing and ϕ_j is increasing.

We identify

$$g_j(x_1, x_2) = \frac{\psi_j(x_j)}{\phi_j(x_j) + a_j x_k}, \qquad j \neq k. \tag{8.32}$$

Let us assume that $a_1 > 0$ (the case $a_2 > 0$ is symmetric to this one). Then

$$F(x) \le y + Dx, \qquad y = \begin{pmatrix} \frac{\sup \psi_1}{a_1} \\ 0 \end{pmatrix}, \quad D = \begin{pmatrix} 0 & 0 \\ \frac{\sup \psi_2}{\inf \phi_2} & 0 \end{pmatrix},$$

and the semiflow induced by F has a compact attractor by Corollary 7.3. It is readily checked that the monotonicity assumptions of Theorem 8.36 are satisfied. The threshold condition for uniform persistence of the total population and of synchronous solutions takes the form

$$1 < \frac{\psi_1(0)\psi_2(0)}{\phi_1(0)\phi_2(0)} = g_1(0,0)g_2(0,0). \tag{8.33}$$

The equations for the boundary two-cycle take the form

$$y^\diamond = \frac{\psi_1(0)z^\diamond}{\phi_1(0) + a_1 z^\diamond}, \qquad z^\diamond = \frac{\psi_2(0)y^\diamond}{\phi_2(0) + a_2 y^\diamond}.$$

We substitute the second equation into the first and divide by $y^\diamond$,

$$1 = \frac{\psi_1(0)}{\phi_1(0) + a_1 z^\diamond} \frac{\psi_2(0)}{\phi_2(0) + a_2 y^\diamond}.$$

We reorganize,

$$\begin{aligned} \psi_1(0)\psi_2(0) =& \phi_1(0)\phi_2(0) + \phi_1(0)a_2 y^\diamond + a_1 z^\diamond(\phi_2(0) + a_2 y^\diamond) \\ =& \phi_1(0)\phi_2(0) + \phi_1(0)a_2 y^\diamond + a_1\psi_2(0)y^\diamond. \end{aligned}$$

We solve for $y^\diamond$,

$$y^\diamond = \frac{\psi_1(0)\psi_2(0) - \phi_1(0)\phi_2(0)}{\phi_1(0)a_2 + a_1\psi_2(0)}.$$

By symmetry,

$$z^\diamond = \frac{\psi_1(0)\psi_2(0) - \phi_1(0)\phi_2(0)}{\phi_2(0)a_1 + a_2\psi_1(0)}.$$

By Theorem 8.36, the condition for uniform persistence of both stages takes the form

$$1 < \frac{\psi_1(y^\diamond)\psi_2(z^\diamond)}{\phi_1(y^\diamond)\phi_2(z^\diamond)}.$$

If this inequality is satisfied, there exists some $\epsilon > 0$ such that

$$\liminf_{t\to\infty} x_j(t) \ge \epsilon, \qquad j = 1, 2,$$

for all solutions with $x_1(0) > 0$ and $x_2(0) > 0$. Further, there exists a compact connected attractor A_1 of compact sets in $(0,\infty)^2$ that is contained in some set $[c^{-1}, c]$, $c > 0$.

Notice that $y^\diamond$ and $z^\diamond$ depend on the intrastage competition coefficients a_1 and a_2 in a decreasing way. If we also assume interstage competition (ψ_j decreasing and ϕ_j increasing), coexistence of the stages is the easier the stronger the competition is within the stages (intrastage competition) When we introduce coefficients b_j of interstage competition by setting $\psi_j(y) = \tilde{\psi}_j(b_j y)$ and $\phi_j(y) = \tilde{\phi}_j(b_j y)$ with increasing $\tilde{\phi}_j$ and decreasing $\tilde{\psi}_j$, we see that coexistence of stages is the easier the weaker the competition is between the two stages (interstage competition). Notice that changing b_j does not affect $y^\diamond$ and $z^\diamond$. So we have a phenomenon familiar from two species competition in standard ecology textbooks: coexistence is easier if intrastage competition is stronger and interstage competition is weaker.

8.6.2. A simple Ricker type model. We assume that the transition from the juvenile to the adult stage is not competitive, but that reproduction obeys Ricker type competition with competitive impact from both juveniles and adults,

$$
\begin{aligned}
x_1(n+1) &= \beta x_2(n) e^{-a x_1(n) - b x_2(n)} = f_1(x_1, x_2), \\
x_2(n+1) &= p x_1(n) = f_2(x_1, x_2).
\end{aligned}
$$

Assume $p\beta > 1$, which is equivalent to $\mathbf{r}(F'(0)) > 1$. We have

$$
f_1(0, z) = \beta z e^{-bz}, \qquad f_2(y, 0) = py.
$$

We use Remark 8.30 to get a handle on the intervals I_1 and I_2,

$$
\phi(y) := f_1(0, f_2(y, 0)) = \beta p y e^{-bpy}, \quad \psi(z) := f_2(f_1(0, z), 0) = p\beta z e^{-bz},
$$

are again of Ricker type, and the dynamics on the boundary can be complicated, even chaotic. We take derivatives,

$$
\phi'(y) = \beta p(1 - pby) e^{-bpy}, \qquad \psi'(z) = p\beta(1 - bz) e^{-bz}.
$$

So ϕ takes its maximum at $y = \frac{1}{pb}$ and $\phi_{\max} = \frac{\beta}{b} e^{-1}$, while ψ takes its maximum at $z = 1/b$ and $\psi_{\max} = \frac{p\beta}{b} e^{-1}$. By Proposition 5.12, $J_1 \subset [\phi(\phi_{\max}), \phi_{\max}]$ and $J_2 \subset [\psi(\psi_{\max}), \psi_{\max}]$.

By Theorem 8.35, with $g_1(y, z) = \beta e^{-ay - bz}$ and $g_2(y, z) = p$, the model has uniform persistence in both stages if

$$
p\beta \exp\Big(-\frac{\beta a}{eb} \Big) > 1.
$$

This is satisfied if β is sufficiently large and a/b sufficiently small. As in the other example, a is a measure of interstage competition, while b is a measure of competition within the adult stage for reproduction.

8.7. An endemic with vaccination and temporary immunity

Vaccination is one of the main means of controlling the spread of infectious diseases. We consider a model where neither the disease nor vaccination leads to permanent immunity. The model is quite basic in so far as we assume that immunity whether it results from the disease or from vaccination is perfect as long as it lasts. For simplicity, we consider a population of constant size which consists of the following epidemiological parts the sizes of which, at time t, are denoted as follows: proportion of susceptible individuals, $S(t)$, of infected individuals, $I(t)$, of removed individuals, $R(t)$, and of vaccinated individuals, $V(t)$,

$$1 = S(t) + I(t) + R(t) + V(t). \tag{8.34}$$

Removed and vaccinated individuals are assumed to be immune to the disease, but eventually return to the susceptible class with per capita rates θ and η, respectively. Individuals are born and die at the same per capita rate μ and are vaccinated at a per capita rate κ. Vaccination, if it occurs on an infected or removed individual, has no effect. The per capita infection rate is σ,

$$\begin{aligned} S' &= \mu - \mu S - \kappa S - \sigma S I + \theta R + \eta V, \\ I' &= \sigma S I - (\gamma + \mu) I, \\ R' &= \gamma I - (\theta + \mu) R, \\ V' &= \kappa S - (\eta + \mu) V. \end{aligned} \tag{8.35}$$

Standard arguments provide local existence of solutions and preservation of nonnegativity. Further, the relation (8.34) is preserved. This implies that solutions whose initial data satisfy (8.34) and are nonnegative remain bounded and nonnegative and exist for all forward times. We use (8.34) to eliminate S,

$$\begin{aligned} I' &= \sigma(1 - I - R - V) I - (\gamma + \mu) I, \\ R' &= \gamma I - (\theta + \mu) R, \\ V' &= \kappa(1 - I - R) - (\kappa + \eta + \mu) V. \end{aligned} \tag{8.36}$$

The solutions to this system induce a continuous semiflow on

$$X = \{(I, R, V) \in \mathbb{R}^3_+;\ I + R + V \le 1\}. \tag{8.37}$$

Since X is a compact metric space, the semiflow has a compact attractor of bounded sets, A. Actually, A is the compact attractor of the whole state space. Using Fourier transform (alias frequency domain) methods (see [**85**] and the references therein), we will show that the attractor has a simple form. We will identify a disease reproduction number $\mathcal{R}$ such that the attractor A is a singleton set formed by a (unique) disease-free equilibrium if

$\mathcal{R} \leq 1$. In particular, the disease-free equilibrium is globally asymptotically stable.

If $\mathcal{R} > 1$, the compact attractor A has three disjoint parts: the disease-free equilibrium that attracts $X_0 = \{(0, R, V) \in X\}$, the endemic equilibrium that attracts all closed subsets Y of X with $I > 0$ whenever $(I, R, V) \in Y$, and a third part formed by all orbits that connect the endemic and the disease-free equilibrium. In particular, the endemic equilibrium, whenever it exists, is locally stable and globally asymptotically stable for $\{(I, R, V) \in X; I > 0\}$. The local stability of the endemic equilibrium also holds in a multigroup model of this type [**217**, Chap.23].

We mention the remarkable paper [**6**] where vaccination may only provide partial immunity. As already observed earlier [**130**], in such a case subcritical bifurcation can lead to multiple endemic equilibria if the reproduction number is smaller than 1. It is shown in [**6**] with different methods than here (compound matrices and geometric ideas) that all solutions converge towards an equilibrium provided the length of the infectious period is not too short and the partial immunity acquired from vaccination lasts at least twice as long as the perfect immunity inferred by the disease. The Fourier transform methods need none of these assumptions, but only work if vaccine protection is perfect. We mention that they were first used for epidemic models in [**208**]. They would still work if disease fatalities were incorporated (similarly as in [**224**]) and/or the durations of immunity were distributed. In the second case, assumptions would be needed on the distributions in order to have global stability of the endemic equilibrium and the existence of the compact global attractor would require a nontrivial proof because the state space becomes infinite dimensional. See [**223**] where the length of prophylactic treatment is distributed.

We mention another remarkable vaccination model [**111**] which shares perfect vaccination with ours, but where the presence of two strains and of super-infection leads to undamped oscillations.

8.7.1. Disease persistence. We choose $\rho(I, R, V) = I$ as disease persistence function and identify

$$X_0 = \{(0, R, V);\ R \geq 0, V \geq 0, R + V \leq 1\},$$

a compact set forward invariant set. The compact attractor of X_0, A_0, which exists by Proposition 5.21, is also the compact attractor of all bounded subsets of X_0. It consists of triples $(0, R, V)$ where (R, V) is the value of a bounded nonnegative solution on $\mathbb{R}$ of

$$\begin{aligned} R' &= -(\theta + \mu)R, \\ V' &= \kappa(1 - R) - (\kappa + \eta + \mu)V. \end{aligned} \tag{8.38}$$

Then R is identically equal to 0; otherwise $R(t) \to \infty$ as $t \to -\infty$. So

$$V' = \kappa - (\kappa + \eta + \mu)V, \qquad t \in \mathbb{R}.$$

For $t > r$,

$$V(t) = V(r)e^{(\kappa+\eta+\mu)(r-t)} + \frac{\kappa}{\kappa + \eta + \mu}\big(1 - e^{(\kappa+\eta+\mu)(r-t)}\big).$$

We take the limit $r \to -\infty$ and find that

$$V(t) = \frac{\kappa}{\kappa + \eta + \mu} =: V^\diamond \quad \text{for all } t \in \mathbb{R}. \tag{8.39}$$

So A_0 is the singleton set formed by the disease-free equilibrium $(0, 0, V^\diamond)$.

In the language of Chapter 8.3, $(0, 0, V^\diamond) \in \Omega \subset A_0$ and so Ω is the singleton set formed by $(0, 0, V^\diamond)$. Since $\Omega = A_0$ contains every compact invariant set in X_0, Ω is isolated in X_0. By Theorem 8.20, the solution semiflow is uniformly weakly ρ-persistent if the disease-free equilibrium is uniformly weakly ρ-repelling. Suppose the second is false. By the contrapositive of Definition 8.15, for every $\epsilon > 0$, there exists a solution with $I(0) > 0$ such that

$$\limsup_{t\to\infty}\Big(I(t) + R(t) + |V(t) - V^\diamond|\Big) < \epsilon.$$

Then $I(t) > 0$ for all $t \geq 0$ and

$$\liminf_{t\to\infty} \frac{I'(t)}{I(t)} > \sigma(1 - \epsilon - V^\diamond) - (\gamma + \mu).$$

We obtain the contradiction $I(t) \to \infty$ as $t \to \infty$ if

$$\sigma(1 - V^\diamond) > \gamma + \mu$$

by choosing $\epsilon > 0$ small enough. This condition can be rewritten in the form of a reproduction number,

$$\mathcal{R} = \frac{\sigma(1 - V^\diamond)}{\gamma + \mu} > 1. \tag{8.40}$$

Notice that $1 - V^\diamond$ is the susceptible fraction of the population at the disease-free equilibrium, $\frac{1}{\gamma+\mu}$ the average duration of the infectious stage (taking natural death into account) and σ is the per capita infection rate.

By Theorem 4.13, the solution semiflow is uniformly ρ-persistent if $\mathcal{R} >$ 1.

Theorem 8.37. *Let $\mathcal{R} > 1$. Then there exists some $\epsilon > 0$ such that $\liminf_{t\to\infty} I(t) \geq \epsilon$ for all solutions of of (8.35 and (8.34) with $I(0) > 0$ and $S(0), R(0), V(0) \geq 0$.*

Below we will show that the disease-free equilibrium is globally asymptotically stable if $\mathcal{R} \leq 1$. The fluctuation method does not seem to work well in this case and we will employ some Fourier transform (alias frequency domain) methods [**85, 151**]. This method will also show that, if $\mathcal{R} > 1$, there exists a unique endemic equilibrium that attracts all solutions with $I(0) > 0$. Actually, we will give a complete characterization of the global attractor in the two scenarios of disease extinction and disease persistence.

8.7.2. Description of the global compact attractor. Since the state space is compact, the solution semiflow has a global compact attractor, $A = \omega(X)$, that attracts the whole state space (Theorem 2.11). If $\mathcal{R} > 1$, the semiflow is uniformly ρ-persistent (Theorem 8.37) and has a persistence attractor, A_1, as described in Theorem 5.7.

Theorem 8.38. (a) *Let $\mathcal{R} \le 1$. Then the compact attractor of bounded sets for Φ is the singleton set formed by the disease-free equilibrium. That is, the disease-free equilibrium attracts the whole state space and is stable.*

(b) *Let $\mathcal{R} > 1$. Then there exists a unique endemic equilibrium and the persistence attractor is the singleton set formed by the endemic equilibrium. That is, the endemic equilibrium attracts all points (I, R, V) with $I > 0$ and all sets $\{(I, R, V) \in X; I \ge \epsilon\}$ with $\epsilon > 0$. Further, the endemic equilibrium is stable.*

In order to prove this theorem, guided by Theorem 1.40, we consider a solution that is defined and bounded for all $t \in \mathbb{R}$. We already know from the considerations at the beginning of Section 8.7.1 that the disease-free equilibrium attracts X_0. We integrate the differential equations for R and V in (8.36),

$$\begin{cases} I' = & I\big[\sigma(1 - I - R - V) - (\gamma + \mu)\big], \\ R(t) = & \displaystyle\int_0^\infty I(t-s)k_1(s)ds, \\ V(t) = & V^\diamond - \displaystyle\int_0^\infty I(t-s)k_2(s)ds - \int_0^\infty R(t-s)k_2(s)ds, \end{cases} \tag{8.41}$$

with $V^\diamond$ in (8.39) and

$$k_1(s) = \gamma e^{-(\theta+\mu)s}, \qquad k_2 = \kappa e^{-(\kappa+\eta+\mu)s}. \tag{8.42}$$

We want to express V completely in terms of I,

$$\begin{aligned} \int_0^\infty R(t-s)k_2(s)ds &= \int_0^\infty k_2(s)ds \int_0^\infty I(t-s-r)k_1(r)dr \\ &= \int_0^\infty k_2(s)ds \int_s^\infty I(t-r)k_1(r-s)dr. \end{aligned}$$

We change the order of integration,

$$\int_0^\infty R(t-s)k_2(s)ds = \int_0^\infty I(t-r)k_3(r)dr$$

with

$$k_3(r) = \int_0^r k_2(s)k_1(r-s)ds =: (k_1 * k_2)(r).$$

So I is the solution of an integro-differential equation,

$$I'(t) = I(t)\Big(\tilde{\sigma} - \int_0^\infty I(t-s)m(ds)\Big) \tag{8.43}$$

with

$$\tilde{\sigma} = \sigma(1-V^\diamond) - (\gamma+\mu) = (\mathcal{R}-1)(\gamma+\mu) \tag{8.44}$$

and a signed measure m on $\mathbb{R}_+$ given by

$$\frac{1}{\sigma}m(ds) = \delta_0(ds) + k_1(s)ds - k_2(s)ds - k_3(s)ds, \tag{8.45}$$

where δ_0 denotes the Dirac measure concentrated at 0.

Lemma 8.39. *The signed measure m satisfies the assumptions of Theorem B.10 and Theorem B.11.*

Proof. The crucial assumption is $\inf_{s\geq 0} \Re \int_0^\infty e^{-ist}m(dt) > 0$. The Fourier transform converts convolutions into products,

$$\begin{aligned}\int_{\mathbb{R}_+} e^{-ist}m(dt) =& 1 + \frac{\gamma}{is+\theta+\mu} - \frac{\kappa}{is+\kappa+\eta+\mu}\\ &- \frac{\gamma}{is+\theta+\mu}\,\frac{\kappa}{is+\kappa+\eta+\mu}.\end{aligned}$$

We make all denominators real by multiplying with the appropriate complex conjugates,

$$\begin{aligned}\int_{\mathbb{R}_+} e^{-ist}m(dt) =& 1 + \frac{\gamma(\theta+\mu-is)}{(\theta+\mu)^2+s^2} - \frac{\kappa(\kappa+\eta+\mu-is)}{(\kappa+\eta+\mu)^2+s^2}\\ &- \gamma\kappa\frac{(\theta+\mu-is)}{(\theta+\mu)^2+s^2}\,\frac{(\kappa+\eta+\mu-is)}{(\kappa+\eta+\mu)^2+s^2}.\end{aligned}$$

We take the real part,

$$\begin{aligned}&\Re\int_{\mathbb{R}_+} e^{-ist}m(dt) = a_1(s) + a_2(s),\\ &a_1(s) = 1 - \frac{\kappa(\kappa+\eta+\mu)}{(\kappa+\eta+\mu)^2+s^2},\\ &a_2(s) = \frac{\gamma(\theta+\mu)}{(\theta+\mu)^2+s^2} - \gamma\kappa\frac{(\theta+\mu)(\kappa+\eta+\mu)-s^2}{[(\theta+\mu)^2+s^2][(\kappa+\eta+\mu)^2+s^2]}.\end{aligned}$$

We use a common denominator for a_1,

$$a_1(s) = \frac{(\kappa+\eta+\mu)^2+s^2-\kappa(\kappa+\eta+\mu)}{(\kappa+\eta+\mu)^2+s^2} = \frac{(\kappa+\eta+\mu)(\eta+\mu)+s^2}{(\kappa+\eta+\mu)^2+s^2}.$$

Notice that $\inf_{s \geq 0} a_1(s) > 0$. $a_2(s)$ has the same sign as

$$1 - \kappa \frac{(\kappa + \eta + \mu) - s^2/(\theta + \mu)}{(\kappa + \eta + \mu)^2 + s^2} \geq \frac{(\kappa + \eta + \mu)^2 - \kappa(\kappa + \eta + \mu)}{(\kappa + \eta + \mu)^2 + s^2} > 0.$$

So,

$$\inf_{s \geq 0} \Re \int_{\mathbb{R}_+} e^{-ist} m(dt) > 0. \qquad \square$$

Let $\mathcal{R} \leq 1$. Then $\tilde{\sigma} \leq 0$ and

$$-I'(t) \geq I(t) \int_0^\infty I(t-s) m(ds).$$

By Theorem B.10, $I(t) = 0$ for all $t \in \mathbb{R}$. By (8.41), $R(t) = 0$ and $V(t) = V^\diamond$ for all $t \in \mathbb{R}$. This shows that the disease-free equilibrium is the compact attractor of bounded sets. By Theorem 2.39, the disease-free equilibrium is stable.

Now let $\mathcal{R} > 1$, i.e., $\tilde{\sigma} > 0$. Then (8.43) has a constant solution, the endemic equilibrium with infectious component

$$I^* = \frac{\tilde{\sigma}}{m(\mathbb{R}_+)}.$$

We substitute this expression into (8.43),

$$\frac{I'(t)}{I(t)} = \int_0^\infty (I^* - I(t-s)) m(ds).$$

Set $u = \ln(I(t)/I^*)$. Then

$$-u'(t) = \int_0^\infty g(u(t-s)) m(ds)$$

with

$$g(u) = I^*(e^u - 1).$$

It is sufficient to consider a total solution that takes values in the persistence attractor A_1 described in Theorem 5.7. In particular, $0 < \inf_{t \in \mathbb{R}} I(t)$. This implies that u is bounded. By Theorem B.11, $u(t) = 0$ for all $t \in \mathbb{R}$, i.e., $I(t) = I^*$ for all $t \in \mathbb{R}$. By (8.41), $R(t) = R^*$ and $V(t) = V^*$ for all $t \in \mathbb{R}$ where R^* and V^* are the proportions of removed and vaccinated individuals at endemic equilibrium, respectively. This shows that the persistence attractor is the singleton set formed by the endemic equilibrium.

By Theorem 5.7, the endemic equilibrium is stable and attracts every point $(\breve{I}, \breve{R}, \breve{V}) \in X$ with $\breve{I} > 0$ and all sets $\{(I, R, V) \in X; I \geq \epsilon\}$ with $\epsilon > 0$.

8.8. Lyapunov exponents and persistence for ODEs and maps

The acyclicity theorem, Theorem 8.17, is one of the most useful tools with which to establish persistence of a dynamical system. It simply requires that we understand the "boundary dynamics" sufficiently well that an acyclic covering $\{M_1, M_2, \dots, M_k\}$ of it can be found. There is considerable flexibility in determining the compact invariant sets M_i. However, we typically need to establish that each M_i is weakly uniformly ρ-repelling. Determining conditions for this to hold can be challenging even for equilibria or periodic orbits as we have seen through numerous examples. So far, essentially ad hoc methods have been used to establish the weak uniform repelling property. For ODES, it usually involves identifying a positive eigenvalue or Floquet exponent corresponding to an eigenvector transverse to the boundary for the linearized system and then comparing solutions starting near M with some perturbed linear system which has similar spectral properties (by continuity) to the linearization about M_i. A similar procedure works for maps. But what if one of the M_i is a more complicated invariant set than a periodic orbit or equilibrium? Can we proceed similarly? We will show that the answer, in principle, is "yes". We just need to work with the normal Lyapunov exponents associated with M_i. In this section, we obtain sufficient conditions for the weak uniform repelling property to hold that apply for general compact invariant sets.

We follow the approach used by Salceanu et al. [**189, 186, 190**], which in turn was inspired by earlier work of [**79, 193**]. Elementary arguments are used here rather than appealing to the multiplicative ergodic theorem as in [**7, 8, 79, 193, 105**]. The use of Lyapunov exponents in the study of biological models was pioneered by Metz et al. [**168**], who proposed that the dominant Lyapunov exponent gives the best measure of invasion fitness, and by Rand et al. [**180**] who used it to characterize the invasion "speed" of a rare species. See more recent reviews by Ferriere and Gatto [**70**], which deals with computational aspects, and by Metz [**167**]. Roughly, a positive dominant Lyapunov exponent corresponding to a potential invading species in the environment set by a resident species attractor implies that the invader can successfully invade.

8.8.1. Co-cycle over a compact boundary invariant set.

The basic setup is that we have a finite dimensional dynamical system (an ODE or a map) on the nonnegative cone in some Euclidean space (a more general domain could be chosen), and our focus is on the behavior of solutions near some compact invariant subset of a boundary face of the cone. We begin by formalizing this. Consider both the discrete dynamical system

$$z(t+1) = F(z(t)),\ z(0) = z_0 \tag{8.46}$$

and the continuous dynamical system

$$z'(t) = F(z(t)),\ z(0) = z_0 \tag{8.47}$$

on the nonnegative cone $Z_+ \equiv \mathbb{R}^m_+$.

Let $|x| = \sum |x_i|$ denote the norm on $\mathbb{R}^m$ and let $|B|$ be the induced matrix norm. We use the notation $x \le y$ when $x_i \le y_i$, $x < y$ when $x \le y$, $x \ne y$, and $x \ll y$ when $x_i < y_i$ for all i; a similar notation applies to matrices.

It is assumed that (8.46) and (8.47) generate a dynamical system on Z_+. In case of (8.46), we assume that $F : Z_+ \to Z_+$ is continuous; in case (8.47), $F : Z_+ \to \mathbb{R}^m$ satisfies

$$z \in Z_+,\ z_i = 0 \Rightarrow F_i(z) \ge 0 \tag{8.48}$$

(see Proposition A.17) and sufficient regularity properties such that solutions $z(t)$ of (8.47) are defined on $t \ge 0$ and are unique. Denote by

$$\Phi_t(z(0)) = \Phi(t, z(0)) = z(t)$$

the associated semiflow $\Phi : J \times Z_+ \to Z_+$ where J is the set of nonnegative integers in case (8.46) and $[0, \infty)$ in case (8.47). We will make further assumptions below.

Assume that $m = p + q,\ p, q \ge 1$ and that Z_+ is decomposed as follows,

$$Z_+ = \mathbb{R}^m_+ = \mathbb{R}^p_+ \times \mathbb{R}^q_+, \quad z = (x, y)$$

and, compatible with this decomposition, that

$$F(z) = (f(z), g(z))$$

where

$$X = \{z = (x, y) \in Z_+ : y = 0\},$$

is forward invariant. In fact, we assume that

$$X \text{ and } Z_+ \setminus X \text{ are forward invariant sets.} \tag{8.49}$$

Forward invariance of X for both (8.46) and (8.47) means that $g(x, 0) = 0$. If $F \in C^1(U)$ where U is some open subset of $\mathbb{R}^m$ containing Z_+, then

$$g(z) = g(x, y) - g(x, 0) = \int_0^1 g_y(x, sy)ds\, y \equiv A(x, y)y,\ z = (x, y) \in Z_+,$$

where $z \to A(z)$ is continuous; here g_y denotes the $q \times q$ partial derivative of g with respect to y. Observe that $A(x, 0) = g_y(x, 0)$. It follows that (8.46) and (8.47) can be expressed as

$$\begin{cases} x(t+1) &= f(z(t)), \\ y(t+1) &= A(z(t))y(t) \end{cases} \tag{8.50}$$

and, respectively, as

$$\begin{cases} x'(t) &= f(z(t)), \\ y'(t) &= A(z(t))y(t), \end{cases} \tag{8.51}$$

where the matrix function $z \to A(z)$ satisfies:

$$A(z) \geq 0, \quad z = (x, 0) \in X \tag{8.52}$$

in case of (8.50), and

$$A_{ij}(z) \geq 0, \ i \neq j, \quad z = (x, 0) \subset X \tag{8.53}$$

in case of (8.51).

Rather than assuming that $F \in C^1(U)$, we just assume that the representations (8.50) and (8.51) hold for (8.46) and (8.47) and that A has the properties just mentioned, i.e., (8.52) and (8.53).

Let $M \subset X$ be compact and forward invariant set for Φ. We envision that in typical applications, M will be an invariant set in the interior of the face X of the cone Z_+. We will focus on the behavior of solutions near M in $Z_+ \setminus X$.

Remark 8.40. The domain Z_+, the nonnegative cone in $\mathbb{R}^m$, is chosen here largely motivated by the applications but it is not the most natural domain from a mathematical perspective. Recall, our interest is in the behavior of solutions near M, a compact, forward invariant "boundary" set. Alternatively, let $W_+ = \mathbb{R}^p \times \mathbb{R}^q_+$ and suppose that (8.46), or (8.47), generates a semiflow Φ on $D = V \cap W_+$, where V is open in $\mathbb{R}^m$ and has nonempty intersection with the lower boundary of W_+ given by $X = \mathbb{R}^p \times \{0\} \subset W_+$. Suppose that $D \setminus X$ and $X \cap D$ are forward invariant. Finally, suppose there is a compact, forward invariant subset M in the boundary $D \cap X$ for Φ. Our focus then is on the behavior of solutions starting near M in $D \setminus X$.

Example 8.41. Consider the discrete time host-parasite model due to Beddington et al. and studied by Kon and Takeuchi [**124**]:

$$\begin{aligned} x(t+1) &= x(t)\exp(r - x(t) - y(t)), \\ y(t+1) &= \theta x(t)(1 - \exp(-y(t))), \end{aligned} \tag{8.54}$$

where $r, \theta > 0$, x denotes host and y denotes parasite. Here,

$$A(x,y) = \begin{cases} \theta x \frac{1-\exp(-y)}{y}, & y \neq 0, \\ \theta x, & y = 0, \end{cases}$$

is continuous. We view system (8.54) as defined on $D = \{(x, y) : x > 0, y \geq 0\}$, taking the point of view of Remark 8.40. Then D and $D \setminus X$ are forward invariant where $X = \{(x, 0) : x > 0\}$.

The dynamics on X are given by the Ricker map

$$f(x) = x \exp(r - x). \tag{8.55}$$

As noted in the previous chapter, the Ricker map $f : [0, \infty) \to [0, \infty)$ has the following properties:

(a) It has a compact attractor A^r of bounded sets.

(b) The Ricker semiflow is uniformly ρ-persistent with $\rho(x) = x$.

(c) A^r can be decomposed as in Theorem 5.7 into the "extinction attractor" $\{0\}$, the "persistence attractor" M, and a set C consisting of total trajectories with α-limit set $\{0\}$ and ω-limit set M. $M \subset (0, \infty)$ is a compact interval, which is invariant, asymptotically stable, and attracts compact subsets of $(0, \infty)$. See Proposition 5.12.

Hereafter, we refer to M as the Ricker attractor.

Remark 8.42. $F : \mathbb{R}^2_+ \to \mathbb{R}^2_+$ defined by $F(x, y) = (x \exp(r - x - y), \theta x(1 - \exp(-y)))$ induces a semiflow that has a compact attractor of bounded sets. The semiflow is uniformly ρ-persistent with $\rho(x, y) = x$ if $r > 0$; the host persists. See Exercise 8.10.

Example 8.43. Consider the classical chemostat model

$$\begin{aligned} S' &= D(S^0 - S) - \sum_i f_i(S) x_i, \\ x_i' &= x_i(f_i(S) - D), \ 1 \leq i \leq n, \end{aligned} \tag{8.56}$$

where $f_i(S)$ is continuously differentiable, $f_i(0) = 0$, $f_i' > 0$, and bounded for $S \geq 0$; $D, S^0 > 0$. If $X = \{(S, x) \in \mathbb{R}^{n+1} : x_i = 0, \ 1 \leq i \leq n\}$, and $M = (S^0, 0)$ is the "washout" equilibrium, then matrix A is diagonal

$$A(S, x) = \operatorname{diag}(f_i(S) - D).$$

This example is trivial but it will be useful by way of pointing to the appropriate hypotheses for our main result.

We introduce below the notation and tools which will be useful for describing the rate of growth of solutions transverse to M in Z_+. The reader may wish to assume that $z \in M$ or is at least near M in the equations below.

The $q \times q$-fundamental matrix solution $P(t, z)$ for

$$v(t + 1) = A(\Phi(t, z)) v(t) \tag{8.57}$$

and for

$$v'(t) = A(\Phi(t,z))v(t) \tag{8.58}$$

are defined by

$$P(t+1,z) = A(\Phi(t,z))P(t,z),\ P(0,z) = I \tag{8.59}$$

for discrete time, and

$$\frac{d}{dt}P(t,z) = A(\Phi(t,z))P(t,z),\ P(0,z) = I \tag{8.60}$$

for continuous time. In either case, it follows that if $z(t) = (x(t), y(t))$ is a solution of (8.46) or (8.47), then

$$y(t) = P(t, z(0))y(0). \tag{8.61}$$

In addition, from (8.52) and (8.53), it follows that P is positive on X:

$$P(t,z) \geq 0,\ z \in X,\ t \in J. \tag{8.62}$$

Following Arnold [**7**], $P(t,z)$ is a matrix cocycle generated by (8.57) (or by (8.58)). It is easy to check (see Exercise 8.13) that P has the cocycle property:

$$P(t_2, \Phi(t_1,z))P(t_1,z) = P(t_1+t_2,z),\ t_1,t_2 \in J. \tag{8.63}$$

A fundamental difference between the discrete time and continuous time systems is that for the former $P(t,z)$ need not be invertible while for the latter it must be. For example, if $F : \mathbb{R}^3_+ \to \mathbb{R}^3_+$ is given by $F(x, y_1, y_2) = (f(x), y_1 + y_2, 0)$, then

$$A(z) = \begin{pmatrix} 1 & 1 \\ 0 & 0 \end{pmatrix},$$

so $P(t,z) = A^t = A$.

For Example 8.41, $P(t,z) = (\theta)^t x(t-1)x(t-2)\cdots x(0)$ for $z = (x(0), 0)$. For Example 8.43, $P(t,z) = \mathrm{diag}(\int_0^t f_i(S(u)) - D)du)$.

8.8.2. Normal Lyapunov exponents.

Definition 8.44. For $z \in M$ and $\eta \in \mathbb{R}^q$, define the *normal* Lyapunov exponent $\lambda(z,\eta)$ by

$$\lambda(z,\eta) = \limsup_{t\to\infty} \frac{1}{t} \ln |P(t,z)\eta|, \tag{8.64}$$

where $t \in J$ and with the convention $\ln(0) = -\infty$.

As M is compact and forward invariant and A is bounded on M, it follows that $\lambda(z,\eta) \in \{-\infty\} \cup \mathbb{R}$.

The Lyapunov exponent $\lambda(z,\eta)$ measures the exponential growth rate of the transverse component of the linearized dynamics about the solution

$\Phi(t,z)$. Indeed, if $z=(x,0)\in M$, then the linearization of (8.46) and (8.47) about $z(t)=\Phi(t,z)$ leads to the Jacobian matrix

$$J(z(t))=\begin{pmatrix} f_x(z(t)) & f_y(z(t)) \\ 0 & g_y(z(t)) \end{pmatrix},$$

partitioned compatible with the decomposition described above. As $A(z(t)) = g_y(z(t))$, the systems (8.57) and (8.58), and the associated fundamental matrix $P(t,z)$, capture only the transverse, or normal, component of the full linearization about the orbit in M.

Ergodic theory can be used to show that the limit superior can be replaced by limit for almost every z, with respect to an invariant measure for the dynamical system [**7, 8, 114**]. We do not make use of ergodic theory here. See [**79, 105, 192, 193**] for the use of ergodic theory ideas in persistence theory.

Among the many properties of Lyapunov exponents, see [**7, 8, 114**], we use only the following two.

Lemma 8.45. *The follow hold:*

$$\lambda(z,\eta_1+\eta_2)\le \max\{\lambda(z,\eta_1),\lambda(z,\eta_2)\}, \tag{8.65}$$

where equality holds if $\lambda(z,\eta_1)\neq\lambda(z,\eta_2)$, *and*

$$\lambda(z,a\eta)=\lambda(z,\eta),\ a\neq 0. \tag{8.66}$$

Proof. Observe that if $y(t)=P(t,z)\eta$ satisfies $|y(t)|\le Me^{at}$, then $\lambda(z,\eta)\le a$ and conversely, if $\lambda(z,\eta)<a$, then there exists M such that $|y(t)|\le Me^{at}$. Let $y_i(t)=P(t,z)\eta_i$, $y(t)=P(t,z)(\eta_1+\eta_2)=y_1(t)+y_2(t)$, and $\lambda_i=\lambda(z,\eta_i)$. Assume that $\lambda_1\le\lambda_2$ and let $a>\lambda_2$. Then $|y_i(t)|\le M_ie^{at}$, $t\ge 0$ for some M_i and

$$|y(t)|\le M_1e^{at}+M_2e^{at}=(M_1+M_2)e^{at},\ t\ge 0.$$

Consequently, $\lambda(z,\eta_1+\eta_2)\le a$ and, as $a>\lambda_2$ is arbitrary, (8.65) holds.

If $\lambda_1<\lambda_2$ and $\lambda(z,\eta_1+\eta_2)<\lambda_2$, choose $a<\lambda_2$ larger than both λ_1 and $\lambda(z,\eta_1+\eta_2)$. Then

$$|y_2(t)|\le|y(t)|+|y_1(t)|\le Me^{at}+Ne^{at}=(M+N)e^{at},\ t\ge 0$$

so $\lambda_2\le a$, a contradiction.

Finally,

$$\lambda(z,a\eta)=\limsup_{t\to\infty}\frac{1}{t}\left[\ln|a|+\ln|P(t,z)\eta|\right]=\limsup_{t\to\infty}\frac{1}{t}\ln|P(t,z)\eta|$$

proving (8.66). □

We now specialize to the case that $\eta \in \mathbb{R}^q_+$. This is primary interest because in that case $z + (0, \eta) \in Z_+$ and $(0, \eta)$ represents a normal vector to M at $z \in M$. By virtue of (8.66), we may as well specialize to unit vectors. Hence, let

$$U_+ = \{\eta \in \mathbb{R}^q_+ : |\eta| = 1\}$$

be the set of nonnegative unit vectors.

The cocycle property (8.63) can be used to show that

$$\lambda(z, \eta) = \lambda(\Phi(s, z), P(s, z)\eta), \; s \geq 0. \tag{8.67}$$

See Exercise 8.14.

Let $\eta, \chi \in U_+$ and define an equivalence relation on U_+ by $\eta \sim \chi$ if there exist $a, b > 0$ such that $a\chi \leq \eta \leq b\chi$. The equivalence classes are the intersection of U_+ with the various faces bounding $\mathbb{R}^q_+$; an equivalence class consists of vectors with common support, the indices of the positive components. If $[\eta]$ and $[\chi]$ denote two such equivalence classes, we write $[\eta] \leqslant [\chi]$ if there exists $a > 0$ such that $a\eta \leq \chi$. Relation $\leqslant$ defines a partial order on the equivalence classes identical to the one induced by support inclusion. An important equivalence class consists of the positive vectors $(\eta \gg 0)$ in U_+. The standard basis vectors for $\mathbb{R}^q$ are denoted by e_i.

Lemma 8.46. *The following hold:*

(a) *If $\eta, \chi \in U_+$ and $\eta \leq \chi$, then $\lambda(z, \eta) \leq \lambda(z, \chi)$.*

(b) *$\lambda(z, \cdot)$ is constant on each equivalence class of relation $\sim$. In fact,*

$$\lambda(z, \eta) = \max\{\lambda(z, e_i) : \eta_i > 0\}. \tag{8.68}$$

(c) *$[\eta] \leqslant [\chi] \Rightarrow \lambda(z, \eta) \leq \lambda(z, \chi)$.*

(d) *$\lambda(z, \hat{1}) = \lambda(z, \eta)$, $\eta \in U_+$, $\eta \gg 0$, where $\hat{1} = (1, 1, \ldots, 1)^T$.*

(e) *If for each i there exists t_i such that $P(t_i, z)e_i \gg 0$ (for example, if $P(T, z) \gg 0$, some $T \in J$), then*

$$\lambda(z, \eta) = \lambda(z, \hat{1}) = \limsup_{t \to \infty} \frac{1}{t} \ln \|P(t, z)\|, \; \eta \in U_+.$$

Proof. Part (a) follows from the fact that $P(t, z) \geq 0$ so $0 \leq P(t, z)\eta \leq P(t, z)\chi$ and $|P(t, z)\eta| \leq |P(t, z)\chi|$. Parts (b) and (c) follow directly from (8.66) and part (a). Part (d) follows from part (b).

Formula (8.68) is proved as follows. If $\eta_i > 0$, then, dropping the z from the notation, $\lambda(e_i) = \lambda(\eta_i e_i) \leq \lambda(\eta)$, which leads to $\lambda(\eta) \geq \max\{\lambda(e_i) : \eta_i > 0\}$. The reverse inequality follows from (8.65).

By (8.67), $\lambda(z, e_i) = \lambda(\Phi(t_i, z), P(t_i, z)e_i)$ and the latter agrees with $\lambda(\Phi(t_i, z), P(t_i, z)\hat{1}) = \lambda(z, \hat{1})$ since $P(t_i, z)\hat{1} \geq P(t_i, z)e_i \gg 0$ by part (d). By (8.68), $\lambda(z, \eta) = \lambda(z, \hat{1})$ for $\eta \in U_+$. Let B^i denote the i-th column of matrix B and $\|B\|_1 = \sum_i |B^i|$. Then $\|B\|_1$ is a matrix norm and $\|B\|_1 = |B\hat{1}|$ if $B \geq 0$. It follows that

$$\lambda(z, \hat{1}) = \limsup_{t \to \infty} \frac{1}{t} \ln \|P(t)\|_1 = \limsup_{t \to \infty} \frac{1}{t} \ln \|P(t)\|$$

since all matrix norms are equivalent. □

Suppose $z \in M$ belongs to a periodic orbit of period $p \in J$: $\Phi(p, z) = z$. Assume $p > 0$ is minimal with this property. Let $O_+(z) = \{\Phi(t, z) : t \geq 0\}$. Our next result shows that under suitable conditions: (1) if $\lambda(z, \eta) > 0$ for all $\eta \in U_+$, then we need not consider $\lambda(w, \eta)$ for other points $w \in O_+(z)$, and (2) if the spectral radius $r(P(p, z)) > 1$, then $\lambda(z, \eta) > 0$ for all $\eta \in U_+$. Recall that since $P(p, z)$ is a nonnegative matrix, the Perron-Frobenius Theorem ensures that its spectral radius is an eigenvalue and the associated eigenvector can be taken nonnegative.

Lemma 8.47. *Let $O_+(z)$ be a p-periodic orbit in M.*

If $\lambda(z, \eta) > 0$ for all $\eta \in U_+$, then $\lambda(w, \eta) > 0$ for all $\eta \in U_+$ and $w \in O_+(z)$ provided that in the discrete time case,

$$P(t, w)\eta \neq 0, \ \ \eta \in U_+, \ \ t \in J, \ \ w \in O_+(z).$$

If $P(T, z) \gg 0$ for some $T \in J$ and $p > 0$, then

$$\lambda(z, \eta) = \frac{\ln(r(P(p, z)))}{p}, \ \ \eta \in U_+. \tag{8.69}$$

In the continuous time case when $O_+(z)$ is an equilibrium, there is $v \in U_+$ such that $A(z)v = \mu v$ and $\lambda(z, v) = \mu$. If $A(z)$ is irreducible, then $P(t, z) \gg 0$ and

$$\lambda(z, \eta) = \mu, \ \ \eta \in U_+.$$

Proof. Let $w = \Phi(s, z)$ and $0 < s < t$. By (8.67),

$$\lambda(w, \eta) = \lambda(\Phi(p - s, w), P(p - s, w)\eta) = \lambda(z, P(p - s, w)\eta), \ \ \eta \in U_+.$$

As $P(p - s, w)\eta \neq 0$ by hypothesis, $P(p - s, w)\eta > 0$ and $\lambda(w, \eta) > 0$.

Recall that $r(P(p, z)) = \lim_n \|P(p, z)^n\|^{1/n}$. Using Lemma 8.46 (e) and $P(np, z) = P(p, z)^n$, $n = 1, 2, \ldots$, we have

$$\lambda(z, \eta) \geq \limsup_{n \to \infty} \frac{1}{np} \ln \|P(p, z)^n\| = \frac{1}{p} \ln \left(\lim_{n \to \infty} \|P(p, z)^n\|^{1/n} \right)$$

On the other hand, writing $t = np + r$ where $n = n(t)$ is a nonnegative integer and $r = r(t)$ satisfies $0 \leq r < p$, we have $P(t, z) = P(r, z)P(p, z)^n$, so

$$\begin{aligned}
\lambda(z, \eta) &= \limsup_{t\to\infty} \frac{1}{t} \ln \|P(r, z)P(p, z)^n\| \\
&\leq \limsup_{t\to\infty} \frac{1}{t} \left(\ln \|P(r, z)\| + \ln \|P(p, z)^n\|\right) \\
&= \limsup_{n\to\infty} \frac{n}{np + r} \frac{1}{n} \ln \|P(p, z)^n\| \\
&= \frac{1}{p} \ln(\lim_{n\to\infty} \|P(p, z)^n\|^{1/n}).
\end{aligned}$$

The final assertion when $O_+(z)$ is an equilibrium follows from $P(t, z)v = e^{\mu t}v$ and the fact that when $A(z)$ is irreducible, then $P(t, z) \gg 0$ and there is a unique $v \gg 0$ in U_+. □

Consider Example 8.41. As noted earlier,

$$P(t, z) = \prod_{i=0}^{t-1} \theta x(i), \quad z = (x(0), 0) \in M$$

where M is the Ricker attractor. Clearly, $U_+ = \{1\}$ and

$$\lambda(z, 1) = \limsup_{t\to\infty} \ln \left(\prod_{i=0}^{t-1} \theta x(i) \right)^{1/t} = \ln \left[\theta \limsup_{t\to\infty} \left(\prod_{i=0}^{t-1} x(i) \right)^{1/t} \right] \tag{8.70}$$

For Example 8.43, $M = \{(S^0, 0)\}$, so

$$P(t, M) = \operatorname{diag}(e^{\lambda_i t}), \quad \lambda_i := f_i(S^0) - D. \tag{8.71}$$

It follows that the Lyapunov exponents are the λ_i.

8.8.3. Uniformly weakly repelling sets via Lyapunov exponents. As above, M is assumed to be a compact, forward invariant subset of X. Define $\rho : Z_+ \to \mathbb{R}_+$ by $\rho(z) = |y| = \sum_i y_i$. Because $\rho^{-1}(0) \cap Z_+ = X$, M is uniformly weakly ρ-repelling if there exists $\epsilon > 0$ such that

$$\limsup_{t\to\infty} d(\Phi(t, z), M) > \epsilon, \ z \in Z_+ \setminus X.$$

Equivalently, in view of (8.49), there exists a neighborhood V of M in Z_+ such that

$$\forall z \in V \setminus X, \ \exists\, t = t(z) > 0, \ \Phi(t, z) \notin V.$$

We stress that M may be an attractor relative to the dynamics restricted to the forward invariant set X, but we are concerned with the behavior of solutions near M initiating from points of the forward invariant set $Z_+ \setminus X$. We seek sufficient conditions for M to be weakly uniformly ρ-repelling.

As motivation for our main result, consider Example 8.43. If $\lambda_i \leq 0$ for some i, then it is easily seen (see [**203**]) that $x_i(t) \to 0$ as $t \to \infty$ for every solution of (8.56). But this means that $M = \{(S^0, 0)\}$ is not weakly uniformly ρ-repelling because M is globally asymptotically stable for the $S - x_i$ subsystem. Therefore, a necessary condition for M to be weakly uniformly ρ-repelling is that all Lyapunov exponents are positive. In the next result we show that if all Lyapunov exponents are positive then M is uniformly weakly ρ-repelling.

Let

$$\Omega(M) = \bigcup_{z \in M} \omega(z)$$

where $\omega(z)$ represents the omega limit set of z.

Theorem 8.48. *Let $M \subset X$ be a nonempty compact forward invariant set. M is uniformly weakly ρ-repelling if*

$$\lambda(z, \eta) > 0, \; z \in M, \; \eta \in U_+. \tag{8.72}$$

Moreover, (8.72) holds if,

$$\lambda(z, \eta) > 0, \; z \in \Omega(M), \; \eta \in U_+ \tag{8.73}$$

and, in the discrete case,

$$A(z)\eta \neq 0, \; z \in M, \; \eta \in U_+. \tag{8.74}$$

Before proceeding to the proof of the theorem, we require the following rather technical lemma which gives an alternative formulation for the positivity of Lyapunov exponents.

Lemma 8.49. *Let $K \subset X$ be compact. Assume that*

$$\forall\, (z, \eta) \in K \times U_+, \; \exists \tau > 0 \;\; \text{such that } |P(\tau, z)\eta| > 1. \tag{8.75}$$

Then there exists $c > 1$ and a bounded neighborhood V of K in Z_+ such that any forward invariant set $L \subset V$ satisfies $L \subset X$ and

$$\forall\, (z, \eta) \in L \times U_+, \; \exists\, \nu_p \nearrow \infty, \; |P(\nu_p, z)\eta| > c^p, \; p \geq 1. \tag{8.76}$$

If, in addition, K is forward invariant, then (8.75) is equivalent to

$$\lambda(z, \eta) > 0, \; (z, \eta) \in K \times U_+. \tag{8.77}$$

Proof. Let $W = K \times U_+$ and $\hat{w} = (\hat{z}, \hat{\eta}) \in W$. Then W is compact. By (8.75), there exists $\hat{\tau} = \hat{\tau}(\hat{z}, \hat{\eta}) > 0$ such that $|P(\hat{\tau}, \hat{z})\hat{\eta}| > 1$ and continuity implies that there exist $c > 1$ and a neighborhood $N(\hat{w})$ of $\hat{w}$ in W such that

$$|P(\hat{\tau}, z)\eta| > c, \; w = (z, \eta) \in N(\hat{w}). \tag{8.78}$$

Since W is compact, there exists a finite set $w^i \in W$, neighborhoods $N(w^i)$, times $\tau^i = \tau(w^i) > 0$, and constants $c^i > 1$, where $1 \le i \le k$ such that $W \subset N := \bigcup_{i=1}^k N(w^i)$ and

$$|P(\tau_i, z)\eta| > c,\ w = (z, \eta) \in N(w^i),\ 1 \le i \le k, \tag{8.79}$$

where $c = \min_i c^i > 1$.

Now let $V \subset Z_+$ be a bounded neighborhood of K such that $V \times U_+ \subset N$ and let $L \subset V$ be forward invariant. We prove that $L \subset X$, arguing by contradiction. If $a = (a_x, a_y) \in L \setminus X$, let $\alpha := a_y/|a_y|$ and note that $\alpha \in U_+$. We will inductively construct the sequence $(\nu_p)_p$ such that $\nu_p \nearrow \infty$ and

$$|P(\nu_p, a)\alpha| > c^p,\ p \ge 1. \tag{8.80}$$

As $(a, \alpha) \in N(w^i)$ for some i, by (8.79), we have $|P(\nu_1, a)\alpha| > c$, where $\nu_1 = \tau_i$. Now suppose $|P(\nu_p, a)\alpha| > c^p$ for some $p \ge 1$. Let

$$\tilde{\alpha} = \frac{P(\nu_p, a)\alpha}{|P(\nu_p, a)\alpha|}.$$

Since $L \setminus X$ is forward invariant, $\Phi(\nu_p, a) \in L \setminus X$, hence $\Phi_y(\nu_p, a) > 0$, where $\Phi(t, z) = (\Phi_x(t, z), \Phi_y(t, z))$. It follows from (8.61) that

$$P(\nu_p, a)\alpha = \frac{1}{|a_y|}P(\nu_p, a)a_y = \frac{1}{|a_y|}\Phi_y(\nu_p, a) > 0.$$

Thus, $\tilde{\alpha} \in U_+$. There exists j such that $(\Phi(\nu_p, a), \tilde{\alpha}) \in N(w^j)$. Then again, from (8.79) we have $|P(\tau_j, \Phi(\nu_p, a))\tilde{\alpha}| > c$ which implies that

$$|P(\tau_j, \Phi(\nu_p, a))P(\nu_p, a)\alpha| > c^{p+1}.$$

Thus, using (8.63), $|P(\nu_{p+1}, a)\alpha| > c^{p+1}$, where $\nu_{p+1} = \nu_p + \tau_j$. The induction is complete and, by construction, $\nu_p \to \infty$ as $p \to \infty$ and (8.80) holds. But then $|\Phi_y(\nu_p, a)| = |P(\nu_p, a)a_y| > c^p|a_y|$, which implies that $|\Phi_y(\nu_p, a)| \to \infty$. This contradiction to the boundedness of L proves that $L \subset X$.

Now we show that (8.76) holds if $L \subset X \cap V$ is forward invariant. If $(z, \eta) \in L \times U_+$, then it belongs to $N(w^i)$ for some i and by (8.79) $|P(\nu_1, z)\eta| > c$ where $\nu_1 = \tau^i$. Now suppose $|P(\nu_p, z)\eta| > c^p$ for some $p \ge 1$. Let $\tilde{\alpha} = P(\nu_p, z)\eta/|P(\nu_p, z)\eta|$. As $P(t, z) \ge 0$ it follows that $\tilde{\alpha} \in U_+$, so $(\Phi(\nu_p, z), \tilde{\alpha}) \in L \times U_+$ and hence it belongs to $N(w^j)$ for some j. By (8.79), $|P(\tau^j, \Phi(\nu_p, z))\tilde{\alpha}| > c$, and we have

$$|P(\nu_p + \tau^j, z)\eta| = |P(\tau^j, \Phi(\nu_p, z))P(\nu_p, z)\eta| > c|P(\nu_p, z)\eta| > c^{p+1}$$

by (8.13). Letting $\nu_{p+1} = \nu_p + \tau^j$, our induction argument is complete.

Now assume that K is also forward invariant. The implication (8.77) $\Rightarrow$ (8.75) is trivial. For the converse, using (8.76) and the fact that $\nu_p \le p\tau$

where $\tau = \max_i \tau^i$, we have $|P(\nu_p, z)\eta|^{1/\nu_p} > c^{p/\nu_p} \geq c^{1/\tau}$ for all $(z, \eta) \in K \times U_+$ and so for $p \geq 1$,

$$\frac{1}{\nu_p} \ln |P(\nu_p, z)\eta| > \frac{1}{\tau} \ln c.$$

Hence,

$$\lambda(z, \eta) = \limsup_{t\to\infty} \frac{1}{t} \ln(|P(t, z)\eta|) \geq \frac{1}{\tau} \ln c > 0.$$

This completes our proof. □

Now we give the proof of Theorem 8.48.

Proof. First we show that (8.72) implies that M is uniformly weakly ρ-repelling. If not, there exists a sequence $(\tilde{z}^m)_m \subset Z_+ \setminus X$ such that

$$\limsup_{t\to\infty} d(\Phi(t, \tilde{z}^m), M) < 1/m, \; m \geq 1.$$

Hence, there exists a sequence $(\tau_m)_m \subset J$ such that, for each $m \geq 1$,

$$d(\Phi(t, \tilde{z}^m), M) < 1/m, \; t \geq \tau_m. \tag{8.81}$$

Let $z^m = (x^m, y^m) = \Phi(\tau_m, \tilde{z}^m)$. Using the positive invariance of $Z_+ \setminus X$, we have that $y^m > 0$. By the semiflow property of Φ and from (8.81) we get

$$d(\Phi(t, z^m), M) < 1/m, \; t \in J, \; m \geq 1. \tag{8.82}$$

Using (8.72), we obtain from Lemma 8.49 (with $K = M$) that there exists a bounded neighborhood V of M in Z_+, having the property that any forward invariant set contained in V is a subset of X. The set $L_m = \{\phi(n, z^m) | n \geq 0\}$ is forward invariant and, according to (8.82), is contained in $V \setminus X$ for large m, contradicting Lemma 8.49. Hence, M is uniformly weakly ρ-repelling.

Now we prove the final assertion. Let $(a, \alpha) \in M \times U_+$. As $\omega(a) \subset M$ is compact and invariant, we can apply Lemma 8.49, with $K = \omega(a)$. Let V be a neighborhood of $\omega(a)$ and $c > 1$ as in the above mentioned lemma. Since $\Phi(t, a) \to \omega(a)$, there exists τ such that $\Phi(t, a) \in V$, $t \geq \tau$, so $L = \{\Phi(t, a) | t \geq \tau\}$ is forward invariant and contained in V. By (8.74), $\tilde{\alpha} = P(\tau, a)\alpha / |P(\tau, a)\alpha| \in U_+$ so, by (8.76), there exists a sequence $\nu_p \to \infty$ such that $|P(\nu_p, \Phi(\tau, a))\tilde{\alpha}| > c^p$, $p \geq 1$. Thus, using the cocycle property (8.63), we may conclude

$$|P(\nu_p + \tau, a)\alpha| > c^p |P(\tau, a)\alpha|, \; p \geq 1.$$

Choosing p large enough that $c^p |P(\tau_a, a)\alpha| > 1$, we have proved that there exists $t > 0$ such that $|P(t, a)\alpha| > 1$. Since $(a, \alpha) \in M \times U_+$ is arbitrary, we are done by the final assertion of Lemma 8.49. □

It is worth pointing out that Lyapunov exponents have been used to characterize when a compact, invariant subset M of an invariant submanifold X, which is asymptotically stable relative to X, is an attractor for the full dynamical system and when it is of saddle-type [**8, 234**]. In [**234**], it is shown that if the "normal" Lyapunov exponents are all negative, then M is an attractor, in a suitable sense (different than that used here).

As a corollary of Theorem 8.48 and Lemma 8.47 we have the result below.

Corollary 8.50. *Assume that $\Omega(M)$ consists of periodic orbits or equilibria and:*

(a) *if $O_+(z) \subset \Omega(M)$ is a p-periodic orbit ($p > 0$), $\exists T \in J$ such that $P(T, z) \gg 0$ and $\mathbf{r}(P(p, z)) > 1$;*

(b) *if $O_+(z) \subset \Omega(M)$ is an equilibrium in the continuous time case, assume that $A(z)$ is irreducible, $A(z)v = \mu v$ with $v \gg 0$ and $\mu > 0$.*

In the discrete case, assume also that $A(z)\eta \neq 0$, $z \in M$, $\eta \in U_+$. Then M is uniformly weakly ρ-repelling.

In order to appreciate the usefulness of Corollary 8.50, consider the effort required to establish the weak uniform ρ-repelling property of the periodic orbit P (note the conflict of notation with our matrix P) of the food chain model in Proposition 8.26. The transverse stability of P is determined by the one-by-one matrix $P(t, z) = e^{\int_0^t e(s)ds}$, which is clearly positive and $P(T, z) > 1$ if $\lambda > 0$ (T here is the period of orbit P). Thus, the weak uniform ρ-repelling property follows directly from Corollary 8.50 (a).

It should be stressed that the results of this section depend sensitively on the relation of Z_+, X and ρ. To see this, consider the May-Leonard system (3.24) where we used $\tilde{\rho}(N) = \min_i N_i$. Then $M = E_1 = (1, 0, 0)$ is weakly uniformly $\tilde{\rho}$-repelling because it is a hyperbolic saddle point and its stable manifold lies in $\tilde{\rho}^{-1}(0)$. However, it is not weakly uniformly ρ-repelling, where $\rho(N) = N_2 + N_3$, because the stable manifold of M contains points N for which $\rho(N) > 0$, namely, points on the heteroclinic orbit from E_3 to E_1. We note that one Lyapunov exponent is negative $(1 - \beta)$ and one $(1 - \alpha)$ is positive.

8.8.4. Host-parasite model.

Recall Example 8.41, the discrete time host-parasite model

$$\begin{aligned} x(t+1) &= x(t)\exp(r - x(t) - y(t)), \\ y(t+1) &= \theta x(t)(1 - \exp(-y(t))). \end{aligned} \tag{8.83}$$

We have noted that the host x persists uniformly but what about the parasite? The following result, due to Takeuchi and Kon [**124**] and formulated in a slightly different way, gives an answer.

Theorem 8.51. *Let M be the Ricker attractor, described in Example 8.41, and suppose that*

$$\limsup_{t\to\infty}\left(\prod_{i=0}^{t-1} x(i)\right)^{1/t} > 1/\theta, \; x(0)\in\Omega(M). \tag{8.84}$$

Then both host and parasite persist: (8.83) is uniformly ρ-persistent with $\rho(x,y)=\min\{x,y\}$.

Remark 8.52. Takeuchi and Kon [**124**] require that

$$\sup_{t\ge 0}\theta^t\prod_{i=0}^{t-1} x(i) > 1, \; x(0)\in\Omega(X).$$

Observe that $\Omega(M)\subset\Omega(X)\subset M$. (8.84) agrees with the condition of Takeuchi and Kon where both are defined.

Proof. In Example 8.41, we noted that the induced semiflow has a compact attractor of bounded sets and that it is uniformly ρ_1-persistent for $\rho_1(x,y)=x$. By Theorem 5.2, it suffices to establish uniform weak ρ-persistence. We apply the acyclicity Theorem 8.17 for this. Using the notation of that result, note that X_0 consists of the two coordinate axes bounding $X=\mathbb{R}^2_+$. Orbits starting on the positive y-axis jump to the origin in one step; orbits starting on the positive x-axis are attracted to M. Compact invariant sets $M_0=\{(0,0)\}$ and M form an acyclic cover of Ω, the union of limit sets of points of X_0. Note that $M_0\mapsto M$ in X_0 but the reverse cannot hold. M_0 is isolated in X because it is a one-step attractor of points on the y-axis and because of uniform ρ_1-persistence; the latter also implies that M_0 is uniformly weakly ρ-repelling. M is locally asymptotically stable in X_0 by Theorem 2.39 and hence it is isolated in X_0. Theorem 8.48 and (8.84) imply that M is weakly uniformly ρ-repelling. Hence, M is isolated in X and we have verified all hypotheses of Theorem 8.17. □

How can we verify (8.84)? Clearly, we need some information about M, the Ricker attractor. A huge amount of information is available for $1D$-maps in general and the Ricker map in particular. It is well known that they undergo a period-doubling cascade to chaos as r is increased and the famous Sarkovski Theorem constrains the allowed periods of the set of coexisting periodic orbits of the Ricker map. Observe that the Ricker map f has fixed point r and critical point 1. Here we content ourselves with a few simple results:

(a) If $0 < r \leq 2$, then $M = \{r\}$.

(b) If $r > 2$, then $M \subset [f^2(1), f(1)]$.

Item (a) is due to Cull [**37**]; see also [**217**, Thm.9.16]. Item (b) can be found in [**124, 186, 229**]; see also Proposition 5.12.

According to Theorem 25 of [**229**], M contains exactly one of the following invariant sets and this invariant set attracts almost all positive initial conditions: (1) a periodic orbit, or (2) a Cantor set, or (3) a finite union of intervals with a dense orbit.

If $0 < r \leq 2$, then (8.84) is equivalent to $\theta r > 1$. The crude estimate (b) implies that (8.84) holds if $\theta f^2(1) > 1$ provided $r > 2$. For more on the dynamics of (8.83), see [**120**].

8.9. Exercises

Exercise 8.1. Verify Remark 8.4.

Exercise 8.2. Let $\Phi : J \times X \to X$ be continuous and A be a compact chain-transitive set in X. Show:

(a) If $J = \mathbb{R}_+$, A is connected.

(b) If $J = \mathbb{Z}_+$, A is flow-connected.

Exercise 8.3. Let $\Phi : J \times X \to X$ be continuous and ϕ a total trajectory such that $\overline{\phi(-J)}$ is compact. Show that $\alpha(\phi)$ is chain-transitive.

Exercise 8.4. Show how the proof of Theorems 8.17 and 8.20 need to be modified if (H) is replaced by

($\tilde{\text{H}}$) ρ is uniformly continuous on X and there exist a set $B \subset X$ and some $c > 0$ such that $d(\Phi_t(x), B) \to 0$ as $t \to \infty$ for all $x \in X$ and $B \cap \{\rho \leq c\}$ has compact closure in X.

Exercise 8.5. Consider the Lotka-Volterra competitive system

$$x' = x(1 - x - 2y - az), \quad y' = y(1 - \frac{1}{2}x - y - bz), \quad z' = z(1 - cx - dy - z),$$

where $a, b, c, d > 0$. You may assume that it generates a continuous semiflow on $\mathbb{R}^3_+$ which reduces to:

$$x' = x(1 - x - 2y), \qquad y' = y(1 - \frac{1}{2}x - y),$$

when $z = 0$. For this reduced system, y drives x to extinction.

(a) Show that there is a compact attractor of bounded subsets of $\mathbb{R}^3_+$. Hint: You may find differential inequalities useful.

(b) Give conditions on a, b, c, d that guarantee that z is uniformly persistent, i.e., that the system is uniformly ρ-persistent for $\rho(x, y, z) = z$. Justify your assertion by appealing to the acyclicity theorem.

Exercise 8.6. Prove the first assertion of Proposition 8.26. Corresponding to $\tilde{\rho}$, $\tilde{X}_0$ is the union of the $y = 0$ face and the $x = 0$ face of the positive octant, and the relevant acyclic decomposition includes E_0 and E_x.

Exercise 8.7. If inequality (8.6) is reversed, then it is known that E_{xy} is globally attracting for the planar system obtained by setting $z = 0$. State a result analogous to Proposition 8.26 in this case.

Exercise 8.8. Show that the host dies out if $(\beta + 1)e^{-\xi} < 1$ for $\xi = \kappa_0 - \alpha - \gamma > 0$ and $(\beta + 1)e^{-\mu - \alpha y_\infty} < 1$ for y_∞ in (8.14) and $I(0) > 0$.

Exercise 8.9. Verify that (8.52) and (8.53) hold. For the latter, use (8.48).

Exercise 8.10. Verify Remark 8.42.

Exercise 8.11. Analyze the two-stage model of Beverton-Holt type

$$x_j(n+1) = \frac{x_k(n)}{a_{j0} + a_{jj}x_j(n) + a_{jk}x_k(n)}, \qquad j, k \in \{1, 2\}, j \neq k.$$

Here $a_{jk} \geq 0$, $a_{j1} + a_{j2} > 0$ for $j = 1$ or $j = 2$, $a_{j0} > 0$.

Exercise 8.12. Verify (8.62). For the continuous case you may use Proposition A.17.

Exercise 8.13. Verify (8.63).

Exercise 8.14. Verify (8.67) but note that $P(s, z)\eta = 0$ is possible in the discrete time case.

Exercise 8.15. Suppose in the food chain model (8.4) we take M_0, M_x and M as our acyclic covering, where M consists of the periodic orbit P and its interior in the $x - y$ plane. Then $\Omega(M) = E_{xy} \cup P$. Use Corollary 8.50 and Theorem 8.48 to show that M is weakly uniformly ρ-repelling if (8.7) and $\lambda > 0$.

Exercise 8.16. Verify Remark 8.52.

Chapter 9

An SI Endemic Model with Variable Infectivity

We consider an infectious disease that leads to either death or permanent immunity. We assume that the infectivity of an infected person depends on its infection age a, the time since infection has occurred. Among other things, this allows to implicitly incorporate a latency and/or removal period by assuming that infectivity is zero or very low at small infection age and high infection age, respectively. Infection age epidemic and endemic models go as far back as to Kermack and McKendrick [**115, 116, 117**] and have received an increased interest through modeling of HIV for which there is indication that there is an early infectivity peak a few weeks after infection and a late infectivity plateau approximately one year before the onset of full AIDS [**4**, 11.3.2] [**172**].

9.1. The model

Let $S(t)$ and $I(t)$ denote the respective amounts of susceptible and infected individuals at time t. The infected individuals are stratified along infection age a,

$$I(t) = \int_0^\infty u(t,a)da, \tag{9.1}$$

where $u(t, \cdot)$ is the infection-age density at time t. The *infective force* is given by

$$J(t) = \int_0^\infty \kappa(a)u(t,a)da. \tag{9.2}$$

The nonnegative function $\kappa \in L^\infty(\mathbb{R}_+)$ describes the infection-age dependent infectivity of an infected individual. The incidence (rate of new disease cases) is given by $S(t)J(t)$. Notice that the incidence is of density-dependent (mass action) type rather than of frequency-dependent (standard) type. A more general incidence which is more appropriate for HIV is considered in [**221**]; our choice is mathematically motivated: a global stability analysis of the endemic equilibrium is easier for mass action than other types of incidence (Section 9.9).

The disease dynamics are described by

$$\begin{aligned} S'(t) =& \Lambda - S(t)J(t) - \mu S(t), \\ (\partial_t + \partial_a)u(t,a) =& -(\mu + \gamma(a))u(t,a), \qquad u(t,0) = S(t)J(t). \end{aligned} \tag{9.3}$$

Further, we have initial conditions

$$S(0) = \breve{S}, \quad u(0,a) = \breve{u}(a), \tag{9.4}$$

where $\breve{S}$ is the initial amount of susceptible individuals and $\breve{u}$ the initial age-density of infected individuals. $\Lambda \geq 0$ is the rate at which individuals are recruited into the epidemiologically relevant part of the population. For childhood diseases, Λ is the population birth rate, for sexually transmitted diseases the recruitment rate into the sexually active part of the population. The parameter $\mu > 0$ is the natural per capita death rate. The nonnegative function γ describes the infection-age dependent rate at which an infected individual ceases to be infected: either by recovering from the disease and being permanently immune or by dying from the disease. This model is the special case of the one considered in [**221**] where a more general incidence is considered which is more appropriate for HIV than the mass action (density-dependent) incidence we have chosen here because it allows a global stability analysis of the endemic equilibrium. The analysis in [**221**] is fairly complete (though terse) with exception of the global stability of the endemic equilibrium which is established by Magal, McCluskey and Webb [**155**]. Here we will give a more detailed presentation of persistence. For global stability, we will use recent insight in how to employ the Volterra Lyapunov function $g(L) = L - 1 - \ln L$ pioneered by Volterra and promoted by Goh [**83**] in epidemic models [**12, 68, 87, 112, 125, 126, 127, 179, 163**]. In particular, we will follow Magal, McCluskey, and Webb [**155**]. We refer to [**68**] for an excellent survey of the Volterra Lyapunov function in epidemic models and to [**145, 146, 149**] as updates.

Persistence theory has been applied to endemic models with other class ages than infection age: progression age [**161**], infectivity age [**184**], removal age [**216**], recovery age [**227**], and treatment age [**222**]. In the cases of treatment age [**223**], conditions for global stability of the endemic equilibrium are derived by frequency domain methods for integro-differential equations [**150**], in the case of recovery age by a variant of the fluctuation method.

9.1.1. Reformulation in the spirit of Lotka. Compared to the exit rates $\gamma(a)$, the probability of still being in the infected class at infection age a (ignoring natural death), $\mathcal{F}(a)$, is epidemiologically more meaningful and more manageable. If $\mathcal{F}$ is positive and locally absolutely continuous (Section A.A.2), the two items are related by

$$\gamma(a) = -\frac{\mathcal{F}'(a)}{\mathcal{F}(a)}, \qquad \mathcal{F}(a) = \exp\Big(-\int_0^a \gamma(s)ds\Big). \tag{9.5}$$

By its interpretation $\mathcal{F}$ has the following properties:

$$\mathcal{F}(0) = 1 \text{ and } \mathcal{F} \text{ decreasing.} \tag{9.6}$$

When we integrate the McKendrick partial differential equation [**165**] for u in (9.3) along characteristics, we obtain

$$u(t,a) = \begin{cases} S(t-a)J(t-a)\mathcal{F}(a)e^{-\mu a}; & t > a \geq 0, \\ \breve{u}(a-t)e^{-\mu t}\dfrac{\mathcal{F}(a)}{\mathcal{F}(a-t)}; & 0 \leq t < a. \end{cases} \tag{9.7}$$

These equations, the first of which was suggested by Lotka [**152**] in the context of demographics, have an interpretation of their own. If, at time t, an infected individual has infection-age $a < t$, then it has been infected at time $t - a$ when the rate of new cases was $S(t-a)J(t-a)$. This term has to be multiplied by the probability $\mathcal{F}(a)$ of still being in the infected class and by the probability $e^{-\mu a}$ of not having died from other causes than the infection. If the infected individual has infection-age $a > t$, then it was already infected at time 0 when it had infection-age $a - t$ and the amount of infected individuals was $\breve{u}(a-t)$. This term has to be multiplied by the probability of not having died from other cases than the infection for a period of length t, $e^{-\mu t}$, and by the conditional probability $\mathcal{F}(a)/\mathcal{F}(a-t)$ of still being in the infected class at infection-age a under the proviso that one already was in that class at infection age $a - t$.

In order to avoid technicalities, we assume that $\mathcal{F}(a) > 0$ for all $a \geq 0$. Our proof of global stability of the endemic equilibrium will actually rely on this assumption.

For future reference, we repeat the complete model,

$$
\begin{aligned}
S' =& \Lambda - SJ - \mu S, \qquad S(0) = \breve{S},\\
J(t) =& \int_0^\infty \kappa(a) u(t,a) da,\\
u(t,a) =& \begin{cases} S(t-a)J(t-a)\mathcal{F}(a)e^{-\mu a}; & t > a \geq 0,\\ \breve{u}(a-t)e^{-\mu t}\dfrac{\mathcal{F}(a)}{\mathcal{F}(a-t)}; & 0 \leq t < a. \end{cases}
\end{aligned}
\tag{9.8}
$$

$\breve{S}$ and $\breve{u}$ are referred to as the initial data of the system.

Disease-free dynamics. If there is no disease, $u \equiv 0$, then the susceptibles obey the equation

$$
S' = \Lambda - \mu S, \qquad S(0) = \breve{S}. \tag{9.9}
$$

This equation can be integrated,

$$
S(t) = S^\diamond + (\breve{S} - S^\diamond)e^{-\mu t}, \qquad t \geq 0, \tag{9.10}
$$

with

$$
S^\diamond = \frac{\Lambda}{\mu}. \tag{9.11}
$$

Notice that $S(t) \to S^\diamond$ as $t \to \infty$ and that $S^\diamond$ is the steady state of (9.9).

9.1.2. Existence and boundedness of solutions. We substitute the third equation into the second and split up the integral at $a = t$,

$$
\begin{aligned}
S'(t) &= \Lambda - S(t)J(t) - \mu S(t),\\
J(t) &= \int_0^t \kappa(a)e^{-\mu a}\mathcal{F}(a)S(t-a)J(t-a)da\\
&\quad + e^{-\mu t}\int_t^\infty \kappa(a)\breve{u}(a-t)\frac{\mathcal{F}(a)}{\mathcal{F}(a-t)}da.
\end{aligned}
\tag{9.12}
$$

We integrate the equation for S and make a substitution in the equation for J,

$$
\begin{aligned}
S(t) =& \breve{S}e^{-\mu t}\exp\Big(-\int_0^t J(s)ds\Big)\\
&+ \Lambda\int_0^t e^{-\mu(t-r)}\exp\Big(-\int_r^t J(s)ds\Big)dr,\\
J(t) =& \int_0^t \kappa(a)e^{-\mu a}\mathcal{F}(a)S(t-a)J(t-a)da\\
&+ e^{-\mu t}\int_0^\infty \kappa(a+t)\breve{u}(a)\frac{\mathcal{F}(a+t)}{\mathcal{F}(a)}da.
\end{aligned}
\tag{9.13}
$$

Let $\breve{u} \in L^1_+(\mathbb{R}_+)$, the set of integrable nonnegative function, and $\|\breve{u}\|_1 = \int_0^\infty |\breve{u}(a)|da$,

$$(9.14) \qquad S^\diamond = \frac{\Lambda}{\mu}, \qquad S^\sharp = \max\{S^\diamond, \breve{S}\}, \qquad \kappa^\sharp = \|\kappa\|_\infty,$$

with the last denoting the essential supremum of $|\kappa(\cdot)|$ on $\mathbb{R}_+$. In order to apply Banach's fixed point theorem, let $\tau > 0$ be a fixed but arbitrary number and

$$Z = \{(S, J);\ S, J \in \mathrm{C}[0, \tau],\ 0 \le S(t) \le S^\sharp, \\ 0 \le J(t) \le \kappa^\sharp \|\breve{u}\|_1 \exp(\kappa^\sharp S^\sharp t)\}.$$

Let $G : Z \to Z$ be given by

$$(9.15) \qquad G(S, J)(t) = (G_1(S, J)(t), G_2(S, J)(t))$$

with $G_1(S, J)(t)$ and $G_2(S, J)(t)$ being the respective right-hand sides of (9.13). The continuity of $G_2(S, J)$ follows from the continuity of translation in $L^1(\mathbb{R}_+)$. It is easily checked that G preserves the estimates incorporated in the definition of Z. Using these estimates, one finds $L_\tau > 0$ such that, if $0 \le t \le \tau$ and $j = 1, 2$,

$$|G_j(S, J)(t) - G_j(\tilde{S}, \tilde{J})(t)| \le L_\tau \int_0^t \big(|S(r) - \tilde{S}(r)| + |J(r) - \tilde{J}(r)|\big)dr.$$

For $\lambda > 0$, we use the norm

$$\|(S, J)\|_\lambda = \sup_{[0,\tau]} e^{-\lambda t}\big(|S(t)| + |J(t)|\big).$$

Then

$$\begin{aligned} & e^{-\lambda t}(|G_j(S, J)(t) - G_j(\tilde{S}, \tilde{J})(t)|) \\ \le & L_\tau \int_0^t e^{-\lambda(t-r)} e^{-\lambda r}(|S(r) - \tilde{S}(r)| + |J(r) - \tilde{J}(r)|)dr. \\ \le & L_\tau \int_0^t e^{-\lambda(t-r)} \|(S, J)\|_\lambda dr, \end{aligned}$$

so

$$\|G(S, J) - G(\tilde{S}, \tilde{J})\|_\lambda \le \frac{2L_\tau}{\lambda} \|(S, J) - (\tilde{S}, \tilde{J})\|.$$

Choosing $\lambda > 0$ large enough makes G a strict contraction on the complete metric space Z endowed with the metric induced by $\|\cdot\|_\lambda$. By Banach's fixed point theorem, we find unique continuous solutions of (9.13) on every interval $[0, \tau]$ and so on $\mathbb{R}_+$. It follows that S is continuously differentiable. Finally, we define u by the third equation in (9.8).

Theorem 9.1. *For any $\breve{S}$ in $\mathbb{R}_+$ and $\breve{u} \in L^1_+(\mathbb{R}_+)$, there exists a unique solution (S, u) of (9.8). S is a nonnegative continuously differentiable function on $\mathbb{R}_+$ and $u(t, \cdot)$ is a continuous function of $t \ge 0$ with nonnegative* *values in the Banach space* $L^1(\mathbb{R}_+)$. *Further, the associated infective force* J *is a nonnegative continuous function on* $\mathbb{R}_+$.

In order to find a global estimate, we notice from (9.1) and (9.7) that

$$\begin{aligned} I(t) = & \int_0^t e^{-\mu a}\mathcal{F}(a)S(t-a)J(t-a)da \\ & + e^{-\mu t}\int_0^\infty \breve{u}(a)\frac{\mathcal{F}(a+t)}{\mathcal{F}(a)}da. \end{aligned} \tag{9.16}$$

By (9.6),

$$\begin{aligned} I(t) \leq & \int_0^t e^{-\mu(t-a)}S(a)J(a)da + e^{-\mu t}\int_0^\infty \breve{u}(a)da \\ =: & \tilde{I}(t). \end{aligned} \tag{9.17}$$

Then $\tilde{I}$ is differentiable and

$$S' + \tilde{I}' = \Lambda - \mu(S + \tilde{I}), \qquad S(0) + \tilde{I}(0) = \breve{S} + \breve{I}, \quad \breve{I} = \int_0^\infty \breve{u}(a)da.$$

We integrate this differential equation and find

$$S(t) + I(t) \leq S(t) + \tilde{I}(t) = (\breve{S} + \breve{I})e^{-\mu t} + S^\diamond(1 - e^{-\mu t}).$$

Since S and I are both nonnegative, they are both bounded. Further, $J(t) \leq \kappa^\sharp I(t)$ is bounded as well. We have the following large-time estimates.

Theorem 9.2. *The solutions of (9.8) are bounded,*

$$S(t) + I(t) \leq \max\{\breve{S} + \|\breve{u}\|_1, S^\diamond\}, \qquad J(t) \leq \kappa^\sharp \max\{\breve{S} + \|\breve{u}\|_1, S^\diamond\}$$

for all $t \geq 0$, *and the upper bounds are eventually uniform,*

$$\limsup_{t\to\infty}(S(t) + I(t)) \leq S^\diamond, \qquad \limsup_{t\to\infty} J(t) \leq \kappa^\sharp S^\diamond.$$

9.2. Host persistence and disease extinction

To find large-time lower bounds for the susceptibles and upper bounds for the infectives, we define

$$S_\infty = \liminf_{t\to\infty} S(t), \qquad J^\infty = \limsup_{t\to\infty} J(t). \tag{9.18}$$

We apply the fluctuation method (Proposition A.14) to (9.12): There exists a sequence $t_k \to \infty$ such that $S(t_k) \to S_\infty$ and $S'(t_k) \to 0$. This implies

$$0 \geq \Lambda - S_\infty J^\infty - \mu S_\infty.$$

We solve this inequality for S_∞,

$$S_\infty \geq S^\diamond \frac{\mu}{\mu + J^\infty}, \tag{9.19}$$

and use the estimate for J^∞ in Theorem 9.2.

Theorem 9.3. *The hosts persist uniformly,*

$$\liminf_{t\to\infty} S(t) \geq \frac{\mu S^\diamond}{\mu + \kappa^\sharp S^\diamond}.$$

To find a condition for disease extinction, we apply a version of the Lebesgue-Fatou lemma [**217**, Sec.B.2] to the J-equation in (9.13),

$$J^\infty \leq \int_0^\infty \kappa(a) e^{-\mu a} \mathcal{F}(a) \limsup_{t\to\infty} S(t-a) J(t-a) da.$$

By Theorem 9.2,

$$J^\infty \leq \int_0^\infty \kappa(a) e^{-\mu a} \mathcal{F}(a) S^\diamond J^\infty da.$$

Define the *basic reproduction number* of the disease,

$$\mathcal{R}_0 = S^\diamond \int_0^\infty \kappa(a) e^{-\mu a} \mathcal{F}(a) da. \tag{9.20}$$

Then $J^\infty \leq \mathcal{R}_0 J^\infty$, and $\mathcal{R}_0 < 1$ implies $J^\infty = 0$. We apply the Lebesgue-Fatou lemma to (9.16) and obtain $I^\infty = 0$ as well.

Theorem 9.4. *If $\mathcal{R}_0 < 1$, the disease goes extinct: $J(t) \to 0$ and $I(t) \to 0$ as $t \to \infty$.*

Notice that $\mathcal{R}_0$ has the usual interpretation of the average number of infections produced by one typical infected individual that is introduced into a completely susceptible population at its steady state $S^\diamond$.

9.3. Uniform weak disease persistence

Assume that $\mathcal{R}_0 > 1$ for the basic reproduction number $\mathcal{R}_0$, i.e., a typical infected individual more than replaces itself. In particular, κ is not zero almost everywhere on $\mathbb{R}_+$ and $\Lambda > 0$. By (9.13), $S(t) > 0$ for all $t > 0$ and

$$\begin{aligned} J(t) =& \int_0^t \kappa(a) e^{-\mu a} \mathcal{F}(a) S(t-a) J(t-a) da + \breve{J}(t), \\ \breve{J}(t) =& e^{-\mu t} \int_0^\infty \kappa(a+t) \breve{u}(a) \frac{\mathcal{F}(a+t)}{\mathcal{F}(a)} da. \end{aligned} \tag{9.21}$$

There is no disease if $\breve{J}(t)$ is zero for all $t \geq 0$ because then $J(t)$ and $I(t)$ are zero for all $t \geq 0$ (by Gronwall's inequality, e.g.). This can happen even if $\breve{u}$ is not zero almost everywhere; for instance, if there is some $b > 0$ such $\kappa(a) = 0$ for all $a \geq b$ and $\breve{u}(a) = 0$ for all $a \leq b$.

In order to rule this case out we assume that $0 < \int_0^\infty \breve{J}(t)dt$. We change the order of integration,

$$\begin{aligned} &0 < \int_0^\infty \breve{u}(a)\mathcal{I}(a)da, \\ (9.22) \qquad &\mathcal{I}(a) = \int_0^\infty e^{-\mu t}\kappa(t+a)\frac{\mathcal{F}(a+t)}{\mathcal{F}(a)}dt. \end{aligned}$$

$\mathcal{I}(a)$ can be interpreted as the *expected remaining infective impact* of an individual at infection-age a. Assumption (9.22) has this interpretation: The initially infected individuals contain some whose expected remaining infective impact is positive.

In the following we will use translations of solutions: for $r \geq 0$, define

$$(9.23) \quad J_r(t) = J(r+t), \quad \breve{J}_r(t) = \breve{J}(r+t), \quad S_r(t) = S(r+t), \qquad t \geq -r.$$

By (9.21),

$$\begin{aligned} J_r(t) &\geq \int_0^t \kappa(a)e^{-\mu a}\mathcal{F}(a)S(t+r-a)J(t+r-a)da + \breve{J}_r(t) \\ (9.24) \qquad &\geq \int_0^t \kappa(a)e^{-\mu a}\mathcal{F}(a)S_r(t-a)J_r(t-a)da + \breve{J}_r(t). \end{aligned}$$

Lemma 9.5. *If (9.22) holds, there exists some $t_0 > 0$ such that $J(t) > 0$ for all $t \geq t_0$. If $\int_0^\infty \breve{u}(a)\mathcal{I}(a)da = 0$, then $J(t) = 0$ for all $t \geq 0$.*

Proof. Assume (9.22). Then, for some small $r > 0$, $\breve{J}$ is not zero almost everywhere on $[r, \infty)$. Since the host persists uniformly by Theorem 9.3, there exists some $\delta > 0$ such that $S_r(t) \geq \delta$ for all $t \geq 0$. By (9.24),

$$J_r(t) \geq \int_0^t \kappa(a)e^{-\mu a}\mathcal{F}(a)\delta J_r(t-a)da + \breve{J}_r(t).$$

Apply Corollary B.6.

As for the second statement, it follows that $\breve{J}(t) = 0$ for all $t \geq 0$ and so $J(t) = 0$ for all $t \geq 0$ by (9.21). □

In order to derive a lower estimate for J^∞, let $\epsilon > 0$. By (9.19), there exists some $r > 0$ such that

$$S_r(t) = S(r+t) \geq \frac{\mu S^\diamond}{\mu + J^\infty + \epsilon}, \qquad t \geq 0.$$

By (9.24),

$$J_r(t) \geq \frac{\mu S^\diamond}{\mu + J^\infty + \epsilon}\int_0^t \kappa(a)e^{-\mu a}\mathcal{F}(a)J_r(t-a)da.$$

We apply the Laplace transform to this equation which converts the convolution into a product. Notice that

$$\widehat{J}_r(\lambda) = \int_0^\infty e^{-\lambda t} J_r(t)dt < \infty$$

for each $\lambda > 0$ because J_r is bounded. Further, $\widehat{J}_r(\lambda) > 0$ by Lemma 9.5. We obtain

$$\widehat{J}_r(\lambda) \geq \frac{\mu S^\diamond}{\mu + J^\infty + \epsilon} \widehat{J}_r(\lambda) \int_0^\infty e^{-\lambda a} \kappa(a) e^{-\mu a} \mathcal{F}(a)da.$$

We divide by $\widehat{J}_r(\lambda) > 0$,

$$1 \geq \frac{\mu S^\diamond}{\mu + J^\infty + \epsilon} \int_0^\infty e^{-\lambda a} \kappa(a) e^{-\mu a} \mathcal{F}(a)da.$$

This holds for all $\lambda, \epsilon > 0$ and so we can take the limit as $\lambda, \epsilon \to 0$,

$$1 \geq \frac{\mu}{\mu + J^\infty} \mathcal{R}_0.$$

We solve this inequality for J^∞.

Theorem 9.6. *If $\mathcal{R}_0 > 1$ and (9.22) holds, then*

$$\limsup_{t\to\infty} J(t) \geq \mu(\mathcal{R}_0 - 1).$$

In a similar way we can derive an upper bound for J_∞.

Corollary 9.7. *Let $\mathcal{R}_0 > 1$ and (9.22) hold. Then*

$$\liminf_{t\to\infty} J(t) \leq \mu(\mathcal{R}_0 - 1) \leq \limsup_{t\to\infty} J(t).$$

9.4. The semiflow

We choose the state space

$$X = \mathbb{R}_+ \times L^1_+(\mathbb{R}_+). \tag{9.25}$$

For the unique solution (S, J) of (9.13) with $(\breve{S}, \breve{u}) \in X$ we define

$$\Phi(t, (\breve{S}, \breve{u})) = (S(t), u(t, \cdot)) \tag{9.26}$$

with

$$u(t,a) = \begin{cases} S(t-a)J(t-a)e^{-\mu a}\mathcal{F}(a), & t > a \geq 0, \\ \breve{u}(a-t)e^{-\mu t}\dfrac{\mathcal{F}(a)}{\mathcal{F}(a-t)}, & 0 \leq t < a. \end{cases} \tag{9.27}$$

Then

$$J(t) = \int_0^\infty \kappa(a)u(t,a)da. \tag{9.28}$$

Theorem 9.8. *Φ is a continuous semiflow on X.*

Proof. For $r, t \geq 0$ define $S_r(t) = S(r+t)$ and $J(r+t) = J_r(t)$. By (9.27), checking the cases $t > a$, $t \leq a < r+t$, and $a > r+t$,

$$u(r+t,a) = \begin{cases} e^{-\mu a}\mathcal{F}(a)S_r(t-a)J_r(t-a), & t > a \geq 0, \\ u(r,a-t)e^{-\mu t}\frac{\mathcal{F}(a)}{\mathcal{F}(a-t)}, & a > t \geq 0. \end{cases}$$

By (9.28), splitting up the integral at $a = t$,

$$\begin{aligned} J_r(t) = & J(r+t) = \int_0^\infty \kappa(a)u(t+r,a)da \\ = & \int_0^t e^{-\mu a}\mathcal{F}(a)S_r(t-a)J_r(t-a)da + \int_t^\infty u(r,a-t)e^{-\mu t}\frac{\mathcal{F}(a)}{\mathcal{F}(a-t)}da. \end{aligned}$$

Further,

$$S_r'(t) = \Lambda - \mu S_r - S_r J_r, \quad t \geq 0, \qquad S_r(0) = S(r).$$

By definition of the semiflow,

$$\Phi_t(S(r), u(r,\cdot)) = (S_r(t), v(t,\cdot))$$

with

$$v(t,a) = \left\{ \begin{array}{ll} e^{-\mu a}\mathcal{F}(a)S_r(t-a)J_r(t-a), & t > a \geq 0 \\ u(r,a-t)e^{-\mu t}\frac{\mathcal{F}(a)}{\mathcal{F}(a-t)}, & a > t \geq 0 \end{array} \right\} = u(t+r,a)$$

because (S_r, J_r) is the unique solution of (9.13) with $(S(r), u(r,\cdot)) \in X$. Hence,

$$\Phi_t(\Phi_r(\breve{S},\breve{u})) = (S(r+t), u(r+t,\cdot)) = \Phi_{r+t}(\breve{S},\breve{u}).$$

The time-continuity of Φ follows from Theorem 9.1. To see that Φ is state-continuous uniformly in finite time, one first establishes that S and J depend continuously on $\breve{S}$ and $\breve{u}$. To this end, one can apply Gronwall's inequality to (9.13) or apply [**31**, Ch.2.Thm.2.2] (with $k = 0$) to the map G in (9.15). See the proof of Theorem 10.11. Continuity of Φ then follows from Lemma 1.34(c). □

9.5. Existence of a global compact attractor

The average length of the infection period (natural death neglected) is given by $\mathcal{D} = \int_0^\infty \mathcal{F}(a)da$ [**217**, 12.2]. It is natural to assume that this average length is finite.

Theorem 9.9. *If* $\mathcal{D} = \int_0^\infty \mathcal{F}(a)da < \infty$, *the semiflow* Φ *has a compact attractor of bounded sets.*

Proof. Following Theorem 2.33, we check that the semiflow Φ is point-dissipative, eventually bounded on bounded sets, and asymptotically smooth.

The first two properties follow from Theorem 9.2 and the definition of the semiflow Φ, (9.26) and (9.27). For asymptotic smoothness, we apply Theorem 2.46. Define

$$\Theta(t, (\breve{S}, \breve{u})) = (0, w(t, \cdot)) \tag{9.29}$$

with

$$w(t,a) = \begin{cases} S(t-a)J_2(t-a)e^{-\mu a}\mathcal{F}(a), & t > a \geq 0, \\ \breve{u}(a-t)e^{-\mu t}\dfrac{\mathcal{F}(a)}{\mathcal{F}(a-t)}, & 0 \leq t < a, \end{cases} \tag{9.30}$$

and

$$\Psi(t, (\breve{S}, \breve{u})) = (S(t), v(t, \cdot)) \tag{9.31}$$

with

$$v(t,a) = \begin{cases} S(t-a)J_1(t-a)e^{-\mu a}\mathcal{F}(a), & t > a \geq 0, \\ 0, & 0 \leq t < a, \end{cases} \tag{9.32}$$

where S and J are solutions of (9.13) and

$$\begin{aligned} J_1(t) =& \int_0^t \kappa(a)e^{-\mu a}\mathcal{F}(a)S(t-a)J(t-a)da, \\ J_2(t) =& e^{-\mu t}\int_0^\infty \kappa(a+t)\breve{u}(a)\frac{\mathcal{F}(a+t)}{\mathcal{F}(a)}da. \end{aligned}$$

Now let C be a bounded subset of initial data in X and let

$$c := \sup\{\breve{S} + \|\breve{u}\|_1; (\breve{S}, \breve{u}) \in C\}, \qquad \tilde{c} = \max\{c, S^\diamond\}.$$

For the solution S and J of (9.13), by Theorem 9.2,

$$S(t) + I(t) \leq \tilde{c}, \qquad J(t) \leq \kappa^\sharp \tilde{c}, \qquad \tilde{c} = \max\{c, S^\diamond\}.$$

Since $\mathcal{F}$ is decreasing,

$$J_2(t) \leq e^{-\mu t}\kappa^\sharp \|\breve{u}\|_1.$$

Now

$$\begin{aligned} \|\Theta(t, (\breve{S}, \breve{u}))\| \leq& e^{-\mu t}\|\breve{u}\|_1 + \int_0^t S(t-a)J_2(t-a)e^{-\mu a}\mathcal{F}(a)da \\ =& e^{-\mu t}\|\breve{u}\|_1 + \tilde{c}\int_0^t J_2(s)e^{-\mu(t-s)}\mathcal{F}(t-s)da \\ \leq& e^{-\mu t}c(1 + \tilde{c}\kappa^\sharp \mathcal{D}) \end{aligned}$$

with $D = \int_0^\infty \mathcal{F}(a)da$. So $\Theta(t, (\breve{S}, \breve{u})) \to 0$ as $t \to \infty$ uniformly for $(\breve{S}, \breve{u}) \in C$ and diam $\Theta(t, C) \to 0$ as $t \to \infty$.

In order to show that $\Phi_t(C)$ has compact closure, we check the criterion in Theorem B.2. All conditions are readily checked except possibly (iii),

$$\begin{aligned}\int_0^{t-h} &|S(t-a-h)J_1(t-a-h)e^{-\mu(a+h)}\mathcal{F}(a+h)\\ &\qquad - S(t-a)J_1(t-a)e^{-\mu a}\mathcal{F}(a)|da\\ &+\int_{t-h}^{t} |S(t-a-h)J_1(t-a-h)e^{-\mu(a+h)}\mathcal{F}(a+h)da\\ &\longrightarrow 0, \qquad h\to 0,\end{aligned}$$

uniformly for initial data from C. We only need to deal with the first summand which can be estimated by

$$\begin{aligned}&\kappa^\sharp\tilde{c}\int_0^{t-h}|S(t-a-h)-S(t-a)|e^{-\mu a}da\\ &+\tilde{c}\int_0^{t-h}|J_1(t-a-h)-J_1(t-a)|e^{-\mu a}da\\ &+\kappa^\sharp\tilde{c}^2\int_0^{t-h}(1-e^{-\mu h})e^{-\mu a}da\\ &+\kappa^\sharp\tilde{c}^2\int_0^{t-h}(\mathcal{F}(a)-\mathcal{F}(a+h))e^{-\mu a}da.\end{aligned} \tag{9.33}$$

It is easily seen that the last two summands tend to 0 as $h\to 0$. As for the first summand, by (9.8),

$$|S'|\le \Lambda + SJ+\mu S\le \Lambda+\tilde{c}(\mu+\kappa^\sharp\tilde{c}).$$

The mean value theorem implies that the first summand in (9.33) converges to 0 as $h\to\infty$, uniformly for initial values in C. Consider the second summand of (9.33) which, after a substitution, becomes

$$\tilde{c}\int_0^{t-h}|J_1(s+h)-J_1(s)|e^{-\mu(s+h-t)}ds. \tag{9.34}$$

After a change of variables,

$$J_1(s)=\int_0^s\tilde{\kappa}(s-r)S(r)J(r)dr$$

with $\tilde{\kappa}(s)=\kappa(s)e^{-\mu s}\mathcal{F}(s)$, $\tilde{\kappa}\in L^1(\mathbb{R}_+)$, $\tilde{\kappa}\le\kappa$. Then

$$\begin{aligned}|J_1(s+h)-J_1(s)|\le& h(\tilde{c}\kappa^\sharp)^2+\int_0^s|\tilde{\kappa}(s+h-r)-\tilde{\kappa}(s-r)|\tilde{c}^2\kappa^\sharp dr\\ \le& h(\tilde{c}\kappa^\sharp)^2+\tilde{c}^2\kappa^\sharp\int_0^\infty|\tilde{\kappa}(r+h)-\tilde{\kappa}(r)|dr.\end{aligned}$$

This converges to 0 as $h \to 0$ uniformly for $s \geq 0$ and uniformly for the initial data in C. Then also the expression in (9.34) converges to 0 as $h \to 0+$ uniformly for initial data in C. □

Total trajectories. Since the compact attractor of bounded sets consists of the orbits of total Φ-trajectories, we will describe the total trajectories for our system.

Let ϕ be a total Φ-trajectory, $\phi(t) = (S(t), u(t,\cdot))$. Then $\phi(r+t) = \Phi(t, \phi(r))$, $t \geq 0$, $r \in R$. Keep $r \in \mathbb{R}$ fixed but arbitrary. Define

$$J(t) = \int_0^\infty \kappa(a) u(t,a) da,$$

$$S_r(t) = S(t+r), \quad J_r(t) = J(t+r), \quad u_r(t,a) = u(t+r,a).$$

Then $(S_r(t), u_r(t,\cdot)) = \Phi(t, (S(r), u(r,\cdot)))$. By definition of the semiflow,

$$S_r' = \Lambda - \mu S_r - S_r J_r, \qquad S_r(0) = S(r),$$

$$u_r(t,a) = \begin{cases} e^{-\mu a}\mathcal{F}(a) S_r(t-a) J_r(t-a), & t > a, \\ u(r, a-t) e^{-\mu t} \frac{\mathcal{F}(a)}{\mathcal{F}(a-t)}, & t < a, \end{cases}$$

and

$$J_r(t) = \int_0^\infty \kappa(a) u_r(t,a) da.$$

So, for all $t \geq 0$,

$$S'(t+r) = \Lambda - \mu S(t+r) - S(t+r) J(t+r),$$

$$u(t+r,a) = \begin{cases} e^{-\mu a}\mathcal{F}(a) S(t+r-a) J(t+r-a), & t > a, \\ u(r, a-t) e^{-\mu t} \frac{\mathcal{F}(a)}{\mathcal{F}(a-t)}, & t < a, \end{cases}$$

$$J(t+r) = \int_0^\infty \kappa(a) u(t+r,a) da.$$

For $s \geq r$, with $t = s - r$,

$$S'(s) = \Lambda - \mu S(s) - S(s) J(s),$$

$$u(s,a) = \begin{cases} e^{-\mu a}\mathcal{F}(a) S(s-a) J(s-a), & s - r > a, \\ u(r, a-(s-r)) e^{-\mu(s-r)} \frac{\mathcal{F}(a)}{\mathcal{F}(a-(s-r))}, & s - r < a, \end{cases}$$

$$\begin{aligned} J(s) &= \int_0^\infty \kappa(a) u(s,a) da \\ &= \int_0^{s-r} \kappa(a) e^{-\mu a}\mathcal{F}(a) S(s-a) J(s-a) da \\ &\quad + \int_{s-r}^\infty \kappa(a) u(r, a-(s-r)) e^{-\mu(s-r)} \frac{\mathcal{F}(a)}{\mathcal{F}(a-(s-r))} da. \end{aligned}$$

This holds for all $r \in \mathbb{R}$, $s \geq r$. We let $r \to -\infty$. Then, for all $s \in \mathbb{R}$,

$$
\begin{aligned}
S'(s) =& \Lambda - \mu S(s) - S(s)J(s),\\
J(s) =& \int_0^\infty \kappa(a) e^{-\mu a} \mathcal{F}(a) S(s-a) J(s-a) da, \qquad (9.35)\\
u(s,a) =& e^{-\mu a} S(s-a) J(s-a), \qquad a \geq 0.
\end{aligned}
$$

We prove some estimates for bounded total trajectories, $\sup_{t\in\mathbb{R}} S(t) < \infty$ and $\sup_{t\in\mathbb{R}} \int_0^\infty u(t,a)da < \infty$. Then J is also bounded on $\mathbb{R}$. Notice that

$$
I(t) = \int_0^\infty u(t,a)da \leq \int_0^\infty S(t-a)J(t-a)e^{-\mu a} dt =: \tilde{I}(t).
$$

After a substitution,

$$
\tilde{I}(t) = \int_{-\infty}^t S(s)J(s)e^{-\mu(t-s)} ds.
$$

Since S and J are continuous, $\tilde{I}$ is differentiable and

$$
\tilde{I}'(t) = S(t)J(t) - \mu \tilde{I}(t), \qquad t \in \mathbb{R}.
$$

This implies

$$
S' + \tilde{I}' = \Lambda - \mu(S + \tilde{I}).
$$

For $t > r$,

$$
S(t) + \tilde{I}(t) = (S(r) + \tilde{I}(r))e^{-\mu(t-r)} + \int_r^t \Lambda e^{-\mu(t-s)} ds.
$$

Since S and $\tilde{I}$ are bounded, we can take the limit as $r \to -\infty$,

$$
S(t) + \tilde{I}(t) = \Lambda/\mu = S^\diamond, \qquad t \in \mathbb{R}.
$$

Since $I \leq \tilde{I}$,

$$
\left.
\begin{aligned}
S(t) + I(t) \leq S^\diamond, \qquad & J(t) \leq \kappa^\sharp S^\diamond\\
u(t,a) \leq \kappa^\sharp (S^\diamond)^2 \mathcal{F}(a) e^{-\mu a}, \qquad & a \geq 0
\end{aligned}
\right\} \quad t \in \mathbb{R}. \qquad (9.36)
$$

Also,

$$
S(t) \geq \int_r^t \Lambda e^{-\mu(t-s)} \exp\Big(- \int_s^t J(\tilde{s}) d\tilde{s} \Big) ds, \qquad t > r.
$$

We take the limit as $r \to -\infty$,

$$
S(t) \geq \frac{\mu S^\diamond}{\mu + \kappa^\sharp S^\diamond}, \qquad t \in \mathbb{R}. \qquad (9.37)
$$

We rephrase these estimates in terms of the compact attractor of bounded sets, A.

Theorem 9.10. *For all* $(\breve{S}, \breve{u}) \in A$,

$$\breve{S} + \int_0^\infty \breve{u}(a)da \leq S^\diamond, \qquad \breve{S} \geq \frac{\mu S^\diamond}{\mu + \kappa^\sharp S^\diamond}$$
$$\int_0^\infty \kappa \breve{u}(a)da \leq \kappa^\sharp S^\diamond$$
$$\breve{u}(a) \leq \kappa^\sharp (S^\diamond)^2 \mathcal{F}(a) e^{-\mu a}, \qquad a \geq 0.$$

9.6. Uniform disease persistence

We define $\rho : X \to \mathbb{R}_+$ by

$$\rho(\breve{S}, \breve{u}) = \int_0^\infty \kappa(a)\breve{u}(a)da. \tag{9.38}$$

Then

$$\rho(\Phi(t,x)) = \int_0^\infty \kappa(a)u(t,a)da = J(t)$$

provides the infective force at time t. By Theorem 9.6, Φ is uniformly weakly ρ-persistent if $\mathcal{R}_0 > 1$. It can happen that $J(0) = 0$ and $J(r) > 0$ and $J(t) = 0$ for some $t > r > 0$. For instance, all initially infected individuals could be in the latency period; at time r some of them are infectious, but at time $t > r$ all of them are in the removed period while those they have infected before are still in the latency period. This rules out the application of most of our theorems which would allow the transition from uniform weak to uniform persistence, and we will use Theorem 5.2. In order to be able to apply this latter Theorem, we have established the existence of a compact attractor of bounded sets under the assumption that $\mathcal{D} < \infty$. (See also [**216**, Thm.2.6] which has been used to overcome a similar difficulty in an endemic model with explicit disease stages of arbitrarily distributed length [**69**].)

We check hypothesis (H1) of Section 5.1.

Let $\phi(t) = (S(t), u(t,\cdot))$ be a total Φ-trajectory with bounded range.

Lemma 9.11. *If* $J(t) = 0$ *for all* $t \leq 0$, *then* $J(t) = 0$ *for all* $t \geq 0$.

Proof. Assume $J(t) = 0$ for all $t \leq 0$. Then

$$J(t) \leq \kappa^\sharp S^\diamond \int_0^t J(s)ds, \qquad t \geq 0.$$

Gronwall's inequality now provides the assertion. □

Lemma 9.12. *The following alternative holds: either* J *is 0 everywhere on* $\mathbb{R}$, *or* J *is positive everywhere on* $\mathbb{R}$.

Proof. By appropriate shifts, we have from the previous lemma for each $r \in \mathbb{R}$ that $J(t) = 0$ for all $t \leq r$ implies that $J(t) = 0$ for all $t \geq r$. This implies the following alternative: either J is identically equal to 0, or there exists a sequence $t_j \to -\infty$ as $j \to \infty$ with $J(t_j) > 0$. Assume the second. Let $J_j(t) = J(t + t_j)$. By (9.37), with $\delta = \inf_{\mathbb{R}} S > 0$,

$$J_j(t) \geq \int_0^t \delta J_j(t-a)\kappa(a)e^{-\mu a}\mathcal{F}(a)da + \breve{J}_j(t)$$

where $\breve{J}_j(0) = J_j(0) = J(t_j) > 0$ and $\breve{J}_j$ is continuous at 0. By Corollary B.6, $J_j(t) > 0$ for $t > b$ with some $b \geq 0$ that does not depend on j. In other words, $J(t) > 0$ for all $t > t_j + b$. Since $t_j \to -\infty$ as $j \to \infty$, $J(t) > 0$ for all $t \in \mathbb{R}$. □

The last lemma proves (H1) and we obtain from Theorem 5.2 that Φ is uniformly ρ-persistent. Recall the formula for the *expected remaining infective impact* of an individual at infection-age a, introduced in (9.22),

$$\mathcal{I}(a) = \int_0^\infty e^{-\mu t}\kappa(t+a)\frac{\mathcal{F}(a+t)}{\mathcal{F}(a)}dt.$$

In the language of Theorem 5.2, Lemma 9.5 translates as

$$X_0 = \Big\{(\breve{S}, \breve{u}) \in X, \int_0^\infty \breve{u}(a)\mathcal{I}(a)da = 0\Big\}.$$

Theorem 9.13. *Let $\mathcal{R}_0 > 1$ and $\mathcal{D} < \infty$. Then there exists some $\epsilon > 0$ such that*

$$\liminf_{t\to\infty} J(t) \geq \epsilon \qquad \textit{and} \qquad \liminf_{t\to\infty} I(t) \geq \epsilon$$

for all solutions with their initial data satisfying $\int_0^\infty \breve{u}(a)\mathcal{I}(a)da > 0$.

Proof. A solution satisfying $\int_0^\infty \breve{u}(a)\mathcal{I}(a)da > 0$ also satisfies $(\breve{S}, \breve{u}) \in X \setminus X_0$, i.e., $\rho(\Phi(s, (\breve{S}, \breve{u})) > 0$ for some $s \geq 0$. Since Φ is uniformly ρ-persistent and a semiflow, the assertion follows for J. By (9.16),

$$I(t) \geq \int_0^t S(t-a)J(t-a)\mathcal{F}(a)e^{-\mu a}da.$$

By a version of the Lebesgue-Fatou lemma [**217**, Sec.B.2],

$$\liminf_{t\to\infty} I(t) \geq S_\infty J_\infty \int_0^\infty \mathcal{F}(a)e^{-\mu a}da.$$

The persistence statement for I now follows from the one for J and the one for S in Theorem 9.3. □

Partition of the compact attractor. Let $\mathcal{R}_0 > 1$ and $\mathcal{D} < \infty$. By Lemma 9.12, the hypothesis (H1) in Theorem 5.7 holds. By Theorem 5.7, the compact attractor A is partitioned into three pairwise disjoint invariant sets, $A = A_0 \cup C \cup A_1$.

A_0 is the compact attractor of bounded sets in X_0 which is the set of those points $(\breve{S}, \breve{u})$ for which $\int_0^\infty \breve{u}(a)\mathcal{I}(a)da = 0$ with $\mathcal{I}(a)$ from (9.22). See Lemma 9.5. For initial data $(\breve{S}, \breve{u}) \in X_0$, $J(t) = 0$ for all $t \geq 0$ and $S' = \Lambda - \mu S$. This implies that A_0 is the singleton set with the element $(S^\diamond, 0)$.

A_1 is uniformly ρ-positive, i.e., there exists some $\delta > 0$ such that

$$\int_0^\infty \kappa(a)u(a)da \geq \delta \text{ for all } (S, u) \in A_1. \tag{9.39}$$

Further, A_1 attracts all solutions whose initial data satisfy (9.22) and all bounded sets on which the solution semiflow is eventually uniformly ρ-positive. Finally, A_1 is stable. Since ρ is linear, (9.38), A_1 is connected by Proposition 5.9. We will need the following estimate later.

Theorem 9.14. *There exist $\breve{\delta} > 0$ such that, for all $(\breve{S}, \breve{u}) \in A_1$,*

$$\breve{u}(a) \geq \breve{\delta}\mathcal{F}(a)e^{-\mu a}, \qquad a \geq 0.$$

Proof. Since A_1 is invariant, there exists a total trajectory $\phi : \mathbb{R} \to A_1$, $\phi(t) = (S, u(t, \cdot))$ with $S(0) = \breve{S}$ and $u(0, a) = \breve{u}(a)$. By (9.39),

$$J(t) = \int_0^\infty \kappa(a)u(t, a)da \geq \delta$$

and

$$u(t, a) = S(t - a)J(t - a)\mathcal{F}(a)e^{-\mu a}.$$

By Theorem 9.10,

$$u(t, a) \geq \frac{\mu S^\diamond}{\mu + \kappa^\sharp S^\diamond}\delta\mathcal{F}(a)e^{-\mu a}, \qquad t \in \mathbb{R}, a \geq 0.$$ □

Since A is the compact attractor of bounded sets in the convex cone of a Banach space, A is connected by Theorem 2.24. Since A_0 and A_1 are compact, C is nonempty. It consists of those points in A through which there exists a total trajectory ϕ with $\phi(t) \to (S^\diamond, 0)$ as $t \to -\infty$ and $\phi(t) \to A_1$ as $t \to \infty$.

9.7. Disease extinction and the disease-free equilibrium

If $\mathcal{R}_0 < 1$, then the number of infectives and the infective force converge to 0 as time tends to infinity by Theorem 9.4. Now that we know the existence of a compact attractor of bounded sets, we can describe the extinction of the disease more precisely. Recall that $\rho : X \to \mathbb{R}_+$ has been defined in (9.38) as

$$\rho(\breve{S}, \breve{u}) = \int_0^\infty \kappa(a)\breve{u}(a)da.$$

Then

$$\rho(\Phi(t, x)) = \int_0^\infty \kappa(a)u(t, a)da = J(t)$$

provides the infective force at time t. X_0 is the set of those points $(\breve{S}, \breve{u})$ for which $\int_0^\infty \breve{u}(a)\mathcal{I}(a)da = 0$ with $\mathcal{I}(a)$ from (9.22). See Lemma 9.5. For initial data $(\breve{S}, \breve{u}) \in X_0$, $J(t) = 0 = I(t) = u(t, \cdot)$ for all $t \geq 0$ and $S' = \Lambda - \mu S$. We integrate this equation,

$$S(t) = S^\diamond + (S(0) - S^\diamond)e^{-\mu t}, \qquad S^\diamond = \frac{\Lambda}{\mu}.$$

This implies that the singleton set formed by $(S^\diamond, u^\diamond)$, $u^\diamond(a) = 0$ for all $a \geq 0$, is the compact attractor of bounded sets in X_0. $(S^\diamond, 0)$ is called the *disease-free equilibrium*.

Theorem 9.15. *Let $\mathcal{R}_0 < 1$ and $\mathcal{D} = \int_0^\infty \mathcal{F}(a)da < \infty$. Then the singleton set formed by the disease-free equilibrium is the compact attractor of bounded sets in X. In particular, it is stable.*

Proof. Recall that the compact attractor of bounded set is the union of bounded total orbits (Proposition 2.34). As in Section 9.6, let $\phi(t) = (S(t), u(t, \cdot))$ be a total Φ-trajectory with bounded range. By (9.35), for $t \in \mathbb{R}$ and $a \geq 0$,

$$\begin{aligned} S' =& \Lambda - SJ - \mu S, \\ u(t, a) =& S(t - a)J(t - a)\mathcal{F}(a)e^{-\mu a}, \\ J(t) =& \int_0^\infty \kappa(a)S(t - a)J(t - a)e^{-\mu a}\mathcal{F}(a)da = \rho(\phi(t)). \end{aligned}$$

For all $t \geq r$, $S(t) \leq S^\diamond + (S(r) - S^\diamond)e^{-\mu(t-r)}$. We take the limit as $r \to -\infty$ and obtain $S(t) \leq S^\diamond$ for all $t \in \mathbb{R}$. We substitute this inequality into the equation for J,

$$J(t) \leq \int_0^\infty \kappa(a)S^\diamond J(t - a)e^{-\mu a}\mathcal{F}(a)da.$$

Let $\tilde{J} = \sup J(\mathbb{R})$. Since the range of ϕ is bounded and J nonnegative, $0 \leq \tilde{J} < \infty$ and, by (9.20),

$$\tilde{J} \leq \mathcal{R}_0\tilde{J}.$$

Since $\mathcal{R}_0 < 1$, $\tilde{J} = 0$ and $J(t) = 0$ for all $t \in \mathbb{R}$. This implies $u(t, \cdot) = 0$ and $S(t) = S^\diamond$ for all $t \in \mathbb{R}$. So the attractor, A, of bounded sets is the singleton set formed by the endemic equilibrium $(S^\diamond, 0)$. The endemic equilibrium is stable by Theorem 2.39. $\square$

9.8. The endemic equilibrium

For this example, it is not clear whether existing theory allows to conclude existence of an endemic equilibrium from the uniform persistence of both host and disease. The estimates in Section 9.5 are not good enough to imply that Φ_t is a condensing map for every $t > 0$ with respect to an appropriate measure of noncompactness [**238**, Thm.1.1.3].

Fortunately, it is easy to determine the endemic equilibrium explicitly.

An endemic equilibrium is a fixed point of the semiflow Φ,

$$\Phi(t, (S^*, u^*)) = (S^*, u^*) \text{ with } u^* \neq 0.$$

From the definition of Φ in (9.26) and (9.27),

$$u^*(a) = \begin{cases} S^* J^* e^{-\mu a} \mathcal{F}(a); & t > a \geq 0, \\ u^*(a-t) e^{-\mu t} \dfrac{\mathcal{F}(a)}{\mathcal{F}(a-t)}; & 0 \leq t < a. \end{cases}$$

Notice that, if u^* is given by the first case, it also satisfies the second case, so

$$u^*(a) = S^* J^* e^{-\mu a} \mathcal{F}(a), \qquad a \geq 0.$$

Since $u^* \neq 0$, also $S^* \neq 0 \neq J^*$. From (9.8),

$$\begin{aligned} 0 =& \Lambda - S^* J^* - \mu S^*, \\ J^* =& \int_0^\infty \kappa(a) u^*(a) da. \end{aligned}$$

We substitute the formula for u^* into the formula for J^* and divide by J^* and use the definition of $\mathcal{R}_0$,

$$S^* = \frac{S^\diamond}{\mathcal{R}_0}. \tag{9.40}$$

We substitute this formula into the equilibrium equation and obtain the following result.

Theorem 9.16. *There is an endemic equilibrium if and only if* $\mathcal{R}_0 > 1$. *It is uniquely determined and satisfies,*

$$\begin{aligned} S^* =& \frac{S^\diamond}{\mathcal{R}_0}, \quad J^* = \frac{\Lambda}{S^*} - \mu = \mu(\mathcal{R}_0 - 1), \\ u^*(a) =& S^* J^* e^{-\mu a} \mathcal{F}(a), \qquad a \geq 0. \end{aligned}$$

9.9. Persistence as a crossroad to global stability

According to the partition of the global compact attractor A as it has been described at the end of Section 9.6, there exists a compact invariant set A_1 which is uniformly ρ-positive and attracts all solutions whose initial data satisfy (9.22) and all bounded sets on which the semiflow is eventually uniformly ρ-positive. In this section we will show that A_1 is the singleton set whose only element is the endemic equilibrium. We use Theorem 2.53 and the same Lyapunov functions as in [**155**].

Theorem 9.17. *Assume $\mathcal{R}_0 > 1$ and $\mathcal{D} = \int_0^\infty \mathcal{F}(a)da < \infty$.*

Then the endemic equilibrium attracts all solutions with their initial data satisfying (9.22). It also attracts all bounded sets on which the semiflow is eventually uniformly ρ-continuous. In particular, it is locally asymptotically stable.

We employ the following nonnegative function which is a well known ingredient for Lyapunov functions in Volterra-Lotka systems [**83**],

$$g(y) = y - 1 - \ln y. \tag{9.41}$$

For $x = (\breve{S}, \breve{u}) \in A_1$, we define

$$\breve{T} = \frac{\breve{S}}{S^*}, \qquad \breve{w}(a) = \frac{\breve{u}(a)}{u^*(a)}. \tag{9.42}$$

The Lyapunov function to be, W, is defined by

$$\begin{aligned} W_1(x) =& g(\breve{T}), \qquad W_2(x) = \int_0^\infty \gamma(a) g(\breve{w}(a))da, \\ W(x) =& W_1(x) + W_2(x), \end{aligned} \tag{9.43}$$

with a function $\gamma \in L^1(\mathbb{R}_+)$ still to be determined. For the ease of the reader we repeat the equations for the endemic equilibrium $x^* = (S^*, u^*)$,

$$\begin{aligned} J^* =& \frac{\Lambda}{S^*} - \mu = \int_0^\infty \kappa(a) u^*(a)da, \\ u^*(a) =& S^* J^* e^{-\mu a} \mathcal{F}(a), \qquad a \geq 0. \end{aligned} \tag{9.44}$$

For $(\breve{S}, \breve{u}) \in A_1$, by Theorem 9.10 and Theorem 9.14, the function $\frac{\breve{u}}{u^*}$ is bounded and bounded away from 0 which implies that W_2 and thus W are well defined.

We check the assumptions of Theorem 2.53. Let $\phi : \mathbb{R} \to A_1$ be a total Φ-trajectory, $\phi(t) = (S(t), u(t, \cdot))$, $S(0) = \breve{S}$ and $u(0, a) = \breve{u}(a)$. We recall the equations for S and u: for all $t \in \mathbb{R}$,

$$\begin{aligned} S' =& \Lambda - SJ - \mu S, \\ u(t,a) =& S(t-a)J(t-a)\mathcal{F}(a)e^{-\mu a}, \quad a \geq 0, \\ J(t) =& \int_0^\infty \kappa(a)u(t,a)da = \rho(\phi(t)). \end{aligned} \tag{9.45}$$

Define

$$T(t) = \frac{S(t)}{S^*}, \qquad w(t,a) = \frac{u(t,a)}{u^*(a)}. \tag{9.46}$$

Then $T(0) = \breve{T}$, $w(0,a) = \breve{w}(a)$, and

$$\begin{aligned} T' =& \frac{\Lambda}{S^*} - TJ - \mu T, \\ J(t) =& \int_0^\infty \kappa(a)u^*(a)w(t,a)da, \\ w(t,a) =& \frac{T(t-a)J(t-a)}{J^*}. \end{aligned} \tag{9.47}$$

By (9.42) and (9.43),

$$\begin{aligned} W_1(\phi(t)) =& g(T(t)), \\ W_2(\phi(t)) =& \int_0^\infty \gamma(a)g(w(t,a))da. \end{aligned}$$

Using the notation $\frac{d}{dt}W_1 = \frac{d}{dt}W_1(\phi(t))$,

$$\frac{d}{dt}W_1 = \Big(1 - \frac{1}{T(t)}\Big)T'(t) = \Big(1 - \frac{1}{T}\Big)\Big(\frac{\Lambda}{S^*} - TJ - \mu T\Big).$$

By (9.44), $\Lambda/S^* = J^* + \mu$ and

$$\frac{d}{dt}W_1(t) = \Big(1 - \frac{1}{T}\Big)\Big(J^* - TJ - \mu(T-1)\Big).$$

After some algebra, we obtain the equation

$$\frac{d}{dt}W_1(t) = \Big(1 - \frac{1}{T}\Big)J^* + (1-T)J - \frac{\mu}{T}(T-1)^2. \tag{9.48}$$

As for W_2, we assume that $\gamma : \mathbb{R}_+ \to \mathbb{R}_+$ is locally absolutely continuous (Section A.A.2) and $\gamma, \gamma' \in L^1(\mathbb{R}_+)$.

Lemma 9.18. *$W_2 \circ \phi$ is locally absolutely continuous and*

$$\frac{d}{dt}W_2 = \gamma(0)g(w(t,0)) + \int_0^\infty \gamma'(a)g(w(t,a))da.$$

Proof. By (9.47), $g(w(t,a)) = \xi(t-a)$ for some continuous bounded function ξ. Let

$$\psi(t) = \gamma(0)\xi(t) + \int_0^\infty \gamma'(a)\xi(t-a)da.$$

Then ψ is continuous and, by Fubini's theorem,

$$\int_r^t \psi(s)ds = \gamma(0)\int_r^t \xi(s)ds + \int_0^\infty \gamma'(a)\Big(\int_{\mathbb{R}} \chi_{[r,t]}(s)\xi(s-a)ds\Big)da.$$

After a substitution,

$$\int_r^t \psi(s)ds = \gamma(0)\int_r^t \xi(s)ds + \int_0^\infty \gamma'(a)\Big(\int_{\mathbb{R}} \chi_{[r,t]}(s+a)\xi(s)ds\Big)da.$$

Using Fubini's theorem again,

$$\int_r^t \psi(s)ds = \gamma(0)\int_r^t \xi(s)ds + \int_{\mathbb{R}}\Big(\int_0^\infty \gamma'(a)\chi_{[r,t]}(s+a)da\Big)\xi(s)ds.$$

Let $a_+ = \max\{a, 0\}$ be the positive part of a real number a. Then

$$\begin{aligned}
\int_r^t \psi(s)ds =&\gamma(0)\int_r^t \xi(s)ds + \int_{\mathbb{R}}\Big(\int_{[r-s]_+}^{[t-s]_+} \gamma'(a)da\Big)\xi(s)ds \\
=&\gamma(0)\int_r^t \xi(s)ds + \int_{\mathbb{R}}\Big(\gamma([t-s]_+) - \gamma([r-s]_+)\Big)\xi(s)ds \\
=&\gamma(0)\int_r^t \xi(s)ds + \int_r^t\Big(\gamma(t-s) - \gamma(0)\Big)\xi(s)ds \\
&+ \int_{-\infty}^r (\gamma(t-s) - \gamma(r-s))\xi(s)ds \\
=&\int_{-\infty}^t \gamma(t-s)\xi(s)ds - \int_{-\infty}^r \gamma(r-s)\xi(s)ds \\
=&W_2(\phi(t)) - W_2(\phi(r)).
\end{aligned}$$

□

We substitute g into the formula of Lemma 9.18,

$$\frac{d}{dt}W_2 = \gamma(0)(w(t,0) - 1 - \ln w(t,0)) + \int_0^\infty \gamma'(a)(w(t,a) - 1 - \ln w(t,a))da.$$

By (9.47),

$$\begin{aligned}
\frac{d}{dt}W_2 =&\gamma(0)\Big(\frac{T(t)J(t)}{J^*} - 1 - \ln w(t,0)\Big) \\
&+ \int_0^\infty \gamma'(a)(w(t,a) - 1 - \ln w(t,a))da.
\end{aligned}$$

In order to combine W_2 with W_1 in an effective way (recall (9.48)), we define

$$\gamma(a) = \int_a^\infty \kappa(a)u^*(a)da \tag{9.49}$$

to obtain

$$\gamma(0) = J^* \qquad \gamma'(a) = -\kappa(a)u^*(a).$$

We substitute this into the formula for $\frac{d}{dt}W_2$,

$$\begin{aligned}\frac{d}{dt}W_2 =& T(t)J(t) - J^* - J^* \ln w(t,0) \\ &- \int_0^\infty \kappa(a)u^*(a)(w(t,a) - 1 - \ln w(t,a))da.\end{aligned}$$

By (9.47) and (9.44),

$$\frac{d}{dt}W_2' = T(t)J(t) - J(t) + \int_0^\infty \kappa(a)u^*(a) \ln \frac{w(t,a)}{w(t,0)} da.$$

We combine $W = W_1 + W_2$,

$$\frac{dW}{dt} = \int_0^\infty \kappa(a)u^*(a)\Big(1 - \frac{1}{T} + \ln \frac{w(t,a)}{w(t,0)}\Big) da - \frac{\mu}{T}(T-1)^2.$$

We reintroduce $g(T(t))$,

$$\begin{aligned}\frac{dW}{dt} =& \int_0^\infty \kappa(a)u^*(a)\Big(-g\Big(\frac{1}{T(t)}\Big) + \ln \frac{w(t,a)T(t)}{w(t,0)}\Big) da \\ &- \frac{\mu}{T}(T-1)^2.\end{aligned}$$

We use g yet another time,

$$\begin{aligned}&\frac{dW}{dt} + \frac{\mu}{T}(T-1)^2 \\ =& -\int_0^\infty \kappa(a)u^*(a)\Big(g\Big(\frac{1}{T(t)}\Big) + g\Big(\frac{w(t,a)T(t)}{w(t,0)}\Big) - \frac{w(t,a)T(t)}{w(t,0)} + 1\Big) da.\end{aligned}$$

By (9.47),

$$\int_0^\infty \kappa(a)u^*(a)\Big(\frac{w(t,a)T(t)}{w(t,0)} - 1\Big) da = \frac{J(t)T(t)}{w(t,0)} - J^* = 0,$$

so

$$\frac{dW}{dt} + \frac{\mu}{T}(T-1)^2 = -J^* g\Big(\frac{1}{T(t)}\Big) - \int_0^\infty \kappa(a)u^*(a) g\Big(\frac{w(t,a)T(t)}{w(t,0)}\Big) da \le 0.$$

Now assume that $\frac{d}{dt}W \circ \phi \equiv 0$. Then $T(t) = 1$ for all $t \in \mathbb{R}$. By (9.47),

$$0 = \frac{\Lambda}{S^*} - J - \mu,$$

so $J = J^*$ by (9.44) and $w(t,a) = 1$ by (9.47). We recall (9.46) and find $\phi \equiv x^*$, the endemic equilibrium.

By Theorem 2.53, the persistence attractor A_1 is the singleton set with the endemic equilibrium x^* as the only element. The statements of Theorem 9.17 now follow from the properties of the persistence attractor A_1 in Theorem 5.7.

Notice that the proof above does not use that the endemic equilibrium is locally stable. In fact, as a consequence of Theorem 5.7, we obtain the stronger statement that the endemic equilibrium attracts a neighborhood of itself (Exercise 2.39).

A simple partition of the global attractor. Under the assumptions of Theorem 9.17, the compact attractor A of bounded sets has a simple three-way partition $A = A_0 \cup C \cup A_1$, where A_0 is the singleton set with the disease-free equilibrium, A_1 is the singleton set with the endemic equilibrium and C consists of total trajectories that connect the disease-free equilibrium to the endemic equilibrium.

9.10. Measure-valued distributions of infection-age

Mathematical convenience has made us assume so far that the infection-age distributions are given by densities rather than measures. More generally than in (9.1), the number of infected individuals can be given as

$$I(t) = m_t(\mathbb{R}_+), \tag{9.50}$$

where, for each $t \geq 0$, m_t is a Borel measure on $\mathbb{R}_+$: for each Borel set Ω in $\mathbb{R}_+$, $m_t(\Omega)$ is the number of infected individuals with infection-age in Ω. The *infective force* is given by

$$J(t) = \int_{\mathbb{R}_+} \kappa(a) m_t(da) \tag{9.51}$$

rather than by (9.2). As before, the nonnegative function κ on $\mathbb{R}_+$ describes the infection-age dependent infectivity of an infected individual. As a first modification we now assume that κ is a bounded Borel measurable function rather than an element in $L^\infty(\mathbb{R}_+)$ which, strictly speaking, is not a space of functions but of equivalence classes. If we have infection-age densities, it does not make a difference for the infective force which representative we choose. This would no longer be valid for infection-age distributions that are measures, and so we must settle on a specific representative.

One of the application-driven reasons to consider infection-age distributions that are measures is to include the case where all individuals that are infected initially have the same infection-age (they form an age-cohort of infected individuals). In this case, the initial distribution is not a density but a point measure.

We have the secondary motivation of presenting an example for a semiflow with time-set $\mathbb{R}_+$ that is state-continuous, uniformly in finite time, but not continuous in time itself. There are certain models of physiologically structured populations where the case for using a state space of measures is more pressing than here: either the solutions becomes measure-valued even for smooth-initial data by some mild form of shock-formation [**3, 213**], or there are equilibria to be taken care of that are measures [**2, 227**]. Developing and analyzing one of these models in a self-contained way would take too much space, though, so we use the infection-age model as an object of demonstration and otherwise refer to additional literature [**55, 56, 57, 58**].

In order to get an idea of how to define the measures m_t we take a look at the formula for J in (9.13) and replace $\breve{u}(a)da$ by $\breve{m}(da)$ where the measure $\breve{m}$ is the initial infection-age distribution,

$$\begin{aligned} J(t) = &\int_0^t \kappa(a)e^{-\mu a}\mathcal{F}(a)S(t-a)J(t-a)da \\ &+ e^{-\mu t}\int_0^\infty \kappa(a+t)\frac{\mathcal{F}(a+t)}{\mathcal{F}(a)}\breve{m}(da). \end{aligned} \tag{9.52}$$

By standard measure theoretic arguments, (9.52) follows from (9.51) and the definition of the measures m_t by

$$\begin{aligned} m_t(\Omega) = &\int_0^t \chi_\Omega(a)e^{-\mu a}\mathcal{F}(a)S(t-a)J(t-a)da \\ &+ e^{-\mu t}\int_0^\infty \chi_\Omega(a+t)\frac{\mathcal{F}(a+t)}{\mathcal{F}(a)}\breve{m}(da), \end{aligned} \tag{9.53}$$

where Ω is an arbitrary Borel set in $\mathbb{R}_+$ and χ_Ω the indicator (or characteristic) function of Ω with is 1 for $a \in \Omega$ and 0 otherwise. Notice that $m_0 = \breve{m}$. The equation for S remains the same,

$$S' = \Lambda - SJ - \mu S, \qquad S(0) = \breve{S}, \tag{9.54}$$

in integrated form,

$$\begin{aligned} S(t) = &\breve{S}e^{-\mu t}\exp\Big(-\int_0^t J(s)ds\Big) \\ &+ \Lambda\int_0^t e^{-\mu(t-r)}\exp\Big(-\int_r^t J(s)ds\Big)dr. \end{aligned} \tag{9.55}$$

The existence of solutions S and J of (9.55) and (9.52) follows mutandis mutatis from Banach's fixed point theorem as in Section 9.1.2. The only difference is that J is Borel measurable rather than continuous and S is locally absolutely continuous on $\mathbb{R}_+$ (Section A.A.2) rather than continuously differentiable. The space Z on which the contraction operates and in which the fixed point is found must be adapted correspondingly, and $\kappa^\sharp$ is now the supremum rather than the essential supremum of κ. We could achieve that J is continuous and S continuously differentiable by assuming that $\kappa \cdot \mathcal{F}$ is continuous, but we would like to avoid this assumption.

Theorem 9.19. *For any $\breve{S}$ in $\mathbb{R}_+$ and $\breve{m} \in \mathcal{M}_+(\mathbb{R}_+)$, there exists a unique solution (S, J) of (9.55) and (9.52). S is a nonnegative locally absolutely continuous function on $\mathbb{R}_+$, and the infective force J is a nonnegative Borel-measurable function on $\mathbb{R}_+$.*

The Borel measures m_t are now defined by (9.53). For each Borel set Ω in $\mathbb{R}_+$, $m_t(\Omega)$ is a Borel measurable function of $t \geq 0$.

The proof and result of the boundedness of solutions remain the same except that $\|\breve{u}\|_1 = \int_0^\infty \breve{u}(a)da$ must be replaced by $\breve{m}(\mathbb{R}_+)$.

The host persistence, disease-extinction and weak disease persistence results in Sections 9.2 and 9.3 also remain the same except that $\breve{u}(a)da$ must be replaced by $\breve{m}(da)$.

In order to choose the right state space, let $\mathcal{M}(\mathbb{R}_+)$ be the space of (signed) finite Borel measures on $\mathbb{R}_+$. The state space is given by $X = \mathbb{R}_+ \times \mathcal{M}_+(\mathbb{R}_+)$ where $\mathcal{M}_+(\mathbb{R}_+)$ is the cone of nonnegative measures. The map $\Phi : \mathbb{R}_+ \times X \to X$ is defined by

$$\Phi(t, (\breve{S}, \breve{m})) = (S(t), m_t),$$

where S and J are the unique solutions of (9.55) and (9.52) and m_t is defined by (9.53). The semiflow property follows as in Section 9.4 by proving an analogous result.

Theorem 9.20. *Let $\breve{S} \in \mathbb{R}_+$ and $\breve{m} \in \mathcal{M}_+(\mathbb{R}_+)$. Then $(S(t), m_t) = \Phi_t(\breve{S}, \breve{m})$ is the unique solution of a weakly formulated semilinear Cauchy problem,*

$$\begin{aligned} S' &= \Lambda - SJ - \mu S, \qquad S(0) = \breve{S}, \\ \Big(\frac{d}{dt} + \mu\Big) \int_0^\infty g(a) m_t(da) &= \int_0^\infty \frac{(g \cdot \mathcal{F})'(a)}{\mathcal{F}(a)} m_t(da) \\ &\quad + g(0) S(t) J(t), \qquad g \in D, \\ m_0 &= \breve{m}, \\ J(t) &= \int_0^\infty \kappa(a) m_t(da), \end{aligned}$$

and Φ is a semiflow on X.

Proof. That $(S(t), m_t) = \Phi_t(\breve{S}, \breve{m})$ is a weak solution of the semilinear Cauchy problem follows as in Theorem 9.8. One only needs to replace

$u(t,a)da$ by $m_t(da)$ and $\breve{u}(a)da$ by $\breve{m}(da)$ and to use that

$$\begin{aligned}\int_{\mathbb{R}_+} g(a)m_t(da) = &\int_{\mathbb{R}_+} \frac{g(a+t)\mathcal{F}(a+t)}{\mathcal{F}(a)}\breve{m}(da)\\ &+ \int_0^t g(t-r)e^{-\mu(t-r)}\mathcal{F}(t-r)S(r)J(r)dr\end{aligned} \tag{9.56}$$

for all bounded Borel measurable functions on $\mathbb{R}_+$. This follows from (9.53) by standard measure-theoretic arguments and by a substitution.

Conversely, let $(S(t), m_t)$ be a weak solution of the semilinear Cauchy problem. The same proof as in Theorem 9.8 with the same replacements provides that (9.56) holds for all $g \in D$. Recall that D is the subspace of those functions g with compact support in $\mathbb{R}_+$ such that $g \cdot \mathcal{F}$ is continuously differentiable on $\mathbb{R}_+$ Equivalently,

$$\begin{aligned}\int_{\mathbb{R}_+} \frac{f(a)}{\mathcal{F}(a)} m_t(da) = &\int_{\mathbb{R}_+} \frac{f(a+t)}{\mathcal{F}(a)}\breve{m}(da)\\ &+ \int_0^t f(t-r)e^{-\mu(t-r)}S(r)J(r)dr\end{aligned}$$

for all continuously differentiable functions f with compact support in $\mathbb{R}_+$. Since continuous functions with compact support can be uniformly approximated by continuously differentiable functions with compact support, the equality holds for all continuous functions f with compact support in $\mathbb{R}_+$. By standard measure-theoretic arguments it then also holds for all bounded Borel measurable functions f with compact support. Setting $f(a) = g(a)\mathcal{F}(a)$, we have (9.56) for all bounded measurable functions g with compact support. By Lebesgue's theorem of dominated convergence, (9.56) holds for all bounded measurable functions g and, in particular, for all functions $g = \chi_\Omega$. So m_t satisfies (9.53) and $(S(t), m_t) = \Phi_t(\breve{S}, \breve{m})$ by definition of the map Φ. □

We finally face the choice of a topology on $\mathcal{M}_+(\mathbb{R}_+)$. As before, we want to work with the persistence function $\rho(\breve{S}, \breve{m}) = \int_{\mathbb{R}_+} \kappa(a)\breve{m}(da)$. Then $\rho(\Phi(t, (\breve{S}, \breve{m}))) = J(t)$. One of the assumptions that seems to be indispensable in Chapter 5 is the state-continuity of $\rho \circ \Phi$; in other words, J needs to depend continuously on $\breve{m}$. The only topology that is strong enough to provide this (via Gronwall's inequality) is the variation of a measure, equivalently,

$$\|\breve{m}\| = \sup\{|\breve{m}(\Omega)|; \Omega \in \mathcal{B}(\mathbb{R}_+)\},$$

where $\mathcal{B}(\mathbb{R}_+)$ denotes the σ-algebra of Borel sets in $\mathbb{R}_+$. The metric induced by this norm on $\mathcal{M}_+(\mathbb{R}_+)$ makes ρ uniformly continuous (even Lipschitz continuous). Once we know that S and J continuously depend on $(\breve{S}, \breve{m})$,

(9.53) implies that Φ is state-continuous, uniformly in finite time. Then $\rho \circ \Phi$ is state-continuous, uniformly in finite time, as well.

The existence of a compact attractor of bounded sets follows as in Section 9.5 with the usual adaptations.

If $\breve{m}$ is the Dirac measure concentrated at $a = 0$, then

$$\begin{aligned}\rho(\Phi(t, (\breve{S}, \breve{m}))) &= J(t) \\ &= \int_0^t \kappa(a) e^{-\mu a} \mathcal{F}(a) S(t-a) J(t-a) da + e^{-\mu t} \kappa(t) \mathcal{F}(t).\end{aligned}$$

Notice that the first term on the right-hand side of this equation is continuous in $t \geq 0$ even if J is not continuous. So J is continuous only if $\kappa \cdot \mathcal{F}$ is continuous, and this condition is necessary (and sufficient) for the time-continuity of $\rho \circ \Phi$. In particular, Φ is not time-continuous because then J would be continuous without this condition. However, by (9.56), $\int_{\mathbb{R}_+} g(a) m_t(da)$ is a continuous function of $t \geq 0$ whenever g is a bounded Borel measurable function such that $g \cdot \mathcal{F}$ is continuous.

We could now take the route of lesser resistance and assume the continuity of $\kappa \cdot \mathcal{F}$. Since $\rho \circ \Phi$ is state-continuous, uniformly in finite time, $\rho \circ \Phi$ would be continuous.

Alternatively, we keep avoiding this assumption and turn to total Φ-trajectories for help. Let $\phi(t) = (S(t), m_t)$ be a total Φ-trajectory with bounded range. For all $t \in \mathbb{R}$, $r \geq 0$, $\phi(t) = \Phi(r, \phi(t-r))$, i.e.,

$$S' = \Lambda - \mu S - SJ \quad \text{on } \mathbb{R},$$

and, for all $t \in \mathbb{R}$ and $r \geq 0$,

$$\begin{aligned}m_t(\Omega) = &\int_0^r \chi_\Omega(a) S(t-a) J(t-a) \mathcal{F}(a) e^{-\mu a} da \\ &+ e^{-\mu r} \int_{\mathbb{R}_+} \chi_\Omega(a+r) \frac{\mathcal{F}(a+r)}{\mathcal{F}(a)} m_{t-r}(da), \\ J(t) = &\int_0^r \kappa(a) S(t-a) J(t-a) e^{-\mu a} \mathcal{F}(a) da \\ &+ e^{-\mu r} \int_{\mathbb{R}_+} \kappa(a) \frac{\mathcal{F}(a+r)}{\mathcal{F}(a)} m_{t-r}(da).\end{aligned}$$

Since $\|m_t\|$ is a bounded function of $t \geq 0$, we can take the limit as $r \to \infty$ and obtain for all $t \in \mathbb{R}$ that

$$
\begin{aligned}
S' =& \Lambda - \mu S - SJ, \\
m_t(\Omega) =& \int_\Omega S(t-a)J(t-a)\mathcal{F}(a)e^{-\mu a}da, \\
J(t) =& \int_0^\infty \kappa(a)S(t-a)J(t-a)e^{-\mu a}\mathcal{F}(a)da,
\end{aligned}
\tag{9.57}
$$

and

$$
J(t) = \int_{\mathbb{R}_+} \kappa(a)m_t(da) = \rho(\phi(t)).
$$

To see that J is continuous, approximate the function $\kappa(a)e^{-\mu a}\mathcal{F}(a)$ by bounded continuous functions η_n in $L^1(\mathbb{R}_+)$. The approximation holds in $L^1(\mathbb{R}_+)$ after multiplication by $e^{-\mu a}$. Then

$$
\begin{aligned}
J(t) =& \lim_{n\to\infty} \int_0^\infty S(t-a)J(t-a)\eta_n(a)da \\
=& \lim_{n\to\infty} \int_{-\infty}^t S(r)J(r)\eta_n(t-r)dr.
\end{aligned}
$$

The first equation shows that the convergence is uniform in $t \in \mathbb{R}$ (S and J are bounded) and the second that we take the limit of continuous functions.

Since J and S are continuous, the total Φ-trajectory ϕ is continuous in spite of Φ being not time-continuous. System (9.57) also implies that the total Φ-trajectories with bounded range are the same as for the state-space $\mathbb{R}_+ \times L^1_+(\mathbb{R}_+)$. So the hypotheses of Theorem 5.2 with Remark 5.3 are satisfied and Φ is uniformly ρ-persistent, i.e., the disease is uniformly persistent. By Theorem 5.7 with Remark 5.8, the compact global attractor has the partition described at the end of Section 9.6. Since equilibria are special total Φ-trajectories with bounded range, we also have the same equilibria as for the state-space $L^1_+(\mathbb{R}_+)$. So Theorem 9.17 holds also for measure-valued infection-age distributions. In particular, the endemic equilibrium is also locally asymptotically stable in the state space with measure-valued infection-age distributions.

Chapter 10

Semiflows Induced by Semilinear Cauchy Problems

Consider a Banach space X. Let the linear operator A in X (mapping its domain $D(A) \subset X$ into X) be closed (Definition B.12) and $X_\circ = \overline{D(A)}$. Let $F : C \to X$ be continuous, $C \subset X_\circ$. We consider the semilinear Cauchy problem

$$u' = Au + F(u), \qquad u(0) = x. \tag{10.1}$$

10.1. Classical, integral, and mild solutions

There are different concepts of solutions of (10.1).

Definition 10.1. Let I be an interval $[0, b)$ with $b \in (0, \infty]$ or an interval $[0, b]$ with $b \in (0, \infty)$.

A function $u : I \to X$ is called a *classical solution* of (10.1) if u is continuous on I and differentiable on $I \setminus \{0\}$, $u(t) \in C \cap D(A)$ for all $t > 0$ and $u' = Au + F(u)$ on $I \setminus \{0\}$, $u(0) = x$.

A function $u : I \to C$ is called an *integral solution* of (10.1) if u is continuous on I, $\int_0^t u(s)ds \in D(A)$ for all $t \in I$, and

$$u(t) = x + A\int_0^t u(s)ds + \int_0^t F(u(s))ds, \qquad t \in I. \tag{10.2}$$

Notice that the natural requirement $\int_0^t u(s)ds \in D(A)$ for all $t > 0$ already implies that $u(t) \in X_\circ = \overline{D(A)}$ because $u(t) = \lim_{h\to 0+} \frac{1}{h}\int_t^{t+h} u(s)ds$. So the requirement that $u(t) \in C \subset X_\circ$ is not as stringent as it may appear.

The relation between the two notions of solution is described in the next result which follows from Corollary B.20 with $f(t) = F(u(t))$.

Proposition 10.2. *Every classical solution of (10.1) is an integral solution. An integral solution u is a classical solution if (and only if) u is differentiable.*

Integral solutions induce a semiflow.

Theorem 10.3. *Assume that for each $x \in C \subset X_\circ$, there exists a unique integral solution $u : \mathbb{R}_+ \to C$. Then $\Phi(t,x) = u(t)$ defines a time-continuous semiflow $\Phi : \mathbb{R}_+ \times C \to C$.*

Proof. From (10.2), $u(0) = x$; this implies $\Phi(0,x) = x$. Let $r \geq 0$ and set $v(t) = \Phi(t+r,x)$. Then, for $t \geq 0$,

$$v(t) = x + A\int_0^{t+r} \Phi(s,x)ds + \int_0^{t+r} F(\Phi(s,x))ds.$$

We split up the integrals,

$$\begin{aligned} v(t) =& x + A\int_0^r \Phi(s,x)ds + A\int_r^{t+r} \Phi(s,x)ds \\ &+ \int_0^r F(\Phi(s,x))ds + \int_r^{t+r} F(\Phi(s,x))ds. \end{aligned}$$

We use the definition of the semiflow and a substitution,

$$v(t) = \Phi(r,x) + A\int_0^t v(s)ds + \int_0^t F(v(s))ds.$$

By definition of the semiflow, $v(t) = \Phi(t, \Phi(r,x))$. □

From now on we assume that A is the generator of a C_0-semigroup. A *C_0-semigroup* is a continuous semiflow $S : \mathbb{R}_+ \times X \to X$ where the maps $S_t = S(t,\cdot)$ are linear operators (which then are automatically bounded). It is common to write $S(t)$ instead of S_t. The *generator* A of S is defined by

$$Ax = \lim_{h\to 0+} \frac{1}{h}(S(h)x - x) \tag{10.3}$$

and its domain $D(A)$ is the set of elements x for which this limit exists. One can show that A is closed and that $D(A)$ is a dense linear subspace of X [**67**, II.1.3], in particular $X_\circ = X$.

Remark 10.4. If S is only state-continuous, uniformly in finite time, it can be shown that $X_\circ = \overline{D(A)}$ equals the continuity space of S introduced in Definition 1.47. In particular, the restriction of S to $X_\circ$ is a C_0-semigroup.

Definition 10.5. A continuous function $u : I \to C \subset X$ is called a mild solution of (10.1) if

$$u(t) = S(t)x + \int_0^t S(t-r)F(u(r))dr, \quad t \in I. \tag{10.4}$$

The next result follows from Theorem B.22 with $f(t) = F(u(t))$.

Proposition 10.6. *Let $u : I \to X$. Then u is a mild solution of (10.1) if and only if it is an integral solution of (10.1).*

Remark 10.7. Integral solutions can also be obtained if A is the generator of a locally Lipschitz continuous integrated semigroup (i.e. a Hille-Yosida operator) [**5**, 3.5] or the generator of an integrated semigroup of locally bounded semivariation [**219**].

Definition 10.8. Let X, Y be subsets of normed vector spaces (or, more generally, let X, Y be metric spaces). Then $F : X \to Y$ is called *Lipschitz (continuous)* if there is a number $\Lambda > 0$ such that $\|F(x_1) - F(x_2)\| \leq \Lambda\|x_1 - x_2\|$ for all $x_1, x_2 \in X$.

If $\tilde{X} \subset X$, we say that F is Lipschitz continuous on $\tilde{X}$ if the restriction of F to $\tilde{X}$ is Lipschitz continuous.

F is called *locally Lipschitz (continuous)* if for every $x \in X$ there exists some $\epsilon > 0$ such that F is Lipschitz continuous on the ϵ-neighborhood of x, $U_\epsilon(x) = \{y \in X; \|y - x\| < \epsilon\}$.

F is called *Lipschitz (continuous)* on compact sets if F is Lipschitz continuous on every compact subset of X.

Theorem 10.9. *Assume in addition that F is Lipschitz continuous on compact sets. Then the integral (mild) solutions of (10.1) are unique.*

Proof. Let u_1 and u_2 be two integral, equivalently mild, solutions. Let $0 < T < \infty$. Since u_1 and u_2 are continuous, $K = u_1([0,T]) \cup u_2([0,T])$ is a compact set. By assumption, there exists some $\Lambda > 0$ such that $\|F(x) - F(y)\| \leq \Lambda\|x - y\|$ for all $x, y \in K$. Then

$$\begin{aligned}\|u_1(t) - u_2(t)\| &\leq \int_0^t \|S(t-s)\|\Lambda\|u_1(s) - u_2(s)\|ds \\ &\leq \int_0^t N_T\Lambda\|u_1(s) - u_2(s)\|ds, \qquad 0 \leq t \leq T,\end{aligned}$$

where $N_T = \max_{t\in[0,T]} \|S(t)\|$. Let $\lambda > 0$. Then

$$e^{-\lambda t}\|u_1(t) - u_2(t)\| \le N_T \int_0^t e^{-\lambda(t-s)} e^{-\lambda s} \Lambda \|u_1(s) - u_2(s)\| ds.$$

Set $\|u\|_\lambda = \max_{0\le t\le T} e^{-\lambda t}\|u(t)\|$ (alternatively we can use a Gronwall argument). Then

$$\|u_1 - u_2\|_\lambda \le N_T \Lambda \sup_{0\le t\le T} \int_0^t e^{-\lambda(t-s)} ds \|u_1 - u_2\|_\lambda \le \frac{N_T\Lambda}{\lambda}\|u_1 - u_2\|_\lambda.$$

We choose λ so large that $\frac{N_T\Lambda}{\lambda} < 1$. This implies that $\|u_1 - u_2\|_\lambda = 0$, i.e., $u_1 = u_2$ on $[0, T]$.

Since T has been arbitrary, $u_1 = u_2$ on $\mathbb{R}_+$. □

Remark 10.10. If F is locally Lipschitz, then F is Lipschitz continuous on compact sets. Actually, for every compact set K there exists a neighborhood U of K such that F is Lipschitz on U.

Proof (by contradiction). Assume that F is locally Lipschitz, but that there is some compact set K such that F is not Lipschitz continuous on any neighborhood of K. In particular, F is continuous on X. Let $U_n = \{x \in X; d(x, K) < 1/n\}$. Let (Λ_n) be a sequence of positive numbers with $\Lambda_n \to \infty$ as $n \to \infty$. Since F is not Lipschitz on U_n, for each $n \in \mathbb{N}$ there exist $x_n, y_n \in U_n$ such that

$$\|F(x_n) - F(y_n)\| > \Lambda_n \|x_n - y_n\|, \qquad n \in \mathbb{N}.$$

For each $n \in \mathbb{N}$, there exists some $w_n, z_n \in K$ with $\|w_n - x_n\| < 1/n$ and $\|z_n - y_n\| < 1/n$. Since K is compact, after choosing subsequences, $w_n \to x$ and $z_n \to z$ for some $x, z \in K$. Then $x_n \to x$ and $y_n \to z$. Since F is continuous, $F(x_n) \to F(x)$ and $F(y_n) \to F(z)$. The inequality above implies that $x_n - y_n \to 0$ and so $x_n, y_n \to x$. Since F is locally Lipschitz, there exist some $\epsilon > 0$ and $\Lambda > 0$ such that

$$\|F(z) - F(y)\| \le \Lambda\|z - y\| \qquad \text{if } z, y \in K, \|z - x\| < \epsilon, \|y - z\| < \epsilon.$$

Choosing subsequences again, we can arrange that $\|x_n - x\| < \epsilon$ and $\|y_n - x\| < \epsilon$ for all $n \in \mathbb{N}$. This implies

$$\Lambda \ge \frac{\|F(x_n) - F(y_n)\|}{\|x_n - y_n\|} \ge \Lambda_n \to \infty,$$

a contradiction. □

10.2. Semiflow via Lipschitz condition and contraction principle

The following is the most rudimentary contraction principle argument that provides a continuous semiflow of integral solutions to (10.1). It can be refined and extended in many directions: see [**160, 175, 214, 233**] and the subsequent sections.

We continue assuming that A is the generator of a C_0-semigroup S. See (10.3).

Theorem 10.11. *Assume that F is (globally) Lipschitz continuous on X. Then there exists a unique integral solution u to (10.1) on $\mathbb{R}_+$. The definition $\Phi(t,x) = u(t)$ provides a continuous semiflow.*

Proof. For $u \in C([0,T],X)$ and $\lambda > 0$, define $\|u\|_\lambda = \sup_{0\le t\le T} e^{-\lambda t}\|u(t)\|$. $C([0,T],X)$ is a Banach space under this norm. One can prove this from scratch or notice that this norm is equivalent to the usual supremum norm. Let Λ be the global Lipschitz constant of F. Define

$$\tilde{G}(v)(t) = S(t)x + \int_0^t S(t-s)F(v(s))ds, \qquad v \in C([0,T],X).$$

The same proof as in Theorem 10.9 shows that $\|\tilde{G}(v_1) - \tilde{G}(v_2)\| \le \frac{N_T\Lambda}{\lambda}\|v_1 - v_2\|_\lambda$. If λ is chosen so large that $\frac{N_T\Lambda}{\lambda} < 1$, $\tilde{G}$ becomes a strict contraction and has a unique fixed point in $C([0,T],X)$ by the contraction mapping theorem (Banach's fixed point theorem) which is a unique mild solution of (10.1) on $[0,T]$. So, for each $n \in \mathbb{N}$, we have a unique mild solution u_n on $[0,n]$. Notice that, if $k,n \in \mathbb{N}$ and $k < n$, then $u_n(t) = u_k(t)$ for all $t \in [0,k]$ because the restriction of u_n to $[0,k]$ is also a solution on $[0,k]$. This implies that the following definition is sound. If $t \ge 0$, choose some $n \ge$t and set

$$u(t) = u_n(t).$$

Obviously, u is a mild solution of (10.1) on $[0,\infty)$. Uniqueness of mild solutions follows from the previous theorem, and the induced semiflow Φ is time-continuous. To see that it is continuous, by Lemma 1.34 it is sufficient to show that Φ is state-continuous uniformly in finite time. To show this, we could simply apply the uniform contraction principle in [**31**, Ch.2.Thm.2.2] to the map $\tilde{G}$. For the convenience of the reader, we illustrate the essence of that principle for the present concrete situation. (Alternatively, we could use Gronwall's inequality.) Let $T > 0$ and $u_j(t) = \Phi(t,x_j)$, $j = 1,2$. By the triangle inequality, for $t \in [0,T]$,

$$\|u_1(t) - u_2(t)\| \le \|S(t)(x_1 - x_2)\| + \int_0^t S(t-s)\Lambda\|u_1(s) - u_2(s)\|ds.$$

With the same notation as before,

$$\|u_1 - u_2\|_\lambda \leq N_T \|x_1 - x_2\| + \frac{N_T \Lambda}{\lambda} \|u_1 - u_2\|_\lambda.$$

Again, we choose $\lambda > 0$ so large that $\frac{N_T\Lambda}{\lambda} < 1$ and obtain

$$\|u_1 - u_2\|_\lambda \leq \frac{N_T}{1 - \frac{N_T\Lambda}{\lambda}} \|x_1 - x_2\|.$$

By definition of the λ-norm and u_j,

$$\|\Phi(t, x_1) - \Phi(t, x_2)\| \leq \frac{e^{\lambda T} N_T}{1 - \frac{N_T\Lambda}{\lambda}} \|x_1 - x_2\|, \qquad 0 \leq t \leq T. \qquad \square$$

10.3. Compactness all the way

We now go easy on the Lipschitz condition for F but assume compactness and boundedness properties to not only obtain a solution semiflow but also a compact attractor.

Lemma 10.12. *Let $S(t)F$ be a compact operator for each $t > 0$ and D a bounded subset of X, $0 < \epsilon < T < \infty$. Then the following hold:*

(a) *The set $\{S(t)F(y); \epsilon \leq t \leq T, y \in D\}$ has compact closure in X.*

(b) *The function $t \mapsto S(t)F(y)$ is uniformly continuous on $[\epsilon, T]$, uniformly for $y \in D$:*
for any $\eta > 0$, there exists $\delta > 0$ such that

$$\sup_{y \in D} \|S(t)F(y) - S(s)F(y)\| \leq \eta \quad \text{if } |t - s| < \delta, \epsilon \leq t, s \leq T.$$

Proof. Let $0 < \epsilon < T < \infty$. Since $S(\epsilon)F$ is compact, $S(\epsilon)F(D)$ has compact closure. Further, the map

$$(t, y) \mapsto S(t - \epsilon)y, \qquad t \in [\epsilon, T], y \in X,$$

is continuous.

(a) This implies that the set

$$\bigcup_{t \in [\epsilon, T]} S(t - \epsilon)S(\epsilon)F(D) = \{S(t)F(y); \epsilon \leq t \leq T, y \in D\}$$

has compact closure.

(b) Further, the map above is uniformly continuous on $S(\epsilon)F(D)$. So $t \mapsto S(t - \epsilon)S(\epsilon)F(y)$ is uniformly continuous on $[\epsilon, T]$, uniformly for $y \in D$. The semigroup property of S implies (b). $\square$

Recall that each C_0-semigroup is exponentially bounded [**67**, V.1.7], i.e., there exist $N \geq 1$ and $\eta \in \mathbb{R}$ such that

$$\|S(t)\| \leq N e^{\eta t}, \qquad t \geq 0.$$

Lemma 10.13. *Let $S(t)F$ be a compact operator for each $t > 0$ and $T > 0$. Further, assume that F maps bounded subsets of X into bounded subsets of X. Then the operator G on $C([0,T],X)$ with the supremum norm defined by*

$$G(w)(t) = \int_0^t S(t-s)F(w(s))ds, \qquad w \in C([0,T],X), t \in [0,T]$$

is compact.

Proof. Let W be a bounded subset of $C([0,T],X)$. We use the Arzela-Ascoli theorem [**164**, Ch.8.3] to show that $G(W)$ has compact closure. Let $\epsilon \in (0,T)$. Let $t \in [\epsilon, T]$. Since the set $D = \{w(s); w \in W, s \in [0,T]\}$ is bounded, by Lemma 10.12 the set

$$K_\epsilon = \{S(t-s)F(w(s));\ s \in [0, t-\epsilon]\} \subset \{S(s)F(y); s \in [\epsilon, T], y \in D\}$$

has compact closure in X, and so its closed convex hull $\overline{\text{co}} K_\epsilon$ is compact by a theorem by Mazur [**62**, V.2.7.Thm.6]. Now

$$\frac{1}{t-\epsilon}\int_0^{t-\epsilon} S(t-s)F(w(s))ds \in \overline{\text{co}} K_\epsilon, \qquad w \in W.$$

This implies that

$$\tilde{K}_\epsilon = \Big\{\int_0^{t-\epsilon} S(t-s)F(w(s))ds;\ w \in W\Big\}$$

has compact closure for each $\epsilon \in (0,t)$. To show that $\{G(w)(t); w \in W\}$ has compact closure (equivalently, is totally bounded), we notice that

$$\{G(w)(t); w \in W\} = \tilde{K}_\epsilon + \breve{K}_\epsilon$$

with

$$\breve{K}_\epsilon = \Big\{\int_{t-\epsilon}^{t} S(t-s)F(w(s))ds;\ w \in W\Big\}.$$

Now every $y \in \breve{K}_\epsilon$ satisfies $\|y\| \leq \epsilon N e^{T\eta_+} \sup_{y \in D} \|F(y)\|)$ where $\eta_+ = \max\{\eta, 0\}$ is the positive part of η. We want to show that, for every $\tilde{\epsilon} > 0$, $\{G(w)(t); w \in W\}$ can be covered by finitely many balls of radius $\tilde{\epsilon}$. If we choose $\epsilon > 0$ small enough, $\breve{K}_\epsilon$ can be covered by one ball of radius $\tilde{\epsilon}/2$ centered at 0. Since $\tilde{K}_\epsilon$ has compact closure, it can be covered by finitely many balls with radius $\tilde{\epsilon}/2$. So $\{G(w)(t); w \in W\}$ can be covered by finitely many balls with radius $\tilde{\epsilon}/2$.

To show that $G(w)$ is continuous uniformly for $w \in W$, let $r \in (0,T]$, $\epsilon \in (0,r)$, $\epsilon \leq p \leq t \leq T$ and $p, t \to r$. Then

$$\begin{aligned} G(w)(t) - G(w)(p) = & \int_p^t S(t-s)F(w(s))ds \\ & + \int_{p-\epsilon}^p [S(t-s)F(w(s)) - S(p-s)F(w(s))]ds \\ & + \int_0^{p-\epsilon} [S(t-s)F(w(s)) - S(p-s)F(w(s))]ds. \end{aligned}$$

By the triangle inequality and a substitution,

$$\begin{aligned} & \|G(w)(t) - G(w)(p)\| \\ & \leq (t-p)Ne^{T\eta_+} \sup_{y \in D} \|F(y)\| + 2\epsilon N e^{T\eta_+} \sup_{y \in D} \|F(y)\| \\ & + \int_\epsilon^p [S(t-p+s)F(w(p-s)) - S(s)F(w(p-s))]ds. \end{aligned}$$

So,

$$\begin{aligned} \|G(w)(t) - G(w)(p)\| \leq & (2\epsilon + t - p)Ne^{T\eta_+} \sup_{y \in D} \|F(y)\| \\ & + r \sup_{\epsilon \leq s \leq r, y \in D} [S(t-p+s)F(y) - S(s)F(y)]. \end{aligned}$$

By Lemma 10.12 (b),

$$\limsup_{p,t \to r} \sup_{w \in W} \|G(w)(t) - G(w)(p)\| \leq 2\epsilon N e^{T\eta_+} \sup_{y \in D} \|F(y)\|.$$

Since $\epsilon \in (0,r)$ has been arbitrary,

$$\limsup_{p,t \to r} \sup_{w \in W} \|G(w)(t) - G(w)(p)\| = 0.$$

So $G(w)$ is continuous at $r \in (0,T]$, uniformly for $w \in W$. The continuity at 0 is shown similarly. So $G(W)$ has compact closure by the Arzela-Ascoli theorem [**164**, Ch.8.3]. □

In addition to the continuity of F, we assume:

- $S(t)F$ is a compact operator for each $t > 0$.
- F grows at most linearly, i.e., there exist $a, b > 0$ such that $\|F(x)\| \leq a + b\|x\|$ for all $x \in X$.

The first assumption is satisfied, e.g., if F is compact or if $S(t)$ is compact for each $t > 0$ and F satisfies the third assumption.

We look for mild solutions,

$$u(t) = S(t)x + \int_0^t S(t-s)F(u(s))ds. \tag{10.5}$$

In order to take advantage of our compactness condition we look for solutions of

$$v(t) = \int_0^t S(t-s)F\big(v(s) + S(s)x\big)ds. \tag{10.6}$$

We want to apply Tychonov's fixed point theorem. For a proof and the original reference see [**49**, Ch.2, Thm.10.1].

Theorem 10.14 (Tychonov). *Let K be a closed convex subset of a locally convex space, $\tilde{G} : K \to K$ continuous, and $\tilde{G}(K)$ have compact closure. Then $\tilde{G}$ has a fixed point, i.e., there exists some $v \in K$ with $\tilde{G}(v) = v$.*

Theorem 10.15. *For each $x \in X$, there exists an integral solution of $u' = Au + F(u)$, $u(0) = x$ on $\mathbb{R}_+$ such that*

$$\|u(t)\| \leq e^{\lambda t}(c_1 + c_2\|x\|), \qquad t \geq 0$$

with constants $\lambda, c_1, c_2 > 0$ that do not depend on x and t.

Proof. We make $\mathrm{C}(\mathbb{R}_+, X)$, the vector space of continuous function on $\mathbb{R}_+$, a locally convex space by endowing it with the topology of uniform convergence on compact intervals. More precisely, we introduce the metric

$$d(u,v) = \sup_{n\in\mathbb{N}} \frac{1}{n}\frac{\|u-v\|_n}{1+\|u-v\|_n}, \qquad \|v\|_n = \sup_{0\leq t\leq n} \|v(t)\|.$$

This metric is compatible with the linear structure of $\mathrm{C}(\mathbb{R}_+, X)$, $d(u,v) = d(u-v,0)$ and $d(\alpha u, 0) \to 0$ as $\alpha \to 0$. This compatibility makes $\mathrm{C}(\mathbb{R}_+, X)$ a locally convex space.

Let $x \in X$ be the initial datum of our Cauchy problem. We set

$$K = \{v \in \mathrm{C}(\mathbb{R}_+, X); \|v(t)\| \leq e^{\lambda t}(c_1 + c_2\|x\|) \quad \forall t \geq 0\}, \tag{10.7}$$

with constants $\lambda > 0$ and $c_1, c_2 > 0$ that are still to be determined. It is easy to see that K is a closed convex subset of $\mathrm{C}(\mathbb{R}_+, X)$. We define $\tilde{G} : K \to X^{\mathbb{R}_+}$ by

$$\tilde{G}(v)(t) = \int_0^t S(t-s)F(v(s) + S(s)x)ds, \qquad v \in K, t \geq 0.$$

With G defined as in Lemma 10.13, but on $\mathrm{C}([0,\infty), X)$,

$$\tilde{G}(v) = G(v + S(\cdot)x).$$

It follows from Lemma 10.13 that $\tilde{G}(v) \in C(\mathbb{R}_+, X)$. Let (v_j) be a sequence in K. The restrictions of v_j to an interval $[0,b]$ form a bounded set in $C([0,n], X)$. By Lemma 10.13, the restrictions of $\tilde{G}(v_j)$ to $[0,n]$ form a set in $C([0,n], X)$ with compact closure. Using a diagonalization procedure, there is a subsequence of (v_j) that converges uniformly on every interval $[0,n]$, $n \in \mathbb{N}$. By construction, the convergence holds with respect to the quasinorm we have chosen. This means that $\tilde{G}(K)$ has compact closure in $C(\mathbb{R}_+, X)$.

We now determine c_1, c_2, and λ such that $\tilde{G}$ maps K into itself. There exist $N \geq 1$ and $\eta \geq 0$ such that

$$\|S(t)\| \leq N e^{\eta t}, \qquad t \geq 0.$$

[**67**, V.1.7] By the triangle inequality,

$$\|\tilde{G}(v)(t)\| \leq \int_0^t N e^{\eta(t-s)}(a + b\|v(s)\| + bNe^{\eta s}\|x\|)ds. \tag{10.8}$$

Since $v \in K$,

$$\|\tilde{G}(v)(t)\| \leq a\frac{N}{\eta}e^{\eta t} + bNte^{\eta t}\|x\| + \int_0^t Ne^{\eta(t-s)}b(c_1 + c_2\|x\|)e^{\lambda s}ds.$$

Choose $\lambda > \eta$. Then

$$\|S(t)x\| + \|\tilde{G}(v)(t)\| \leq a\frac{N}{\eta}e^{\eta t} + N(1 + bt)e^{\eta t}\|x\| + (c_1 + c_2\|x\|)\frac{Nb}{\lambda - \eta}e^{\lambda t}.$$

Choosing λ, c_1, and c_2 large enough, we achieve that

$$\|S(t)x\| + \|\tilde{G}(v)(t)\| \leq (c_1 + c_2\|x\|)e^{\lambda t}, \qquad t \geq 0. \tag{10.9}$$

In particular, by definition of K in (10.8), $\tilde{G}$ maps K into K. Standard estimates show that the continuity of F implies the continuity of $\tilde{G}$.

Tychonov's fixed point theorem now provides a solution v of (10.6). We set $u(t) = S(t)x + v(t)$ and obtain a mild solution of $u' = Au + F(u)$, $u(0) = x$, which is also an integral solution. By (10.9),

$$\|u(t)\| \leq (c_1 + c_2\|x\|)e^{\lambda t}.$$ □

Theorem 10.16. *Assume, in addition, that F is Lipschitz continuous on compact sets. Then the integral solutions are unique, and the induced semiflow is continuous and asymptotically smooth.*

Proof. Uniqueness of integral (or mild) solutions follows from Theorem 10.9. We define the semiflow $\Phi : \mathbb{R} \times X \to X$ by

$$\Phi(t, x) = u(t)$$

with u being the integral (or mild) solution of $u' = Au + F(u)$ on $(0, \infty)$, $u(0) = x$. Since an integral solution u is continuous by definition, the semiflow Φ is time-continuous. By Lemma 1.34, for Φ to be continuous, it is sufficient to show that Φ is state-continuous uniformly in finite time. Suppose not. Then there exist some $\epsilon > 0$, $T > 0$, $x \in X$ and sequences (x_j) in X and (t_j) in $[0, T]$ such that $x_j \to x$ as $j \to \infty$ and

$$\|\Phi(t_j, x_j) - \Phi(t_j, x)\| \geq \epsilon, \qquad j \in \mathbb{N}.$$

Without restriction of generality, we can assume that $\|x_j\| \le \|x\| + 1$. By Theorem 10.15,

$$\|\Phi(t, x_j)\| \le \big(c_1 + c_2(\|x\| + 1)\big)e^{\lambda t}, \qquad t \ge 0.$$

Let $u_j(t) = \Phi(t, x_j)$. Then

$$u_j(t) = S(t)x_j + \int_0^t S(t-s)F(u_j(s))ds, \qquad t \ge 0.$$

By Lemma 10.13, the sequence of functions

$$u_j(t) - S(t)x_j = \int_0^t S(t-s)F(u_j(s))ds = G(u_j)(t)$$

has compact closure in $C([0,T], X)$. After choosing subsequences,

$$\tilde{u}(t) - S(t)x = \lim_{j\to\infty} (u_j(t) - S(t)x_j)$$

exists for each $t \ge 0$, with the convergence being uniform for $t \in [0,T]$. We take the limit as $j \to \infty$ in the integral equation and find that $\tilde{u}$ is a mild solution of $u' = Au + F(u)$ on $(0,T]$, $u(0) = x$. By uniqueness, $u = \tilde{u}$ on $[0,T]$ and $\Phi(t, x_j) \to \Phi(t,x)$ as $j \to \infty$ uniformly for $t \in [0,T]$, a contradiction.

To show that Φ is asymptotically smooth, notice that

$$\Phi(t,x) = S(t)x + \Psi(t,x)$$

with $\|S(t)x\| \le Ne^{-\eta t}\|x\| =: \theta(t, \|x\|)$ and

$$\Psi(t,x) = \int_0^t S(t-s)F(\Phi(s,x))ds = G(\Phi(\cdot, x))(t).$$

By Lemma 10.13, $\Psi_t(D)$ has compact closure whenever D is bounded and forward invariant under Φ. Theorem 2.46 implies that Φ is asymptotically smooth. □

While it would be possible to give abstract conditions for Φ to be point-dissipative and eventually bounded on bounded sets (and thus to have a compact attractor), these conditions are rarely met in concrete applications where point-dissipativity and eventual boundedness rather follow from inherent balance laws.

10.4. Total trajectories

Hypothesis (H1) of Theorem 5.2 and a condition in Theorem 4.21 are formulated in terms of total trajectories. Therefore, we describe total trajectories in the case that the semiflow is induced by a semilinear Cauchy problem.

Theorem 10.17. *Let A be a closed linear operator in X and let $F : X \to X$ be continuous. Assume that, for each $x \in X_\circ = \overline{D(A)}$, there exists a unique integral solution of (10.1) on $\mathbb{R}_+$. Then $\phi : \mathbb{R} \to X_\circ$ is a total trajectory of the solution semiflow Φ if and only if $\int_r^t u(s)ds \in D(A)$ and*

$$\phi(t) - \phi(r) = A \int_r^t \phi(s)ds + \int_r^t F(\phi(s))ds, \qquad r \leq t.$$

The proof is left as an exercise.

Theorem 10.18. *Let A be the generator of a C_0-semigroup S and let $F : X \to X$ be continuous. Assume that, for each $x \in X$, there exists a unique integral (or mild) solution of (10.1) on $\mathbb{R}_+$. Let $\phi : \mathbb{R} \to X$.*

Then ϕ is a total trajectory of the solution semiflow Φ if and only if

$$\phi(t) = S(t-r)\phi(r) + \int_r^t S(t-s)F(\phi(s))ds, \qquad r \leq t. \tag{10.10}$$

If, in addition, $\|S(t)\| < 1$ for some $t > 0$ and $\phi(\mathbb{R})$ has compact closure, ϕ is a total trajectory if and only if

$$\phi(t) = \int_{-\infty}^t S(t-s)F(\phi(s))ds, \qquad t \in \mathbb{R}. \tag{10.11}$$

Proof. Let $r \in \mathbb{R}$ be fixed but arbitrary. Define $\phi_t(r) = \phi(t+r)$ for $t \in \mathbb{R}$. Then the following chain of equivalences holds:

$$\phi(t) = S(t-r)\phi(r) + \int_r^t S(t-s)F(\phi(s))ds, \qquad -\infty < r \leq t < \infty,$$

if and only if (by a time shift)

$$\phi(t+r) = S(t)\phi(r) + \int_r^{t+r} S(t+r-s)F(\phi(s))ds, \qquad t \geq 0,$$

if and only if (by substitution)

$$\phi_r(t) = S(t)\phi(r) + \int_0^t S(t-s)F(\phi_r(s))ds, \qquad t \geq 0,$$

if and only if (by definition of the semiflow)

$$\phi_r(t) = \Phi(t, \phi(r)), \qquad t \geq 0,$$

if and only if (by definition of ϕ_r)

$$\phi(t+r) = \Phi(t, \phi(r)), \qquad t \geq 0.$$

This proves the first description of total trajectories. Assume $\|S(t)\| < 1$ for some $t > 0$. Then there are $N \geq 1$ and $\eta > 0$ such that $\|S(t)\| \leq Ne^{-\eta t}$ for all $t \geq 0$ [**67**, V.1.7]. Also, assume that $\phi(\mathbb{R})$ had compact closure. Then $F(\phi(\mathbb{R}))$ has compact closure as well and is a bounded set, and (10.11) follows from (10.10) by taking the limit as $r \to -\infty$. In turn assume (10.11). Then, for $r \le t$, we can split the integral at r,

$$\phi(t) = \int_{-\infty}^{r} S(t-s)F(\phi(s))ds + \int_{r}^{t} S(t-s)F(\phi(s))ds.$$

By the semigroup property,

$$\phi(t) = \int_{-\infty}^{r} S(t-r)S(r-s)F(\phi(s))ds + \int_{r}^{t} S(t-s)F(\phi(s))ds.$$

Since $S(t-r)$ is a bounded linear operator, it interchanges with the integral,

$$\phi(t) = S(t-r)\int_{-\infty}^{r} S(r-s)F(\phi(s))ds + \int_{r}^{t} S(t-s)F(\phi(s))ds.$$

We use (10.11) for $t = r$,

$$\phi(t) = S(t-r)\phi(r) + \int_{r}^{t} S(t-s)F(\phi(s))ds,$$

and obtain that ϕ satisfies (10.11), and conclude that ϕ is a total trajectory of Φ. □

10.5. Positive solutions: The low road

For applications in population dynamics, one is interested in solutions with positive values. In an abstract setting, let us say a Banach space X, this is captured by a positive closed convex cone $X_+ \subset X$:

(a) If $x, y \in X_+$, then $x + y \in X_+$.

(b) If $x \in X_+$, $\alpha \ge 0$, then $\alpha x \in X_+$.

(c) If $x, -x \in X_+$, then $x = 0$.

(d) X_+ is a closed subset of X.

More generally, we are interested in integral solutions to the semilinear Cauchy problem (10.1) which take values in a certain closed convex set $C \subset X$.

Let A be the generator of a positive C_0-semigroup S and C the closed convex set of interest; we assume

$$S(t)C \subset C, \qquad t \ge 0.$$

The high road to positive solutions assumes a subtangential condition for F with respect to C,

$$\frac{1}{h} d(x + hF(x), C) \to 0, \qquad h \to 0+, x \in C.$$

From there one can take two routes.

One constructs the solutions in such a way that they take values in C right away. This typically precludes the application of fixed point theorems and requires use of variants of the Euler polygon method or other approximation schemes [**160, 214**]. Alternatively, one can try to prove the forward invariance of C a posteriori [**135**, 3.9, 4.4]. Both approaches are tedious.

Here we take a low a posteriori road using a stronger condition than the subtangential condition. We call $F : D \to X$, $C \subset D \subset X$, *locally uniformly subtangential to* C if for each $x \in C$ there exist $\delta > 0$ and $\epsilon > 0$ such that

$$y + hF(y) \in C \qquad \text{if } y \in C, \|y - x\| < \epsilon, 0 \le h \le \delta. \tag{10.12}$$

Proposition 10.19. *Let A be the generator of a C_0-semigroup S, C a closed convex subset of X with $S(t)C \subset C$ for all $t \ge 0$, and let $F : C \to X$ be locally Lipschitz continuous and locally uniformly subtangential to C. Then, for each $x \in C$, there exists some $\tau > 0$ and a unique integral solution of (10.1) defined on $[0, \tau]$ with values in C.*

Proof. Let $x \in X_+$. Choose ϵ and $\delta > 0$ according to the locally uniform subtangential condition of F. Choose $\lambda > 1/\delta$ and $\lambda > \eta$ where

$$\|S(t)\| \le Ne^{\eta t}, \qquad t \ge 0.$$

Notice that every integral solution u to (10.1) also satisfies

$$u(t) = x + (A - \lambda) \int_0^t u(s)ds + \int_0^t \big(F(u(s)) + \lambda u(s)\big)ds.$$

In turn, every solution of this equation is an integral solution of (10.1). It is easy to see that $A - \lambda$ is the generator of the C_0-semigroups $e^{-\lambda t}S(t)$. So u is an integral solution of (10.1) if and only if it solves

$$u(t) = e^{-\lambda t}S(t)x + \int_0^t e^{-\lambda(t-s)}S(t-s)[F(u(s)) + \lambda u(s)]ds, \qquad t \ge 0. \tag{10.13}$$

Let $\tilde{G}(u)(t)$ denote the right-hand side of this equation and

$$K_\tau = \{u \in C([0, \tau], C); \|u(t) - x\| \le \epsilon, 0 \le t \le \tau\}.$$

K_τ is a complete metric space when endowed with the supremum norm. The aim is to choose τ so small that $\tilde{G}$ becomes a strict contraction from K_τ to K_τ and apply Banach's fixed point theorem. Standard estimates show that $\tilde{G}(u)$ is continuous on $[0, \tau]$. Since F is locally Lipschitz continuous, there exists some $\Lambda > 0$ such that $\|F(y) - F(z)\| \le \Lambda\|y - z\|$ whenever $y, z \in X$ and $y, z \in U_\epsilon(x)$ (possibly after making $\epsilon > 0$ smaller). Here $U_\epsilon(x)$ is the

ϵ-neighborhood of x. By the triangle inequality,

$$\|\tilde{G}(u)(t) - x\| \leq \|e^{-\lambda t}S(t)x - x\| + \int_0^t e^{-\lambda(t-s)}\|S(t-s)\|\Big(\|F(u(s)) - F(x) + F(x)\| + \lambda\|u(s)\|\Big)ds.$$

We use the Lipschitz property of F, and the estimates for S,

$$\|\tilde{G}(u)(t) - x\| \leq \|e^{-\lambda t}S(t)x - x\| + \int_0^t e^{-(\lambda-\eta)(t-s)}N\Big((\Lambda+\lambda)\|u(s) - x\| + \|F(x)\| + \lambda\|x\|\Big)ds.$$

Since $u \in K_\tau$, $\|u(s) - x\| \leq \epsilon$ for $s \in [0,\tau]$ and so, for $t \in [0,\tau]$,

$$\|\tilde{G}(u)(t) - x\| \leq \|e^{-\lambda t}S(t)x - x\| + \int_0^t e^{-(\lambda-\eta)(t-s)}N\Big((\Lambda+\lambda)\epsilon + \|F(x)\| + \lambda\|x\|\Big)ds.$$

We evaluate the integral, for $t \in [0,\tau]$,

$$\|\tilde{G}(u)(t) - x\| \leq \|e^{-\lambda t}S(t)x - x\| + \frac{1 - e^{-(\lambda-\eta)t}}{\lambda - \eta}N\Big((\Lambda+\lambda)\epsilon + \|F(x)\| + \lambda\|x\|\Big).$$

The right-hand side of this equation converges to 0 as $t \to 0$. So we achieve $\|\tilde{G}(u)(t) - x\| \leq \epsilon$ for all $t \in [0,\tau]$ by choosing $\tau > 0$ small enough.

To see that $\tilde{G}(u)(t) \in C$, we notice that

$$\tilde{G}(u)(t) = e^{-\lambda t}S(t)x + (1 - e^{-\lambda t})z$$

with

$$z = \frac{1}{1 - e^{-\lambda t}}\int_0^t e^{-\lambda(t-s)}S(t-s)(F(u(s)) + \lambda u(s))ds.$$

Since $S(t)x \in C$, it is sufficient to show that $z \in C$. Since u and $F \circ u$ are continuous, $z = \lim_{n\to\infty} z_n$ with

$$z_n = \frac{1}{1 - e^{-\lambda t}}\sum_{j=0}^{n-1}\Big(\int_{s_j}^{s_{j+1}} e^{-\lambda(t-s)}ds\Big)S(t-s_j)\Big(F(u(s_j)) + \lambda u(s_j)\Big),$$

where $s_j = \frac{tj}{n}$. To see that z_n is a convex combination of elements in C,

$$z_n = \sum_{j=0}^{n-1}\frac{\lambda}{1 - e^{-\lambda t}}\Big(\int_{s_j}^{s_{j+1}} e^{-\lambda(t-s)}ds\Big)S(t-s_j)\Big(\frac{1}{\lambda}F(u(s_j)) + u(s_j)\Big).$$

By choice of λ, $(1/\lambda)F(u(s_j)) + u(s_j) \in C$ and so

$$S(t-s_j)((1/\lambda)F(u(s_j)) + u(s_j)) \in C.$$

Further,

$$\sum_{j=0}^{n-1} \frac{\lambda}{1-e^{-\lambda t}} \int_{s_j}^{s_{j+1}} e^{-\lambda(t-s)} ds = \frac{\lambda}{1-e^{-\lambda t}} \int_0^t e^{-\lambda(t-s)} ds = 1.$$

Since C is convex and closed, $z_n \in C$ and $z \in C$. This finishes the proof that $\tilde{G}$ maps K_τ into itself.

For $u_1, u_2 \in K_\tau$, $t \in [0, \tau]$,

$$\begin{aligned} &\big\|\tilde{G}(u_1)(t) - \tilde{G}(u_2)(t)\big\| \\ &\le \int_0^t e^{-\lambda(t-s)} \|S(t-s)\| \,\big(\|F(u_1)(s) - F(u_2)(s)\| + \lambda\|u_1(s) - u_2(s)\|\big) ds. \end{aligned}$$

Here we have used the linearity of the integral and of $S(t-s)$. By the Lipschitz condition for F and the estimate for S,

$$\big\|\tilde{G}(u_1)(t) - \tilde{G}(u_2)(t)\big\| \le \int_0^t e^{-(\lambda-\eta)(t-s)} N(\Lambda + \lambda)\|u_1(s) - u_2(s)\| ds.$$

We take the supremum over t and evaluate the integral,

$$\begin{aligned} &\sup_{0\le t\le\tau} \|\tilde{G}(u_1)(t) - \tilde{G}(u_2)(t)\| \\ &\le \frac{1-e^{-(\lambda-\eta)\tau}}{\lambda-\eta} N(\Lambda+\lambda) \sup_{0\le s\le\tau} \|u_1(s) - u_2(s)\|. \end{aligned}$$

By choosing $\tau > 0$ as small as needed, $\tilde{G}$ becomes a strict contraction, and $\tilde{G}$ has a fixed point in K_τ which is a solution of (10.13) on $[0, \tau]$. By our previous consideration, it is an integral solution of (10.1) on $[0, \tau]$. □

Proposition 10.20. *Let C be a closed convex subset of X and $F : C \to X$ be locally Lipschitz continuous and locally uniformly subtangential to C. Further, assume that F maps bounded subsets of C into bounded subsets of X. Then, for each $x \in C$, there exists some $\tau > 0$ and a unique integral solution $u : [0, \tau) \to C$ of (10.1) such that $\tau = \infty$ or u is not bounded on $[0, \tau)$.*

Proof. Let $x \in C$. Then there exists some $\tilde{\tau} > 0$ and a unique integral solution $u : [0, \tilde{\tau}] \to C$ of (10.1). Let τ be the supremum to those numbers $\tilde{\tau} > 0$ for which there is an solution $u : [0, \tilde{\tau}) \to C$ of (10.1). By uniqueness, there exists a unique integral solution $u : [0, \tau) \to C$ of (10.1). Suppose that

$\tau < \infty$ and u is bounded on $[0,\tau)$. Let $\delta \in (0,\tau)$. Then, for $\tau-\delta < r < t < \tau$,

$$\begin{aligned}\|u(t)-u(r)\| \leq &\|S(t)x - S(r)x\| + \int_{\tau-\delta}^{t} \|S(t-s)F(u(s))\|ds \\ &+ \int_{\tau-\delta}^{r} \|S(r-s)F(u(s))\|ds \\ &+ \int_{0}^{\tau-\delta} \|S(t-s)F(u(s)) - S(r-s)F(u(s))\|ds.\end{aligned}$$

Let Q be the compact set $F(u([0,\tau-\delta]))$ and $c = \sup_{s\in[0,\tau]} \|F(u(s))\|$ and $N_\tau = \sup_{s\in[0,\tau]} \|S(s)\|$. After a substitution,

$$\begin{aligned}\|u(t)-u(r)\| \leq &\|S(t)x - S(r)x\| + 2\delta N_\tau c \\ &+ \int_0^\tau \sup_{y\in Q} \|S(t-r+s)y - S(s)y\|ds.\end{aligned}$$

The map $(s,y) \to S(s)y$ is continuous and uniformly continuous on $[0,\tau + \delta] \times Q$. This implies that

$$\limsup_{r,t\to\tau} \|u(t)-u(r)\| \leq 2\delta N_\tau c.$$

Since this holds for any $\delta \in (0,\tau)$,

$$\limsup_{r,t\to\tau} \|u(t)-u(r)\| = 0.$$

Since X is complete and C is closed, the limit $z = \lim_{t\to\tau-} u(t)$ exists and is an element of C. Since A is closed, by (10.2),

$$z = x + A\int_0^\tau u(s)ds + \int_0^\tau F(u(s))ds.$$

This means that the definition $u(\tau) = z$ provides an integral solution on $[0,\tau]$. By Proposition 10.19, there exists some $\delta > 0$ and a unique solution v of

$$v(t) = z + A\int_0^t v(s)ds + \int_0^t F(v(s))ds, \qquad 0 \leq t \leq \delta.$$

We define

$$w(t) = \begin{cases} u(t); & 0 \leq t < \tau, \\ v(t-\tau); & \tau \leq t \leq \tau+\delta.\end{cases}$$

Since $u(\tau) = z = v(0)$, w is continuous. A similar consideration as in Theorem 10.3 shows that w is an integral solution of (10.1) on $[0,\tau+\delta]$, a contradiction to the choice of τ. □

Theorem 10.21. *Let C be a closed convex subset of X and $F : C \to X$ be locally Lipschitz continuous on C and locally uniformly subtangential to C. Further, assume that F maps bounded subsets of C into bounded subsets of X. Finally, assume that all integral solutions of (10.1) on finite intervals*

$[0, \tau)$, $\tau \in (0, \infty)$, with values in C are bounded (e.g. if C is bounded). Then there exists a continuous semiflow $\Phi : \mathbb{R}_+ \times C \to C$ such that $\Phi(\cdot, x)$ is an integral solution of (10.1) on $\mathbb{R}_+$.

Remark 10.22. More precisely, Φ is time-continuous and locally Lipschitz state-continuous in the following sense: For each $x \in C$ and $T > 0$ there exist some $\epsilon > 0$ and $N > 0$ (which depend on x and T) such that

$$\|\Phi(t, y) - \Phi(t, z)\| \le N\|y - z\| \qquad \text{if } t \in [0, T], y, z \in U_\epsilon(x),$$

where $U_\epsilon(x) = \{y \in C; \|y - x\| < \epsilon\}$.

Proof. By Proposition 10.20, for each $x \in C$ there exists an integral solution of (10.1) on $\mathbb{R}_+$. Since F is locally Lipschitz and thus Lipschitz on compact sets (Remark 10.10), the solutions are unique and induce a time-continuous solution semiflow Φ by Theorem 10.3. Let $T > 0$. To show that Φ is state-continuous, uniformly for $t \in [0, T]$, let $x \in X$ and $K = \Phi([0, T] \times \{x\})$. Since Φ is time-continuous, K is compact. By Remark 10.10, there exists some $\delta > 0$ such that F is Lipschitz on the neighborhood $\bar{U}_\delta(K) = \{y \in C; d(y, K) \le \delta\}$ of K; let Λ be a respective Lipschitz constant. Let $y \in U_\delta(K)$ and τ the supremum of those numbers $\tilde{\tau}$ such that $\Phi(t, y) \in U_\delta(K)$ for all $t \in [0, \tilde{\tau}]$. Then, for all $t \in [0, \tau)$, $\Phi(t, y) \in U_\delta(K)$. Since $\bar{U}_\delta(K)$ is closed and Φ time-continuous, for all $t \in [0, \tau]$, $\Phi(t, y) \in \bar{U}_\delta(K)$ and

$$\begin{aligned}\|\Phi(t, y) - \Phi(t, x)\| \le &\|S(t)y - S(t)x\| \\ &+ \int_0^t \|S(t - s)\|\Lambda\|\Phi(s, y) - \Phi(s, x)\|ds.\end{aligned}$$

Let $c = \sup_{t\in[0,T]} \|S(t)\|$. For all $t \in [0, \tau] \cap [0, T]$,

$$\|\Phi(t, y) - \Phi(t, x)\| \le c\|y - x\| + \Lambda c \int_0^t \|\Phi(s, y) - \Phi(s, x)\|ds.$$

By Gronwall's inequality,

$$\|\Phi(t, y) - \Phi(t, x)\| \le c\|y - x\|e^{c\Lambda t}, \qquad t \in [0, \tau] \cap [0, T].$$

Choose $\epsilon \in (0, \delta)$ such that $c\epsilon e^{c\Lambda T} < \delta$. Then, if $t \in [0, \tau] \cap [0, T]$ and $\|x - y\| < \epsilon$, we have $\|\Phi(t, y) - \Phi(t, x)\| < \delta$ and $\Phi(t, y) \in U_\delta(K)$ by choice of K. By definition of τ, $\tau \ge \max\{\tau, T\}$ and so $\tau \ge T$ and $\Phi(t, y) \in U_\delta(K)$ for all $t \in [0, T]$ and $y \in U_\epsilon(x)$. Now let $y, z \in U_\epsilon(x)$. Then $\Phi(t, y)$ and $\Phi(t, z)$ are elements of $U_\delta(K)$ for all $t \in [0, T]$ and we can perform the same estimate as before with z replacing x. This provides

$$\|\Phi(t, y) - \Phi(t, z)\| \le c\|y - z\|e^{c\Lambda t} \quad \text{if } t \in [0, T].$$

This estimate shows that Φ is state-continuous, uniformly for $t \in [0, T]$ and Φ is continuous by Lemma 1.34. □

10.6. Heterogeneous time-autonomous boundary conditions

Consider a semilinear Cauchy problem of the form

$$\begin{aligned} u'(t) =&\tilde{A}u + F(u), \qquad z = Bu(t), \\ u(0) =&x, \end{aligned} \tag{10.14}$$

where $\tilde{A}$ is a linear operator from $D(\tilde{A}) \subset X$ to X, B is a linear operator from $D(\tilde{A}) \subset X$ into a Banach space Z, and $z \in Z$ given. Let A be the restriction of $\tilde{A}$ to

$$D(A) = \{x \in D(\tilde{A}); Bx = 0\} \tag{10.15}$$

and assume that A is the generator of a C_0-semigroup S.

Further, assume that there exists some $\tilde{x} \in D(\tilde{A})$ such that

$$0 = \tilde{A}\tilde{x}, \qquad z = B\tilde{x}. \tag{10.16}$$

Define

$$v(t) = u(t) - \tilde{x}.$$

Then u is a solution of (10.14) if and only if v is a solution of

$$\begin{aligned} v' =&Av + F(v + \tilde{x}), \\ v(0) =&y, \qquad y = x - \tilde{x}. \end{aligned} \tag{10.17}$$

Then $\tilde{F}$ with $\tilde{F}(v) = F(v + \tilde{x})$ inherits the respective Lipschitz properties from F, and we can use the appropriate theorems derived before to find a solution semiflow Ψ induced by the integral solutions to (10.17) with initial datum y. Define

$$\Phi(t, x) = \Psi(t, x - \tilde{x}) + \tilde{x}. \tag{10.18}$$

Lemma 10.23. *If Ψ is a semiflow, so is Φ.*

Proof. By definition of Φ, used several times,

$$\begin{aligned} \Phi(t + r, x) =&\Psi(t + r, x - \tilde{x}) + \tilde{x} = \Psi(t, \Psi(r, x - \tilde{x})) + \tilde{x} \\ =&\Psi(t, \Phi(r, x) - \tilde{x}) + \tilde{x} = \Phi(t, \Phi(r, x)). \end{aligned}$$

□

We now assume that X is an ordered Banach space with a closed convex cone X_+ and $S(t)$ a positive semigroup, i.e., $S(t)X_+ \subset X_+$ for all $t > 0$.

Theorem 10.24. *Assume that $S(t)$ is a positive semigroup, $\tilde{x} \in X_+$, and $S(t)\tilde{x} \leq \tilde{x}$ for all $t \geq 0$. Assume that $F : X_+ \to X$ is locally Lipschitz continuous and locally uniformly subtangential to X_+. Further, assume that F maps bounded subsets of X_+ into bounded subsets of X.*

Then, for each $x \in X_+$, there exists some $\tau > 0$ and a unique solution $u : [0, \tau) \to X_+$ of

$$u(t) = x + A \int_0^t (u(s) - \tilde{x})ds + \int_0^t F(u(s))ds, \tag{10.19}$$

$t \in [0, \tau)$, such that $\tau = \infty$ or u is not bounded on $[0, \tau)$.

If all solutions of (10.19) on finite intervals $[0, \tau)$, $\tau \in (0, \infty)$, with values in X_+ are bounded, then there exists a continuous semiflow $\Phi : \mathbb{R}_+ \times X_+ \to X_+$ such that $u(t) = \Phi(t, x)$ solves (10.19) for all $t \in \mathbb{R}_+$.

If X_+ has nonempty interior $\breve{X}_+$ and $S(t)x \in \breve{X}_+$ for every $x \in X_+ \setminus \{0\}$ and $t > 0$, then $\Phi(t, x) \in \breve{X}_+$ for every $x \in X_+ \setminus \{0\}$ and $t > 0$.

Equation (10.19) tacitly assumes that $\int_0^t (u(s) - \tilde{x})ds \in D(A)$ which implies by (10.15) that $B \int_0^t (u(s) - \tilde{x})ds = 0$.

Proof. Set $F_\lambda(z) = F(z) + \lambda z$. Then u solves (10.19) if and only if it solves

$$u(t) = x + (A - \lambda) \int_0^t (u(s) - \tilde{x})ds + \int_0^t F_\lambda(u(s))ds - \lambda \int_0^t \tilde{x}ds.$$

Define $v(t) = u(t) - \tilde{x}$ as above. Then u solves (10.19) if and only

$$v(t) = x - \tilde{x} + (A - \lambda) \int_0^t v(s)ds + \int_0^t F_\lambda(v(s) + \tilde{x})ds - \lambda \int_0^t \tilde{x}ds.$$

Since $A - \lambda$ is the generator of $e^{-\lambda t}S(t)$, by Proposition 10.6 this holds if and only if

$$\begin{aligned} v(t) =& e^{-\lambda t}S(t)(x - \tilde{x}) + \int_0^t e^{\lambda(t-s)}S(t - s)F_\lambda(v(s) + \tilde{x})ds \\ & - \int_0^t \lambda e^{\lambda(t-s)}S(t - s)\tilde{x}ds. \end{aligned}$$

In terms of u,

$$\begin{aligned} u(t) =& \tilde{x} + e^{-\lambda t}S(t)(x - \tilde{x}) + \int_0^t e^{-\lambda(t-s)}S(t - s)F_\lambda(u)ds \\ & - \int_0^t \lambda e^{-\lambda(t-s)}S(t - s)\tilde{x}ds. \end{aligned}$$

Since $S(t)\tilde{x} \leq \tilde{x}$ by assumption,

$$u(t) = \breve{u}(t)x + \int_0^t e^{-\lambda(t-s)}S(t - s)F_\lambda(u)ds$$

with

$$\begin{aligned}\breve{u}(t) =& \tilde{x} + e^{-\lambda t} S(t)(x - \tilde{x}) - \int_0^t S(s)\lambda e^{-\lambda s} \tilde{x} ds \\ \geq & \, e^{-\lambda t} S(t) x \geq 0.\end{aligned}$$

Notice that $\breve{u}(t) \to \tilde{x} - \lambda(\lambda - A)^{-1}\tilde{x}$ as $t \to \infty$. Similarly, as in Proposition 10.19, we find a unique solution $u : [0, \tau] \to X_+ \cap U_\epsilon(x)$ where $\epsilon > 0$ is such that $F_\lambda(X_+ \cap U_\epsilon(x)) \subset X_+$.

If $S(t)x \in \breve{X}_+$ for $t > 0$, then $\breve{u}(t) \in \breve{X}_+$ and $\breve{u}(t) \leq u(t) \in \breve{X}_+$.

u is also a unique solution of (10.19) and, by Theorem B.22, of

$$u(t) = \tilde{x} + S(t)(x - \tilde{x}) + \int_0^t S(t - s) F(u(s)) ds. \tag{10.20}$$

Let $\tilde{\tau}$ be the supremum of all those τ for which there is a (unique) solution of this integral equation on $[0, \tau)$. By uniqueness, there is a solution on $[0, \tilde{\tau})$. Suppose that $\tilde{\tau} < \infty$ and u is bounded on $[0, \tau)$. Similarly, as in Proposition 10.20, the solution can be extended to $[0, \tilde{\tau}]$. Now there exists some $\delta > 0$ and a solution $\tilde{u} : [0, \delta] \to X_+$ of

$$\tilde{u}(t) = u(\tilde{\tau}) + A \int_0^t (\tilde{u}(s) - \tilde{x}) ds + \int_0^t F(\tilde{u}(s)) ds. \tag{10.21}$$

We extend u to $[0, \tilde{\tau} + \delta]$ by

$$u(t + \tilde{\tau}) = \tilde{u}(t), \qquad t \in [0, \delta].$$

For $t \in [0, \delta]$,

$$u(t + \tilde{\tau}) = u(\tilde{\tau}) + A \int_0^t (\tilde{u}(s) - \tilde{x}) ds + \int_0^t F(\tilde{u}(s)) ds.$$

We substitute for $u(\tilde{\tau})$ and use the definition of u,

$$\begin{aligned}u(t + \tilde{\tau}) =& x + A \int_0^{\tilde{\tau}} (u(s) - \tilde{x}) ds + \int_0^{\tilde{\tau}} F(u(s)) ds \\ &+ A \int_0^t (u(s + \tilde{\tau}) - \tilde{x}) ds + \int_0^t F(u(s + \tilde{\tau})) ds.\end{aligned}$$

We make a change of variables and combine the integrals,

$$u(t + \tilde{\tau}) = x + A \int_0^{t+\tilde{\tau}} (u(s) - \tilde{x}) ds + \int_0^{t+\tilde{\tau}} F(u(s)) ds. \tag{10.22}$$

This contradicts the choice of $\tilde{\tau}$ and implies that $\tilde{\tau} = \infty$. We define $\Phi(t, x) = u(t)$. On the one hand, by (10.21), $\tilde{u}(t) = \Phi(t, \Phi(\tilde{\tau}, x))$. On the other hand, by (10.22), $\Phi(t + \tilde{\tau}) = u(t + \tilde{\tau}) = \tilde{u}(t)$. So Φ is a semiflow. The continuity of Φ is shown similarly as in the proof of Theorem 10.21 using (10.20). □

Corollary 10.25. *Let $\lambda > 0$ be such that $e^{-\lambda t}\|S(t)\| \to 0$ as $t \to \infty$ and u a solution of*

$$u(t) = x + A\int_0^t (u(s) - \tilde{x})ds - \lambda \int_0^t u(s)ds.$$

Then $u(t) \to v$ as $t \to \infty$ with $v \in D(\tilde{A})$ and $(\tilde{A} - \lambda)v = 0$, $Bv = z$.

Proof. From the proof of the previous theorem, with $F_\lambda = 0$,

$$u(t) = \breve{u}(t) = \tilde{x} + e^{-\lambda t}S(t)(x - \tilde{x}) - \int_0^t S(s)\lambda e^{-\lambda s}\tilde{x}ds.$$

Since $e^{-\lambda t}\|S(t)\| \to 0$ as $t \to \infty$, there exist $\delta > 0$ and $N \geq 1$ such that $e^{-\lambda t}\|S(t)\| \leq Ne^{-\delta t}$ for all $t \geq 0$ [**67**, Sec.V.1]. Using the relation between the Laplace transform of a semigroup and the resolvent of its generator [**67**, II.1.10],

$$u(t) \to \tilde{x} - \int_0^\infty S(s)\lambda e^{-\lambda s}\tilde{x}ds = \tilde{x} - \lambda(\lambda - A)^{-1}\tilde{x}.$$

Set $y = \lambda(\lambda - A)^{-1}\tilde{x}$. Then $y \in D(\tilde{A})$, $By = 0$ and

$$(\lambda - \tilde{A})y = \lambda\tilde{x}.$$

Set $v = \tilde{x} - y$. Then $Bv = B\tilde{x} = z$ and $\tilde{A}v = \tilde{A}\tilde{x} - \tilde{A}y = \lambda v$. $\square$

Chapter 11

Microbial Growth in a Tubular Bioreactor

11.1. Model description

Imagine a thin fluid-filled tube of length L and cross sectional area A containing motile bacteria feeding on a substrate which enters the tube at the left end. Both substrate and bacteria leak out of the right end of the tube at a rate depending on their local concentration. Substrate diffuses and bacteria do a "tumble and run" motion that can be modeled by diffusion provided it is not biased by substrate density. Assume that the tube occupies the interval $0 < \xi < L$ and $B(\tau, \xi)$ denotes the biomass concentration of bacteria (milligrams/cm^3) and $N(\tau, \xi)$ denotes the concentration of nutrient (milligrams/cm^3) at time τ. Notice that we assume that there is no variation of concentration of either biomass or nutrient transverse to the axis of the tube. Our definitions imply that the weight in milligrams of bacteria in the tube at time τ is given by

$$A \int_0^L B(\tau, \xi) d\xi$$

and the weight of nutrient is given by replacing B by N.

The partial differential equations for B and N are

$$\begin{aligned} N_\tau &= d_N N_{\xi\xi} - \frac{1}{\gamma} H(N) B, \\ B_\tau &= d_B B_{\xi\xi} + H(N) B, \quad 0 < \xi < L, \ \tau > 0, \end{aligned}$$

with boundary conditions

$$\begin{aligned}
-d_N N_\xi(\tau, 0) &= F > 0, \\
d_B B_\xi(\tau, 0) &= 0, \\
d_N N_\xi(\tau, L) + r_N N(\tau, L) &= 0, \\
d_B B_\xi(\tau, L) + r_B B(\tau, L) &= 0, \quad \tau > 0,
\end{aligned}$$

and initial conditions

$$\begin{aligned}
N(0, \xi) &= N_0(\xi) \geq 0, \\
B(0, \xi) &= B_0(\xi) \geq 0, \quad 0 < \xi < L.
\end{aligned}$$

The specific growth rate $H(N)$ is a continuously differentiable function of nutrient concentration satisfying $H(0) = 0$ and $H'(N) \geq 0$; no growth when no nutrient, more growth when more nutrient. Let $m = H(\infty) < \infty$ the maximum growth rate and assume it is finite. The nutrient level at which growth is half its maximum value will be denoted by $N_{1/2}$; it is called the half-saturation constant. It satisfies $H(N_{1/2}) = m/2$. Growth parameters m and $N_{1/2}$ for a given bacterial population are commonly measured from growth experiments in the chemostat [**203**]. A typical choice of H is the Monod function

$$H(N) = \frac{mN}{N_{1/2} + N}.$$

Constant $\gamma > 0$ is the yield constant reflecting the assumed proportionality of bacterial consumption and growth, d_N denotes the substrate diffusion constant, d_B the "motility constant" for the bacteria, that is, the effective diffusion coefficient for cell motility. F is the flux (milligrams per square centimeter per hour) of fresh substrate into the tube at $\xi = 0$ and $r_N, r_B > 0$ are leak rates of nutrient and biomass at the right end $\xi = L$.

For a thorough discussion of the modeling issues we strongly recommend the monographs of Berg [**14**] and Segel [**198**]. Here we follow the treatment in [**138**]. See [**203**] for the basics of microbial growth.

In order to better understand the model, let us see how the total weight of bacteria and nutrient change as a function of time. Starting with the nutrient

$$\begin{aligned}
\frac{d}{d\tau}A\int_0^L N(\tau,\xi)d\xi &= A\int_0^L N_\tau(\tau,\xi)d\xi \\
&= A\int_0^L \left(d_N N_{\xi\xi} - \frac{1}{\gamma}H(N)B\right)d\xi \\
&= Ad_N N_\xi|_0^L - A\int_0^L \frac{1}{\gamma}H(N)Bd\xi \\
&= A[F - r_N N(\tau,L)] - A\int_0^L \frac{1}{\gamma}H(N)Bd\xi,
\end{aligned}$$

where we have used the boundary conditions in the last line. This says that total nutrient increases at the rate FA due to fresh nutrient entering at the left end of the tube and decreases due to nutrient leaking out at rate $Ar_N N(\tau, L)$ at the right end of the tube and decreases due to consumption of nutrient at rate $A\int_0^L \frac{1}{\gamma}H(N)Bd\xi$ throughout the tube.

As for bacteria, we have:

$$\begin{aligned}
\frac{d}{d\tau}A\int_0^L B(\tau,\xi)d\xi &= A\int_0^L B_\tau(\tau,\xi)d\xi \\
&= A\int_0^L (d_B B_{\xi\xi} + H(N)B)\,d\xi \\
&= Ad_B B_\xi|_0^L + A\int_0^L H(N)Bd\xi \\
&= -Ar_B B(\tau,L) + A\int_0^L H(N)Bd\xi.
\end{aligned}$$

Bacterial biomass increases at the rate $A\int_0^L H(N)Bd\xi$ due to growth in response to consuming nutrient and decreases due to biomass leaking out at rate $Ar_B B(\tau, L)$ at the right end of the tube.

These calculations confirm that our equations are consistent with our verbal description of how nutrient and bacteria enter and leave the tube. It also confirms that the signs of terms in the equations and boundary conditions are accurate. Clearly, the flux F of nutrient into the tube will be an important parameter since $F = 0$ results in a decreasing level of nutrient in the tube which will ultimately starve out the bacteria.

It is useful to nondimensionalize parameters and variables. This can be done in several ways. We choose to scale time using the biological time scale of bacterial growth. The maximum growth parameter m has units of hour^{-1}; an organism growing at rate m has doubling time $\ln(2)/m$. We scale nutrient concentration by $N_{1/2}$. Our scaling is as follows:

$$x = \xi/L,\ t = m\tau,\ S = N/N_{1/2},\ u = B/(\gamma N_{1/2}). \tag{11.1}$$

Further, we define nondimensional parameters

$$d_S = d_N/mL^2,\ d_u = d_B/mL^2,\ r_S = r_N/mL,\ r_u = r_B/mL \tag{11.2}$$

and the growth function

$$f(S) = \frac{1}{m}H(N_{1/2}S). \tag{11.3}$$

Our scaling ensures that $f(\infty) = 1$ and $f(1/2) = 1/2$. Finally, we define our scaled nutrient flux by

$$F_0 = \frac{F}{mLN_{1/2}}. \tag{11.4}$$

Then the scaled equations become

$$\begin{aligned} S_t &= d_S S_{xx} - f(S)u, \\ u_t &= d_u u_{xx} + f(S)u, \quad 0 < x < 1,\ t > 0, \end{aligned} \tag{11.5}$$

with boundary conditions

$$\begin{aligned} -d_S S_x(t,0) &= F_0, \\ u_x(t,0) &= 0, \\ d_S S_x(t,1) + r_S S(t,1) &= 0, \\ d_u u_x(t,1) + r_u u(t,1) &= 0, \quad t > 0, \end{aligned} \tag{11.6}$$

and initial conditions

$$\begin{aligned} S(0,x) &= S_0(x) \geq 0, \\ u(0,x) &= u_0(x) \geq 0, \quad 0 < x < 1, \end{aligned} \tag{11.7}$$

where

$$S_0(x) = N_0(\eta)/N_{1/2},\quad u_0(x) = B_0(\eta)/(\gamma N_{1/2}).$$

An appropriate state space for (11.5)-(11.7) is

$$X = \{(S,u) \in C([0,1],\mathbb{R})^2 : S, u \geq 0\}$$

with norm

$$\|(S,u)\| = \|S\|_\infty + \|u\|_\infty$$

where $\|u\|_\infty = \sup_{0\leq x\leq 1}|u(x)|$. We expect that (11.5)-(11.7) generates a semiflow

$$\Phi(t, S_0, u_0)(x) = (S(t,x), u(t,x)),\ x \in [0,1],\ t > 0.$$

This will be shown later.

11.2. The no-bacteria invariant set

When no bacteria are initially present ($u_0 = 0$), then no bacteria can arise in the future (a basic fact of biology since immigration/contamination is not modeled), so $u \equiv 0$. Therefore, the set

$$X_0 = \{(S, 0) \in X\}$$

is forward invariant. The dynamics on X_0 are described by the equations

$$\begin{aligned} S_t &= d_S S_{xx}, \\ -d_S S_x(t,0) &= F_0, \\ d_S S_x(t,1) + r_S S(t,1) &= 0, \\ S(0,x) &= S_0(x). \end{aligned} \tag{11.8}$$

The steady state solution of (11.8) is given by

$$S = \hat{S}(x) = \frac{F_0}{d_S}(1-x) + \frac{F_0}{r_S}.$$

As expected, the nutrient level is highest near the source $x = 0$. The equilibrium solution $(\hat{S}, 0)$ of (11.5)-(11.7) is traditionally called the "washout equilibrium" since the bacteria have been "washed out".

Consider the eigenvalue problem

$$\begin{aligned} d_S \psi_{xx} &= \lambda \psi, \\ \psi_x(0) &= 0, \\ d_S \psi_x(1) + r_S \psi(1) &= 0, \end{aligned} \tag{11.9}$$

for the eigenvalue λ and eigenvector $\psi \neq 0$.

Lemma 11.1. *Eigenvalue problem (11.9) has countably many solutions $\{(\psi_n, -\mu_n)\}_{n \geq 1}$, satisfying*

$$0 > -\mu_1 > -\mu_2 > -\mu_3 > \cdots$$

In fact, $\mu_n = d_S z_n^2$ where z_n are the positive roots of

$$\tan(z) = \frac{r_S/d_S}{z}$$

listed in increasing order; they satisfy $(n-1)\pi < z_n < (n-1)\pi + \pi/2$, $n \geq 1$. The corresponding eigenfunctions, given by

$$\psi_n(x) = b_n \cos(z_n x), \; 2 = b_n^2 \left[1 + \frac{r_S/d_S}{(r_S/d_S)^2 + z_n^2}\right],$$

where $b_n > 0$, form an orthonormal basis for $L^2(0,1)$. The eigenfunction ψ_1 is positive on $[0,1]$. μ_1 satisfies

$$0 < z_1 < \min\{\frac{\pi}{2}, (r_S/d_S)^{1/2}\}. \tag{11.10}$$

The first eigenfunction ψ_1 is called the principal eigenfunction and its corresponding eigenvalue $-\mu_1$ is the principal eigenvalue of (11.9).

We briefly remind the reader how to solve (11.8). First, one removes the inhomogeneity in the boundary condition (the nonzero term F_0 on the left side of the first boundary condition) by setting $R(t,x) = S(t,x) - \hat{S}(x)$. Then R satisfies (11.8) but with homogeneous boundary conditions ($S_x(t,0) = 0$); hereafter we refer to these as the *homogeneous boundary conditions.* Of course, the initial condition must be changed as well by replacing S_0 by $R_0 \equiv S_0 - \hat{S}$. By a standard separation of variables argument, R is given by

$$\begin{aligned} R(t,x) &= \sum_{n\geq 1} \left(\int_0^1 R_0(y)\psi_n(y)dy \right) e^{-\mu_n t}\psi_n(x) \\ &= \int_0^1 G(t,x,y)R_0(y)dy \end{aligned}$$

where G is the Green's function given by

$$G(t,x,y) = \sum_{n\geq 1} e^{-\mu_n t}\psi_n(x)\psi_n(y). \tag{11.11}$$

See e.g., Chapter 5, Theorem 3.1 in [**86**] for details and conditions on R_0 for $R(t,x)$ to be a solution. S is then given by $S(t,x) = \hat{S}(x) + R(t,x)$.

The Green's function G can easily be shown to be a C^∞ function for $t > 0$, and to satisfy the partial differential equation (11.8) with homogeneous boundary conditions for $t > 0$. See e.g. Chapt. 5, sec. 3 of [**86**].

Proposition 11.2. *Define*

$$[V(t)\phi](x) = \int_0^1 G(t,x,y)\phi(y)dy, \ x \in [0,1], \ t > 0.$$

Then $\{V(t)\}_{t\geq 0}$*, with* $V(0) = I$ *the identity operator, defines a family of bounded linear operators on* $C([0,1],\mathbb{R})$ *satisfying the semigroup property*

$$V(t)V(s) = V(t+s), \ t,s \geq 0.$$

There exists $M > 0$ *such that*

$$\|V(t)\phi\|_\infty \leq Me^{-\mu_1 t}\|\phi\|_\infty, \ t > 0, \ \phi \in C([0,1],\mathbb{R}). \tag{11.12}$$

$V(t)$ *is a compact operator for each* $t > 0$ *and it is positive, that is,*

$$\phi \geq 0 \Rightarrow V(t)\phi \geq 0. \tag{11.13}$$

The unique solution of (11.8) is given by $S(t,x) = \hat{S}(x) + V(t)[S_0 - \hat{S}](x)$ *for* $S_0 \in C([0,1],\mathbb{R})$*. In particular, it is nonnegative. Finally, for each* $\phi \in C([0,1],\mathbb{R})$*, the map* $t \to V(t)\phi$ *is continuous.*

Proof. Let D be the subspace of twice continuously differentiable functions on $[0,1]$ which satisfy the homogeneous boundary conditions. For $\phi \in D$, $S(t,x) = \int_0^1 G(t,x,y)\phi(y)dy$ is the solution of (11.8) with the homogeneous boundary conditions and $S_0 = \phi$. See e.g. Theorem 1-1, Chapt. 5 of ([**86**]). Since $\psi_1 > 0$, $m^{-1} = \inf \psi_1 > 0$ and

$$-\|\phi\|_\infty m\psi_1(x) \leq \phi(x) \leq \|\phi\|_\infty m\psi_1(x), \ x \in [0,1].$$

Now $w_\pm(t,x) = \pm m\|\phi\|_\infty e^{-\mu_1 t}\psi_1(x)$ is a solution of the same partial differential equation and boundary conditions as S and $w_-(0,x) \leq S(0,x) \leq w_+(0,x)$, $x \in [0,1]$. By Theorem A.23 it follows that

$$w_-(t,x) \leq S(t,x) \leq w_+(t,x), \ x \in [0,1], \ t > 0,$$

which implies that

$$\|S(t,x)\| \leq Me^{-\mu_1 t}\|\phi\|_\infty, \quad t > 0,$$

where $M = m\|\psi_1\|_\infty$. Hence, (11.12) holds for $\phi \in D$. Since D is a dense subset of $C([0,1],\mathbb{R})$, $V(t)$ can be extended to be a bounded linear operator on $C([0,1],\mathbb{R})$ enjoying the estimate (11.12).

By a similar argument as above, if $\phi \in D$ satisfies $\phi(x) \geq c\psi_1(x)$ for some $c \geq 0$ then Theorem A.23 implies that $S(t,x) \geq c\psi_1(x)e^{-\mu_1 t} \geq 0$. By extension, $V(t)\phi \geq c\psi_1(x)e^{-\mu_1 t}$ for every $\phi \in C([0,1],\mathbb{R})$ satisfying $\phi \geq c\psi_1(x)$. Indeed, if $\phi(x) > c\psi_1(x)$, $x \in [0,1]$, then the above holds because ϕ can be uniformly approximated by functions $\chi_n \in D$ satisfying the inequality $\chi_n(x) \geq c\psi_1(x)$ and $V(t)\phi = \lim_n V(t)\chi_n$ uniformly; the general case follows from this. Therefore, $V(t)\phi \geq 0$ for every $\phi \in C([0,1],\mathbb{R})$ satisfying $\phi \geq 0$. As a consequence, we may conclude $G(t,x,y) \geq 0, t > 0, x, y \in [0,1]$. In fact, since $G(t,x,y)$ is C^∞ and satisfies the boundary conditions for $t > 0$, we may conclude from Theorem A.23 that $G(t,x,y) > 0$ for $t > 0, x, y \in [0,1]$.

To prove the semigroup property, it suffices to prove that $V(t)V(s)\phi = V(t+s)\phi$ for $\phi \in D$. Now $S_1(t,x) = V(t)[V(s)\phi](x)$ and $S_2(t,x) = V(t+s)[\phi](x)$ satisfy (11.8) with homogeneous boundary conditions and with $S_0 = V(s)\phi \in D$. As solutions of this equation are unique (a simple consequence of the maximum principle), $S_1 = S_2$.

As

$$|G_x(t,x,y)| \leq \sum_{n\geq 1} b_n^2 e^{-\mu_n t} z_n \leq \sum_{n \geq 1} 2z_n e^{-dsz_n^2 t} \equiv m(t),$$

the series on the right converging, we may justify that

$$[V(t)\phi]_x = \int_0^1 G_x(t,x,y)\phi(y)dy$$

and

$$|[V(t)\phi]_x| \le \int_0^1 |G_x(t,x,y)||\phi(y)|dy \le m(t)\|\phi\|_\infty.$$

This, and the Ascoli-Arzela Theorem [**164**, Ch.8.3] proves that $V(t)$ is compact for $t > 0$.

The assertion that $S(t,x) = \hat{S}(x) + V(t)[S_0 - \hat{S}](x)$ satisfies (11.8) is proved exactly as in Theorem 3-1, Chapt. 5 [**86**]. Assertion (11.13) and the nonnegativity of the solution of (11.8) follow from the maximum principle Theorem A.23.

In order to establish the continuity of the map $t \to V(t)\phi$ for each ϕ, it suffices, by the semigroup property, to show that it is continuous at $t = 0$. Let $Z = \{\phi \in C([0,1],\mathbb{R}) : \|V(t)\phi - \phi\|_\infty \to 0, t \to 0\}$. It is clear that Z is a subspace and, in fact, it is closed. For if $\epsilon > 0$ and $\phi \in \overline{Z}$, we find $\psi \in Z$ such that $\|\phi - \psi\|_\infty < \epsilon/2(M+1)$, so

$$\begin{aligned} \|V(t)\phi - \phi\|_\infty &= \|V(t)(\phi - \psi) + V(t)\psi - \psi + \psi - \phi\|_\infty \\ &\le (M+1)\|\phi - \psi\|_\infty + \|V(t)\psi - \psi\|_\infty \\ &\le \epsilon/2 + \epsilon/2 = \epsilon, \ 0 < t < \delta, \end{aligned}$$

where δ corresponds to $\epsilon/2$ in the definition of continuity of $t \to V(t)\psi$ at $t = 0$. The set of eigenfunctions $K = \{\psi_n\}_{n\ge1} \subset Z$ since $V(t)\psi_n = e^{-\mu_n t}\psi_n$. As every function in D is the uniform limit of its Fourier expansion in the eigenfunctions (see e.g., Theorem 10.1 in Chapt.7 [**86**]), the span of K is dense in $C([0,1],\mathbb{R})$. Therefore, $Z = C([0,1],\mathbb{R})$. □

Remark 11.3. The semigroup V is symmetric and strongly positive. The Green's function G has the following properties:

(a) If $0 < r < t$, then $\inf\{G(s,x,y); r \le s \le t, 0 \le x, y \le 1\} > 0$.

(b) $e^{\mu_1 t}\int_0^1 G(t,x,y)dy$ is bounded for $t \ge 0$ and $x \in [0,1]$.

(c) If $0 < r$, $e^{\mu_1 t}G(t,x,y)$ is bounded for $t \ge r$ and $x \in [0,1]$.

Proof. By construction, $G(t,x,y) = G(t,y,x)$. Thus V is symmetric,

$$\int_0^1 V(u)(x)v(x)dx = \int_0^1 u(x)V(v)(x)dx.$$

(a) This is immediate from $G(t,x,y) > 0, t > 0$, and continuity.

(b) This follows from $\int_0^1 G(t,\cdot,y)dy = V(t)\mathbf{1}$ and (11.12), where $\mathbf{1}$ is the function that is identically equal to 1 on $[0,1]$.

(c) Since the family of functions ψ_n is uniformly bounded, $G(t,x,y) \le \tilde{c}\sum_{n=1}^\infty e^{-\mu_n t}$. Since μ_n increases like n^2, the assertion follows. □

Remark 11.4. If $F_0 = 0$, that is, if the boundary conditions are homogeneous, then $\hat{S} = 0$, in which case $S(t,x) = [V(t)S_0](x)$ is the solution of (11.8).

Corollary 11.5. *Every solution of (11.8) converges to $\hat{S}$ as $t \to \infty$, uniformly for $x \in [0,1]$. Consequently, every solution of (11.5)-(11.7) with $u_0 = 0$ converges to the washout equilibrium in X.*

Proof. This follows immediately from the representation $S(t,x) = \hat{S}(x) + R(t,x)$ and Proposition 11.2. □

11.3. The solution semiflow

We now tackle the problem of showing that (11.5)-(11.7) generates a semiflow on X. A mild solution is a continuous function $(S,u) : [0,T] \to X$, for some $T > 0$, satisfying

$$\begin{aligned} S(t) &= \hat{S} + V(t)[S_0 - \hat{S}] - \int_0^t V(t-s)F(S(s),u(s))ds, \\ u(t) &= U(t)u_0 + \int_0^t U(t-s)F(S(s),u(s))ds, \end{aligned} \tag{11.14}$$

where $U(t)$ is the semigroup corresponding to $V(t)$ on changing d_S, r_S to d_u, r_u and $F : X \to C([0,1],\mathbb{R})$ is defined by

$$[F(S,u)](x) = f(S(x))u(x), \quad 0 \le x \le 1.$$

Alternatively, an integral solution, as in (10.19), satisfies:

$$\begin{aligned} S(t) &= S_0 + A_V \int_0^t (S(s) - \hat{S})ds - \int_0^t F(S(s),u(s))ds, \\ u(t) &= u_0 + A_U \int_0^t u(s)ds + \int_0^t F(S(s),u(s))ds, \end{aligned} \tag{11.15}$$

where A_U is the generator of semigroup U and A_V is the generator of semigroup V.

Theorem 11.6. *Equations (11.15) generate a continuous semiflow Φ on X given by*

$$\Phi(t, S_0, u_0) = (S(t), u(t)), \quad t > 0. \tag{11.16}$$

If $(S_0, u_0) \in X$ satisfies $u_0 \neq 0$, then $u(t,x) = u(t)(x) > 0$, $x \in [0,1], t > 0$.

Proof. We use Theorem 10.24. (11.15) takes the form of (10.19) with $x = (S,u)$, $\tilde{x} = (\hat{S}, 0)$, $A = (A_V, A_U)$ and $\tilde{F} = (-F, F)$. Note that $(V(t), U(t))\tilde{x} = (V(t)\hat{S}, 0) \le (\hat{S}, 0)$, $t \ge 0$ where we have used that $\hat{S} + V(t)(S_0 - \hat{S}) \ge 0$, $t \ge 0$, for every $S_0 \ge 0$. Nonlinearity $\tilde{F}$ is clearly locally Lipschitz and maps bounded subsets of X to bounded subsets of $C([0,1],\mathbb{R}^2)$; we verify the local uniform subtangential condition relative to X (note our X is X_+ in the notation of Theorem 10.24). Given $(\bar{S}, \bar{u}) \in X$ there exists $m > 0$ such that $f(S) \leq mS$ for all S belonging to $[0, \|\bar{S}\|_\infty + 1]$. Choose $\epsilon = 1$ and $\delta = \frac{1}{m(\|\bar{u}\|_\infty+1)}$. If $(S, u) \in X$ satisfies $\|S - \bar{S}\|_\infty + \|u - \bar{u}\|_\infty < \epsilon$ and $0 \leq h < \delta$, then

$$S - hF(S, u) = S - hf(S)u \geq S(1 - mhu) \geq S(1 - mh(\|\bar{u}\|_\infty + 1)) \geq 0.$$

Since $u + hF(S, u) \geq 0$ for $(S, u) \in X$, the local uniform subtangential condition is verified.

Theorem 10.24 implies the existence of a maximally defined solution $(S, u) : [0, \tau) \to X$. In order to conclude that $\tau = +\infty$ we must show that the solution does not blow up as $t \to \tau$. While we obtain strong bounds on solutions in the next sections, weaker ones suffice for our purposes. A solution of (11.15) is also a solution of (11.14) by arguments similar to those of Proposition 10.6. From (11.14), we have $S(t) \leq \hat{S} + V(t)(S_0 - \hat{S})$, so S is bounded on $[0, \tau)$. This ensures that $F(S(t), u(t)) \leq Mu(t)$, $t \in [0, \tau)$ so, from (11.14), we have

$$u(t) \leq U(t)u_0 + M \int_0^t U(t - s)u(s)ds.$$

Easy Gronwall estimates establish boundedness of u on $[0, \tau)$. We conclude that $\tau = \infty$. It follows from Theorem 10.24 that Φ is a continuous semiflow.

The final assertion follows from the fact that $[U(t)\phi](x) > 0$, $x \in [0, 1], t > 0$ for every $\phi \in C([0, 1], \mathbb{R}_+)$ satisfying $\phi \neq 0$ (similarly for V in place of U) by Remark 11.3. Since $u(t) \geq U(t)u_0$ it follows that $u(t)(x) > 0$, $x \in [0, 1], t > 0$. □

A solution of (11.15) is also a solution of (11.14) by arguments similar to those of Proposition 10.6. It can be shown that mild solutions of (11.14) are, in fact, classical solutions of (11.5)-(11.7). By this, we mean that for $w = S, u$, derivatives w_t, w_x, w_{xx} exist and are continuous for $(t, x) \in (0, \infty) \times (0, 1)$, the derivative w_x is continuous for $(t, x) \in (0, \infty) \times [0, 1]$, $w(t, x)$ is continuous for $(t, x) \in [0, \infty) \times [0, 1]$, $w(0, x) = w_0(x), x \in [0, 1]$ and $(S(t, x), u(t, x))$ satisfies (11.5) and (11.6) for $t > 0$.

See [**153, 159, 174**] for full details. We will use this fact in the following in order to make use of the comparison Theorem A.23.

11.4. Bounds on solutions and the global attractor

We begin by establishing a biologically important a priori bound on solutions of the system (11.5)-(11.7).

Proposition 11.7. *Let $\psi > 0$ be the principle eigenfunction corresponding to the largest eigenvalue $-\mu < 0$ of*

$$\begin{aligned} -\psi_{xx} &= \mu\psi, \\ 0 &= \psi_x(0), \\ 0 &= \psi_x(1) + r\psi(1), \end{aligned}$$

where $r = \min\{r_S/d_S, r_u/d_u\}$, normalized such that $\psi(0) = 1$. Let

$$W(t) = \int_0^1 \psi(x)(S(t,x) + u(t,x))dx,$$

which can be viewed as a measure of the total nutrient (bacterial biomass can be viewed as "processed nutrient") in the tube. Then for every solution (S,u) of (11.5)-(11.7) with $(S_0, u_0) \in X$ we have the estimate

$$W(t) \leq W(0)e^{-\mu D t} + \frac{F_0}{\mu D}(1 - e^{-\mu D t}) \tag{11.17}$$

where $D = \min\{d_S, d_u\}$.

Proof. It suffices to prove (11.17) for smooth $(S_0, u_0) \in X$, i.e., for $(S_0, u_0) \in D(A_V) \times D(A_U)$ since this set is dense in X. By the mild formulation (11.14), it follows that, if $(S_0, u_0) \in D(A_V) \times D(A_U)$, then $(S(t), u(t)) \in D(A_V) \times D(A_U)$ for $t > 0$ implying that $S_{xx}(t,x), u_{xx}(t,x)$ are continuous for $x \in [0,1], t > 0$. It follows from the classical formulation that $S_t(t,x), u_t(t,x)$ are continuous for $x \in [0,1], t > 0$. This justifies the following argument.

We differentiate W for $t > 0$, integrate by parts, and apply the boundary conditions,

$$\begin{aligned} W'(t) &= \int_0^1 \psi(S_t + u_t)dx \\ &= \int_0^1 \psi(d_S S_{xx} + d_u u_{xx})dx \\ &= -S(t,1)[d_S\psi_x(1) + r_S\psi(1)] - u(t,1)[d_u\psi_x(1) + r_u\psi(1)] \\ &\quad + S(t,0)d_S\psi_x(0) + d_u u(t,0)\psi_x(0) + F_0\psi(0) \\ &\quad + \int_0^1 (d_u u + d_S S)\psi_{xx}dx \\ &= -S(t,1)\psi(1)(r_S - r d_S) - u(t,1)\psi(1)(r_u - r d_u) + F_0 \\ &\quad - \mu \int_0^1 (d_S S + d_u u)\psi dx \\ &\leq F_0 - D\mu W(t). \end{aligned}$$

Integration of this differential inequality leads to (11.17). □

The total amount of nutrient in the tube is ultimately limited by the influx rate of nutrient into the tube.

Our next result gives a pointwise bound on the nutrient.

Proposition 11.8. *For any solution of (11.5)-(11.7) with $(S_0, u_0) \in X$ we have*

$$S(t,x) \le \hat{S}(x) + V(t)(S_0 - \hat{S})(x) \tag{11.18}$$

where $V(t)$ is the semigroup described in Proposition 11.2 related to (11.8). In particular, given $\epsilon > 0$, there exists $T > 0$ depending on initial data, such that

$$0 \le S(t,x) \le \hat{S}(x) + \epsilon, \ 0 \le x \le 1, \ t \ge T. \tag{11.19}$$

Proof. This is a corollary to the proof of Theorem 11.6. □

Observe that (11.18) says that the nutrient level is lower when there are consumers than when there are no consumers since the right-hand side is just the solution of (11.8). This should seem obvious on biological grounds.

We now wish to show that the semiflow Φ is point dissipative and eventually bounded on every bounded subset of X.

Proposition 11.9. *There exists a continuous function $K : [0,\infty) \to [0,\infty)$ and $M > 0$ such that for every solution (S,u) of (11.5)-(11.7) with $(S_0, u_0) \in X$,*

$$\|(S(t,\cdot), u(t,\cdot)\| \le K(\|(S_0, u_0)\|), \quad t \ge 0,$$

and

$$\limsup_{t\to\infty} \|(S(t,\cdot), u(t,\cdot))\| \le M.$$

Proof. Since $f(S) \le c, \ S \ge 0$ for some $c > 0$,

$$u(t,x) \le \int_0^1 G(t,x,y)u(0,y)dy + c\int_0^t \int_0^1 G(s,x,y)u(t-s,y)dsdy. \tag{11.20}$$

Here, abusing notation, we again use G for the Green's function with r_u, d_u replacing r_S, d_S. Let $\epsilon \in (0,1)$, to be determined later. We estimate the first integral using Remark 11.3 (b) and split up the second integral,

$$\begin{aligned} u(t,x) \le & e^{-\mu_1 t}\tilde{c}\|u(0,\cdot)\|_\infty + c\int_\epsilon^t \int_0^1 G(s,x,y)u(t-s,y)dsdy \\ & + c\int_0^\epsilon \int_0^1 G(s,x,y)u(t-s,y)dsdy, \ t \ge \epsilon. \end{aligned}$$

By Remark 11.3 (c), for every $\epsilon > 0$ there exists some $c_\epsilon > 0$ such that

$$cG(s,x,y) \le c_\epsilon e^{-\mu_1 s}, \qquad s \ge \epsilon, \quad 0 \le x,y \le 1.$$

Therefore, since u is bounded on $[0,\tau]\times[0,1]$ for every $\tau\in(0,\infty)$,

$$\begin{aligned}u(t,x) \le& e^{-\mu_1 t}\tilde{c}\|u(0,\cdot)\|_\infty + \int_\epsilon^t c_\epsilon e^{-\mu_1 s}ds \sup_{0\le r\le t}\int_0^1 u(r,y)dy \\ &+ c\int_0^\epsilon\int_0^1 G(s,x,y)dsdy \sup_{t-\epsilon\le r\le t}\|u(r,\cdot)\|_\infty,\ t\ge\epsilon.\end{aligned}\tag{11.21}$$

By Remark 11.3 (b),

$$c\int_0^\epsilon\int_0^1 G(s,x,y)dsdy \le \frac{1}{2}, \qquad 0\le x\le 1,\tag{11.22}$$

if $\epsilon>0$ is chosen small enough. There exists $M>0$, depending only on $\|u_0\|_\infty$ and $\|S_0\|_\infty$, such that

$$\int_0^1 u(t,y)dy \le M, \quad t\ge 0.\tag{11.23}$$

This follows from (11.17) of Proposition 11.7 and the fact that ψ is strictly positive on $[0,1]$. By (11.23), (11.21), and (11.22),

$$\|u(t,\cdot)\|_\infty \le e^{-\mu_1 t}\tilde{c}\|u(0,\cdot)\|_\infty + M_\epsilon + \frac{1}{2}\sup_{t-\epsilon\le r\le t}\|u(r,\cdot)\|_\infty, \quad t\ge\epsilon,$$

with $M_\epsilon=\frac{c_\epsilon M}{\mu_1}$. A straightforward Gronwall estimate starting with (11.20) gives $Q_\epsilon>0$ such that

$$\|u(t,\cdot)\|_\infty \le Q_\epsilon\|u(0,\cdot)\|_\infty, \qquad 0\le t\le\epsilon.$$

We combine the two estimates.

$$\|u(t,\cdot)\|_\infty \le (e^{-\mu_1 t}\tilde{c}+Q_\epsilon)\|u(0,\cdot)\|_\infty + M_\epsilon + \frac{1}{2}\sup_{0\le r\le t}\|u(r,\cdot)\|_\infty, \quad t\ge 0.\tag{11.24}$$

Dropping the exponential on the right side makes it monotone increasing in t, so we can replace the left side of (11.24) by $\sup_{0\le r\le t}\|u(r,\cdot)\|_\infty$, which leads to

$$\frac{1}{2}\sup_{0\le t\le\tau}\|u(t,\cdot)\|_\infty \le (\tilde{c}+Q_\epsilon)\|u(0,\cdot)\|_\infty + M_\epsilon.$$

This implies that u is bounded on $\mathbb{R}_+\times[0,1]$ and, together with Proposition 11.8, gives the first estimate.

In order to get the second estimate, we use the semigroup property and (11.17) to replace M in (11.23) by a bound that is independent of the initial data (related to $F_0/\mu D$). Then by taking limit superior of both sides of (11.24), we find

$$\limsup_{t\to\infty}\|u(t,\cdot)\|_\infty \le M_\epsilon + \frac{1}{2}\limsup_{t\to\infty}\|u(t,\cdot)\|_\infty$$

and

$$\limsup_{t\to\infty} \|u(t,\cdot)\|_\infty \le 2M_\epsilon. \qquad \square$$

Proposition 11.10. *Every orbit* $\{(S(t,\cdot),u(t,\cdot)) : t \ge 0\}$ *has compact closure in* X*. Furthermore, there exists a compact, connected, invariant attractor for* Φ *which attracts bounded sets in* X*.*

Proof. By Theorem 2.33 we must show that Φ is point dissipative, asymptotically smooth, and eventually bounded on every bounded set in X. Proposition 11.9 gives the first and last of these. Now we can argue that Φ is asymptotically smooth as in Theorem 10.16. $\square$

11.5. Stability of the washout equilibrium

The stability of the washout equilibrium $(\hat{S},0)$ can be determined by linearization. The formal procedure is to change variables by letting

$$S(t,x) = \hat{S}(x) + \epsilon Y(t,x), \;\; u(t,x) = 0 + \epsilon U(t,x)$$

where ϵ is a small parameter, insert these into (11.5)-(11.6), and drop terms of higher than first order in ϵ. This leads to the so-called "variational equation" associated with the washout equilibrium given by

$$\begin{aligned} Y_t &= d_S Y_{xx} - f(\hat{S})U, \\ U_t &= d_u U_{xx} + f(\hat{S})U, \quad 0 < x < 1, \; t > 0, \end{aligned} \tag{11.25}$$

with homogeneous boundary conditions (i.e., (11.6) with $F_0 = 0$). Observe that this is a linear system of partial differential equations together with linear homogeneous boundary conditions.

In a similar manner to that of the stability of an equilibrium for an ordinary differential equation, we will show that the stability of the washout equilibrium depends on the sign of the real part of the eigenvalues of an eigenvalue problem. Seeking solutions of (11.25) of the form

$$Y = e^{\lambda t} X(x), \quad U = e^{\lambda t} V(x)$$

we arrive at the associated eigenvalue problem

$$\begin{aligned} \lambda X &= d_S X_{xx} - f(\hat{S}(x))V, \\ \lambda V &= d_u V_{xx} + f(\hat{S}(x))V, \end{aligned} \tag{11.26}$$

with boundary conditions

$$\begin{aligned} X_x(0) &= V_x(0) = 0, \\ 0 &= d_S X_x(1) + r_S X(1), \\ 0 &= d_u V_x(1) + r_u V(1). \end{aligned} \tag{11.27}$$

The eigenvalue problem decouples since the equation for V does not contain X. This part of the eigenvalue problem

$$\begin{aligned} \lambda V &= d_u V_{xx} + f(\hat{S}(x))V, \\ 0 &= d_u V_x(1) + r_u V(1), \\ 0 &= V_x(0), \end{aligned} \tag{11.28}$$

is the key, since once V is known, X can be computed.

Lemma 11.11. *Eigenvalue problem (11.28) has countably many eigenvalues satisfying*

$$\cdots \lambda_n < \lambda_{n-1} < \cdots < \lambda_2 < \lambda_1.$$

The eigenfunction V_1 associated to λ_1 is positive on $[0,1]$ and we have the estimate

$$f(\hat{S}(1)) - d_u \hat{z}_1^2 \leq \lambda_1 \leq f(\hat{S}(0)) - d_u \hat{z}_1^2 \tag{11.29}$$

where $\hat{z}_1$ is defined as in Lemma 11.1 on replacing r_S by r_u.

The eigenvalues of (11.26) are a countable set of real numbers contained in the set

$$\{\lambda_n\}_{n\geq 1} \cup \{-\mu_n\}_{n\geq 1}$$

where $\{-\mu_n\}_{n\geq 1}$ are eigenvalues of (11.9).

If $\lambda_1 > -\mu_1$, in particular, if $\lambda_1 \geq 0$, then it is a simple eigenvalue of (11.26) and all other eigenvalues of (11.26) satisfy $\lambda < \lambda_1$.

λ_1 is a continuous, strictly increasing function of F_0 satisfying $\lambda_1 \to 1 - d_u \hat{z}_1^2$ as $F_0 \to \infty$ and $\lambda_1 \to -d_u \hat{z}_1^2$ as $F_0 \to 0$.

Proof. The first assertion is a well known result from Sturm-Liouville eigenvalue theory. See Chapt. 5 of [**232**]. If λ is an eigenvalue of (11.26) with eigenvector (X, V) and $V = 0$, then λ must be one of the $\{-\mu_n\}_{n\geq 1}$ and observe that $\lambda = -\mu_n$, $n \geq 1$ is an eigenvalue of (11.26) with eigenvector $(X, V) = (\psi_n, 0)$ where ψ_n is the corresponding eigenvector of (11.9) for eigenvalue μ_n. On the other hand, if λ is an eigenvalue of (11.26) with eigenvector (X, V) and $V \neq 0$, then λ must be one of the $\{\lambda_n\}_{n\geq 0}$. This implies the second assertion.

If λ_m is not one of the $\{-\mu_n\}_{n\geq 1}$, then it is an eigenvalue of (11.26) with corresponding eigenvector (X_m, V_m) where V_m is an eigenvector of (11.28) corresponding to λ_m and X_m is the unique solution of

$$\lambda_m X = d_S X_{xx} - f(\hat{S}(x))V_m$$

which satisfies the appropriate boundary conditions. See Theorem 2-1, Chapter 7 of [**86**]. Existence and uniqueness of X_m follows since λ_m is not one of the eigenvalues $\{-\mu_n\}_{n\geq 1}$. If $\lambda_1 > -\mu_1$, then it is not one of the $\{-\mu_n\}_{n\geq 1}$ since they are all negative by Lemma 11.1.

If λ does not belong to $\{\lambda_n\}_{n\geq 1} \cup \{-\mu_n\}_{n\geq 1}$, then it belongs to the resolvent set for (11.26). The easy proof uses Theorem 2-1, Chapter 7 of [**86**].

The estimate (11.29) follows immediately from Lemma 11.1, modified by replacing r_S by r_u, and using $f(\hat{S}(1)) \leq f(\hat{S}(x)) \leq f(\hat{S}(0))$, if one knows that the principle eigenvalue is a monotone function of the "potential function" $f(\hat{S}(x))$. But this follows from the well known variational characterization of the eigenvalue

$$(11.30) \qquad \lambda_1 = \max_{\phi\in W^1(0,1),\ \int_0^1 \phi^2 dx=1} \left[\int_0^1 (f(\hat{S})\phi^2 - d_u\phi_x^2)dx - d_u r_u \phi(1)^2 \right],$$

where equality holds when ϕ is the principle eigenfunction of (11.28). See e.g., Theorem 2.1 and Corollary 2.2 [**25**] or Chapter 7 of [**166**]. $W^1(0,1)$ is the space consisting of absolutely continuous functions on $[0,1]$ which have square integrable derivatives.

The continuity of λ_1 as a function of F_0 follows from the continuity of the principle eigenvalue on the potential function $f(\hat{S}(\cdot))$ and the continuity of $F_0 \to \hat{S}(x) = F_0\left(\frac{1}{d_S}(1-x) + \frac{1}{r_S}\right)$. The strict monotonicity of λ_1 as a function of F_0 follows from (11.30). The limiting values of λ_1 as F_0 approach zero and infinity follow from (11.29). □

It is reasonable to expect that the stability of the washout equilibrium depends on the sign of λ_1: If $\lambda_1 < 0$ then it is asymptotically stable; if $\lambda_1 > 0$, then it is unstable. We do not formalize this assertion since we will show below that the washout equilibrium is globally attracting when $\lambda_1 < 0$ and that the bacteria uniformly persist when $\lambda_1 > 0$.

By Lemma 11.11, if $d_u\hat{z}^2 \geq 1$, then $\lambda_1 < 0$ for all $F_0 > 0$. If $d_u\hat{z}^2 < 1$, then there exists $F_0^* > 0$ such that $\lambda_1 < 0$ for $0 < F_0 < F_0^*$ and $\lambda_1 > 0$ for $F_0 > F_0^*$.

Remark 11.12. A consequence of (11.10) is that $d_u\hat{z}^2 < 1$ holds if

$$\frac{d_B}{mL}\min\{\pi^2/4, Lr_N/d_N\} < 1.$$

The inequality $d_u\hat{z}^2 < 1$ holds when bacterial motility d_B is sufficiently small, or when the length L of the tube is sufficiently large, or when the maximum bacterial growth rate m is sufficiently large. As noted above, when this holds there exists a threshold value for the nutrient influx rate F_0^* above which the washout equilibrium is unstable.

Remark 11.13. Formula (11.30) has many important implications. As noted in [**25**], among these is the continuity of λ_1 with respect to the "potential function" $m(x) \equiv f(\hat{S}(x))$. Indeed, the formula holds where

$f(\hat{S}(x))$ in (11.28) is replaced by a general bounded measurable function $m(x)$. It implies that given $\epsilon > 0$, there exists $\delta > 0$ such that, if $m_i,\ i = 1, 2$, are two such functions satisfying $\sup_{0<x<1} |m_1(x) - m_2(x)| < \delta$, then $|\lambda_1(m_1) - \lambda_1(m_2)| < \epsilon$. This result will be used frequently below.

11.5.1. The basic reproduction number. By virtue of (11.29) and the continuity and monotonicity of $\hat{S}$, there exists a unique value $\eta \in [0, 1]$ such that

$$\lambda_1 = f(\hat{S}(\eta)) - d_u \hat{z}_1^2. \tag{11.31}$$

In terms of the unscaled parameters, this becomes

$$\lambda_1 = \frac{1}{m}\left[H(\hat{N}(\eta L)) - \frac{d_B}{L^2}\hat{z}_1^2\right]$$

where $\hat{N}(\xi) = \frac{F}{d_N}(L - \xi) + \frac{F}{r_N}$ is the unscaled washout equilibrium. As the sign of λ_0 is determined by the term in brackets, we are motivated to define a basic reproduction number by

$$\mathcal{R}_0 = H(\hat{N}(\eta L))\frac{L^2}{d_B}\frac{1}{\hat{z}_1^2}. \tag{11.32}$$

The washout equilibrium is stable when $\mathcal{R}_0 < 1$ and unstable when $\mathcal{R}_0 > 1$. The factor $H(\hat{N}(\eta L))$ of $\mathcal{R}_0$ is the rate at which a unit of bacterial biomass would increase if it were to be placed in the tube (at $\xi = \eta L$) in which nutrient levels are at the washout equilibrium level. It is a function of d_S, r_S, F, L and the specific growth rate H. We claim that the other factor, $\frac{L^2}{d_B}\frac{1}{\hat{z}_1^2}$ depending on d_B, r_B, L, is the *mean residence time* of a bacterium in the tube. Therefore, $\mathcal{R}_0$ measures the amount of bacterial biomass produced by a unit biomass before it leaks out of the tube.

To justify our claim concerning the mean residence time of bacteria in the tube, consider a unit inoculum $B(0, \xi)$ ($\int_0^L B(0, \xi)d\xi = 1$) of bacteria occupying a nutrient-free tube at $\tau = 0$. How long will they remain in the tube before exiting at the right-hand end $\xi = L$? $B(\tau, \xi)$ satisfies

$$\begin{aligned} B_\tau &= d_B B_{\xi\xi}, \\ d_B B_\xi(\tau, 0) &= 0, \\ d_B B_\xi(\tau, L) + r_B B(\tau, L) &= 0, \quad \tau > 0. \end{aligned}$$

It is natural to take $B(0, \xi) = B_1(\xi)$ to be the normalized positive principle eigenfunction of

$$\begin{aligned} d_B B_{\xi\xi} &= \rho B, \\ d_B B_\xi(0) &= 0, \\ d_B B_\xi(L) + r_B B(L) &= 0. \end{aligned} \tag{11.33}$$

Then an easy calculation shows that

$$B(\tau, \xi) = B_1(\xi)e^{\rho_1 \tau}$$

where ρ_1 is the corresponding principle (largest) eigenvalue. We will soon show that it is negative. The amount of bacteria at time τ is given by

$$\text{total bacteria at time } \tau = \int_0^L B(\tau, \xi)d\xi = e^{\rho_1 \tau},$$

where we have used the normalization of B_1. It follows that the mean residence time of bacteria in the tube (for this inoculum) is $-1/\rho_1$.

To see that $-1/\rho_1$ agrees with $d_B \hat{z}_1^2$, let $\psi_1(x) = B_1(xL)$, $0 \leq x \leq 1$, and observe that $\psi_1 > 0$ satisfies (11.9), where we must replace d_S, r_S by r_u, d_u, corresponding to eigenvalue parameter $\lambda = \rho_1/m$. It follows that $\rho_1/m = -\mu_1$, so the mean residence time is

$$-1/\rho_1 = 1/m\mu_1 = 1/md_u\hat{z}_1^2 = L^2/d_B\hat{z}_1^2.$$

11.5.2. Global stability of the washout equilibrium. Our next result implies that the washout equilibrium is globally attracting when $\lambda_1 < 0$.

Theorem 11.14. *If $\lambda_1 < 0$, equivalently, if $\mathcal{R}_0 < 1$, then every solution $(S(t,x), u(t,x))$ of (11.5)-(11.7) satisfies*

$$S(t,x) \to \hat{S}(x), \; u(t,x) \to 0, \; t \to \infty,$$

uniformly for $x \in [0,1]$. Actually, $(\hat{S}, 0)$ attracts all bounded subsets of the state space.

Proof. We apply Proposition 5.22. To this end, we define

$$\rho(S_0, u_0) = \|u_0\|_\infty = \max_{0 \leq x \leq 1} u_0(x), \quad (S_0, u_0) \in X.$$

Then $X_0 = \{(S_0, 0); S_0 \in C([0,1], \mathbb{R})\}$ in the language of Proposition 5.22. By Proposition 11.2, if $u_0 = 0$,

$$\|S(t) - \hat{S}\|_\infty \leq \|S_0 - \hat{S}\|_\infty Me^{-\mu_1 t}, \qquad t \geq 0.$$

This implies that $(\hat{S}, 0)$ attracts every bounded subset of X_0.

The largest eigenvalue λ_1 of (11.28) depends continuously on the "potential function" $f(\hat{S}(x))$, as noted in Remark 11.13. This implies that there exists $\epsilon > 0$ such that the largest eigenvalue $\hat{\lambda}_1$ of (11.28) with potential $f(\hat{S}(x))$ replaced by $f(\hat{S}(x) + \epsilon)$, satisfies $\hat{\lambda}_1 < 0$. Let $\hat{V}_1(x)$ denote the corresponding eigenfunction, which we may assume is positive on $[0,1]$ (see [**232**]).

Let B be a bounded subset of the state space.

By (11.18), there exists $\tau > 0$ such that $S(t,x) \leq \hat{S}(x)+\epsilon,\ x \in [0,1],\ t \geq \tau$, for all solutions with initial data in B. Consequently, $u(t,x)$ satisfies

$$u_t \leq du_{xx} + f(\hat{S}(x) + \epsilon)u, \quad x \in [0,1],\ t \geq \tau$$

together with the boundary conditions. The corresponding linear differential equation, with equality rather than inequality, has as one of its solutions $w(t,x) = Ke^{\hat{\lambda}_1(t-\tau)}\hat{V}_1(x)$ for arbitrary scalar K.

We choose K such that $u(\tau,x) \leq K,\ x \in [0,1]$, for all solutions with initial data in B. This is possible by Proposition 11.9. Then we have $u(\tau,x) \leq w(\tau,x),\ x \in [0,1]$. Theorem A.23 implies that

$$u(t,x) \leq w(t,x), \qquad x \in [0,1],\ t \geq \tau.$$

As $w(t,x) \to 0$ as $t \to \infty$ uniformly in $x \in [0,1]$, $\rho(S(t,\cdot),u(t,\cdot)) = \|u(t,\cdot)\|_\infty \to 0$ as $t \to \infty$ uniformly for all solutions with initial data in B.

By Proposition 5.22, $(\hat{S},0)$ attracts all bounded subsets of the state space. □

11.6. Persistence of the microbial population

As $\int_0^1 u(t,x)dx$ denotes the total biomass in the tube at time t, it is natural to define $\rho : X \to [0,\infty)$ by

$$\rho(S(t),u(t)) = \int_0^1 u(t,x)dx.$$

We would like to show that when $\lambda_1 > 0$, then the system is uniformly ρ-persistent. We begin with a preliminary result taking

$$\rho(S(t),u(t)) = \|u(t,\cdot)\|_\infty.$$

Proposition 11.15. *If $\lambda_1 > 0$, there exists $\eta > 0$ such that for all solutions of (11.5)-(11.7) with $u_0 \neq 0$,*

$$\limsup_{t\to\infty} \max_{0\leq x\leq 1} u(t,x) \geq \eta. \tag{11.34}$$

Proof. The largest eigenvalue λ_1 of (11.28) depends continuously on the "potential function" $f(\hat{S}(x))$ which implies that there exists $\epsilon > 0$ such that the largest eigenvalue $\hat{\lambda}_1$ of (11.28) with potential $f(\hat{S}(x))$ replaced by $f(\hat{S}(x) - \epsilon)$, satisfies $\hat{\lambda}_1 > 0$. Let $\hat{V}_1(x)$ denote the corresponding eigenfunction, which we may assume is positive on $[0,1]$.

For every $\delta > 0$, the equation

$$(11.35)\qquad \begin{aligned} S_t &= d_S S_{xx} - \delta, \\ F_0 &= -d_S S_x(t,0), \\ 0 &= d_S S_x(t,1) + r_S S(t,1), \end{aligned}$$

has a unique equilibrium $\hat{S}_\delta$ which attracts all solutions. This can be seen by an argument similar to that given for (11.8). The equilibrium is easily computed using superposition to be $\hat{S}_\delta = \hat{S} + \delta[\frac{x^2}{2d_S} - (\frac{1}{2d_S} + \frac{1}{r_S})]$. We choose $\delta > 0$ so small that

$$\|\hat{S} - \hat{S}_\delta\| < \epsilon/2.$$

Choose $\eta > 0$ such that

$$2\eta f(\hat{S}(x) + 1) < \delta,\ 0 \le x \le 1.$$

Let (S_0, u_0) be such that $u_0 \neq 0$ and $(S(t,x), u(t,x))$ be the corresponding solution. Suppose that (11.34) is false. Then there exists $\tau > 0$ such that

$$u(t,x) \le 2\eta,\ 0 \le x \le 1,\ t \ge \tau.$$

We may assume that τ is so large that $S(t,x) \le \hat{S}(x) + 1$ for $t \ge \tau$ by (11.19). Consequently,

$$u(t,x) f(S(t,x)) < \delta,\ 0 \le x \le 1, \quad t \ge \tau,$$

so S satisfies

$$S_t \ge d_S S_{xx} - \delta,\ t \ge \tau.$$

By the comparison theorem, comparing $S(t,x)$ with the solution of (11.35) satisfying the same initial conditions at $t = \tau$, we find that $S(t,x)$ exceeds that solution, which converges uniformly to $\hat{S}_\delta$. Hence, there exists $T \ge \tau$ such that $S(t,x) \ge \hat{S}_\delta(x) - \epsilon/2 \ge \hat{S}(x) - \epsilon$ for $t \ge T$.

But then

$$u_t \ge d_u u_{xx} + f(\hat{S} - \epsilon) u, \quad t \ge T.$$

The corresponding linear differential equation has the family of solutions $u_c(t,x) = c\hat{V}_1(x) e^{\hat{\lambda}_1(t-T)}$ for each real c. Since $u(T,x) > 0$ for all $x \in [0,1]$ there exists $c > 0$ such that $u(T,x) \ge c\hat{V}_1(x)$, $0 \le x \le 1$. Thus by the comparison theorem

$$u(t,x) \ge c\hat{V}_1(x) e^{\hat{\lambda}_1(t-T)}, \quad t \ge T.$$

This is impossible since the right-hand side is unbounded. This proves (11.34). □

We can now prove the main result of this section.

Theorem 11.16. *If $\lambda_1 > 0$, then there exists $\epsilon > 0$ such that for all solutions of (11.5)-(11.7) with $u_0 \neq 0$,*

$$\liminf_{t\to\infty} \int_0^1 u(t,x)dx \geq \epsilon \tag{11.36}$$

and

$$\liminf_{t\to\infty} \min_{0\leq x\leq 1} u(t,x) \geq \epsilon. \tag{11.37}$$

Proof. Proposition 11.15 establishes uniform weak $\|\cdot\|_\infty$-persistence. By Theorem 4.13 with $B = A$, the compact attractor of Theorem 11.9, we have uniform $\|\cdot\|_\infty$-persistence. We use Corollary 4.22 with $\rho = \|\cdot\|_\infty$. Note that both $\hat{\rho}(S,u) = \int_0^1 u(x)dx$ and $\tilde{\rho}(S,u) = \min_{0\leq x\leq 1} u(x)$ are continuous on X.

Given a total trajectory $\{(S(t),u(t));\, t \in \mathbb{R}\}$ with pre-compact range we must show that

$$\inf_{t\in\mathbb{R}} \|u(t,\cdot)\|_\infty > 0 \implies \min_{0\leq x\leq 1} u(0,x) > 0.$$

Since Theorem 11.6 implies that a solution $\hat{u}(t,x) > 0$ if $\hat{u}_0 \neq 0$, we conclude that $u(0,x) > 0,\ x \in [0,1]$, by applying Theorem 11.6 to $\hat{u}_0 = u(-1,\cdot)$. The required conclusion follows. By Corollary 4.22, (11.37) follows and it implies (11.36). □

We can summarize the results of this chapter as follows. The key "invasion eigenvalue" λ_1, alternatively, $\mathcal{R}_0$, completely determines the essential features of long term dynamics of (11.5)-(11.7). If $\mathcal{R}_0 < 1$, then the bacteria are washed out; if $\mathcal{R}_0 > 1$, then the bacteria persist if they are initially present.

It is natural to expect the existence of an equilibrium solution characterized by a positive density of bacteria when $\lambda_1 > 0$. Such an equilibrium $(S^*(x), u^*(x))$ must satisfy

$$\begin{aligned} 0 &= d_S S_{xx} - f(S)u, \\ 0 &= d_u u_{xx} + f(S)u, \quad 0 < x < 1, \end{aligned} \tag{11.38}$$

and the boundary conditions

$$\begin{aligned} -d_S S_x(0) &= F_0, \\ u_x(0) &= 0, \\ d_S S_x(1) + r_S S(1) &= 0, \\ d_u u_x(1) + r_u u(1) &= 0. \end{aligned} \tag{11.39}$$

The following is a special case of a result in [**138**]. The uniqueness assertion is most remarkable.

Theorem 11.17. *If* $\lambda_1 > 0$*, there exists a unique equilibrium solution* $(S^*(x), u^*(x)) \in X$ *with* $u^* \neq 0$*. Indeed,*

$$S^*(x) < \hat{S}(x), \quad u^*(x) > 0, \qquad x \in [0,1].$$

Proof. We prove existence of a positive equilibrium point using Theorem 6.2; for uniqueness, see [**138**]. Clearly, X is a closed subset of a Banach space, $\rho(S,u) = \int_0^1 u dx$ is continuous and concave, and Φ is uniformly ρ-persistent and has a compact attractor of compact sets. Hypothesis (H1) of Theorem 6.2 is satisfied by arguments similar to those used to prove Theorem 11.16. We need only establish that Φ_t is condensing for $0 < t \leq 1$ to conclude from Theorem 6.2 that there exists an equilibrium $(S^*(x), u^*(x))$ with $\int_0^1 u^*(x)dx > 0$. Then, by Theorem 11.6, $u^*(x) > 0$, $0 \leq x \leq 1$; the bounds on S^* follow from Proposition 11.8 and the maximum principle.

We sketch the argument that Φ_t is compact for each $t \in (0,1]$. By (11.14) and compactness of the semigroups $U(t), V(t)$ it suffices to focus on the two integral terms. According to Lemma 10.13, it is easily seen that these integral terms, taken together, define a compact operator L on $C([0,1], X)$. By Theorem 11.6 and Proposition 11.9, the map H: $(S_0, u_0) \to \Phi(t, S_0, u_0)$, $0 \leq t \leq 1$, is a continuous map from X to $C([0,1], X)$ which maps bounded sets to bounded sets; continuity uses that Φ is state continuous, uniformly in finite time by Lemma 1.34. As the projection $P : C([0,1], X) \to X$, defined by $P(S,u) = (S(1), u(1))$, is continuous, $P \circ L \circ H$ is compact. Hence, it follows that Φ_1 is compact and the same argument works for each $t \in (0,1]$. $\square$

Numerical simulations suggest that the equilibrium (S^*, u^*) attracts all solutions with initial data (S_0, u_0) satisfying $u_0 \neq 0$.

The paper [**138**] extends much of this analysis to three-dimensional bioreactors of general shape and multiple competing bacterial populations.

11.7. Exercises

Exercise 11.1. What are the units of d_N and r_N?

Exercise 11.2. Check that the scaling of parameters and variables given in (11.1)-(11.4) yields (11.5)-(11.7). Also, verify that scaled variables and parameters are unitless.

Exercise 11.3. Give the complete proof of Theorem 11.16 by verifying one set of hypotheses of Theorem 4.13.

Exercise 11.4. Verify Lemma 11.1. See [**86, 232**].

Exercise 11.5. Verify the semigroup property directly using the formula for the Green's function. Hint: Interchange order of integration and use the orthogonality of the eigenfunctions.

Exercise 11.6. Is $\Psi(t,\phi) \equiv V(t)\phi$ a continuous semiflow on $C([0,1], \mathbb{R})$? Justify your answer.

Exercise 11.7. Use the maximum principle Theorem A.23 to show that classical solutions of (11.5)-(11.7) corresponding to nonnegative initial data are nonnegative.

Exercise 11.8. Show that the mean residence time for bacteria in the tube decreases with increasing bacterial motility d_B if $d_B < Lr_B$.

Exercise 11.9. In the special case that $d_S = d_u$ and $r_S = r_u$ system (11.5)-(11.7) can be simplified. Add the S and u equations to show that $S+u$ satisfies (11.8) and therefore, by Corollary 11.5, $S(t,x) + u(t,x) \to \hat{S}(x)$ as $t \to \infty$, uniformly in $x \in [0,1]$. Setting $S + u = \hat{S}$ allows the reduction of (11.5)-(11.7) to a single equation for u. The asymptotic behavior of this "limiting system" can be completely determined using either variational methods or by applying the theory of monotone dynamical systems [**200**]. Unfortunately, the motility d_B of a bacterium is typically smaller than the diffusion constant d_N for a nutrient molecule, making this assumption inappropriate. Berg [**14**] lists the diffusivity of a small molecule in water as $10^{-5} cm^2/sec$ whereas the motility of an *E. Coli* cell is $4 \times 10^{-6} cm^2/sec$.

Chapter 12

Dividing Cells in a Chemostat

We consider a chemostat model where the consumer population consists of cells. Since cells of different age a (time since they split for the last time) may have different resource uptake rates $\kappa(a) \geq 0$, $\kappa \in L^\infty(\mathbb{R}_+)$, we stratify the biomass of the cell population according to age,

$$C(t) = \int_0^\infty c(t,a)da. \tag{12.1}$$

In a preliminary modeling attempt we assume that there is a per unit cell-biomass division rate β (which is a nonnegative function of cell age). As before, R is the mass of the resource (nutrient) supporting cell growth,

$$\begin{aligned} R' =& D(R^\diamond - R(t)) - f(R(t)) \int_0^\infty \kappa(a) c(t,a)da, \\ (\partial_t + \partial_a) c(t,a) =& \Big(\kappa(a) f(R(t)) - D - \beta(a)\Big) c(t,a), \\ c(t,0) =& \int_0^\infty \beta(a) c(t,a)da, \end{aligned} \tag{12.2}$$

with initial data $R(0) = \check{R}$, $c(0,a) = \check{c}(a)$. As for the boundary condition, notice that cell division preserves cell biomass. The two daughter cells together have the same biomass as their mother cell had before the division.

$R^\diamond$ is the input nutrient mass, D is the dilution rate of the chemostat, and $\kappa(a)f(R)$ is the specific biomass growth rate of cells of age a at resource level R. We assume that $f : \mathbb{R}_+ \to \mathbb{R}_+$ is

(12.3) Lipschitz continuous and monotone increasing, $f(0) = 0, f(\infty) < \infty$.

In this model we have implicitly assumed that the yield parameter which describes how resource mass is converted into cell biomass is independent of cell age. This has allowed us to remove the yield parameter by scaling.

We introduce a cell's probability of not dividing before age a,

$$\mathcal{F}(a) = \exp\Big(-\int_0^a \beta(s)ds\Big). \tag{12.4}$$

By differentiation we see that

$$\beta(a) = -\frac{\mathcal{F}'(a)}{\mathcal{F}(a)}.$$

In order to include probability distributions $\mathcal{F}$ that are not differentiable we assume that there is

(12.5) a decreasing function $\mathcal{F} : \mathbb{R}_+ \to \mathbb{R}_+$ with $\mathcal{F}(0) = 1$

such that $\mathcal{F}(a)$ is a cell's probability of not dividing before age a. We introduce the Lebesgue-Stieltjes measure associated with $\mathcal{F}$: There exists a uniquely determined nonnegative regular Borel measure m on $\mathbb{R}_+$ such that

$$m([0, a)) = 1 - \mathcal{F}(a) \tag{12.6}$$

whenever $a > 0$ and $\mathcal{F}$ is continuous at a [**164**]. For $0 < a_1 < a_2$ the probability that a cell divides at an age in $[a_1, a_2)$ is given by $m([a_1, a_2)) = \mathcal{F}(a_1-) - \mathcal{F}(a_2-)$, while the probability of dividing at an age in $(a_1, a_2]$ is given by $m((a_1, a_2]) = \mathcal{F}(a_1+) - \mathcal{F}(a_2+)$. Here the attached symbols $+$ and $-$ mean that the right-hand or the left-hand limit are taken respectively.

Our model now takes the form

$$\left\{\begin{aligned} R' &= D(R^\diamond - R(t)) - f(R(t)) \int_0^\infty \kappa(a)c(t,a)da, \\ \frac{\partial}{\partial t}c(t,a) &= -\mathcal{F}(a)\frac{\partial}{\partial a}\frac{c(t,a)}{\mathcal{F}(a)} + \Big(\kappa(a)f(R(t)) - D\Big)c(t,a), \\ c(t,0) &= \int_{\mathbb{R}_+} \frac{c(t,a)}{\mathcal{F}(a)} m(da), \end{aligned}\right. \tag{12.7}$$

with initial data $R(0) = \breve{R}$, $c(0,a) = \breve{c}(a)$.

The form of the equation for $c(t,a)$ is partially motivated by observing that for smooth c and $\mathcal{F}$ we have

$$\frac{\partial}{\partial t}c(t,a) + \mathcal{F}(a)\frac{\partial}{\partial a}\frac{c(t,a)}{\mathcal{F}(a)} = \frac{\partial}{\partial t}c(t,a) + \frac{\partial}{\partial a}c(t,a) - \beta(a)c(t,a).$$

We begin by tackling the linear problem

$$
\begin{aligned}
\frac{\partial}{\partial t}c(t,a) =& - \mathcal{F}(a)\frac{\partial}{\partial a}\frac{c(t,a)}{\mathcal{F}(a)},\\
c(t,0) =& \int_{\mathbb{R}_+} \frac{c(t,a)}{\mathcal{F}(a)} m(da),\\
c(0,a) =& g(a)
\end{aligned}
\tag{12.8}
$$

in the next section.

Remark 12.1. Let $\mathcal{F}$ be positive and set $u(t,a) = c(t,a)/\mathcal{F}(a)$. In Exercise 12.1 we ask the reader to verify that u solves

$$
\begin{aligned}
\frac{\partial u}{\partial t} + \frac{\partial u}{\partial a} =& 0,\\
u(t,0) =& b(t),\\
u(0,a) =& g(a)/\mathcal{F}(a)
\end{aligned}
\tag{12.9}
$$

where b is the cell (population) division rate

$$
b(t) = \int_0^\infty c(t,a) m(da). \tag{12.10}
$$

In addition:

a) The method of characteristics can be used to show that u can be written as

$$
u(t,a) = \begin{cases} b(t-a), & t > a, \\ \frac{g(a-t)}{\mathcal{F}(a-t)}, & t < a. \end{cases} \tag{12.11}
$$

b) c can be written as

$$
c(t,a) = \begin{cases} b(t-a)\mathcal{F}(a), & t > a, \\ g(a-t)\frac{\mathcal{F}(a)}{\mathcal{F}(a-t)}, & t < a. \end{cases} \tag{12.12}
$$

c) If c satisfies (12.8) and b is given by (12.10), then b satisfies the integral equation

$$
b(t) = \int_{[0,t)} b(t-a) m(da) + \int_{[t,\infty)} \frac{g(a-t)}{\mathcal{F}(a-t)} m(da). \tag{12.13}
$$

12.1. An integral equation

We assume that $\mathcal{F}$ is strictly positive on $\mathbb{R}_+$ because many of the usual probability distributions on $\mathbb{R}_+$ like the Γ distribution or the lognormal distribution have this property. Further, we assume that

(12.14) for every $a_1 > 0$ there is an $a_2 > a_1$ with $\mathcal{F}(a_2) < \mathcal{F}(a_1)$.

As usual, $L^1(\mathbb{R}_+)$ denotes the Banach space of Lebesgue integrable functions on $\mathbb{R}_+$. More precisely, each element in $L^1(\mathbb{R})$ is an equivalence class of integrable functions which are equal to each other a.e. $L^1_+(\mathbb{R}_+)$ is the cone of functions with nonnegative representatives.

For given $g \in L^1(\mathbb{R}_+)$, we study the following integral equation for the cell-division rate b,

$$b(t) = \int_{[0,t)} b(t-a)m(da) + \int_{[t,\infty)} \frac{g(a-t)}{\mathcal{F}(a-t)} m(da). \tag{12.15}$$

This equation is motivated by Remark 12.1. At the first glance, it is not clear whether the equation is meaningful because g is an equivalence class of functions that differ on sets of Lebesgue measure 0.

We introduce the *expected remaining time till cell division* at age $s \geq 0$,

$$\mathcal{D}(s) = \int_s^\infty \frac{\mathcal{F}(a)}{\mathcal{F}(s)} da = \int_0^\infty \frac{\mathcal{F}(a+s)}{\mathcal{F}(s)} da. \tag{12.16}$$

Notice that $\mathcal{F}(a|s) = \frac{\mathcal{F}(a+s)}{\mathcal{F}(s)}$ is the conditional probability that a cell does not divide before age $a+s$, provided it has not divided at age s. See [**217**, Sec.12.4]. The following relations follow from [**217**, Prop.12.1] and its proof.

Lemma 12.2. *The following statements are equivalent:*

(a) *$\mathcal{D}(a)$ is a bounded function of $a \in \mathbb{R}_+$.*

(b) *There exist some $\delta \in (0,1)$ and $t_0 > 0$ such that $\frac{\mathcal{F}(a+t)}{\mathcal{F}(a)} \leq (1-\delta)$ for all $a \geq 0$, $t \geq t_0$.*

(c) *There exist $\epsilon > 0$, $N \geq 1$ such that $\frac{\mathcal{F}(a+t)}{\mathcal{F}(a)} \leq Ne^{-\epsilon t}$ for all $a \geq 0$, $t \geq 0$.*

Further, we have the following equivalence.

Lemma 12.3. *The following are equivalent:*

(i) *For every $a > 0$, $\dfrac{\mathcal{F}(a+s)}{\mathcal{F}(s)} \to 0$ as $s \to \infty$.*

(ii) *For every $\lambda > 0$, $\displaystyle\int_0^\infty \frac{\mathcal{F}(a+s)}{\mathcal{F}(s)} e^{-\lambda a} da \longrightarrow 0$, as $s \to \infty$.*

Proof. Let $\lambda, h > 0$. Since $\mathcal{F}$ is decreasing,

$$\int_0^\infty \frac{\mathcal{F}(a+s)}{\mathcal{F}(s)} e^{-\lambda a} da \geq \int_0^h \frac{\mathcal{F}(a+s)}{\mathcal{F}(s)} e^{-\lambda a} da \geq \frac{1-e^{-\lambda h}}{\lambda} \frac{\mathcal{F}(h+s)}{\mathcal{F}(s)},$$

and (ii) implies (i).

Now assume (i) and let $h > 0$ be arbitrary. Since $\mathcal{F}$ is decreasing,

$$\int_0^\infty \frac{\mathcal{F}(a+s)}{\mathcal{F}(s)} e^{-\lambda a} da \leq \int_0^h e^{-\lambda a} da + \int_h^\infty \frac{\mathcal{F}(h+s)}{\mathcal{F}(s)} e^{-\lambda a} da$$

$$= \frac{1-e^{-\lambda h}}{\lambda} + \frac{\mathcal{F}(h+s)}{\mathcal{F}(s)} \frac{e^{-\lambda h}}{\lambda}.$$

By (i),

$$\limsup_{s\to\infty} \int_0^\infty \frac{\mathcal{F}(a+s)}{\mathcal{F}(s)} e^{-\lambda a} da \leq \frac{1-e^{-\lambda h}}{\lambda}.$$

This holds for every $h > 0$ and so

$$\limsup_{s\to\infty} \int_0^\infty \frac{\mathcal{F}(a+s)}{\mathcal{F}(s)} e^{-\lambda a} da = 0. \qquad \square$$

In the following we will make heavy use of the Laplace transform. If $g : \mathbb{R}_+ \to \mathbb{R}$ is bounded, then $\hat{g} = \mathcal{L}\{g\}$ with

$$\hat{g}(\lambda) = \mathcal{L}\{g\}(\lambda) = \int_0^\infty e^{-\lambda t} g(t) dt, \qquad \lambda > 0,$$

is the Laplace transform of g. $\hat{g}(\lambda)$ is also defined if $g \in L^1(\mathbb{R}_+)$ and $\lambda \geq 0$, and then $|\hat{g}(\lambda)| \leq \|g\|$ with $\|g\| = \int_0^\infty |g(t)| dt$ being the norm in $L^1(\mathbb{R}_+)$. If g is nonnegative and Borel measurable, we can still take the Laplace transform but may obtain a function with values ∞.

Proposition 12.4. *For given $g \in L^1(\mathbb{R}_+)$, equation (12.15) is meaningful and has a unique solution b with*

$$\mathcal{L}\{|b(\cdot)|\}(\lambda) := \int_0^\infty e^{-\lambda t} |b(t)| dt < \infty \qquad \textit{for all } \lambda > 0.$$

If $g \geq 0$, then $b \geq 0$, and $b = 0$ a.e. if and only if $g = 0$ a.e. Further,

$$\frac{\|g\|}{\lambda \hat{\mathcal{F}}(\lambda)} \geq \mathcal{L}\{|b(\cdot)|\}(\lambda).$$

Finally, there exists some $\delta_\lambda > 0$ (independent of g) such that

$$\delta_\lambda \frac{\|g\|}{\lambda \hat{\mathcal{F}}(\lambda)} \leq \hat{b}(\lambda)$$

if one of the following two conditions is satisfied:

(i) *$\mathcal{D}(\cdot)$ is bounded,*

(ii) *$\frac{\mathcal{F}(a+s)}{\mathcal{F}(s)} \to 0$ as $s \to \infty$ for every $a > 0$.*

Proof. Let h denote

$$h(t) = \int_{[t,\infty)} \frac{|g(a-t)|}{\mathcal{F}(a-t)} m(da). \tag{12.17}$$

To show that h is well defined as a function in $L^1_{\mathrm{loc}}(\mathbb{R}_+)$, we take its Laplace transform,

$$\hat{h}(\lambda) = \int_0^\infty e^{-\lambda t} h(t) dt = \int_0^\infty e^{-\lambda t} \Big(\int_{[t,\infty)} \frac{|g(a-t)|}{\mathcal{F}(a-t)} m(da) \Big) dt.$$

By Tonelli's theorem, we can change the order of integration even if $\hat{h}(\lambda) = \infty$,

$$\hat{h}(\lambda) = \int_{\mathbb{R}_+} \Big(\int_0^a e^{-\lambda t} \frac{|g(a-t)|}{\mathcal{F}(a-t)} dt \Big) m(da).$$

We change the variables,

$$\hat{h}(\lambda) = \int_{\mathbb{R}_+} \Big(\int_0^a e^{\lambda s} \frac{|g(s)|}{\mathcal{F}(s)} ds \Big) e^{-\lambda a} m(da).$$

We change the order of integration again,

$$\hat{h}(\lambda) = \int_0^\infty e^{\lambda s} \frac{|g(s)|}{\mathcal{F}(s)} \Big(\int_{[s,\infty)} e^{-\lambda a} m(da) \Big) ds. \tag{12.18}$$

We integrate the interior integral by parts ([**99**, Thm. 21.67], or combine Lemma B.27 and Corollary B.11 in [**217**]),

$$\begin{aligned}\hat{h}(\lambda) =& \int_0^\infty e^{\lambda s} \frac{|g(s)|}{\mathcal{F}(s)} \Big(e^{-\lambda s} \mathcal{F}(s) - \lambda \int_s^\infty e^{-\lambda a} \mathcal{F}(a) da \Big) ds \\ =& \int_0^\infty |g(s)| \Big(1 - \lambda \int_s^\infty e^{-\lambda(a-s)} \frac{\mathcal{F}(a)}{\mathcal{F}(s)} da \Big) ds.\end{aligned} \tag{12.19}$$

This also yields

$$\hat{h}(\lambda) = \|g\| - \int_0^\infty e^{\lambda s} \frac{|g(s)|}{\mathcal{F}(s)} \lambda \int_s^\infty e^{-\lambda a} \mathcal{F}(a) da \Big) ds. \tag{12.20}$$

If g is zero Lebesgue almost everywhere, this expression is 0. Since h is nonnegative, for every $t > 0$,

$$\|g\| \geq \hat{h}(\lambda) \geq e^{-\lambda t} \int_0^t h(s) ds.$$

By (12.17), the forcing term in (12.15) is well defined as a function in $L^1_{\mathrm{loc}}(\mathbb{R}_+)$.

To solve the integral equation, we will use a fixed point argument. To this end, we consider the Banach space

$$\begin{aligned} E_\lambda =& \{ b \in L^1_{\mathrm{loc}}(\mathbb{R}_+); \|b\|_\lambda < \infty \}, \\ \|b\|_\lambda =& \int_0^\infty e^{-\lambda t} |b(t)| dt. \end{aligned} \tag{12.21}$$

Define $F : E_\lambda \to E_\lambda$ by

$$F(b)(t) = \int_{[0,t)} b(t-a) m(da), \qquad t \geq 0.$$

This way, the integral equation (12.15) becomes the fixed point problem $b = F(b) + h$. Since the Laplace transform converts convolutions into products,

$$\|F(b)\|_\lambda \le \|b\|_\lambda \hat{m}(\lambda),$$

where

$$\hat{m}(\lambda) = \int_{\mathbb{R}_+} e^{-\lambda a} m(da)$$

is the Laplace transform of the measure m. In particular F is well defined. Choose $\lambda > 0$. We integrate by parts,

$$\hat{m}(\lambda) = 1 - \lambda \int_0^\infty e^{-\lambda a} \mathcal{F}(a) da < 1.$$

Existence of solutions now follows from the contraction mapping theorem (Banach's fixed point theorem) or the theory of Neumann series. Notice that the unique solution b is a linear function of g. Since F maps nonnegative functions to nonnegative functions, b is nonnegative whenever g is nonnegative.

We note that

$$\|b\|_\lambda \le \frac{\hat{h}(\lambda)}{1 - \hat{m}(\lambda)} = \frac{\hat{h}(\lambda)}{\lambda \hat{\mathcal{F}}(\lambda)}, \tag{12.22}$$

with equality holding if $h \ge 0$. To show that $b = 0$ a.e. implies that $g = 0$ a.e., assume that $\|g\| > 0$. Since $\mathcal{F}$ is decreasing and we assume (12.14), there exists a strictly increasing sequence (a_k) in $\mathbb{R}_+$ such that $a_k \to \infty$ as $k \to \infty$, $\mathcal{F}$ is continuous at a_k, and $\mathcal{F}(a_{k+1}) < \mathcal{F}(a_k)$. There exists some $k \in \mathbb{N}$ such that $\int_0^{a_k} |g(a)| da > 0$. By (12.18),

$$\begin{aligned}
\hat{h}(\lambda) &\ge \int_0^{a_k} e^{\lambda s} \frac{|g(s)|}{\mathcal{F}(s)} ds \Big(\int_{[a_k, \infty)} e^{-\lambda a} m(da) \Big). \\
&\ge \int_0^{a_k} \frac{|g(s)|}{\mathcal{F}(s)} ds \sum_{j=k}^{\infty} \int_{[a_j, a_{j+1})} e^{-\lambda a} m(da). \\
&\ge \int_0^{a_k} |g(s)| ds \sum_{j=k}^{\infty} e^{-\lambda a_{j+1}} (\mathcal{F}(a_j) - \mathcal{F}(a_{j+1})) > 0.
\end{aligned}$$

By (12.22), $\|b\|_\lambda > 0$ and, by (12.20),

$$\|b\|_\lambda = \frac{\hat{h}(\lambda)}{\lambda \hat{\mathcal{F}}(\lambda)} \le \frac{1}{\lambda \hat{\mathcal{F}}(\lambda)} \|g\|.$$

Assume (i) that $\mathcal{D}$ is bounded. By Lemma 12.2, there exist some $\delta_0 \in (0, 1)$ and $t_0 > 0$ such that

$$\frac{\mathcal{F}(a+t)}{\mathcal{F}(a)} \le 1 - \delta_0, \qquad a \ge 0, t \ge t_0.$$

Since $\mathcal{F}$ is decreasing, for all $s \geq 0$,

$$\begin{aligned} 1-\lambda\int_s^\infty e^{-\lambda(a-s)}\frac{\mathcal{F}(a)}{\mathcal{F}(s)}da &= 1-\lambda\int_0^\infty e^{-\lambda a}\frac{\mathcal{F}(a+s)}{\mathcal{F}(s)}da \\ = 1-\lambda\int_0^{t_0} e^{-\lambda a}\frac{\mathcal{F}(a+s)}{\mathcal{F}(s)}da &- \lambda\int_{t_0}^\infty e^{-\lambda a}\frac{\mathcal{F}(a+s)}{\mathcal{F}(s)}da \\ \geq 1-\lambda\int_0^{t_0} e^{-\lambda a}da &- \lambda\int_{t_0}^\infty e^{-\lambda a}(1-\delta_0)da = \delta_0 e^{-\lambda t_0}. \end{aligned}$$

We set $\delta_\lambda = \delta_0 e^{-\lambda t_0}$. By (12.22) and (12.19),

$$\|b\|_\lambda = \frac{\hat{h}(\lambda)}{\lambda\hat{\mathcal{F}}(\lambda)} \geq \frac{\delta_\lambda}{\lambda\hat{\mathcal{F}}(\lambda)}\|g\|,$$

and δ_λ does not depend on g.

Assume (ii). Revisit (12.19). It follows from Lemma 12.3 that there is some $\epsilon > 0$ such that

$$1-\lambda\int_s^\infty e^{-\lambda(a-s)}\frac{\mathcal{F}(a)}{\mathcal{F}(s)}da \geq \epsilon > 0$$

for sufficiently large $s > 0$. By (12.14), this expression is positive for every $s \geq 0$. Since this expression also continuously depends on $s \geq 0$, it is bounded away from 0 for all $s \geq 0$. □

12.2. A C_0-semigroup

Our strategy of analyzing (12.7) consists in rewriting it as a semilinear Cauchy problem (Chapter 10). To this end, we show that the linear problem (12.8) can be solved by a C_0-semigroup.

For $t, a \geq 0$ and $g \in L^1(\mathbb{R}_+)$, we define

$$[S(t)g](a) = \begin{cases} b(t-a)\mathcal{F}(a), & t > a, \\ g(a-t)\frac{\mathcal{F}(a)}{\mathcal{F}(a-t)}, & t < a, \end{cases} \tag{12.23}$$

where b is the unique solution of (12.15),

$$b(t) = \int_{[0,t)} b(t-a)m(da) + \int_{[t,\infty)} \frac{g(a-t)}{\mathcal{F}(a-t)}m(da). \tag{12.24}$$

Recall Proposition 12.4. See also (12.12).

Lemma 12.5. *S is a semigroup of bounded linear operators.*

Proof. To show that $S(t)$ maps $L^1(\mathbb{R}_+)$ into itself and find $\|S(t)\, g\|$, we note that

$$\|S(t)g\| \leq \int_0^t |b(t-a)|\mathcal{F}(a)da + \int_t^\infty |g(a-t)|\frac{\mathcal{F}(a)}{\mathcal{F}(a-t)}da.$$

Since $\mathcal{F}(a)$ is decreasing and $\mathcal{F}(a) \leq 1$ by (12.5),

$$\|S(t)g\| \leq \int_0^t |b(s)|ds + \|g\| \leq e^t \mathcal{L}\{|b|\}(1) + \|g\|.$$

By Proposition 12.4,

$$\|S(t)g\| \leq \Big(\frac{e^t}{\mathcal{F}(1)} + 1\Big)\|g\|.$$

We will improve on this estimate dramatically in the next lemma.

To show the semigroup property, for $t, r \geq 0$, by (12.23),

$$[S(t)S(r)g](a) = \begin{cases} \tilde{b}(t-a)\mathcal{F}(a), & t > a, \\ [S(r)g](a-t)\frac{\mathcal{F}(a)}{\mathcal{F}(a-t)}, & t < a, \end{cases} \tag{12.25}$$

where $\tilde{b}$ is the solution of

$$\tilde{b}(t) = \int_{[0,\,t)} \tilde{b}(t-a)m(da) + \int_{[t,\,\infty)} \frac{[S(r)g](a-t)}{\mathcal{F}(a-t)} m(da). \tag{12.26}$$

Again by definition of S in (12.23), after some simplification,
(12.27)

$$[S(t)S(r)g](a) = \begin{cases} \tilde{b}(t-a)\mathcal{F}(a), & t > a, \\ b(r-(a-t))\mathcal{F}(a), & t+r > a > t, \\ g(a-t-r)\dfrac{\mathcal{F}(a)}{\mathcal{F}(a-t-r)}, & t+r < a. \end{cases}$$

We claim that $\tilde{b}(t-a) = b(r+t-a)$ for $t \geq a$. To this end we show that $\tilde{b}$ and $b(r+\cdot)$ solve the same integral equation and invoke uniqueness of solutions. By (12.26) and (12.23),

$$\begin{aligned}\tilde{b}(t) = &\int_{[0,t)} \tilde{b}(t-a)m(da) + \int_{[t,t+r)} \frac{b(r+t-a)\mathcal{F}(a-t)}{\mathcal{F}(a-t)} m(da) \\ &+ \int_{[t+r,\infty)} \frac{g(a-t-r)\mathcal{F}(a-t)}{\mathcal{F}(a-t-r)\mathcal{F}(a-t)}.\end{aligned}$$

We simplify,

$$\begin{aligned}\tilde{b}(t) = &\int_{[0,t)} \tilde{b}(t-a)m(da) + \int_{[t,t+r)} b(r+t-a)m(da) \\ &+ \int_{[t+r,\infty)} \frac{g(a-t-r)}{\mathcal{F}(a-t-r)} m(da).\end{aligned} \tag{12.28}$$

As for b, by (12.24),

$$b(t+r) = \int_{[0,\,t+r)} b(t+r-a)m(da) + \int_{[t+r,\infty)} \frac{g(a-t-r)}{\mathcal{F}(a-t-r)} m(da).$$

We split the integral,

$$
(12.29)\qquad \begin{aligned} b(t+r) = & \int_{[0,t)} b(t+r-a)m(da) + \int_{[t,t+r)} b(t+r-a)m(da) \\ & + \int_{[t+r,\infty)} \frac{g(a-t-r)}{\mathcal{F}(a-t-r)} m(da). \end{aligned}
$$

By (12.28) and (12.29), $\tilde{b}$ and $b(r+\cdot)$ solve the same integral equation. By uniqueness, $\tilde{b}(t) = b(r+t)$ for a.a. $t \geq 0$. We substitute this result into (12.27),

$$
\begin{aligned} [S(t)S(r)g](a) &= \begin{cases} b(r+t-a)\mathcal{F}(a), & t+r > a \geq 0, \\ g(a-t-r)\dfrac{\mathcal{F}(a)}{\mathcal{F}(a-t-r)}, & t+r < a, \end{cases} \\ &\overset{(12.23)}{=} [S(t+r)g](a). \end{aligned}
$$

□

Lemma 12.6. *S is a C_0-semigroup of positive contractions. Moreover, $\|S(t)g\| = \|g\|$ for all $g \in L^1_+(\mathbb{R}_+)$. Finally, the unique solution b of (12.24) satisfies $b(t) = \int_{\mathbb{R}_+} [S(t)g](a) \frac{m(da)}{\mathcal{F}(a)}$.*

Proof. Let $g \in L^1(\mathbb{R}_+)$. If g is nonnegative, so is b by Proposition 12.4, and $S(t)g$ is a nonnegative function as well. Further,

$$
\begin{aligned} \|S(t)g - g\| \leq & \int_0^t |b(t-a)|\mathcal{F}(a)da + \int_0^t |g(a)|da \\ & + \int_t^\infty \Big| g(a-t)\frac{\mathcal{F}(a)}{\mathcal{F}(a-t)} - g(a) \Big| da. \end{aligned}
$$

Since $\mathcal{F}(a) \leq 1$, after some substitutions,

$$
\|S(t)g - g\| \leq \int_0^t |b(a)|da + \int_0^t |g(a)|da + \int_0^\infty \Big| g(a)\frac{\mathcal{F}(a+t)}{\mathcal{F}(a)} - g(a+t) \Big| da.
$$

Since $b, g \in L^1([0,1])$, and by the triangle inequality,

$$
\begin{aligned} \limsup_{t\to 0} \|S(t)g - g\| &\leq \limsup_{t\to 0} \int_0^\infty \Big| g(a)\frac{\mathcal{F}(a+t)}{\mathcal{F}(a)} - g(a+t) \Big| da \\ &\leq \limsup_{t\to 0} \Big(\int_0^\infty |g(a)| \Big| \frac{\mathcal{F}(a+t)}{\mathcal{F}(a)} - 1 \Big| da + \int_0^\infty |g(a) - g(a+t)|da \Big). \end{aligned}
$$

Since $g \in L^1(\mathbb{R}_+)$, $\int_0^\infty |g(a) - g(a+t)|da \to 0$ as $t \to 0$ [**100**, Thm.3.8.3] and, by Fatou's lemma,

$$
\limsup_{t\to 0} \|S(t)g - g\| \leq \int_0^\infty |g(a)| \limsup_{t\to 0} \Big| \frac{\mathcal{F}(a+t)}{\mathcal{F}(a)} - 1 \Big| da.
$$

Since $\mathcal{F}$ is decreasing and nonnegative, it is continuous almost everywhere. This implies that $\limsup_{t\to 0}\left|\frac{\mathcal{F}(a+t)}{\mathcal{F}(a)} - 1\right| = 0$ for almost all $a \geq 0$. Thus $\limsup_{t\to 0} \|S(t)g - g\| = 0$.

Let $g \in L^1_+(\mathbb{R}_+)$. We consider the Laplace transform of $S(\cdot)g$ using (12.23),

$$\int_0^\infty e^{-\lambda t}\|S(t)g\|dt = \|b\|_\lambda \hat{F}(\lambda) + \int_0^\infty e^{-\lambda t}\Big(\int_t^\infty |g(a-t)|\frac{\mathcal{F}(a)}{\mathcal{F}(a-t)}da\Big)dt,$$

with $\|b\|_\lambda$ given by (12.21). By (12.22) and (12.20),

$$\|b\|_\lambda \hat{F}(\lambda) = \frac{1}{\lambda}\hat{h}(\lambda) = \frac{1}{\lambda}\|g\| - \int_0^\infty e^{\lambda s}\frac{|g(s)|}{\mathcal{F}(s)}\int_s^\infty e^{-\lambda a}\mathcal{F}(a)da\Big)ds.$$

We combine the last two equations, change the order of integration and perform a substitution,

$$\int_0^\infty e^{-\lambda t}\|S(t)g\|dt = \frac{1}{\lambda}\|g\| = \int_0^\infty e^{-\lambda t}\|g\|dt.$$

This equality holds for all $\lambda > 0$. Since $\|S(t)g\|$ is a continuous function of t, the uniqueness properties of the Laplace transform imply $\|S(t)g\| = \|g\|$.

For $g \in L^1(\mathbb{R}_+)$, let $|g|(a) = |g(a)|$ be the modulus or absolute value of g. Since S is a positive semigroup, $|S(t)g| \leq S(t)|g|$. So $\|S(t)g\| \leq \|S(t)|g|\,\| \leq \|\,|g|\,\|$.

The last statement follows from (12.23) and (12.24). □

The next result prepares connecting the C_0-semigroup S to the linear problem (12.8) and rewriting (12.7) as a semilinear Cauchy problem.

Since S is a contraction semigroup, we can define the Laplace transform of S by

$$\hat{S}(\lambda)g = \widehat{S(\cdot)g} = \int_0^\infty e^{-\lambda t}S(t)gdt, \qquad \lambda > 0.$$

Lemma 12.7. *Let $g \in L^1(\mathbb{R}_+)$. Then $\frac{[\hat{S}(\lambda)g](a)}{\mathcal{F}(a)}$ is a locally absolutely continuous function of $a \geq 0$ and*

$$\Big(\lambda + \frac{d}{da}\Big)\frac{[\hat{S}(\lambda)g](a)}{\mathcal{F}(a)} = \frac{g(a)}{\mathcal{F}(a)} \qquad \textit{for a.a. } a > 0,$$

$$[\hat{S}(\lambda)g](0) = \int_{\mathbb{R}_+}\frac{[\hat{S}(\lambda)g](a)}{\mathcal{F}(a)}m(da).$$

Further,

$$\frac{\|g\|}{\lambda\hat{\mathcal{F}}(\lambda)} \geq [\hat{S}(\lambda)g](0).$$

Finally, there exists some $\delta_\lambda > 0$ such that

$$[\hat{S}(\lambda)g](0) \ge \frac{\delta_\lambda}{\lambda\hat{\mathcal{F}}(\lambda)}\|g\| \text{ for all } g \in L^1_+(\mathbb{R}_+)$$

if one of the following two conditions is satisfied:

(i) $\mathcal{D}(\cdot)$ *is bounded,*

(ii) $\frac{\mathcal{F}(a+s)}{\mathcal{F}(s)} \to 0$ *as* $s \to \infty$ *for every* $a > 0$.

See Definition A.6.

Proof. By (12.23),

$$[\hat{S}(\lambda)g](a) = \int_0^a e^{-\lambda t} g(a-t)\frac{\mathcal{F}(a)}{\mathcal{F}(a-t)}dt + \int_a^\infty e^{-\lambda t} b(t-a)\mathcal{F}(a)dt.$$

After a change of variables,

$$\frac{[\hat{S}(\lambda)g](a)}{\mathcal{F}(a)} = \int_0^a e^{-\lambda(a-s)}\frac{g(s)}{\mathcal{F}(s)}ds + e^{-\lambda a}\int_0^\infty e^{-\lambda s} b(s)ds.$$

This implies the assertion. In particular, $[\hat{S}(\lambda)g](0) = \hat{b}(\lambda)$. Recall Lemma 12.6 and Proposition 12.4. □

12.3. A semilinear Cauchy problem

If A is the generator of the semigroup S, (10.3), we have $\hat{S}(\lambda) = (\lambda - A)^{-1}$ ([**67**, Thm.1.10] or any other book on operator semigroups). Let $\tilde{g} \in D(A)$ and set $g = (\lambda - A)\tilde{g}$. Then $\tilde{g} = \hat{S}(\lambda)g$. By Lemma 12.7, $\tilde{g}$ satisfies the boundary condition

$$\tilde{g}(0) = \int_{\mathbb{R}_+} \frac{\tilde{g}(a)}{\mathcal{F}(a)} m(da).$$

Further, $\tilde{g}/\mathcal{F}$ is absolutely continuous and, since $\tilde{g} = \hat{S}(\lambda)g$,

$$[(\lambda - A)\tilde{g}](a) = g(a) = \mathcal{F}(a)\Big(\lambda + \frac{d}{da}\Big)\frac{\tilde{g}(a)}{\mathcal{F}(a)}.$$

So, if $\tilde{g}$ in $D(A)$, then $\frac{\tilde{g}(a)}{\mathcal{F}(a)}$ is a locally absolutely continuous function of $a \ge 0$, and

$$[A\tilde{g}](a) = -\mathcal{F}(a)\frac{d}{da}\frac{\tilde{g}(a)}{\mathcal{F}(a)}.$$

We conjecture that this is actually the accurate description of A.

However this may be, since $v(t) = S(t)g$ is an integral solution of $v'(t) = Av(t)$, $v(0) = g$, by Theorem B.19, $c = S(\cdot)g$ can be considered a generalized solution of the linear problem (12.8).

Our partial insight also justifies rewriting the full cell division model (12.7) in an abstract form. Let $u(t) = c(t, \cdot)$ and $\breve{u} = \breve{c}(\cdot)$, and let A be the generator of S and B the bounded linear operator

$$[Bu](a) = \kappa(a)u(a). \tag{12.30}$$

The differential system (12.7) takes the form

$$\begin{aligned} R' &= D(R^\diamond - R) - f(R)\|Bu\|, & R(0) &= \breve{R}, \\ u' &= Au - Du + f(R)Bu, & u(0) &= \breve{u}. \end{aligned} \tag{12.31}$$

We apply the results in Chapter 10.5. To this end we rewrite the system (12.31) as a semilinear Cauchy problem. Let Y be the Banach space $\mathbb{R} \times L_1(\mathbb{R}_+)$. Define

$$\tilde{A}(R, u) = (-DR, Au - Du), \qquad R \in \mathbb{R}, u \in D(A).$$

$\tilde{A}$ is the generator of the C_0-semigroup

$$\tilde{S}(t)(R, u) = \big(e^{-Dt}R, e^{-Dt}S(t)u\big).$$

We define the nonlinearity F by

$$F(R, u) = \big(DR^\diamond - f(R)\|Bu\|, f(R)Bu\big).$$

Then (12.31) takes the form $(R, u)' = \tilde{A}(R, u) + F(R, u)$. We check the assumptions of Theorem 10.21. Let C be the closed convex cone $\mathbb{R}_+ \times L^1_+(\mathbb{R}_+)$. Then $F : C \to Y$ is locally Lipschitz continuous and maps bounded subsets of C into bounded subsets of Y. To see that F is locally uniformly subtangential to C, let $\Lambda > 0$ be a Lipschitz constant for f,

$$|f(R) - f(\tilde{R})| \le \Lambda|R - \tilde{R}|, \qquad R, \tilde{R} \in \mathbb{R}_+.$$

Then

$$\begin{aligned} (R, u) + hF(R, u) =& (R + hDR^\diamond - hf(R)\|Bu\|, u + hf(R)Bu) \\ \ge& (R - h\Lambda R\|B\|\,\|u\| + hDR^\diamond, u + hf(R)Bu) \ge (0, 0) \end{aligned}$$

if $h \le 1/(\Lambda\|B\|\,\|u\|)$. This implies (10.12). Obviously, $\tilde{S}(t)C \subset C$ for all $t \ge 0$. To complete checking the assumptions of Theorem 10.21, let $(R, u) : [0, \tau) \to C$ be an integral (or mild) solution of $(R, u)' = \tilde{A}(R, u) + F(R, u)$,

$$\begin{aligned} R(t) =& \breve{R}e^{-Dt} + \int_0^t e^{-D(t-s)}(DR^\diamond - f(R(s))\|Bu(s)\|)ds, \\ u(t) =& e^{-Dt}S(t)\breve{u} + \int_0^t e^{-D(t-s)}S(t - s)f(R(s))Bu(s)ds. \end{aligned} \tag{12.32}$$

From the first equation,

$$R(t) \le \breve{R}e^{-Dt} + R^\diamond.$$

In $L^1(\mathbb{R}_+)$ we have the abstract L-space property $\|x+y\| = \|x\| + \|y\|$ for $x, y \in L^1_+(\mathbb{R}_+)$; so for nonnegative solutions, by Lemma 12.6,

$$
\begin{aligned}
\|u(t)\| =& \|S(t)\breve{u}\|e^{-Dt} + \int_0^t e^{-D(t-r)} f(R(r))\|S(t-r)Bu(r)\|dr \\
=& \|\breve{u}\|e^{-Dt} + \int_0^t e^{-D(t-r)} f(R(r))\|Bu(r)\|dr.
\end{aligned} \tag{12.33}
$$

So $\|u(t)\|$ is differentiable and

$$
\frac{d}{dt}\|u(t)\| = -D\|u(t)\| + f(R(t))\|Bu(t)\|. \tag{12.34}
$$

R in (12.32) is differentiable as well and

$$
R'(t) = DR^\diamond - f(R(t))\|Bu(t)\| - DR(t).
$$

Let $M(t) = R(t) + \|u(t)\|$ be the total mass. We add the two differential equations, $M' = D(R^\diamond - M)$, and integrate,

$$
M(t) = M(0)e^{-Dt} + R^\diamond(1 - e^{-Dt}),
$$

which implies that M is bounded and so are R and $\|u(\cdot)\|$ because they are nonnegative. The following global existence result follows from Theorem 10.21.

Theorem 12.8. *There exists a continuous semiflow $\Phi : \mathbb{R}_+ \times X \to X$, $X = \mathbb{R}_+ \times L^1_+(\mathbb{R}_+)$ such that $(R(t), u(t)) = \Phi(t, (\breve{R}, \breve{u}))$ is a solution of (12.32) defined for all $t \ge 0$.*

Further, we have the following relations,

$$
\|u(t)\| \ge e^{-Dt}\|\breve{u}\|, \qquad t \ge 0,
$$

and, for $M(t) = R(t) + \|u(t)\|$,

$$
M(t) = M(0)e^{-Dt} + R^\diamond(1 - e^{-Dt}), \qquad t \ge 0.
$$

The lower estimate for $\|u(t)\|$ follows from (12.33). In particular, the cell biomass stays positive if it is positive at the beginning.

12.4. Extinction and weak persistence via Laplace transform

To formulate and prove extinction and persistence results, we introduce

$$
\begin{aligned}
\mathcal{P}(R, \lambda) =& \int_{\mathbb{R}_+} \mathcal{G}(R, a)e^{-\lambda a} m(da), \\
\mathcal{P}_0 = \mathcal{P}(R^\diamond, 0) =& \int_{\mathbb{R}_+} \mathcal{G}(R^\diamond, a) m(da)
\end{aligned} \tag{12.35}
$$

with

$$
\mathcal{G}(R, a) = \exp\Big(\int_0^a \kappa(s) f(R) ds - Da\Big). \tag{12.36}
$$

$\mathcal{P}(R,0)$ is a measure of a cell's biomass production at resource level R, and $\mathcal{P}_0$ the *basic biomass production number* since it gives the expected amount of biomass produced by a unit of biomass of age zero introduced into the chemostat at the washout equilibrium where $R = R^\diamond,\ c = 0$. The expression $\exp\Big(\int_0^a \kappa(s) f(R^\diamond)ds\Big)$ is the amplification factor of a cell's biomass from age 0 to age a provided the cell has neither been washed out nor divided up to that age. The factor e^{-Da} is the probability this unit biomass has not washed out before age a and $m(da)$ is the probability of cell division in $[a, a + da)$.

Theorem 12.9. *Let $\mathcal{P}_0 < 1$ and*

$$\operatorname*{ess-sup}_{s\geq 0} \int_s^\infty \frac{\mathcal{G}(R^\diamond, a)\mathcal{F}(a)}{\mathcal{G}(R^\diamond, s)\mathcal{F}(s)} da < \infty. \tag{12.37}$$

Then the cell population goes extinct in the sense that $\int_0^\infty \int_0^\infty c(t,a)dtda < \infty$ and $\int_0^\infty c(t,a)da \to 0$ as $t \to \infty$.

Condition (12.37) means that, at high cell-age, cell division is faster than biomass acquisition.

Proof. For $\lambda > 0$, we take the vector-valued Laplace transform of (12.32),

$$\hat{u}(\lambda) = \mathcal{L}\{u\}(\lambda) = \int_0^\infty e^{-\lambda t} u(t)dt,$$

which exists because $\lambda > 0$ and u is bounded,

$$\hat{u}(\lambda) = \hat{S}(\lambda + D)\breve{u} + \hat{S}(\lambda + D)\mathcal{L}\{f(R)Bu\}(\lambda). \tag{12.38}$$

By Lemma 12.7, $\frac{[\hat{u}(\lambda)](a)}{\mathcal{F}(a)}$ is locally absolutely continuous and

$$\begin{aligned} \Big(\lambda + D + \frac{d}{da}\Big)\frac{[\hat{u}(\lambda)]}{\mathcal{F}} &= \frac{\breve{u}}{\mathcal{F}} + \frac{\mathcal{L}\{f(R)Bu\}(\lambda)}{\mathcal{F}}, \\ [\hat{u}(\lambda)](0) &= \int_{\mathbb{R}_+} [\hat{u}(\lambda)](a)\frac{m(da)}{\mathcal{F}(a)}. \end{aligned} \tag{12.39}$$

Since, by Theorem 12.8,

$$R(t) \leq M(t) = R^\diamond + (\breve{R} - R^\diamond)e^{-Dt}$$

and f is increasing and locally Lipschitz continuous, there exists some $N > 0$ such that

$$f(R(t)) \leq f(R^\diamond) + Ne^{-Dt}$$

and

$$\mathcal{L}\{f(R)Bu\}(\lambda) \leq f(R^\diamond)B\hat{u}(\lambda) + NB\hat{u}(\lambda + D).$$

We substitute this inequality into (12.39) and spell out the new inequality more concretely,

$$
\begin{aligned}
&\Big(\lambda + D + \frac{d}{da}\Big)\frac{[\hat{u}(\lambda)](a)}{\mathcal{F}(a)}\\
&\le \frac{\breve{u}(a)}{\mathcal{F}(a)} + f(R^\diamond)\kappa(a)\frac{[\hat{u}(\lambda)](a)}{\mathcal{F}(a)} + N\frac{[B\hat{u}(\lambda+D)](a)}{\mathcal{F}(a)},\\
&[\hat{u}(\lambda)](0) = \int_{\mathbb{R}_+}[\hat{u}(\lambda)](a)\frac{m(da)}{\mathcal{F}(a)}.
\end{aligned}
\tag{12.40}
$$

We integrate this differential inequality and drop $\lambda > 0$ wherever possible. By (12.36),

$$
\begin{aligned}
\frac{[\hat{u}(\lambda)](a)}{\mathcal{F}(a)} \le & [\hat{u}(\lambda)](0)\mathcal{G}(R^\diamond, a) + g_\lambda(a),\\
g_\lambda(a) = & \int_0^a \Big(\frac{\breve{u}(s)}{\mathcal{F}(s)} + N\frac{[B\hat{u}(\lambda+D)](s)}{\mathcal{F}(s)}\Big)\frac{\mathcal{G}(R^\diamond, a)}{\mathcal{G}(R^\diamond, s)}ds.
\end{aligned}
\tag{12.41}
$$

We substitute the inequality (12.41) into the initial condition of (12.39). By (12.35),

$$
[\hat{u}(\lambda)](0) \le [\hat{u}(\lambda)](0)\ \mathcal{P}_0 + \int_{\mathbb{R}_+} g_\lambda(a)m(da).
$$

Since $\lambda > 0$ and by assumption, $\mathcal{P}(R^\diamond, \lambda) \le \mathcal{P}_0 < 1$ and we can solve for $[\hat{u}(\lambda)](0)$,

$$
[\hat{u}(\lambda)](0) \le \frac{1}{1-\mathcal{P}_0}\int_{\mathbb{R}_+} g_\lambda(a)m(da).
\tag{12.42}
$$

We aim at taking the limit of this inequality for $\lambda \to 0$. As for g_λ, we change the order of integration,

$$
\begin{aligned}
&\int_{\mathbb{R}_+} g_\lambda(a)m(da)\\
&= \int_0^\infty \Big(\frac{\breve{u}(s)}{\mathcal{F}(s)} + N\frac{[B\hat{u}(\lambda+D)](s)}{\mathcal{F}(s)}\Big)\Big(\int_{[s,\infty)}\frac{\mathcal{G}(R^\diamond, a)}{\mathcal{G}(R^\diamond, s)}m(da)\Big)ds.
\end{aligned}
$$

We use Beppo Levi's theorem of monotone convergence, integrate by parts, and obtain for almost all $s \ge 0$,

$$
\begin{aligned}
&\int_{[s,\infty)}\frac{\mathcal{G}(R^\diamond, a)}{\mathcal{G}(R^\diamond, s)}m(da) = \lim_{\alpha\to\infty}\int_{[s,\alpha]}\frac{\mathcal{G}(R^\diamond, a)}{\mathcal{G}(R^\diamond, s)}m(da)\\
&= \mathcal{F}(s) + \lim_{\alpha\to\infty}\Big(-\frac{\mathcal{G}(R^\diamond, \alpha)}{\mathcal{G}(R^\diamond, s)}\mathcal{F}(\alpha) + \int_s^\alpha \frac{\partial_a\mathcal{G}(R^\diamond, a)}{\mathcal{G}(R^\diamond, s)}\mathcal{F}(a)da\Big).
\end{aligned}
$$

Here $\partial_a \mathcal{G}(R^\diamond, a)$ is the partial derivative of $\mathcal{G}$ with respect to a. We take the partial derivative with respect to a in (12.36) and drop a nonpositive term,

$$\frac{1}{\mathcal{F}(s)} \int_{[s,\infty)} \frac{\mathcal{G}(R^\diamond, a)}{\mathcal{G}(R^\diamond, s)} m(da) \leq 1 + \int_s^\infty \big[\kappa(a) f(R^\diamond) + D\big] \frac{\mathcal{G}(R^\diamond, a)}{\mathcal{G}(R^\diamond, s)} \frac{\mathcal{F}(a)}{\mathcal{F}(s)} da.$$

By assumption (12.37), this is a bounded function of s and so there exists some constant $k_1 > 0$ such that, for all $\lambda > 0$,

$$\int_{\mathbb{R}_+} g_\lambda(a) m(da) \leq k_1 \big(\|\breve{u}\| + N \|B\hat{u}(D+\lambda)\| \big). \tag{12.43}$$

By Beppo Levi's theorem of monotone convergence and (12.39),

$$\begin{aligned}
[\hat{u}(0)](0) &= \int_{\mathbb{R}_+} \Big(\int_0^\infty c(t,a) dt \Big) \frac{m(da)}{\mathcal{F}(a)} \\
&= \lim_{\lambda \to 0} \int_{\mathbb{R}_+} \Big(\int_0^\infty e^{-\lambda t} c(t,a) dt \Big) \frac{m(da)}{\mathcal{F}(a)} = \lim_{\lambda \to 0} [\hat{u}(\lambda)](0).
\end{aligned}$$

We take the limit of (12.42) for $\lambda \to 0$ and use (12.43),

$$[\hat{u}(0)](0) \leq \big(\|\breve{u}\| + N \|B\hat{u}(D)\| \big) \frac{k_1}{1 - \mathcal{P}_0} < \infty.$$

We also take the limit $\lambda \to 0$ in (12.41),

$$\begin{aligned}
[\hat{u}(0)](a) \leq & [\hat{u}(0)](0)\, \mathcal{G}(R^\diamond, a) \mathcal{F}(a) \\
& + \int_0^a \big(\breve{u}(s) + N[B\hat{u}(D)](s) \big) \frac{\mathcal{F}(a)}{\mathcal{F}(s)} \frac{\mathcal{G}(R^\diamond, a)}{\mathcal{G}(R^\diamond, s)} ds.
\end{aligned}$$

By (12.37) and a change in the order of integration,

$$\int_{\mathbb{R}_+^2} c(t,a) dt da = \int_0^\infty \|u(t)\| ds = \|\hat{u}(0)\| < \infty.$$

By (12.33) and a version of the Lebesgue-Fatou theorem [**217**, B.2],

$$\|u(t)\| \leq e^{-Dt} \|\breve{u}\| + \int_0^t e^{-D(t-r)} f(R(r)) \|B\| \|u(r)\| dt \longrightarrow 0, \qquad t \to \infty.$$

□

The Laplace transform also helps us to establish a condition for uniform weak persistence of the cell population. Recall (12.35).

Theorem 12.10. *Let*

$$1 < \mathcal{P}_0 = \int_{\mathbb{R}_+} \mathcal{G}(R^\diamond, a) m(da).$$

Then the cell population persists uniformly weakly, i.e., there exists some $\epsilon > 0$ such that

$$\limsup_{t \to \infty} \int_0^\infty c(t,a) da \geq \epsilon$$

for every solution with $\int_0^\infty c(0,a)da > 0$.

Proof. Since $\mathcal{P}_0 > 1$, we can choose $\epsilon \in (0, R^\diamond)$ such that

$$1 < \int_{\mathbb{R}_+} \mathcal{G}(R^\diamond - \epsilon, a) m(da). \tag{12.44}$$

By Theorem 12.8, there exists some $t_0 > 0$ such that $M(t) \geq R^\diamond - (\epsilon/2)$ for all $t \geq t_0$. If the cell population does not persist weakly uniformly, then there exists a solution with $0 \neq u(0) \in L^1_+(\mathbb{R}_+)$ and $\|u(t)\| \leq \epsilon/2$ for all $t \geq t_1$ with an appropriate t_1. By translating forward in time, we can assume that $\|u(t)\| \leq \epsilon/2$ for all $t \geq 0$ and $R(t) \geq R^\diamond - \epsilon$. As in the proof of Theorem 12.9, $\frac{[\hat{u}(\lambda)](a)}{\mathcal{F}(a)}$ is locally absolutely continuous and satisfies (12.39),

$$\Big(\lambda + D + \frac{d}{da}\Big)\frac{[\hat{u}(\lambda)]}{\mathcal{F}} = \frac{\breve{u}}{\mathcal{F}} + \frac{\mathcal{L}\{f(R)Bu\}(\lambda)}{\mathcal{F}},$$
$$[\hat{u}(\lambda)](0) = \int_{\mathbb{R}_+} [\hat{u}(\lambda)](a)\frac{m(da)}{\mathcal{F}(a)}.$$

Since $R(t) \geq R^\diamond - \epsilon$ for all $t \geq 0$, $\mathcal{L}\{f(R)Bu\}(\lambda) \geq f(R^\diamond - \epsilon)B\hat{u}(\lambda)$. We substitute this inequality into the previous equation and spell out the resulting inequality more concretely,

$$\begin{aligned} &\Big(\lambda + D + \frac{d}{da}\Big)\frac{[\hat{u}(\lambda)](a)}{\mathcal{F}(a)} \geq \frac{\breve{u}(a)}{\mathcal{F}(a)} + \kappa(a) f(R^\diamond - \epsilon)\frac{[\hat{u}(\lambda)](a)}{\mathcal{F}(a)}, \\ &[\hat{u}(\lambda)](0) = \int_{\mathbb{R}_+} [\hat{u}(\lambda)](a)\frac{m(da)}{\mathcal{F}(a)}. \end{aligned} \tag{12.45}$$

We integrate this differential inequality. Since $0 \neq \breve{u} \in L^1_+(\mathbb{R}_+)$, $\hat{u}(\lambda) \neq 0$ and

$$\frac{[\hat{u}(\lambda)(a)]}{\mathcal{F}(a)} \geq [\hat{u}(\lambda)](0)\mathcal{G}(R^\diamond - \epsilon, a)e^{-\lambda a}. \tag{12.46}$$

Notice from (12.38) and Lemma 12.6 that

$$[\hat{u}(\lambda)](0) \geq \int_{\mathbb{R}_+} [\hat{S}(\lambda + D)\breve{u}](a)\frac{m(da)}{\mathcal{F}(a)} = \hat{b}(\lambda + D), \tag{12.47}$$

where b is the solution of (12.24) with $\breve{u}$ replacing g. By Proposition 12.4, $\hat{b}(\lambda + D)$ is a positive number if $0 \neq \breve{u} \in L^1_+(\mathbb{R}_+)$. We substitute the inequality (12.46) into the initial condition of (12.40), divide by $[\hat{u}(\lambda)](0)$ and obtain

$$1 \geq \int_{\mathbb{R}_+} \mathcal{G}(R^\diamond - \epsilon, a)e^{-\lambda a} m(da).$$

This holds for all $\lambda > 0$, and thus for $\lambda = 0$ by Beppo Levi's theorem of monotone convergence, a contradiction to (12.44). □

We announce the following condition for uniform persistence of the cell population.

Theorem 12.11. *Let $\mathcal{P}_0 > 1$. Then the cell population persists uniformly (i.e., there exists some $\epsilon > 0$ such that*

$$\liminf_{t\to\infty} \int_0^\infty c(t,a)da \geq \epsilon$$

for every solution with $\int_0^\infty c(0,a)da > 0$*),*

provided that one of the following two assumptions are satisfied:

(i) $\dfrac{\mathcal{F}(a+s)}{\mathcal{F}(s)} \to 0$ *as* $s \to \infty$ *for every* $a > 0$*,*

or

(ii) $\mathcal{D}$ *is bounded.*

Because it is difficult to establish the existence of a compact attractor for the cell division model, the proof of this theorem is postponed to Section 13.4 in Chapter 13 after we will have learned about yet another way in which uniform weak persistence implies uniform persistence. This approach will also work for time-heterogeneous problems.

12.5. Exercises

Exercise 12.1. Verify Remark 12.1.

Chapter 13

Persistence for Nonautonomous Dynamical Systems

All natural systems are subject to time-dependent forces, like the diurnal rhythm of day and night or the yearly seasons or weather patterns that change from year to year. Time-dependent forces lead to nonautonomous semiflows which have been neglected so far in our development of persistence theory.

Even in laboratory systems one may be interested in simulating time-heterogeneous environmental conditions. The simple chemostat is an example that is easy enough to establish an explicit large-time lower bound for the concentration of the consumer species. When this is possible, one sometimes speaks about *practical persistence.* The upcoming example does not require an elaborate dynamical systems theory. For more sophisticated manifestations of practical persistence see [**24, 25, 26, 28, 27, 35**].

13.1. The simple chemostat with time-dependent washout rate

A resource of mass R is fed into and washed out of a chemostat at time-dependent rates which are proportional to each other. The target mass, without a consumer, is denoted by $R^\diamond > 0$. On this resource grows a consumer with biomass C which is removed from the chemostat at the same rate as the resource. The washout (or dilution) rate $D : [r_0, \infty) \to \mathbb{R}_+$ is a given function of time,

$$\left\{\begin{array}{rcl} R' &=& D(t)(R^\diamond - R) - f(R)C \\ C' &=& (f(R) - D(t))C \end{array}\right\}, \quad t \geq r, \tag{13.1}$$

with initial data

$$r \geq r_0 \text{ and } R(r) = R_0,\ C(r) = C_0.$$

The growth function $f : \mathbb{R}_+ \to \mathbb{R}$ is strictly increasing and locally Lipschitz continuous, $f(0) = 0$. We have scaled out the yield constant, a positive number multiplying $f(R)$ in the equation for R.

For reasons that will become clear later, we assume that D is Borel measurable, locally integrable, and

$$\int_{r_0}^\infty D(t)dt = \infty. \tag{13.2}$$

This means that R and C are locally absolutely continuous (Definition A.6) and that the differential equations are satisfied for almost all times. The solutions are found in the usual way, by applying the contraction mapping theorem (Banach's fixed point theorem) to the integral equations obtained by integrating (13.1).

We consider the total biomass

$$M = R + C, \tag{13.3}$$

which satisfies the differential equation

$$M' = D(t)(R^\diamond - M). \tag{13.4}$$

We obtain the solution

$$M(t) = R^\diamond + (R_0 - R^\diamond)\frac{\mathcal{F}(t)}{\mathcal{F}(r)}, \tag{13.5}$$

where

$$\mathcal{F}(t) = \exp\Big(-\int_{r_0}^t D(s)ds\Big), \qquad t \geq r_0, \tag{13.6}$$

and $\mathcal{F}(t)/\mathcal{F}(r) = \exp\big(-\int_r^t D(s)ds\big)$ is the probability of not being washed out between time r and time t. By (13.2),

$$M(t) \to R^\diamond \quad \text{as } t \to \infty. \tag{13.7}$$

We integrate the consumer equation

$$\frac{1}{t}\ln\frac{C(t+r)}{C_0} = \frac{1}{t}\int_0^t [f(R(r+s)) - D(r+s)]ds.$$

Since $R(t) \le M(t)$ and f is increasing,

$$\limsup_{t\to\infty} \frac{1}{t} \ln \frac{C(t+r)}{C_0} \le f(R^\diamond) - \liminf_{t\to\infty} \frac{1}{t} \int_0^t D(r+s)]ds.$$

We obtain the following condition for consumer extinction.

Theorem 13.1. *Let*

$$f(R^\diamond) < \liminf_{t\to\infty} \frac{1}{t} \int_0^t D(r+s)ds.$$

Then $C(t) \to 0$ as $t \to \infty$.

An example for practical persistence. For the simple chemostat it is possible to derive an explicit though complicated large-time lower bound.

Theorem 13.2. *Assume that there exists a strictly increasing sequence (t_j) such that $t_0 = r_0$, $t_j \to \infty$ as $j \to \infty$, and*

$$\sup_{j\in\mathbb{N}}(t_j - t_{j-1}) < \infty, \qquad \inf_{j\in\mathbb{N}}\Big(f(R^\diamond)(t_j - t_{j-1}) - \int_{t_{j-1}}^{t_j} D(s)ds\Big) > 0.$$

Then the consumer is uniformly persistent in the following sense: There exists some $\epsilon_0 > 0$ such that for all $r \ge r_0$ and all solutions (R, C) of (13.1) on $[r, \infty)$ with $R(r) \ge 0, C(r) > 0$,

$$\liminf_{t\to\infty} C(t) \ge \epsilon_0.$$

For the value of ϵ_0 see the last line of the subsequent proof.

Proof. By (13.1) and (13.3),

$$\frac{C'}{C} = f(M - C) - D(t), \qquad t \ge r.$$

In the following,

$$f^\diamond := f(R^\diamond) > 0.$$

Let $\epsilon \in (0, f^\diamond)$. By (13.7), there exists some $\tilde r \ge r$ such that $|M(s) - R^\diamond| < \epsilon/\Lambda$ for all $s \ge \tilde r$ where $\Lambda > 0$ is the Lipschitz constant of f on $[0, R^\diamond]$. Since f is increasing,

$$\frac{C'}{C} \ge f\Big(R^\diamond - \frac{\epsilon}{\Lambda} - C\Big) - D(t) \ge f^\diamond - \epsilon - \Lambda C - D(t), \qquad t \ge \tilde r.$$

This is a differential inequality of Bernoulli type to which we apply the Leibniz transformation $N = 1/C$,

$$N' \le \Lambda + N(D(t) + \epsilon - f^\diamond), \qquad t \ge \tilde r.$$

We use an integrating factor,

$$\begin{aligned} N(t) \leq & N(\tilde{r}) \exp\Big(\int_{\tilde{r}}^{t} D(s)ds + (\epsilon - f^{\diamond})(t - \tilde{r})\Big) \\ & + \Lambda \int_{\tilde{r}}^{t} \exp\Big(\int_{u}^{t} D(s)ds + (\epsilon - f^{\diamond})(t - u)\Big) du. \end{aligned}$$

Let

$$c := \sup_{j \in \mathbb{N}} (t_j - t_{j-1}), \qquad \delta := \inf_{j \in \mathbb{N}} \Big(f^{\diamond}(t_j - t_{j-1}) - \int_{t_{j-1}}^{t_j} D(s)ds \Big).$$

By assumption, $c < \infty$ and $\delta > 0$. Choose k and n such that $t_k \leq \tilde{r} < t_{k+1}$ and $t_n \leq t \leq t_{n+1}$. As $t \to \infty$, $n \to \infty$. We enlarge the integrals by changing the integration limits,

$$\begin{aligned} N(t) \leq & N(\tilde{r}) \exp\Big(\int_{t_k}^{t_{n+1}} D(s)ds + (\epsilon - f^{\diamond})(t_{n+1} - t_k)\Big) \\ & \exp\Big(f^{\diamond}(\tilde{r} - t_k) + f^{\diamond}(t_{n+1} - t)\Big) \\ & + \Lambda \int_{t_k}^{t_{n+1}} \exp\Big(\int_{u}^{t_{n+1}} D(s)ds + (\epsilon - f^{\diamond})(t_{n+1} - u)\Big) \\ & \exp(f^{\diamond}(t_{n+1} - t)) du. \end{aligned}$$

We split up some of the integrals and define

$$a_{n,k} = \int_{t_k}^{t_{n+1}} D(s)ds + (\epsilon - f^{\diamond})(t_{n+1} - t_k).$$

This yields

$$\begin{aligned} N(t) \leq & N(\tilde{r}) e^{a_{n,k}} e^{2f^{\diamond}c} \\ & + \Lambda \sum_{j=k}^{n} \int_{t_j}^{t_{j+1}} \exp\Big(\int_{u}^{t_{n+1}} D(s)ds + (\epsilon - f^{\diamond})(t_{n+1} - u)\Big) e^{f^{\diamond}c} du. \end{aligned}$$

We replace u by t_j in the second summand and make the appropriate adjustments,

$$N(t) \leq N(\tilde{r}) e^{a_{n,k}} e^{2f^{\diamond}c} + \Lambda e^{2f^{\diamond}c} \sum_{j=k}^{n} c e^{a_{n,j}}.$$

In the definition of $a_{n,k}$, we split up the integrals and telescope,

$$a_{n,k} = \sum_{j=k}^{n} \Big(\int_{t_j}^{t_{j+1}} D(s)ds + (\epsilon - f^{\diamond})(t_{j+1} - t_j) \Big).$$

By the definitions of c and δ above, $a_{n,k} \le (n-k)(\epsilon c - \delta)$. We substitute these inequalities into the ones for $N(t)$,

$$N(t) \le N(\tilde{r})e^{(n-k)(\epsilon c-\delta)} + \Lambda e^{2f^\diamond c} c \sum_{j=k}^{n} e^{(n-j)(\epsilon c-\delta)}.$$

Let $c\epsilon < \delta$. Since $n \to \infty$ as $t \to \infty$, with $N^\infty = \limsup_{t\to\infty} N(t)$,

$$N^\infty \le \frac{\Lambda c}{1 - e^{c\epsilon-\delta}} e^{2f^\diamond c}.$$

Since this holds for any sufficiently small $\epsilon > 0$, $N^\infty \le \frac{\Lambda c}{1-e^{-\delta}} e^{2f^\diamond c}$. We recall $C = 1/N$ and $f^\diamond = f(R^\diamond)$,

$$\liminf_{t\to\infty} C(t) \ge \frac{1-e^{-\delta}}{\Lambda c e^{2f(R^\diamond)c}}. \qquad \square$$

The following condition compares more readily to the consumer extinction condition in Theorem 13.1.

Corollary 13.3. *Assume that there is some $t > 0$ such that*

$$f(R^\diamond) > \frac{1}{t} \sup_{r\ge r_0} \int_0^t D(s+r)ds.$$

Then the consumer persists uniformly.

Proof. Define $t_n = r_0 + nt$. $\square$

If the consumer persists, then its large-time behavior is independent of the initial conditions.

Theorem 13.4. *Let the assumptions of Theorem 13.2 be satisfied. Let $r_1, r_2 \ge r_0$ and R_j, C_j, $j = 1,2$, be solutions of (13.1) on $[r_j, \infty)$ with respective initial data $\breve{R}_j > 0, \breve{C}_j > 0$ at r_j. Then $\frac{C_1(t)}{C_2(t)} \to 1$ as $t \to \infty$.*

Proof. As we have seen before, with $M_j = R_j + C_j$,

$$\frac{C_j'}{C_j} = f(M_j(t) - C_j) - D(t), \qquad M_j(t) \overset{t\to\infty}{\longrightarrow} R^\diamond.$$

Further, $C_j^\infty \le R^\diamond$. We define $y = C_1/C_2$. Then $0 < y(t)$ for all $t \ge s = \max\{r_j\}$ and

$$\frac{y'}{y} = \frac{C_1'}{C_1} - \frac{C_2'}{C_2} = f(M_1 - C_1) - f(M_2 - C_2) = f(M_1 - yC_2) - f(M_2 - C_2).$$

Notice that the right-hand side of this differential equation is continuous. So y is continuously differentiable, though C_1 and C_2 may be only locally absolutely continuous. Since C_2 is bounded away from 0 by Theorem 13.2, we can assume that $C_2(t) \geq \delta > 0$ for all $t \geq s$. If $y^\infty < 1$, then

$$\liminf_{t\to\infty} \frac{y'(t}{y(t)} \geq \inf_{\delta \leq C \leq R^\diamond} [f(R^\diamond - y^\infty C) - f(R^\diamond - C)] > 0$$

and y grows exponentially. So $y^\infty \geq 1$. If $y_\infty < 1$, by one version of the fluctuation method (Lemma A.13), there exists a sequence $s_n \to \infty$ with $y'(s_n) = 0$ and $y(s_n) \to y_\infty$. So

$$0 = \lim_{n\to\infty} [f(R^\diamond - y_\infty C_2(t_n)) - f(R^\diamond - C_2(t_n))] > 0,$$

again a contradiction. This implies that $\liminf_{t\to\infty} C_1(t)/C_2(t) \geq 1$. By symmetry $\liminf_{t\to\infty} C_2(t)/C_1(t) \geq 1$ as well. □

A similar result for a slightly different chemostat model has been proved in [**66**].

13.2. General time-heterogeneity

Even for the simple chemostat, explicit large-time lower bounds for the consumer species are complicated. While they may provide some insight into how the lower bound depends on key parameters of the model, the quantitative information that is obtained is presumably very rough. For general systems, we therefore restrict ourselves to qualitative persistence results.

One possible line of approach is trying to embed the time-heterogeneous problem into a time-homogeneous one using skew product flows; this makes it possible to use compact attractors and acyclicity methods [**169, 238**]. Here it will be shown that uniform weak implies uniform persistence in a similar though technically more onerous way as in the time-homogeneous case [**216**]. Because of the technicalities involved we restrict ourselves to the easier continuous time situation where we do not need to deal with the extra complications created by jumps.

Let X be a set, $r_0 \in \mathbb{R}$, and

$$\Delta = \{(t,s) \in \mathbb{R}^2; r_0 \leq s \leq t < \infty\}.$$

A mapping $\Psi : \Delta \times X \to X$ is called a *(nonautonomous) semiflow* on X (anchored at r_0) if

$$\left.\begin{aligned} \Psi(t,s,\Psi(s,r,x)) &= \Psi(t,r,x) \\ \Psi(r,r,x) &= x \end{aligned}\right\} \left\{ \begin{aligned} t \geq s \geq r \geq r_0, \\ x \in X. \end{aligned}\right.$$

If X is a topological space and the mapping Ψ is continuous and a semiflow, Ψ is called a *continuous semiflow*. The semiflows we will consider are not necessarily continuous.

In this context, the semiflow we have considered in the previous chapters are *autonomous semiflows*: $\Psi(t+r, r, x)$ does not depend on $r \geq r_0$ for $t \geq 0, x \in X$.

Let

$$\rho : X \to [0, \infty)$$

be a nonnegative function on X.

Definition 13.5. The semiflow Ψ is called *uniformly ρ-persistent* if there exists some $\epsilon > 0$ such that

$$\liminf_{t\to\infty} \rho(\Psi(t, r, x)) > \epsilon \quad \text{whenever} \quad r \geq r_0, x \in X, \rho(x) > 0.$$

Ψ is called *uniformly weakly ρ-persistent* if there exists some $\epsilon > 0$ such that

$$\limsup_{t\to\infty} \rho(\Psi(t, r, x)) > \epsilon \quad \text{whenever} \quad r \geq r_0, x \in X, \rho(x) > 0.$$

Theorem 13.6. *The semiflow Ψ is uniformly ρ-persistent if it is uniformly weakly ρ-persistent and the subsequent conditions are satisfied.*

$\diamondsuit$ *For all $x \in X$ and $r \geq r_0$, $\rho(\Psi(t, r, x))$ is continuous in $t \geq r$.*

Further, there exist some $\epsilon_0 > 0$ and a sequence (B_k) of subsets of X with the following properties:

$\heartsuit$ *If $x \in X$, $\rho(x) > 0$, and $k \in \mathbb{N}$, $r \geq r_0$, then there exists some $t_k \geq r$ such that $\Psi(t, r, x) \in B_k$ for all $t \geq t_k$.*

$\spadesuit$ *Whenever (y_k) is a sequence in X and (s_k) is a sequence in $\mathbb{R}$ such that $r_0 \leq s_k \to \infty$ as $k \to \infty$ and $\Psi(t + s_k, s_k, y_k) \in B_k$ for all $k \in \mathbb{N}$ and $t \geq 0$ and $0 < \rho(y_k) = \rho(y_1) \leq \epsilon_0$ for all $k \in \mathbb{N}$ we have the following:*

- $\spadesuit_1$ *the continuity of $\rho(\Psi(t + s_k, s_k, y_k))$ in $t \geq 0$ holds uniformly in k (possibly after choosing subsequences).*
- $\spadesuit_2$ *if $\tau \in (0, \infty)$ and $\sigma(t) = \lim_{k\to\infty} \rho(\Psi(t + s_k, s_k, y_k))$ exists uniformly for t in $[0, \tau]$ and $\sigma(t) \leq \sigma(0) = \sigma(\tau)$ for all $t \in [0, \tau]$, then $\sigma(s) > 0$ for all $s \in (0, \tau)$.*
- $\spadesuit_3$ *if $\sigma(t) = \lim_{k\to\infty} \rho(\Psi(t + s_k, s_k, y_k))$ exists uniformly for t in all bounded subintervals of $\mathbb{R}_+$, then $\sigma(s) > \sigma(0)$ for some $s \geq 0$.*

Remark 13.7. Notice that, when checking $\spadesuit_{2,3}$, we automatically have $\sigma(0) = \rho(y_1) \in (0, \epsilon_0]$.

$\spadesuit_1$ means that for all $\epsilon > 0$ and $t \geq 0$ there exists some $\delta > 0$ such that $|\rho(t + s_k, s_k, y_k) - \rho(r + s_k, s_k, y_k)| < \epsilon$ for all $k \in \mathbb{N}$ and all $r \geq 0$ with $|t - r| < \delta$. This property is sometimes referred to as equicontinuity.

Proof. Choose $\epsilon_0 > 0$ according to the assumptions. Since Ψ is uniformly weakly ρ-persistent, there exists some $\epsilon_1 > 0$ such that

$$\limsup_{t\to\infty} \rho(\Psi((t+r,r,x)) > \epsilon_1 \text{ whenever } r \ge r_0, x \in X, \rho(x) > 0.$$

Let $\epsilon_2 \in (0, \min\{\epsilon_0, \epsilon_1\})$. Suppose that Ψ is not uniformly ρ-persistent. Then, for every $k \in \mathbb{N}$, there exists some $r_k \ge r_0$ and some $x_k \in X$ such that

$$\liminf_{t\to\infty} \rho(\Psi(t+r_k, r_k, x_k)) < \frac{\epsilon_2}{k}, \qquad \rho(x_k) > 0.$$

This implies that, for each k, $\rho(\Psi(t + r_k, r_k, x_k))$ as a function of t keeps oscillating between ϵ_2/k and ϵ_1. So we can find numbers $s_k \ge r_k$ and $t_k, u_k \ge 0$ such that, for all $k \in \mathbb{N}$,

$$\begin{aligned} \rho(\Psi(s_k, r_k, x_k)) &= \epsilon_2 = \rho(\Psi(u_k + t_k + s_k, r_k, x_k)), \\ \rho(\Psi(t_k + s_k, r_k, x_k)) &= \epsilon_2/k, \\ \rho(\Psi(s + s_k, r_k, x_k)) &\le \epsilon_2, \ 0 \le s \le t_k + u_k. \end{aligned}$$

Since we can choose s_k arbitrarily large, we can arrange that $s_k \to \infty$ and $\Psi(t + s_k, r_k, x_k) \in B_k$ for all $t \ge 0$. We set $y_k = \Psi(s_k, r_k, x_k)$. By the semiflow property, for $k \in \mathbb{N}$

$$\begin{aligned} \Psi(t + s_k, s_k, y_k) &\in B_k, \qquad t \ge 0, \\ \rho(y_k) &= \epsilon_2 = \rho(\Psi(u_k + t_k + s_k, s_k, y_k)), \\ \rho(\Psi(t_k + s_k, s_k, y_k)) &= \epsilon_2/k, \\ \rho(\Psi(s + s_k, s_k, y_k)) &\le \epsilon_2, \qquad 0 \le s \le t_k + u_k. \end{aligned}$$

We claim that $t_k + u_k \to \infty$ as $k \to \infty$. If not, after choosing subsequences, $u_k \to \tilde{u}$ and $t_k \to \tilde{t}$ as $k \to \infty$. By $\spadesuit_1$ and the Arzela-Ascoli Theorem [**164**, Ch.8.3],

$$\sigma(s) = \lim_{k\to\infty} \rho(\Psi(s + s_k, s_k, y_k))$$

exists uniformly in $s \in [0, \tilde{u} + \tilde{t} + 1]$. This implies that $\sigma(0) = \epsilon_2$, $\sigma(\tilde{t}) = 0$, $\sigma(\tilde{t} + \tilde{u}) = \epsilon_2$, and $\sigma(s) \le \epsilon_2 < \epsilon_0$ for all $s \in [0, \tilde{s} + \tilde{t}]$, a contradiction to $\spadesuit_2$.

So $u_k + t_k \to \infty$ as $k \to \infty$. Again by the Arzela-Ascoli Theorem [**164**, Ch.8.3], after choosing subsequences, the limit

$$\sigma(s) = \lim_{k\to\infty} \rho(\Psi(s + s_k, s_k, y_k))$$

exists uniformly for s in every bounded subinterval of $\mathbb{R}_+$. This implies that $\sigma(0) = \epsilon_2 > 0$ and $\sigma(s) \le \epsilon_2$ for all $s \in \mathbb{R}_+$. This contradicts $\spadesuit_3$. □

13.3. Periodic and asymptotically periodic semiflows

Definition 13.8. A semiflow $\Psi : \Delta \times X \to X$ is called periodic, with period $\eta > 0$, if

$$\Psi(t+\eta, r+\eta, x) = \Psi(t, r, x), \qquad t \ge r \ge r_0. \tag{13.8}$$

Theorem 13.9. *Let $\Psi : \Delta \times X \to X$ be a periodic semiflow, X a metric space with metric d and Ψ continuous. Let $\rho : X \to \mathbb{R}_+$ be uniformly continuous.*

Then Ψ is uniformly ρ-persistent whenever it is uniformly weakly ρ-persistent and a closed subset B of X exists with the following properties:

(a) *For all $x \in X$, $\rho(x) > 0$, $r \ge r_0$, $d(\Psi(t, r, x), B) \to 0$ as $t \to \infty$.*
(b) *If $0 < \epsilon_1 < \epsilon_2 < \infty$, the intersection $B \cap \{\epsilon_1 \le \rho(x) \le \epsilon_2\}$ is compact.*
(c) *If $y \in B$ and $\rho(y) > 0$, there exist no $s \ge r_0$, $t > r > 0$ such that $\rho(\Psi(r+s, s, y)) = 0$ and $\rho(\Psi(t+s, s, y)) > 0$.*

Proof. Since Ψ is uniformly weakly ρ-persistent, there exists some $\epsilon_0 > 0$ such that $\limsup_{t\to\infty} \rho(\Psi(t, r, x)) > \epsilon_0$ for all $x \in X$, $\rho(x) > 0$ and $r \ge r_0$.

For $k \in \mathbb{N}$, we set $B_k = \{x \in X; d(x, B) < 1/k\}$.

We check the assumptions of Theorem 13.6. $\diamondsuit$ and $\heartsuit$ are satisfied. To check $\spadesuit$, let (y_k) be a sequence in X and (s_k) a sequence in $\mathbb{R}$ such that $r_0 \le s_k \to \infty$ as $k \to \infty$ and $y_k \in B_k$ and $0 < \rho(y_k) = \rho(y_1) \le \epsilon_0$ for all $k \in \mathbb{N}$. By choice of B_k, there exists a sequence (z_k) in B such that $d(y_k, z_k) < 1/k$ for all $k \in \mathbb{N}$. Since ρ is uniformly continuous, $|\rho(y_k) - \rho(z_k)| \to 0$ as $k \to \infty$. By Assumption (b), (z_k) has a convergent subsequence with limit in the closed set B. So, after choosing subsequences, there exists some $y \in B$ such that $y_k \to y$ as $k \to \infty$.

In considering $\rho(\Psi(t + s_k, s_k, y_k))$, by Definition 13.8 we can assume that $s_k \in [r_0, r_0 + \eta]$. Let $K = \{y_k : k \in \mathbb{N}\} \cup \{y\}$, $T > 0$. Then $(t, s, z) \mapsto \rho(\Psi(t+s, s, z))$ is uniformly continuous on the compact set $[0, T] \times [0, \eta] \times K$. This implies $\spadesuit_1$.

As for $\spadesuit_{2,3}$ we have that

$$\sigma(t) = \rho(\Psi(t + s, s, y))$$

for some $s \in [r_0, r_0 + \eta]$, $y \in B$. $\spadesuit_2$ follows from (c) and $\spadesuit_3$ from $\sigma(0) = \rho(y) = \rho(y_1) \in (0, \epsilon_0)$ and $\limsup_{t\to\infty} \sigma(t) > \epsilon_0$. ☐

Definition 13.10. A semiflow $\tilde{\Psi} : \Delta \times X \to X$ is called asymptotically periodic if there exists a periodic semiflow $\Psi : \Delta \times X \to X$ such that, whenever $r_0 \le s_k \to \infty$ and $t_k \to t \ge 0$ and $x_k \to x$ as $k \to \infty$, then

$$d(\tilde{\Psi}(t + s_k, s_k, x_k), \Psi(t + s_k, s_k, x)) \to 0, \qquad k \to \infty.$$

Ψ is called a *limit semiflow* of $\tilde{\Psi}$.

Remark 13.11. Let $\Psi, \tilde{\Psi} : \Delta \times X \to X$ be semiflows, X a metric space and Ψ and $\tilde{\Psi}$ continuous. Let $\tilde{\Psi}$ be asymptotically periodic and Ψ a periodic limit-semiflow of $\tilde{\Psi}$.

Let $\rho : X \to \mathbb{R}_+$ be uniformly continuous.

Then Ψ and $\tilde{\Psi}$ are both uniformly ρ-persistent if they are both uniformly weakly ρ-persistent and a subset B of X exists with the following property:

(a) For all $x \in X$, $\rho(x) > 0$, $r \geq r_0$, $d(\Psi(t,r,x), B) \to 0$ as $t \to \infty$.

(b) If $0 < \epsilon_1 < \epsilon_2 < \infty$, the intersection $B \cap \{\epsilon_1 \leq \rho(x) \leq \epsilon_2\}$ has compact closure.

(c) If $y \in B$ and $\rho(y) > 0$, there exist no $s \geq r_0$, $t > r > 0$ such that $\rho(\Psi(r+s,s,x)) = 0$ and $\rho(\Psi(t+s,s,x)) > 0$.

The reader is asked to prove Remark 13.11 in Exercise 13.1.

13.4. Uniform persistence of the cell population

The persistence results for nonautonomous semiflow can also be useful for autonomous semiflows for which it is difficult to establish the existence of a compact attracting set like for the cell division model in Chapter 12.

Assume that $\mathcal{P}_0 = \mathcal{P}(R^\diamond, 0) > 1$ with $\mathcal{P}_0$ from (12.35). In order to prove uniform persistence of the cell population (Theorem 12.11), we choose the state space

$$X = \{(R, g); R \in \mathbb{R}_+, g \in L^1_+(\mathbb{R}_+)\}$$

and the function $\rho(x,g) = \|g\|$ for $(x,g) \in X$. The semiflow Φ is induced by the solutions (R,u) of (12.32). See Theorem 12.8. So $\rho(\Phi(t,(\breve{R},\breve{u})) = \|u(t)\|$ with u from (12.32). By Theorem 12.10, Φ is uniformly weakly $\|\cdot\|$-persistent.

In order to make the connection to the language in Section 13.2, we choose $r_0 = 0$ and set $\Psi(t,r,x) = \Phi(t-r,x)$ for $t \geq r \geq 0$ and $x \in X$. This implies $\Psi(s+r,r,x) = \Phi(s,x)$ for all $s, r \geq 0$.

We check the assumptions of Theorem 13.6. Since $\|u(t)\|$ is a locally absolutely continuous function of $t \geq 0$ by (12.33), $\diamondsuit$ is satisfied. We define

$$\tilde{B} = \{(R,g) \in X; R + \|g\| = R^\diamond\}, \quad \tilde{B}_k = \Big\{(R,g) \in X; \big|R + \|g\| - R^\diamond\big| < \frac{1}{k}\Big\}.$$

Since $R(t) + \|u(t)\| \to R^\diamond$ as $t \to \infty$ by Theorem 12.8, $\heartsuit$ is satisfied. (We use $\tilde{B}$ and $\tilde{B}_k$ because B already denotes the bounded linear operator $[Bu](a) = \kappa(a)u(a)$.) Since $\mathcal{P}_0 = \mathcal{P}(R^\diamond, 0) > 1$ we can choose $\epsilon_0 \in (0, R^\diamond)$ such that

$$\mathcal{P}(R^\diamond - 2\epsilon_0, 0) > 1.$$

Recall (12.35). To check ♠, let

$$\left.\begin{array}{ll} y_k = (\breve{R}_k, \breve{u}_k) \in X, & 0 < \|\breve{u}_k\| = \|\breve{u}_1\| \le \epsilon_0 \\ \Phi(t, y_k) \in \tilde{B}_k, & t \ge 0 \end{array}\right\} \quad k \in \mathbb{N}. \tag{13.9}$$

Notice that the sequence (s_k) in ♠ drops out. Set

$$(R_k(t), u_k(t)) = \Phi(t, y_k), \qquad t \ge 0.$$

By (12.32), for $t \ge 0$,

$$\begin{aligned} u_k(t) =& e^{-Dt} S(t)\breve{u}_k + \int_0^t e^{-D(t-r)} S(t-r) B v_k(r) dr, \\ v_k(t) =& f(R_k(r)) u_k(r). \end{aligned} \tag{13.10}$$

Since $\Phi(t, y_k) \in \tilde{B}_k$,

$$R^\diamond - \frac{1}{k} \le \|u_k(t)\| + R_k(t) \le R^\diamond + \frac{1}{K}, \qquad t \ge 0. \tag{13.11}$$

By Lemma 12.6, cf. (12.33),

$$\|u_k(t)\| = e^{-Dt}\|\breve{u}_k\| + \int_0^t e^{-D(t-r)} f(R_k(r)) \|B u_k(r)\| dr.$$

Since $\|\breve{u}_k\| = \|\breve{u}_1\|$, the sequences of functions (R_k) and $\|u_k\|$ are equibounded and equicontinuous. In particular ♠$_1$ is satisfied. Since $\|u_k(t)| \ge e^{-Dt}\|\breve{u}_1\|$, ♠$_2$ is also satisfied.

To check ♠$_3$ we take the Laplace transforms of u_k in (13.10), cf. (12.38),

$$\hat{u}_k(\lambda) = \hat{S}(\lambda + D)\breve{u}_k + \hat{S}(\lambda + D) B \hat{v}_k(\lambda). \tag{13.12}$$

Since $L^1(\mathbb{R}_+)$ has the abstract L space property $\|u + v\| = \|u\| + \|v\|$ for all $u, v \in L^1_+(\mathbb{R}_+)$, by (13.11),

$$\|\hat{u}_k(\lambda)\| = \mathcal{L}\{\|u_k(\cdot)\|\}(\lambda) \le \frac{R^\diamond + (1/k)}{\lambda}. \tag{13.13}$$

By Lemma 12.7, $\hat{u}_k(\lambda)$ is locally absolutely continuous (in particular well defined and continuous at $a = 0$),

$$\hat{u}_k(\lambda)(0) = \int_{\mathbb{R}_+} \hat{u}_k(\lambda)(a) \frac{m(da)}{\mathcal{F}(a)}, \tag{13.14}$$

and

$$\Big(\lambda + D + \frac{d}{da}\Big) \frac{\hat{u}_k(\lambda)(a)}{\mathcal{F}(a)} = \frac{\breve{u}_k(a)}{\mathcal{F}(a)} + \frac{\kappa(a)\hat{v}_k(\lambda)(a)}{\mathcal{F}(a)}. \tag{13.15}$$

Here we have used the definition of B, (12.30).

Suppose ♠$_3$ does not hold. Then (see Remark 13.7),

$$\lim_{k\to\infty} \|u_k(t)\| \le \|u_1(0)\| \le \epsilon_0, \qquad t \ge 0,$$

with the limit being uniform for t in bounded subintervals of $\mathbb{R}_+$. Since $R_k(t) \geq R^\diamond - (1/k) - \|u_k(t)\|$ by (13.11), there exists a sequence (t_k) with $t_k \to \infty$ such that

$$R_k(t) \geq R^\diamond - \frac{3}{2}\epsilon_0 - \frac{1}{k}, \qquad 0 \leq t \leq t_k, k \in \mathbb{N}.$$

Then there exists some $k_0 \in \mathbb{N}$ such that

$$R_k(t) \geq R^\diamond - 2\epsilon_0, \qquad 0 \leq t \leq t_k, \ k \geq k_0.$$

From now on, $k \geq k_0$. Since f is increasing and $v_k(t) = f(R_k(t))u_k(t)$,

$$\begin{aligned}\hat{v}_k(\lambda) \geq & \int_0^{t_k} e^{-\lambda t} f(R^\diamond - 2\epsilon_0)u_k(t)dt \\ = & f(R^\diamond - 2\epsilon_0)\Big(\hat{u}_k(\lambda) - \int_{t_k}^\infty e^{-\lambda t/2}e^{-\lambda t/2}u_k(t)dt\Big) \\ \geq & f(R^\diamond - 2\epsilon_0)\hat{u}_k(\lambda) - f(R^\diamond)e^{-\lambda t_k/2}\hat{u}_k(\lambda/2).\end{aligned}$$

By (13.13),

$$\hat{v}_k(\lambda) \geq f(R^\diamond - 2\epsilon_0)\hat{u}_k(\lambda) - \frac{c}{\lambda}e^{-\lambda t_k/2}$$

for some $c > 0$. We substitute this inequality into (13.15),

$$\Big(\lambda + D + \frac{d}{da}\Big)\frac{\hat{u}_k(\lambda)(a)}{\mathcal{F}(a)} \geq \frac{\breve{u}_k(a)}{\mathcal{F}(a)} + f(R^\diamond - 2\epsilon_0)\kappa(a)\frac{\hat{u}_k(\lambda)(a)}{\mathcal{F}(a)} - \frac{ce^{-\lambda t_k/2}}{\lambda\mathcal{F}(a)}.$$

Here we have absorbed an upper bound for κ into c. We integrate this differential inequality and use $\mathcal{G}$ from (12.36),

$$\begin{aligned}\frac{\hat{u}_k(\lambda)(a)}{\mathcal{F}(a)} \geq & \hat{u}_k(\lambda)(0)e^{-\lambda a}\mathcal{G}(R^\diamond - 2\epsilon_0, a) \\ & - \frac{e^{-\lambda t_k/2}}{\lambda}\int_0^a \frac{c}{\mathcal{F}(r)}e^{-\lambda(a-r)}\frac{\mathcal{G}(R^\diamond - 2\epsilon_0, a)}{\mathcal{G}(R^\diamond - 2\epsilon_0, r)}dr.\end{aligned}$$

Notice that we can drop $e^{-\lambda(a-r)}$ in this estimate. Recall (12.35). We fix $\lambda > 0$ such that $\mathcal{P}(R^\diamond - 2\epsilon_0, \lambda) > 1$ and choose $a^\diamond > 0$ such that

$$\mathcal{P}^\diamond = \int_{[0,a^\diamond)} e^{-\lambda a}\mathcal{G}(R^\diamond - 2\epsilon_0, a)m(da) > 1.$$

From the boundary condition (13.14),

$$\begin{aligned}\hat{u}_k(\lambda)(0) \geq & \int_{[0,a^\diamond)} \hat{u}_k(\lambda)(a)\frac{m(da)}{\mathcal{F}(a)} \\ \geq & \hat{u}_k(\lambda)(0)\mathcal{P}^\diamond - \frac{e^{-\lambda t_k/2}}{\lambda}\int_{[0,a^\diamond)} m(da)\int_0^a \frac{c}{\mathcal{F}(r)}\frac{\mathcal{G}(R^\diamond - 2\epsilon_0, a)}{\mathcal{G}(R^\diamond - 2\epsilon_0, r)}dr.\end{aligned}$$

$\mathcal{G}(R^\diamond - \epsilon_0, \cdot)$ is a continuous strictly positive function. So there exists some $\tilde{c} > 0$ such that

$$\frac{\mathcal{G}(R^\diamond - 2\epsilon_0, a)}{\mathcal{G}(R^\diamond - 2\epsilon_0, r)} \le \tilde{c}, \qquad 0 \le r \le a \le a^\diamond.$$

We substitute this estimate into the previous inequality, reorganize, and change the order of integration,

$$\begin{aligned}
\hat{u}_k(\lambda)(0)(\mathcal{P}^\diamond - 1) \le & \frac{e^{-\lambda t_k/2}}{\lambda} \int_{[0,a^\diamond)} m(da) \int_0^a \frac{c\tilde{c}}{\mathcal{F}(r)} dr \\
\le & \frac{e^{-\lambda t_k/2}}{\lambda} \int_0^{a^\diamond} \frac{c\tilde{c}}{\mathcal{F}(r)} dr \int_{[r,a^\diamond)} m(da) \\
\le & \frac{e^{-\lambda t_k/2}}{\lambda} \int_0^{a^\diamond} dr \frac{c\tilde{c}}{\mathcal{F}(r)} (\mathcal{F}(r) - \mathcal{F}(a^\diamond)) \le \frac{e^{-\lambda t_k/2}}{\lambda} c\tilde{c}a^\diamond.
\end{aligned}$$

So $\hat{u}_k(\lambda)(0) \to 0$ as $k \to \infty$. However, by (13.12), Lemma 12.7, and (13.9),

$$\begin{aligned}
\hat{u}_k(\lambda)(0) \ge [\hat{S}(\lambda + D)\breve{u}_k](0) \ge & \frac{\delta_{\lambda+D}}{(\lambda + D)\hat{\mathcal{F}}(\lambda + D)} \|\breve{u}_k\| \\
= & \frac{\delta_{\lambda+D}}{(\lambda + D)\hat{\mathcal{F}}(\lambda + D)} \|\breve{u}_1\| > 0,
\end{aligned}$$

a contradiction.

13.5. Exercises

Exercise 13.1. Prove Remark 13.11.

Chapter 14

Forced Persistence in Linear Cauchy Problems

Let X be a Banach space and A the generator of a C_0-semigroup $S = \{S(t); t \geq 0\}$. Further, let $f : \mathbb{R}_+ \to X$ be bounded and continuous. We consider the inhomogeneous Cauchy problems,

$$u'(t) = Au(t) + f(t), \quad t > r, \qquad u(r) = x, r \geq 0. \tag{14.1}$$

The variation of constants formula

$$u(t) = S(t-r)x + \int_r^t S(t-s)f(s)ds \tag{14.2}$$

provides unique integral solution,

$$u(t) - x = A\int_r^t u(s)ds + \int_r^t f(s)ds, \qquad t \geq r, \tag{14.3}$$

with the implicit understanding that $\int_r^t u(s)ds \in D(A)$. (This follows from Proposition B.22 after a shift in time.)

We assume that $\|S(t)\| \to 0$ as $t \to \infty$. This means that, if the forcing function is zero, the solutions converge to 0.

We are interested in finding conditions on f and A such that the solutions remain bounded away from 0. If u is an integral solution with initial data (x, r) and $\tilde{r} > r$, then the restriction of u to $[\tilde{r}, \infty)$ is an integral solution with initial data $(u(\tilde{r}), \tilde{r})$. This can easily been seen by manipulating the integrals (cf. proof of Theorem 10.3). So the definition

$$\Psi(t,r,x) = u(t), \qquad 0 \le r \le t, x \in X, \tag{14.4}$$

where u is the solution of (14.3), defines a continuous nonautonomous semiflow anchored at 0. Our task can now be rephrased as finding conditions for Ψ to be uniformly ρ-continuous with $\rho(x) = \|x\|$.

14.1. Uniform weak persistence and asymptotic Abel-averages

Since $\|S(t)\| \to 0$ as $t \to \infty$, there exists $M \ge 1$ and $\omega < 0$ such that $S(t) \le Me^{\omega t}$ [**67**, Prop.V.1.7]. We introduce an equivalent norm such that [**67**, Sec.II.3]

$$\|S(t)\| \le e^{\omega t}, \qquad t \ge 0. \tag{14.5}$$

This implies that $\lambda - A$ has bounded everywhere defined inverses for all $\lambda \ge 0$ which are given by the Laplace transform of the semigroup S,

$$(\lambda - A)^{-1} = \int_0^\infty e^{-\lambda t} S(t) dt. \tag{14.6}$$

The following condition for uniform weak persistence involves the asymptotic Abel-averages of $A^{-1}f(\cdot)$.

Theorem 14.1. *For all $r \ge 0$, $x \in X$,*

$$\limsup_{t\to\infty} \|\Psi(t,r,x)\| \ge \limsup_{\lambda \to 0} \Big\| \int_0^\infty \lambda e^{-\lambda t} A^{-1} f(t) dt \Big\|.$$

In particular, Ψ is uniformly weakly $\|\cdot\|$-persistent if this number is positive.

Proof. Let η be the right-hand side of the inequality,

$$\eta := \limsup_{\lambda \to 0} \Big\| A^{-1} \hat{f}(\lambda) \Big\|. \tag{14.7}$$

Here $\hat{f}(\lambda) = \int_0^\infty e^{-\lambda t} f(t) dt$ is the Laplace transform of f. Suppose that there are $r \ge 0$, $x \in X$ such that $\limsup_{t\to\infty} \|\Psi(t,r,x)\| < \eta$. By the semiflow property, we can assume that there is some $\epsilon \in (0,\eta)$ such that $\|\Psi(t,r,x)\| < \eta - \epsilon$ for all $t \ge r$. Set $u(t) = \Psi(t+r,r,x)$ and $g(t) = f(r+t)$. We make the appropriate change of variables in (14.2),

$$u(t) = S(t)x + \int_0^t S(t-s) g(s) ds.$$

For all $\lambda \ge 0$, $\hat{u}(\lambda)$ is defined and

$$\|\lambda \hat{u}(\lambda)\| \le \eta - \epsilon.$$

Since the Laplace transform converts convolutions into compositions and by (14.6),

$$\lambda\hat{u}(\lambda) = \lambda(\lambda - A)^{-1}x + (\lambda - A)^{-1}\lambda\hat{g}(\lambda).$$

By the triangle inequality,

$$\begin{aligned}\eta - \epsilon &\geq \limsup_{\lambda\to 0} \|\lambda\hat{u}(\lambda)\| \\ &\geq \limsup_{\lambda\to 0}\Big(-\|\lambda(\lambda - A)^{-1}x\| + \|(\lambda - A)^{-1}\lambda\hat{g}(\lambda)\|\Big)\end{aligned}$$

Since $(\lambda - A)^{-1} \stackrel{\lambda\to 0}{\longrightarrow} (-A)^{-1}$ with respect to the operator norm,

$$\eta - \epsilon \geq \limsup_{\lambda\to 0} \|(-A)^{-1}\lambda\hat{g}(\lambda)\|. \tag{14.8}$$

Since $g(t) = f(r+t)$,

$$\lambda\hat{g}(\lambda) = \lambda\int_r^\infty e^{-\lambda(t-r)}f(t)dt = e^{\lambda r}\Big(\lambda\hat{f}(\lambda) - \lambda\int_0^r e^{-\lambda t}f(t)dt\Big).$$

This implies that $\lambda\hat{g}(\lambda) - \lambda\hat{f}(\lambda) \to 0$ as $\lambda \to 0$. By (14.8),

$$\eta - \epsilon \geq \limsup_{\lambda\to 0} \|\lambda A^{-1}\hat{f}(\lambda)\|,$$

a contradiction to (14.7). □

14.2. A compact attracting set

We assume f is bounded and uniformly continuous and that A^{-1} is compact. By the resolvent identity,

$$(\mu - A)^{-1} - (\lambda - A)^{-1} = (\lambda - \mu)(\mu - A)^{-1}(\lambda - A)^{-1}, \qquad \lambda, \mu \geq 0,$$

all operators $(\lambda - A)^{-1}$, $\lambda > 0$, are compact as well.

Proposition 14.2. *If f is bounded and uniformly continuous and A^{-1} is compact, then $K = \{\Psi(t, r, 0); t \geq r \geq 0\}$ has compact closure.*

Since $\Psi(t, r, x) = S(t-r)x + \Psi(t, r, 0)$, this and (14.5) imply that $d(\Psi(t, r, x), K) \to 0$ as $t \to \infty$.

Proof. We set

$$\begin{aligned} u_r(t) =& \Psi(t + r, r, 0), \qquad t, r \geq 0 \\ f_r(t) =& f(t + r) \\ J_h =& (\mathbb{I} - hA)^{-1}, \qquad h > 0. \end{aligned} \tag{14.9}$$

Then J_h is compact of every $h > 0$ and

$$u_r(t) = \int_0^t S(s)f_r(t - s)ds. \tag{14.10}$$

By (14.5),

$$\|u_r(t)\| \leq \frac{1}{|\omega|} \sup_{s\geq 0} \|f(s)\|,$$

and K is a bounded set. Since f is uniformly continuous by assumption, $u_r(t)$ is uniformly continuous in $t \geq 0$, uniformly for $r \geq 0$. The following estimate follows from [**226**, L.2.2],

$$\begin{aligned} \|J_h u_r(t) - u_r(t)\| \leq & 4\int_t^{t+h} \|f_r(s)\| \frac{h+t-s}{h} ds \\ & +4\int_0^1 \|u_r(t+hs) - u_r(t)\| ds. \end{aligned} \tag{14.11}$$

This implies that $\|J_h u_r(t) - u_r(t)\| \to 0$ as $h \to 0$ uniformly in $t, r \geq 0$. Further, $\{J_h u_r(t); t, r \geq 0\} = J_h K$ has compact closure for each $h > 0$. This implies that K has compact closure. □

14.3. Uniform persistence in ordered Banach space

We further assume that X is an ordered Banach space with normal cone X_+ [**49**], S is a strictly positive semigroup, $S(t)(X_+ \setminus \{0\}) \subset X_+ \setminus \{0\}$ for all $t \geq 0$, and $f(t) \in X_+$ for all $t \geq 0$.

Theorem 14.3. *Assume that f is bounded and uniformly continuous on $\mathbb{R}_+$ and A^{-1} is compact. Ψ is uniformly ρ-persistent on X_+ if*

$$\limsup_{\lambda\to 0} \Big\|(-A^{-1})\lambda \int_0^\infty e^{-\lambda t} f(t)dt\Big\| > 0,$$

and

$$\limsup_{\lambda\to 0} \liminf_{r\to\infty} \Big\|(-A^{-1})\lambda \int_0^\infty e^{-\lambda t} f(r+t)dt\Big\| > 0.$$

Proof. We check the assumptions of Theorem 13.6. Notice that

$$\Psi(t+r, r, x) = S(t)x + \int_0^t S(s)f(r+t-s)ds.$$

With the compact set K from Proposition 14.2, define

$$B_k = \{x \in X_+; d(x, K) < 1/k\}.$$

Then ♡ follows from Proposition 14.2. Let (y_k) be a sequence in X_+ such that $y_k \in B_k$ for all $k \in \mathbb{N}$ and $\|y_k\| = \|y_1\| > 0$. For every $k \in \mathbb{N}$ there exists some $z_k \in K$ such that $\|y_k - z_k\| \leq 1/k$. Let $\epsilon > 0$. Choose $j \in \mathbb{N}$ such that $\|y_k - z_k\| < \epsilon/4$ for $k \geq j$. Since S is strongly continuous, there exists some $\delta > 0$ such that $\|S(t)y_k - S(r)y_k\| < \epsilon$ if $\|t - r\| < \delta$ and $k = 1, \dots, j$. Since $(y, t) \to S(t)y$ is continuous, it is uniformly continuous on $[0, r+1] \times K$. So there exists some $\delta > 0$ such that $\|S(t)z_k + S(r)z_k\| < \epsilon/4$ if $|t - r| < \delta$, $k \in \mathbb{N}$. For $k \geq j$,

$$\begin{aligned} &\|S(t)y_k - S(r)y_k\| \\ \leq &\|S(t)y_k - S(t)z_k\| + \|S(t)z_k - S(r)z_k\| + \|S(r)z_k - S(r)y_k\| \\ \leq &2\|y_k - z_k\| + \|S(t)z_k - S(r)z_k\| < \epsilon. \end{aligned}$$

This shows that the continuity of $S(t)y_k$ as a function of t is uniform in k. Since f is uniformly continuous, the continuity of $\Psi(t+s_k, s_k, y_k)$ is uniform in k for any sequence (s_k). This proves ($\spadesuit_1$).

Since $f(t) \in X_+$ and S is a positive semigroup, $\Psi(t+s_k, s_k, y_k) \geq S(t)y_k$. Assume that

$$\sigma(t) = \lim_{k\to\infty} \|\Psi(t + s_k, s_k, y_k)\|$$

exists for $t \in [0, \tau]$. Since X_+ is a normal cone [**49**], there exists some $\eta > 0$ such that

$$\sigma(t) \geq \eta \limsup_{k\to\infty} \|S(t)y_k\|.$$

As before $\|y_k - z_k\| < 1/k$ for suitable $z_k \in K$. Since K is compact, $y_k \to y$ for some $y \in K$ after choosing a subsequence. Then $\|y\| = \|y_1\| > 0$ and $\sigma(t) \geq \eta\|S(t)y\| > 0$ because S is a strictly positive semigroup by assumption. This shows $\spadesuit_2$.

To establish $\spadesuit_3$ by contradiction, assume that

$$\sigma(t) = \lim_{k\to\infty} \|\Psi(t + s_k, s_k, y_k)\|$$

exists uniformly for $t \geq 0$ in all bounded subintervals of $\mathbb{R}_+$ and $\sigma(t) \leq \epsilon_0$ where $\epsilon_0 > 0$ can be chosen as small as wanted. Set $u_k(t) = \Psi(t + s_k, s_k, y_k)$ and $f_k(t) = f(s_k + t)$. Then

$$\epsilon_0 \geq \sigma(t) = \lim_{k\to\infty} \|u_k(t)\|$$

uniformly for $t \geq 0$ in all bounded subintervals of $\mathbb{R}_+$. Further,

$$u_k(t) = S(t)y_k + \int_0^t S(t - s)f_k(s)ds.$$

We take Laplace transforms for $\lambda > 0$,

$$\lambda\hat{u}_k(\lambda) = \lambda(\lambda - A)^{-1}y_k + (\lambda - A)^{-1}\lambda\hat{f}_k(\lambda).$$

By (14.5),

$$\|u_k(t)\| \leq \|y_k\| + \frac{1}{|\omega|}\sup_{s\geq 0}\|f(r_k + s)\| \leq \|y_1\| + \frac{1}{|\omega|}\sup_{s\geq 0}\|f(s)\|.$$

By Lebesgue's theorem of dominated convergence, for all $\lambda > 0$,

$$\begin{aligned}\epsilon_0 \geq & \lambda\hat{\sigma}(\lambda) = \lambda \int_0^\infty e^{-\lambda t} \lim_{k\to\infty} \|u_k(t)\| dt \\ = & \lim_{k\to\infty} \lambda \int_0^\infty e^{-\lambda t} \|u_k(t)\| dt \geq \limsup_{k\to\infty} \|\lambda \hat{u}_k(\lambda)\| \\ \geq & \limsup_{k\to\infty} \Big(-\|\lambda(\lambda - A)^{-1} y_k\| + \|(\lambda - A)^{-1} \lambda \hat{f}_k(\lambda)\| \Big).\end{aligned}$$

Since $\|(\lambda - A)^{-1} - (-A)^{-1}\| \to 0$ as $\lambda \to 0$ and $\|y_k\| = \|y_1\|$,

$$\epsilon_0 \geq \limsup_{\lambda\to 0} \limsup_{k\to\infty} \|(-A)^{-1} \lambda \hat{f}_k(\lambda)\|.$$

Since the last number is greater than or equal to

$$\limsup_{\lambda\to 0} \liminf_{r\to\infty} \Big\|(-A)^{-1} \lambda \int_0^\infty e^{-\lambda t} f(r+t) dt \Big\|,$$

which is positive by assumption, we obtain a contradiction by choosing $\epsilon_0 > 0$ smaller than this number. □

Lemma 14.4. *Let Y be a separable subset of $X_+ \setminus \{0\}$. Then there exists some $x^* \in X_+^*$ such that $\langle y, x^* \rangle > 0$ for all $y \in Y$.*

Proof. Let $\{y_j; j \in \mathbb{N}\}$ be a dense countable subset of Y. By the Hahn-Banach theorem, there exist $x_j^* \in X^*$ such that $\langle y_j, x_j^* \rangle = \|y_j\|$ and $\|x_j^*\| \leq 1$. Since X_+ is normal, X^* is reproducing (generating), $X^* = X_+^* - X_+^*$, and there exists some $\eta \geq 1$ such that, for all $x^* \in X^*$, $x^* = y^* - z^*$ with suitable $y^*, z^* \in X_+^*$, $\|y^*\|, \|z^*\| \leq \eta \|x^*\|$ [**49**, Prop.19.1(d)]. So we can assume that $x_j^* \in X_+^*$, $\langle y_j, x_j^* \rangle \geq \|y_j\|$ and $\|x_j^*\| \leq \eta$. Define

$$x^* = \sum_{j=1}^\infty 2^{-j} x_j^*.$$

Let $y \in Y$. Then there exists some $n \in \mathbb{N}$ such that $\|y - y_n\| \leq \frac{\|y\|}{4\eta}$. Since $x_j^* \in X_+^*$ for all j,

$$\begin{aligned}&\langle y, x^* \rangle \geq 2^{-n} \langle y, x_n^* \rangle = 2^{-n} (\langle y - y_n, x_n^* \rangle + \langle y_n, x_n^* \rangle) \\ \geq & 2^{-n}(\|y_n\| - \eta\|y - y_n\|) \geq 2^{-n}(\|y\| - 2\eta\|y - y_n\|) \geq 2^{-n} \frac{\|y\|}{2} > 0.\end{aligned}$$ □

Corollary 14.5. *Assume that f is bounded and uniformly continuous on $\mathbb{R}_+$ and A^{-1} is compact. Further, assume that there are some $\tau > 0$, $m \in \mathbb{N}$ and $\delta > 0$ such that*

$$\Big\| \int_{n\tau}^{(n+1)\tau} (-A)^{-1} f(s) ds \Big\| \geq \delta, \qquad n \geq m.$$

Then Ψ is uniformly $\|\cdot\|$-persistent.

Proof. Let $Y := \overline{(-A)^{-1}X_+} \setminus \{0\}$. Since $(-A)^{-1}$ is a compact positive operator, Y is a separable subset of $X_+ \setminus \{0\}$. By Lemma 14.4, there exists some $x^* \in X_+^*$ such that $\langle y, x^* \rangle > 0$ for all $y \in Y$. We can assume that $\|x^*\| = 1$. Let

$$v_n = \int_{n\tau}^{(n+1)\tau} f(s)ds.$$

Since $(-A)^{-1}$ is compact and $\|Av_n\| \geq \delta$, $n \geq m$, and (v_n) is bounded, $\{(-A)^{-1}v_n; n \geq m\}$ has compact closure in Y. So there exists some $\delta_0 > 0$ such that

$$\langle (-A)^{-1}v_n, x^* \rangle \geq \delta_0, \qquad n \in \mathbb{N}, n \geq m. \tag{14.12}$$

For every $r \geq 0$ there are unique $n_r \in \mathbb{Z}_+$ and $s_r \in [0, \tau)$ such that $r = n_r\tau + s_r$. In particular, $n_0 = 0$ and $s_0 = 0$. As $r \to \infty$, $n_r \to \infty$. To check the assumptions of Theorem 14.3, for $\lambda \geq 0$,

$$\begin{aligned}
\xi(\lambda, r) :=& \Big\|(-A)^{-1}\lambda \int_0^\infty e^{-\lambda t} f(r+t)dt\Big\| \\
=& \Big\|\lambda \int_0^\infty e^{-\lambda t}(-A)^{-1} f(\tau n_r + s_r + t)dt\Big\| \\
\geq& \lambda\Big\|\int_0^\infty e^{-\lambda(t-s_r)}(-A)^{-1} f(\tau n_r + t)dt\Big\| \\
&- \lambda \int_0^{s_r} e^{-\lambda t}\|(-A)^{-1} f(\tau n_r + t)\|dt.
\end{aligned}$$

Since $0 \leq s_r \leq \tau$ and $\|x^*\| \leq 1$,

$$\begin{aligned}
&\limsup_{\lambda \to 0} \inf_{r \geq 0} \xi(\lambda, r) \\
\geq& \limsup_{\lambda \to 0} \inf_{r \geq 0} \lambda \Big\|\int_0^\infty e^{-\lambda t}(-A)^{-1} f(\tau n_r + t)dt\Big\| \\
\geq& \limsup_{\lambda \to 0} \inf_{r \geq 0} \lambda \Big\langle \sum_{n=0}^\infty \int_{n\tau}^{(n+1)\tau} e^{-\lambda(n+1)\tau}(-A)^{-1} f(\tau n_r + t)dt, x^* \Big\rangle \\
\geq& \limsup_{\lambda \to 0} \inf_{r \geq 0} \lambda \sum_{n=m}^\infty e^{-\lambda(n+1)\tau} \big\langle (-A)^{-1} v_{n_r+n}, x^* \big\rangle.
\end{aligned}$$

By (14.12) and $n_r \in \mathbb{Z}_+$,

$$\begin{aligned}
\limsup_{\lambda \to 0} \inf_{r \geq 0} \xi(\lambda, r) \geq& \limsup_{\lambda \to 0} \lambda \sum_{n=m}^\infty e^{-\lambda(n+1)\tau} \delta_0 \\
=& \limsup_{\lambda \to 0} \delta_0 \frac{\lambda e^{-(m+1)\lambda\tau}}{1 - e^{-\lambda\tau}} = \frac{\delta_0}{\tau} > 0.
\end{aligned}$$

The assertion now follows from Theorem 14.3. □

Chapter 15

Persistence via Average Lyapunov Functions

As in the time homogeneous case [**80, 82, 104, 106, 109, 110**], average Lyapunov functions can be used to prove persistence in time-heterogeneous situations. Burton and Hutson [**18**] combine average Lyapunov functions with skew product flows, but skew product flows are not needed in this context if the concept of ρ-persistence is fully exploited.

Let X be a set, $\rho : X \to [0, \infty)$ a nonnegative function on X, and $\Psi : \Delta \times X \to X$ a *(nonautonomous) semiflow* on X (anchored at r_0). See Chapter 13.2.

For each function $\theta : [r_0, \infty) \times X \to \mathbb{R}$, we define a function $\tilde{\theta} : \mathbb{R}_+ \times [r_0, \infty) \times X \to \mathbb{R}$ by

$$\tilde{\theta}(t, r, x) := \theta(t + r, \Psi(t + r, r, x)). \tag{15.1}$$

By the semiflow property,

$$\tilde{\theta}(t, s + r, \Psi(s + r, r, x)) = \tilde{\theta}(t + s, r, x), \qquad t, s \geq 0, r \geq r_0. \tag{15.2}$$

The ˜ notation also applies to ρ. We assume that

◇ $\tilde{\rho}(t, r, x)$ is a continuous function of $t \geq 0$ for all $r \geq r_0$, $x \in X$.

We further assume that Ψ has a pseudo-attracting set.

Definition 15.1. A subset B of X is called a pseudo-attracting set if there exists a sequence (B_k) of subsets of X with $B \subset B_k$ and the following properties:

(1) if $x \in X$, $\rho(x) > 0$, and $k \in \mathbb{N}$, $r \geq r_0$, then there exists some $t_k \geq r$ such that $\Psi(t, r, x) \in B_k$ for all $t \geq t_k$.

(2) Whenever (y_k) is a sequence in X and (s_k) is a sequence in $\mathbb{R}$ such that

$$r_0 \leq s_k \to \infty \text{ as } k \to \infty,$$
$$0 < \rho(y_k) = \rho(y_1) \text{ for all } k \in \mathbb{N},$$
$$\Psi(t + s_k, s_k, y_k) \in B_k \text{ for all } k \in \mathbb{N}, t \geq 0,$$

then the continuity of $\tilde{\rho}(t, s_k, y_k)$ as a function of $t \geq 0$ holds uniformly in k (possibly after choosing subsequences).

(3) If (y_k) is a sequence in X with $y_k \in B_k$ for all $k \in \mathbb{N}$ and $\sup_{k \in \mathbb{N}} \rho(y_k) < \infty$, then there exists some $y \in B$ such that, after choosing a subsequence, $\rho(y_k) \to \rho(y)$ as $k \to \infty$.

Cf. Theorem 13.6.

15.1. Weak average Lyapunov functions

We define the *orbital derivative* of ρ as

$$\begin{aligned} \dot{\rho}(r, x) &= \limsup_{h \to 0+} \frac{\rho(\Psi(r + h, r, x)) - \rho(x)}{h} \\ &= \limsup_{h \to 0+} \frac{\tilde{\rho}(h, r, x) - \rho(x)}{h}, \qquad r \geq r_0. \end{aligned} \tag{15.3}$$

Definition 15.2. Assume that the semiflow Ψ has a pseudo-attracting set B as in Definition 15.1.

ρ is called a *weak average Lyapunov function* for Ψ if there exists a function $\theta : [r_0, \infty) \times X \to \mathbb{R}$ and a positive continuous function $\xi : (0, 1] \to (0, \infty)$ with the following properties:

(i) $\displaystyle\int_0^1 \frac{du}{\xi(u)} = \infty.$

(ii) If $x \in X$ and $0 < \rho(x) \leq 1$, then

$$\dot{\rho}(t, x) \geq \xi(\rho(x))\theta(t, x).$$

(iii) If $r \geq r_0$, then the function $s \mapsto \tilde{\theta}(s, r, x)$ is continuous on $\mathbb{R}_+$.

(iv) There are no element $x \in B$ and no sequence (s_j) in $[r_0, \infty)$ such that

$$\rho(x) = 0 \quad \text{and} \quad \limsup_{j \to \infty} \int_0^T \tilde{\theta}(t, s_j, x)dt \leq 0 \text{ for all } T > 0.$$

(v) If (x_n) is a sequence in X with $x_n \in B_n$ and $\rho(x_n) \leq 1$ for all $n \in \mathbb{N}$ and (s_n) is a sequence in $[r_0, \infty)$, then there exists some $x \in B$ such that, after choosing a subsequence, $\rho(x_n) \to \rho(x)$ and

$$\int_0^T \tilde{\theta}(t, s_n, x_n)dt - \int_0^T \tilde{\theta}(t, s_n, x)dt \to 0, \qquad n \to \infty, T > 0.$$

Cf. [**82**] and [**106**, Thm.12.2.1].

Remark 15.3. Condition (iii) can be replaced by

(iii)' If $t > 0$, $r \geq r_0$, then the function $s \mapsto \tilde{\theta}(s, r, x)$ is Borel measurable and integrable on $[0, t]$ and the function $s \mapsto \tilde{\rho}(s, r, x)$ is absolutely continuous on $[0, t]$. See Section A.A.2.

Lemma 15.4. *Let ρ be a weak average Lyapunov function. Then there exists some $k \in \mathbb{N}$ with the following property:*

If $s \geq r_0$, $y \in X$, $y \in B_k$ and $\rho(y) < 1/k$, then there exists some $T > 0$ such that

$$\int_0^T \tilde{\theta}(t, s, y)dt > 1/k.$$

Proof. Suppose that the statement of the Lemma is false. Then we find sequences (s_j) in $[r_0, \infty)$, $(y_j) \in X$ such that, for all $j \in \mathbb{N}$, $y_j \in B_j$, $\rho(y_j) < 1/j$ and

$$\int_0^T \tilde{\theta}(t, s_j, y_j)dt \leq \frac{1}{j} \quad \forall T \geq 0.$$

Because of Definition 15.2 (v), there exists some $y \in B$ such that $\rho(y) = 0$ and, after choosing subsequences,

$$\limsup_{j \to \infty} \int_0^T \tilde{\theta}(t, s_j, y)dt \leq 0 \quad \forall T \geq 0.$$

This contradicts Definition 15.2 (iv). □

Lemma 15.5. *If $r \geq r_0$, $x \in X$, $\rho(x) > 0$, then $\tilde{\rho}(s, r, x) > 0$ for all $s \geq 0$. If in addition $\tilde{\rho}(s, r, x) < 1$ for all $0 \leq s \leq t$, then*

$$\int_{\rho(x)}^{\tilde{\rho}(s,r,x)} \frac{du}{\xi(u)} \geq \int_0^s \tilde{\theta}(u, r, x)du \qquad \forall s \in [0, t].$$

Proof. Since $\tilde{\rho}(\cdot, r, x)$ is continuous, $\tilde{\rho}(s, r, x) = \rho(\Psi(r + s, r, x)) > 0$ for $s > 0$ and s close to 0. Suppose that $\tilde{\rho}(s, r, x) = 0$ for some $s \in (0, t]$. Then there exists $\tilde{t} \in (0, t)$ such that $\tilde{\rho}(s, r, x) > 0$ for $s \in [0, \tilde{t})$ and $\tilde{\rho}(\tilde{t}, r, x) = 0$. Since $\tilde{\rho}(\cdot, r, x)$ is continuous, there exists some $\hat{t} \in (0, \tilde{t})$ such that, for $s \in [0, \tilde{t} - \hat{t})$,

$$\begin{aligned} 1 >& \tilde{\rho}(s + \hat{t}, r, x) = \rho(\Psi(s + \hat{t} + r, r, x)) \\ =& \rho(\Psi(s + \hat{t} + r, \hat{t} + r, \Psi(\hat{t} + r, r, x)) = \tilde{\rho}(s; \hat{t} + r, \Psi(\hat{t} + r, r, x)). \end{aligned}$$

Renaming $\Psi(\hat{t} + r, r, x)$ by x, $\hat{t} + r$ by r and $\tilde{t} - \hat{t}$ by $\tilde{t}$, we can assume that

$$\tilde{\rho}(\tilde{t}, r, x) = 0, \qquad 0 < \tilde{\rho}(s, r, x) = \rho(\Psi(s + r, r, x)) < 1 \qquad \forall s \in [0, \tilde{t}).$$

By the semiflow property and (15.3),

$$\limsup_{h\to 0+} \frac{1}{h}[\rho(\Psi(u+h,r,x) - \rho(\Psi(u,r,x))] = \dot{\rho}(u,\Psi(u,r,x))$$
$$\geq \xi(\rho(\Psi(u,r,x)))\theta(u,\Psi(u,r,x))$$

for $r \leq u < r + \tilde{t}$, with the inequality following from Definition 15.2 (ii). Substituting $u = r + s$, using the definition of $\tilde{\rho}$ and $\tilde{\theta}$ in (15.1), and the definition

$$\partial_s^+ \tilde{\rho}(s,r,x) := \limsup_{h\to 0+} \frac{1}{h}[\tilde{\rho}(s+h,r,x) - \tilde{\rho}(s,r,x)],$$

we have

$$\partial_s^+ \tilde{\rho}(s,r,x) \geq \xi(\tilde{\rho}(s,r,x))\tilde{\theta}(s,r,x) \qquad \forall s \in [0,\tilde{t}].$$

Set

$$\psi(s) = \int_{\rho(x)}^{\tilde{\rho}(s,r,x)} \frac{dv}{\xi(v)} - \int_0^s \tilde{\theta}(v,r,x)dv.$$

By the chain rule in Exercise A.2 (recall that ξ is positive) and the fundamental theorem of calculus,

$$\liminf_{h\to 0+} \frac{1}{h}[\psi(s+h) - \psi(s)] \geq 0, \qquad \psi(0) = 0.$$

By Corollary A.2 and Remark A.3, ψ is increasing and $\psi(s) \geq 0$. By definition of ψ,

$$\int_{\rho(x)}^{\tilde{\rho}(s,r,x)} \frac{du}{\xi(u)} \geq \int_0^s \tilde{\theta}(u,r,x)du, \qquad s \in (0,\tilde{t}).$$

Taking the limit $s \to \tilde{t}$, we have $\tilde{\rho}(s,r,x) \to 0$ and, by Definition 15.2 (i),

$$\int_0^s \tilde{\theta}(u,r,x)du \to -\infty, \qquad s \to \tilde{t}-$$

a contradiction to (iii). So $\tilde{\rho}(s,r,x) > 0$ for all $s \in (0,t]$ and the other claim of the lemma follows by the same considerations. □

Remark 15.6. If assumption (iii) in Definition 15.2 is replaced by (iii)' in Remark 15.3, the proof of Lemma 15.5 is to be modified as follows:

Proof. Define $G(w) = \int_{\rho(x)}^{w} \frac{du}{\xi(u)}$ for $u \in (0,1]$. Since $\xi : (0,1] \to (0,\infty)$ is continuous, G is continuously differentiable on $(0,1]$ and Lipschitz continuous on each closed subinterval of $(0,1]$ by the mean value theorem. Since $\tilde{\rho}(s,r,x)$ is an absolutely continuous function of $s \in [0,\tilde{t}]$, $G(\tilde{\rho}(s,r,x))$ is an absolutely continuous function of s in every closed subinterval of $[0,\tilde{t})$ by

Proposition A.5. Thus ψ is absolutely continuous on every closed subinterval of $[0,\tilde{t})$ and ψ is differentiable at a.e. $s \in [0,\tilde{t})$ with

$$\begin{aligned}\psi'(s) =&\xi(\tilde{\rho}(s,r,x))\partial_s\tilde{\rho}(s,r,x) - \tilde{\theta}(s,r,x)\\ =&\xi(\tilde{\rho}(s,r,x))\partial_s^+\tilde{\rho}(s,r,x) - \tilde{\theta}(s,r,x) \ge 0.\end{aligned}$$

This implies that ψ is increasing. □

Theorem 15.7. *If ρ is a weak average Lyapunov function for the semiflow Ψ, then Ψ is uniformly weakly ρ-persistent.*

Proof. Assume that Ψ is not uniformly weakly ρ-persistent. Then, for $k \in \mathbb{N}$ chosen from Lemma 15.4, there exists some $r_1 \ge r_0, x \in X, \rho(x) > 0$ such that

$$\limsup_{t\to\infty} \rho(\Psi(t,r_1,x)) < 1/k.$$

Without restriction we can choose $k > 1$. Let B_k be as in Definition 15.1. We find some $r \ge r_1$ such that

$$0 < \rho(\Psi(t+r,r_1,x)) \le 1/k, \quad \Psi(t+r,r_1,x) \in B_k, \quad t \ge 0.$$

We set $x_0 = \Psi(r,r_1,x)$, use the semiflow property and (15.1),

$$\tilde{\rho}(t,r,x_0) \le 1/k, \qquad \Psi(t+r,r,x_0) \in B_k \qquad \forall t \ge 0. \tag{15.4}$$

By Lemma 15.4, there exists some $T_1 > 0$ such that

$$\int_0^{T_1} \tilde{\theta}(t,r,x_0)dt > 1/k.$$

Set $x_1 = \Psi(r+T_1,r,x_0)$. By Lemma 15.4, there exists some $T_2 > 0$ such that

$$\int_0^{T_2} \tilde{\theta}(t,r+T_1,x_1)dt > 1/k.$$

By (15.2) and a change of variables,

$$\int_{T_1}^{T_1+T_2} \tilde{\theta}(t,r,x_0)dt > 1/k.$$

Hence

$$\int_0^{T_1+T_2} \theta(t,r,x_0)dt > 2/k.$$

Continuing this way we find numbers T_m such that

$$\int_0^{\bar{T}_m} \tilde{\theta}(t,r,x_0)dt > m/k$$

where $\bar{T}_m = T_1 + \cdots + T_m$. By Lemma 15.5,

$$\int_{\rho(x_0)}^{\tilde{\rho}(\bar{T}_m,r,x_0)} \frac{du}{\xi(u)} \ge \int_0^{\bar{T}_m} \tilde{\theta}(u,r,x_0)du > m/k \quad \forall m \in \mathbb{N}.$$

By choice of r and x_0, $\tilde{\rho}(\bar{T}_m, r, x_0) < 1$ by (15.4), and

$$\int_{\rho(x_0)}^{1} \frac{du}{\xi(u)} > m/k \quad \forall m \in \mathbb{N},$$

a contradiction because $\rho(x_0) > 0$. □

15.2. Strong average Lyapunov functions

Definition 15.8. A weak average Lyapunov function ρ is called a strong average Lyapunov function if

$$\liminf_{s\to\infty} \int_0^t \tilde{\theta}(u, s, y)du > -\infty$$

whenever $t \geq 0, y \in B, \rho(y) > 0$.

Theorem 15.9. *If ρ is a strong average Lyapunov function, then Ψ is uniformly ρ-persistent.*

Proof. We apply Theorem 13.6. Ψ is uniformly weakly ρ-persistent because ρ is a weak average Lyapunov function (Theorem 15.7). Recall that $\diamondsuit$ is an overall assumption for this chapter. $\heartsuit$ is part of Definition 15.1 (i). $\spadesuit_1$ follows from Definition 15.1 (ii).

Suppose that $\spadesuit_2$ is false. Then, for sufficiently small $\epsilon > 0$ and some $t > 0$

$$\sigma(s) = \lim_{j\to\infty} \tilde{\rho}(s, s_j, y_j)$$

uniformly on some interval $[0, t]$ with $s_j \to \infty$, $\Psi(s + s_j, s_j, y_j) \in B_j$ for all $j \in \mathbb{N}$,

$$0 \leq \sigma(s) \leq \sigma(0) = \epsilon, \quad \sigma(t) = 0.$$

Notice that $\rho(y_j) = \tilde{\rho}(0, s_j, y_j) \to \sigma(0) = \epsilon$. If we choose $\epsilon < 1$, for sufficiently large j, we have $\tilde{\rho}(s, s_j, y_j) < 1$ for all $s \in [0, t]$. By Lemma 15.5,

$$\int_{\rho(y_j)}^{\tilde{\rho}(t,s_j,y_j)} \frac{du}{\xi(u)} \geq \int_0^t \tilde{\theta}(u, s_j, y_j)du. \tag{15.5}$$

Since $\rho(y_j) \to \epsilon$ and $\tilde{\rho}(t, s_j, y_j) \to \sigma(t) = 0$, we have from property (i) of Definition 15.2 that

$$\int_0^t \tilde{\theta}(u, s_j, y_j)du \to -\infty, \qquad j \to \infty.$$

By property (v) of Definition 15.2, there exists some $y \in B$ such that $\rho(y) = \epsilon$ and

$$\int_0^t \tilde{\theta}(u, s_j, y)du \to -\infty, \qquad j \to \infty,$$

a contradiction because ρ is a strong Lyapunov function.

Suppose that ♠$_3$ is false. Then, for any $\epsilon > 0$, we find sequences (y_j) and (s_j) such that

$$\sigma(s) = \lim_{j\to\infty} \tilde{\rho}(s, s_j, y_j)$$

locally uniformly on $[0, \infty)$ with $s_j \to \infty$, $y_j \in B_j$, $\rho(y_j) = \rho(y_1) = \epsilon$, and

$$\sigma(0) = \rho(y_1) > 0, \qquad 0 < \sigma(s) \leq \sigma(0) \quad \forall s \geq 0.$$

Let $t \geq 0$. By a similar argument as before, we have (15.5). Taking the limit for $j \to \infty$ and noticing that $\rho(y_j) = \sigma(0)$, we have

$$\int_{\sigma(0)}^{\sigma(t)} \frac{du}{\xi(u)} \geq \limsup_{j\to\infty} \int_0^t \tilde{\theta}(u, s_j, y_j)du.$$

Since $\sigma(t) \leq \sigma(0)$,

$$0 \geq \limsup_{j\to\infty} \int_0^t \tilde{\theta}(u, s_j, y_j)du \qquad \forall t \geq 0.$$

By Definition 15.2 (v), there exists some $y \in B$ such that $\rho(y) = \epsilon$ and, after choosing subsequences,

$$0 \geq \limsup_{j\to\infty} \int_0^t \tilde{\theta}(u, s_j, y)du \qquad \forall t \geq 0.$$

Since this holds for every $\epsilon > 0$, there exist sequences $\epsilon_k \to 0, k \to \infty$, $y_k \in B$, $\rho(y_k) = \epsilon_k$ and sequences $s_j^k \to \infty$ $(j \to \infty)$, such that

$$\limsup_{j\to\infty} \int_0^t \tilde{\theta}(r, s_j^k, y_k)dr \leq 0 \quad \forall t \geq 0.$$

Again by Definition 15.2 (v) and Definition 15.1 (2), after choosing a subsequence of y_k and applying a diagonalization procedure to the sequences (s_j^k), we find a sequence $r_j \to \infty$ and some $y \in B$ such that $\rho(y) = 0$ and

$$\limsup_{j\to\infty} \int_0^t \tilde{\theta}(s, r_j, y)ds \leq 0 \quad \forall t \geq 0.$$

This contradicts Definition 15.2 (iv). □

15.3. The time-heterogeneous hypercycle equation

To illustrate the combination of time-heterogeneous persistence theory and average Lyapunov functions, we extend the persistence results in [**106**], Chapter 12 (or in their incarnation [**196**] at the dawn of persistence theory) to the hypercycle equation with time-dependent catalysis rates.

This equation describes the interaction between n types of RNA molecules the frequencies of which are denoted by x_j, $j = 1, \ldots, n$.

Let $r_0 \in \mathbb{R}$ be the earliest time we may ever want to observe this interaction. The catalysis rates are described by continuous bounded functions $\gamma_j : [r_0, \infty) \to \mathbb{R}$.

Define $f : [r_0, \infty) \times \mathbb{R}^n \to \mathbb{R}$ by

$$f(t,x) = \sum_{k=1}^{n} \gamma_k(t) x_k x_{k-1}, \quad x_0 := x_n,\ x = (x_1, \dots, x_n). \tag{15.6}$$

With this f, the time-heterogeneous hypercycle equation takes the form

$$\begin{aligned} x_j' &= x_j \big[\gamma_j(t) x_{j-1} - f(t,x)\big], \qquad j = 1, \dots, n,\ t \ge r, \\ x_0(t) &= x_n(t), \qquad x(r) = x^\circ, \end{aligned} \tag{15.7}$$

on the simplex

$$X = \{x \in \mathbb{R}^n; x_j \ge 0, x_1 + \cdots + x_n = 1\},$$

with initial data $r \ge r_0, x^\circ \in X$. Notice that X is invariant under the solutions of (15.7) which induce a nonautonomous semiflow Ψ on X. There is a trivial pseudo-attracting set, namely $B = X = B_k$ (Definition 15.1).

Following [**106, 196**] we try the function

$$\rho(x) = x_1 \cdots x_n \tag{15.8}$$

as average Lyapunov function. Its orbital derivative satisfies

$$\dot\rho(t,x) = \rho(x)\Big[\sum_{j=1}^{n} \gamma_j(t) x_{j-1} - n f(t,x)\Big]. \tag{15.9}$$

In view of Definition 15.2, this suggests the choice

$$\xi(u) = u, \qquad \theta(t,x) = \sum_{j=1}^{n} \gamma_j(t) x_{j-1} - n f(t,x). \tag{15.10}$$

Notice that

$$\tilde\theta(s, r, x^\circ) = \theta(s + r, y(s))$$

where y is the solution of

$$y_j'(s) = y_j(s)\big[\gamma_j(s+r) y_j(s) - f(s+r, y(s))\big], \quad s \ge 0, \qquad y(0) = x^\circ.$$

By (15.10),

$$|\theta(t,x)| \le (n+1) \max_{j=1}^{n} |\gamma_j(t)|, \qquad t \ge r_0, x \in X,$$

which implies that $\tilde\theta(s, r, x)$ is a bounded function of all three variables. By Definition 15.8, ρ is a strong average Lyapunov function once it has been shown to be a weak average Lyapunov function.

Most of the properties in Definition 15.2 are easily verified; (v) follows from the Arzela-Ascoli Theorem [**164**, Ch.8.3] and the compactness of X. To find conditions that make ρ a weak average Lyapunov function, we suppose that Definition 15.2 (iv) does not hold. Then there are a sequence (s_m) in $[r_0, \infty)$ and an element $x^\circ \in X$ with $\rho(x^\circ) = 0$ and a sequence of solutions x^m such that

$$\limsup_{m\to\infty} \int_0^t \theta(s + s_m, x^m(s))ds \le 0 \quad \forall t \ge 0, \tag{15.11}$$

with x^m solving

$$\frac{d}{ds}x_j^m(s) = x_j^m \left(\gamma_j(s + s_m)x_{j-1}^m(s) - f(s + s_m, x^m(s))\right), \quad x^m(0) = x^\circ.$$

From (15.10) and (15.11),

$$0 \ge \limsup_{m\to\infty} \int_0^t \sum_{j=1}^n \gamma_j(s + s_m)x_{j-1}^m(s)ds - n \int_0^t f(s + s_m, x^m(s))ds, \tag{15.12}$$

and either x_j^m identically equal to 0, or $x_j^\circ > 0$ and

$$\ln x_j^m(t) - \ln x_j^\circ = \int_0^t \left(\gamma_j(s + s_m)x_{j-1}^m(s) - f(s + s_m, x^m(s))\right) ds. \tag{15.13}$$

Recall that the functions γ_j are bounded. Then the derivatives of x_j^m are bounded with a bound independent of m. If follows from the Arzela-Ascoli Theorem [**164**, Ch.8.3] that, after choosing a subsequence, $x^m \to y$ locally uniformly for some continuous function $y : \mathbb{R}_+ \to X$. By a corollary to Alaoglu's theorem [**164**, Thm.10.13], after replacing (s_j) by a subsequence, $\gamma_j(t + s_m) \to \bar\gamma_j(t)$ in the weak* topology with L^∞ considered as the dual space of L^1. More precisely,

$$\int_0^t \gamma_j(s + s_m)\zeta(s)ds \overset{m\to\infty}{\longrightarrow} \int_0^t \bar\gamma_j(s)\zeta(s)ds$$

for every $t > 0$ and every $\zeta \in L^1[0, t]$.

The limit functions $\bar\gamma_j$ are essentially bounded on $[0, \infty)$. We take the limit of (15.12) and (15.13) as $m \to \infty$,

$$0 \ge \sum_{j=1}^n \int_0^t \bar\gamma_j(s)y_{j-1}(s)ds - n \int_0^t \bar f(s, y(s))ds, \tag{15.14}$$

and, unless $x_j^\circ = 0$ and y_j identically equal to 0,

$$\ln y_j(t) - \ln x_j^\circ = \int_0^t \left(\bar\gamma_j(s)y_{j-1}(s) - \bar f(s, y(s))\right) ds. \tag{15.15}$$

Here $\bar f$ is given by (15.6) with $\bar\gamma_k$ replacing γ_k. We combine (15.14) and (15.15) into

$$\ln y_j(t) - \ln x_j^\circ \le \int_0^t \bar\gamma_j(s)y_{j-1}(s)ds - \frac{1}{n}\sum_{k=1}^n \int_0^t \bar\gamma_j(s)y_{k-1}(s)ds. \tag{15.16}$$

Since $\rho(x^\circ) = 0$, there is some $j \in \{1, \ldots, n\}$ such that $x_j^\circ = 0$. Then y_j is identically equal to 0. We now derive assumptions such that, for every $j \in \{1, \ldots, n\}$, $y_{j-1}(t) \to 0$ as $t \to \infty$ implies $y_j(t) \to 0$ as $t \to \infty$. This will yield a contradiction because it will imply that $y_j(t) \to 0$ as $t \to \infty$ for all $j = 1, \ldots, n$, while $\sum_{j=1}^n y_j(t) = 1$ for all $t \geq 0$. Without restriction of generality we can assume that $x_j^\circ > 0$ and $y_j(t) > 0$. If $y_{j-1}(t) \to 0$ as $t \to \infty$, then

$$\limsup_{t\to\infty} \frac{1}{t} \ln y_j(t) \leq -\frac{1}{n} \liminf_{t\to\infty} \frac{1}{t} \sum_{k=1}^n \int_0^t \bar{\gamma}_k(s) y_{k-1}(s) ds.$$

We obtain $y_j(t) \to 0$ as $t \to \infty$ if $\limsup_{t\to\infty} \frac{1}{t} \ln y_j(t) < 0$, i.e., if

$$\begin{aligned} 0 &< \liminf_{t\to\infty} \frac{1}{t} \sum_{k=1}^n \int_0^t \bar{\gamma}_k(s) y_{k-1}(s) ds \\ &= \liminf_{t\to\infty} \frac{1}{t} \sum_{k=1}^n \lim_{m\to\infty} \int_0^t \gamma_k(s + s_m) y_{k-1}(s) ds. \end{aligned}$$

Define

$$\gamma_\diamond(t) = \min_{k=1}^n \gamma_k(t). \tag{15.17}$$

Since $\sum_{k=1}^n y_{k-1}(t) = 1$,

$$\frac{1}{t} \sum_{k=1}^n \lim_{m\to\infty} \int_0^t \gamma_k(s + s_m) y_{k-1}(s) ds \geq \frac{1}{t} \limsup_{m\to\infty} \int_0^t \gamma_\diamond(s + s_m) ds,$$

and ρ is a weak average Lyapunov function if

$$\liminf_{t\to\infty} \frac{1}{t} \inf_{r\geq r_0} \int_0^t \gamma_\diamond(s + r) ds > 0.$$

Since we already checked Definition 15.8, under this condition ρ is a strong Lyapunov function and, by Theorem 15.5, the semiflow generated by the hypercycle equations is uniformly ρ-persistent. This can be reformulated as follows:

Theorem 15.10. *Let*

$$\gamma_\diamond(t) := \min_{k=1}^n \gamma_k(t) \quad \textit{and} \quad \liminf_{t\to\infty} \inf_{r\geq r_0} \frac{1}{t} \int_r^{r+t} \gamma_\diamond(s) ds > 0. \tag{15.18}$$

Then there exists some $\epsilon > 0$ such that, for all $r \geq r_0$, $x^\circ \in X$, $x_j^\circ > 0$, $j = 1, \ldots, n$, the solutions of the hypercycle equations (15.6), (15.7) satisfy

$$\liminf_{t\to\infty} x_j(t) \geq \epsilon \qquad \forall j = 1, \ldots, n.$$

Corollary 15.11. *Let*

$$\gamma_\diamond(t) := \min_{k=1}^{n} \gamma_k(t), \qquad \gamma_\diamond^+(t) = \max\{\gamma_\diamond(t), 0\}.$$

Assume there exists a strictly increasing sequence (t_j) in $[r_0, \infty)$ such that $t_j \to \infty$ as $j \to \infty$ and

$$\sup_{j\in\mathbb{N}} \int_{t_j}^{t_{j+1}} \gamma_\diamond^+(s)ds < \infty, \qquad \inf_{j\in\mathbb{N}} \int_{t_j}^{t_{j+1}} \frac{\gamma_\diamond(s)}{t_{j+1}-t_j} ds > 0.$$

Then there exists some $\epsilon > 0$ such that, for all $r \ge r_0$, $x^\circ \in X$, $x_j^\circ > 0$, $j = 1, \dots, n$, the solutions of the hypercycle equations (15.6), (15.7) satisfy

$$\liminf_{t\to\infty} x_j(t) \ge \epsilon \qquad \forall j = 1, \dots, n.$$

Proof. Let $\delta > 0$ and $c > 0$ such that

$$\int_{t_j}^{t_{j+1}} \gamma_\diamond^+(s)ds \le c, \qquad \int_{t_j}^{t_{j+1}} \frac{\gamma_\diamond(s)}{t_{j+1}-t_j} ds > \delta, \qquad j \in \mathbb{N}. \tag{15.19}$$

Let $r \ge t_1$ and $t \ge 0$. Choose j and k such that

$$t_j \le r < t_{j+1} \text{ and } t_k \le r + t < t_{k+1}. \tag{15.20}$$

Then $k \ge j$ and

$$\frac{1}{t}\int_r^{r+t} \gamma_\diamond(s)ds \ge \frac{1}{t}\int_{t_j}^{t_{k+1}} \gamma_\diamond(s)ds - \frac{1}{t}\int_{t_k}^{t_{k+1}} \gamma_\diamond^+(s)ds - \frac{1}{t}\int_{t_j}^{t_{j+1}} \gamma_\diamond^+(s)ds.$$

By (15.19) and (15.20)

$$\frac{1}{t}\int_r^{r+t} \gamma_\diamond(s)ds \ge \frac{1}{t}\sum_{i=j}^{k}\int_{t_i}^{t_{i+1}} \gamma_\diamond(s)ds - \frac{2c}{t}$$

$$\ge \frac{1}{t}\sum_{i=j}^{k} \delta(t_{i+1}-t_i) - \frac{2c}{t} \ge \delta\frac{t_{k+1}-t_j}{t} - \frac{2c}{t} \ge \delta - \frac{2c}{t}.$$

Notice that the right-hand side does not depend on $r \ge t_1$,

$$\inf_{r\ge t_1} \frac{1}{t}\int_0^t \gamma_\diamond(s+r)ds \ge \delta - \frac{2c}{t}.$$

If $r < t_1$,

$$\frac{1}{t}\int_0^t \gamma_\diamond(s+r)ds = \frac{1}{t}\int_r^{r+t} \gamma_\diamond(s)ds \ge \frac{1}{t}\int_{t_1}^{t_1+t} \gamma_\diamond(s)ds - \frac{2}{t}(t_1 - r)\sup\gamma_\diamond.$$

We combine this inequality with the previous one,

$$\inf_{r\ge r_0} \frac{1}{t}\int_0^t \gamma_\diamond(s+r)ds \ge \delta - \frac{2}{t}(c + (t_1 - r_0)\sup\gamma_\diamond).$$

This implies (15.18). □

If the catalysis rates $\gamma_j(t)$ are periodic with the same period τ, we can have strictly positive periodic solutions.

Theorem 15.12. *Let $\gamma_j : \mathbb{R} \to \mathbb{R}$, $j = 1, \dots, n$, be periodic with the same period $\tau > 0$,*

$$\gamma_\diamond(t) := \min_{k=1} \gamma_k(t), \qquad \textit{and} \qquad \int_0^\tau \gamma_\diamond(s)ds > 0.$$

Then there exists some $\epsilon > 0$ such that, for all $r \in \mathbb{R}$, $x^\circ \in X$, $x_j^\circ > 0$, $j = 1, \dots, n$, the solutions of the hypercycle equations (15.6), (15.7) satisfy

$$\liminf_{t \to \infty} x_j(t) \geq \epsilon \qquad \forall j = 1, \dots, n.$$

Further, there exists a strictly positive periodic solution.

Proof. For the persistence part, we apply Corollary 15.11. Choose $\ell \in \mathbb{N}$ large enough that $\ell\tau \geq r$ and $t_j = (\ell + j - 1)\tau$ for $j \in \mathbb{N}$. Since $\gamma_\diamond$ is also periodic with period τ, $\int_{t_j}^{t_{j+1}} \gamma^\diamond(s)ds = \int_0^\tau \gamma^\diamond(s)ds$. The same holds for $\gamma_\diamond^+$. Thus the assumptions of Corollary 15.11 are satisfied.

For the existence of the periodic solution, let $\Phi_1 : X \to X$ be the period map, $\Phi_1 = \Psi(\tau, 0, \cdot)$ and Φ the discrete semiflow $\Phi_t = \Phi_1^t = \Psi(t\tau, 0, \cdot)$, $t \in \mathbb{Z}_+$. We can choose $r_0 = 0$.

To apply Theorem 6.2, we first observe that the simplex X is convex. By Theorem 15.10, Φ is uniformly $\tilde{\rho}$-persistent for $\tilde{\rho}(x) = \min_j x_j$. The state space X is compact itself, so there trivially is a compact attractor of the whole state space and Φ_1 is a compact map, in particular, condensing. Assumption (H1) of Theorem 6.2 is satisfied because of the form of the differential system 15.7, and ρ is concave. By Theorem 6.2, Φ_1 has a fixed point $x^\diamond$ with $x_j^\diamond > 0$, $j = 1, \dots, n$. The solution with $x^\diamond$ at $t = 0$ is a periodic solution. By Corollary 15.11, this solution is strictly positive. □

Remark 15.13. Continuity and boundedness of the catalysis rates γ_j can be replaced by Borel measurability and essential boundedness.

In this case, assumption (iii)' in Remark 15.3 is satisfied.

The solution x is Lipschitz continuous and the differential equations (15.7) hold almost everywhere. Since ρ in (15.8) is Lipschitz continuous on X, $\tilde{\rho}(s, r, x)$ is a Lipschitz continuous function of s and in particular absolutely continuous on every bounded interval in $\mathbb{R}_+$. See Chapter A.2. $\theta(t, x)$ is Borel measurable in t and continuous in x and thus Borel measurable in both variables. This implies that $\tilde{\theta}(s, r, x)$ is a Borel measurable and essentially bounded function of $s \geq 0$.

For further-reaching results for the time-autonomous hypercycle equation we refer to [**108**].

15.4. Exercises

Exercise 15.1. Check Definition 15.2, (i), (ii), (iii), (v), for the hypercycle equation in Section 15.3.

Exercise 15.2. Prove Corollary 13.3 using an average Lyapunov function.

Appendix A

Tools from Analysis and Differential Equations

A.1. Lower one-sided derivatives

Let $f : (a, b) \to \mathbb{R}$ be continuous, $a < b$ real numbers. Then the lower left and right derivatives of f are defined as

$$f'_-(t) = \liminf_{0<h\to 0} \frac{f(t) - f(t-h)}{h}, \tag{A.1}$$
$$f'_+(t) = \liminf_{0<h\to 0} \frac{f(t+h) - f(t)}{h}.$$

Theorem A.1. *Let $f'_-(t) < 0$ for all $t \in (a, b)$ or $f'_+(t) < 0$ for all $t \in (a, b)$. Then f is strictly decreasing.*

Proof. Suppose that f is not strictly decreasing. Then there exist $a < r < t < b$ such that $f(r) \le f(t)$. Since f is continuous, f takes its minimum on the interval $[r, t]$ at some point s. We can choose $r \le s < t$. Then $f(s+h) \ge f(s)$ for all sufficiently small $h > 0$ and $f'_+(s) \ge 0$, a contradiction. f also takes its maximum on the interval $[r, t]$ at some point s. We can choose $r < s \le t$. Then $f(s - h) \le f(s)$ for all sufficiently small $h > 0$ and $f'_-(s) \ge 0$, a contradiction. □

Corollary A.2. *Let $f'_-(t) \le 0$ for all $t \in (a, b)$ or $f'_+(t) \le 0$ for all $t \in (a, b)$. Then f is decreasing on (a, b).*

Proof. Let $\epsilon > 0$ and define $g(t) = f(t) - \epsilon t$. Then $g'_+(t) = f'_+(t) - \epsilon < 0$. By Theorem A.1, $g(t)$ is strictly decreasing for all $\epsilon > 0$. This implies that f is decreasing. □

Remark A.3. Analogous results hold when "$\leq$" is replaced by "$\geq$", lim inf by lim sup in the definitions of f'_- and f'_+, and "decreasing" by "increasing".

One then speaks about *upper one-sided derivatives.*

A.2. Absolutely continuous functions

The following result in definition can be found in most advanced analysis books ([**164**], Definition 6.8 and Theorem 6.6, e.g.).

Theorem A.4 (and Definition). *Let $-\infty < a < b < \infty$ and $f : [a,b] \to \mathbb{R}$.*

Then the following two statements are equivalent, and f is called absolutely continuous *if it satisfies one (and then both) of them.*

(i) *f is differentiable almost everywhere on $[a,b]$, $f' \in L^1[a,b]$ and*

$$f(t) - f(a) = \int_a^t f(s)ds, \qquad t \in [a,b].$$

(ii) *For every $\epsilon > 0$ there exists some $\delta > 0$ such that, if $(a_1, b_1), \dots, (a_n, b_n)$, $n \in \mathbb{N}$, are disjoint open intervals, then*

$$\sum_{i=1}^n (b_i - a_i) < \delta \implies \sum_{i=1}^n |f(b_i) - f(a_i)| < \epsilon.$$

Obviously, every absolutely continuous function is continuous and every continuously differentiable function absolutely continuous. The easiest non-trivial example of an absolutely continuous function is a Lipschitz continuous function. In that case, $f' \in L^\infty[a,b]$.

Proposition A.5. *Let $-\infty < a < b < \infty$ and $f : [a,b] \to \mathbb{R}$ be absolutely continuous and $g : f([a,b]) \to \mathbb{R}$ be Lipschitz continuous.*

Proof. Let $\epsilon > 0$ and Λ a Lipschitz constant for g. Choose $\eta = \epsilon/(\Lambda + 1)$. Since f is absolutely continuous, there exists some $\delta > 0$ such that Theorem A.4 (ii) holds with η replacing ϵ. Let $(a_1, b_1), \dots, (a_n, b_n)$ be disjoint intervals with $\sum_{i=1}^n (b_i - a_i) < \delta$. Then

$$\sum_{i=1}^n |g(f(b_i)) - g(f(a_i))| \leq \Lambda \sum_{i=1}^n |f(b_i) - f(a_i)| \leq \Lambda\eta < \epsilon.$$ □

Definition A.6. If $\mathcal{I}$ is an open or half-open interval in $\mathbb{R}$ (including $\mathbb{R}_+$ or $\mathbb{R}$ itself), then $f : \mathcal{I} \to \mathbb{R}$ is called locally absolutely continuous if f is absolutely continuous on every compact subinterval of $\mathcal{I}$.

Corollary A.7. *Let $\mathcal{I}$ be as in Definition A.6. Then f is locally absolutely continuous on $\mathcal{I}$ if and only if f is differentiable a.e. on $\mathcal{I}$, f' is L^1 on every compact subinterval of $\mathcal{I}$ and*

$$f(t) - f(r) = \int_r^t f'(s)ds, \qquad r, t \in \mathcal{I}.$$

A.3. The method of fluctuation

Let $f : [b, \infty) \to \mathbb{R}$. Then the *limit superior* and the *limit inferior* of f as $t \to \infty$ are defined as

$$\begin{aligned} f^\infty &:= \limsup_{t\to\infty} f(t) = \inf_{r\geq b} \sup\{f(t); t \geq r\}, \\ f_\infty &:= \liminf_{t\to\infty} f(t) = \sup_{r\geq b} \inf\{f(t); t \geq r\}. \end{aligned} \tag{A.2}$$

Proposition A.8. *The following hold:*

(a) *If $f^\infty > -\infty$, for every $\eta > 0$ there exists $r > b$ such that*
$$f(t) < f^\infty + \eta \qquad \forall t \in [r, \infty).$$
(b) *If $f^\infty < \infty$, for every $\eta > 0$ and every $r > b$, there exists some $t \in [r, \infty)$ such that $f(t) > f^\infty - \eta$.*

(c) *If $f_\infty < \infty$, for every $\eta > 0$ there exists $r \geq b$ such that*
$$f(t) > f_\infty - \eta \qquad \forall t \in [r, \infty).$$
(d) *If $f_\infty > -\infty$, for every $\eta > 0$ and every $r > b$, there exists some $t \in [r, \infty)$ such that $f(t) < f_\infty + \eta$.*

(e) *$f_\infty \leq f^\infty$.*

Proof. Set

$$f^r = \sup\{f(t); t \geq r\}, \qquad f_r = \inf\{f(t); t \geq r\}.$$

Let $\eta > 0$.

(a) By definition of the infimum, there exists some $r > 0$ such that $f^r < f^\infty + \eta$. Otherwise $f^\infty + \eta$ were a lower bound of the f^r larger than f^∞. By definition of the supremum, $f(t) \leq f^r < f^\infty + \eta$ for all $t \geq r$.

(b) By definition of the infimum, $f^\infty \leq f^r$ for all $r > 0$. By definition of the supremum, for all $r > 0$ there exists some $t \geq r$ such that $f(t) > f^\infty - \eta$. Otherwise $f^\infty - \eta$ were an upper bound of the set $\{f(t); t \geq r\}$ that is smaller than f^r.

(c) and (d) are proved similar.

(e) Suppose that $f_\infty > f^\infty$. Let $\eta = \frac{f_\infty - f^\infty}{4}$. By (a) and (c), there exists some $t > 0$ such that

$$f(t) > f_\infty - \eta > f^\infty + \eta > f(t), \qquad \text{a contradiction.}$$ □

Example A.9. $\limsup_{t\to\infty} \cos t = 1$, $\liminf_{t\to\infty} \cos t = -1$.

Proposition A.10. *Let $f : [b, \infty) \to \mathbb{R}$. Then $f(t)$ converges as $t \to \infty$ if and only if $f^\infty = f_\infty$. Moreover, if one and then also the other of these two statements hold, $\lim_{t\to\infty} f(t) = f^\infty = f_\infty$.*

Proof. Suppose that $L = \lim_{t\to\infty} f(t)$ exists. Then, for all $\eta > 0$, there exists some $r \geq b$ such that $L - \eta < f(t) < L + \eta$ for all $t \geq r$. Hence $L - \eta \leq f_r$ and $f^r \leq L + \eta$. By definition of the supremum and the infimum and Proposition A.8 (e), $L - \eta \leq f_\infty \leq f^\infty \leq L + \eta$. Since this holds for all $\eta > 0$, $L = f_\infty = f^\infty$.

Suppose that $f_\infty = f^\infty =: L$. Let $\eta > 0$. By Proposition A.8, there exists some $r > b$ such that

$$L - \eta = f_\infty - \eta < f(t) < f^\infty + \eta = L + \eta \qquad \forall t \geq r.$$

This implies $f(t) \to L$ as $t \to \infty$. □

Definition A.11. Let $f : [b, \infty) \to \mathbb{R}$ be bounded. Define the ω-limit set of f, $\omega(f)$ as the collection of those $x \in \mathbb{R}$ such that there exists a sequence (t_k) with $t_k \to \infty$ as $k \to \infty$ and $x = \lim_{k\to\infty} f(t_k)$.

Lemma A.12. *Let $f : [b, \infty) \to \mathbb{R}$ be bounded. Then $f^\infty = \max \omega(f)$ and $f_\infty = \min \omega(f)$.*

The preceding definitions and results carry over with little change in the case that $f : \{b, b+1, b+2, \cdots\} \to \mathbb{R}$. Of course, this case is usually treated as a sequence $\{f_n = f(b+n)\}_{n\geq 0}$.

If we imagine that $f(t)$ is the density of some population at time t, then necessarily $f(t) \geq 0$ and we will be concerned with whether or not the population ultimately survives. A very weak statement of survivability is that $f^\infty > 0$; this just says that the population exceeds some positive threshold at arbitrarily large times. A stronger condition is $f_\infty > 0$ which implies that there is a positive "floor", or threshold (e.g., $f_\infty/2$) such that $f(t)$ exceeds this threshold for all sufficiently large time.

Lemma A.13 (Fluctuation method 1 [**217**, Lemma A.20]). *Let $f : [b, \infty) \to \mathbb{R}$ be bounded and differentiable and have no limit as $t \to \infty$. Then there exist sequences $s_k, t_k \to \infty$ as $k \to \infty$ with the following properties:*

$$f(s_k) \to f_\infty, \quad k \to \infty, \qquad f'(s_k) = 0 \quad \forall k \in \mathbb{N},$$
$$f(t_k) \to f^\infty, \quad k \to \infty, \qquad f'(t_k) = 0 \quad \forall k \in \mathbb{N}.$$

If f is twice differentiable, one has in addition that

$$f''(s_k) \geq 0, \quad f''(t_k) \leq 0 \qquad \forall k \in \mathbb{N}.$$

Proposition A.14 (fluctuation method 2). *Let $f : [b, \infty) \to \mathbb{R}$ be bounded and differentiable. Then there exist sequences $s_k, t_k \to \infty$ such that*

$$\left.\begin{array}{ll} f(s_k) \to f_\infty, & f'(s_k) \to 0 \\ f(t_k) \to f^\infty, & f'(t_k) \to 0 \end{array}\right\} k \to \infty.$$

Proof. By Lemma A.13, we can assume that $f(t)$ has a finite limit as $t \to \infty$. Assume that the statement of this proposition does not hold. Then there exists some $r > b, \epsilon > 0$ such that $|f'(t)| \geq \epsilon$ for all $t \geq r$. By the intermediate value theorem for derivatives [**118**, Exc. 5-2.18], $f'(t)$ has the same sign for all $t \geq r$. In one case, $f'(t) \geq \epsilon$ for all $t \geq r$. Then $f(t)$ increases to infinity. In the second case, $f'(t) \leq -\epsilon$ for all $t \geq r$. Then $f(t)$ decreases to $-\infty$ as $t \to \infty$. In either case, we obtain a contradiction to the boundedness of f. □

Lemma A.15 ([**217**, Lemma A.6]). *Let $f : [b, c) \to \mathbb{R}$ be continuous and differentiable on (b, c). Let $t \in (b, c)$ and $\bar{f} = \max_{[b,t]} f$. Then $\bar{f} = f(b)$ or there exists $s \in (b, t]$ such that $\bar{f} = f(s)$ and $f'(s) \geq 0$.*

Lemma A.16 (Barbalat [**217**, Cor.A.19]). *Let $f : [b, \infty) \to \mathbb{R}$ be differentiable and f' be uniformly continuous on $[b, \infty)$. If $\lim_{t\to\infty} f(t)$ exists and is finite, then $f'(t) \to 0$ as $t \to \infty$.*

A.4. Differential inequalities and positivity of solutions

A.4.1. ODEs. Systems of ordinary differential equations arising in biology typically involve components which are inherently nonnegative (they may be densities or concentrations) and therefore it becomes important to establish that solutions which start off with nonnegative components continue to satisfy this property. We begin by establishing some useful notation. The nonnegative cone in $\mathbb{R}^n$, denoted by $\mathbb{R}^n_+$, is the set of all n-tuples with nonnegative coordinates. One can define a partial order on $\mathbb{R}^n$ by $y \leq x$ if $y_i \leq x_i$ for all i. We write $x \ll y$ if $x_i < y_i$ for all i. The same notation will be used for matrices with a similar meaning.

In this section we consider the nonautonomous system

$$x' = f(t, x) \tag{A.3}$$

where $f : \mathbb{R} \times D \to \mathbb{R}^n$ is continuous, D is an open subset of $\mathbb{R}^n$, and $f = (f_1, f_2, \ldots, f_n)$.

Proposition A.17. *Suppose that f in (A.3) has the property that solutions of initial value problems $x(t_0) = x_0 \geq 0$ are unique and, for all i, $f_i(t, x) \geq 0$ whenever $x \geq 0$ satisfies $x_i = 0$. If $x(t_0) \geq 0$ then $x(t) \geq 0$ for all $t \geq t_0$ for which it is defined.*

Proof. The assertion is obvious when f satisfies the stronger condition that $f_i(t,x) > 0$ whenever $x \geq 0$ satisfies $x_i = 0$. The general case can be treated by a limiting argument. For $s > 0$, define $f_s(t,x) = f(t,x) + sv$ where v is the vector with all entries equal to 1. Then f_s satisfies the stronger condition and so the solution $x_s(t)$ of $x' = f_s(t,x)$, $x(t_0) = x_0 \geq 0$ satisfies $x_s(t) \geq 0$ for all $t \geq t_0$ for which it is defined. By continuity of solutions of differential equations with respect to parameters (Lemma 3.1, Chapt. 1 of [**90**]), the solution $x(t)$ of (A.3) and $x(t_0) = x_0 \geq 0$ can be approximated at any fixed $t \geq t_0$ by $x_s(t)$, i.e., $x_s(t) \to x(t)$, $s \to 0$. It follows that $x(t) \geq 0$. □

The assumption that solutions of initial value problems for (A.3) are unique is satisfied, for example, if f satisfies a Lipschitz condition in the x variable on each compact subset of $\mathbb{R} \times D$.

Extracting information from differential inequalities is an important tool in applied dynamics. We follow the presentation in Coppel [**34**] and Smith and Waltman [**203**]. The idea is to compare solutions of (A.3) with solutions of the system of differential inequalities

$$x' \leq f(t,x) \tag{A.4}$$

or

$$x' \geq f(t,x) \tag{A.5}$$

on an interval. We assume that solutions of initial value problems for (A.3) are unique.

For general vector-valued functions f it is not possible to make any comparisons between solutions of (A.3) and solutions of (A.4) or (A.5). For example, $(x(t), y(t)) = (1, -t)$ satisfies

$$\begin{aligned} x' &\leq -y, \\ y' &\leq x, \end{aligned} \tag{A.6}$$

for $t \geq 0$ and $(\bar{x}(t), \bar{y}(t)) = (\cos(t), \sin(t))$ satisfies the corresponding differential equation with $(\bar{x}(0), \bar{y}(0)) = (x(0), y(0))$ but it is not true that $(x(t), y(t)) = (1, -t) \leq (\cos(t), \sin(t)) = (\bar{x}(t), \bar{y}(t))$ for $t \geq 0$.

Definition A.18. Function f is said to be *quasimonotone* in D if for each i and all t, $f_i(t,a) \leq f_i(t,b)$ for any two points a and b in D satisfying $a \leq b$ and $a_i = b_i$.

In words, f_i must be increasing in x_j for $j \neq i$. Observe that for $n = 1$, that is, for the case of a scalar-valued function f, the quasimonotone condition is automatically satisfied. The quasimonotone condition is quite restrictive for $n > 1$, but the result below is still very useful in this case.

Theorem A.19. *Let f be continuous and be quasimonotone on D. Let $x(t)$ be a solution of (A.3) defined on $[a,b]$. If $z(t)$ is a continuous function on $[a,b]$ satisfying (A.4) on (a,b) with $z(a) \le x(a)$, then $z(t) \le x(t)$ for all t in $[a,b]$. If $y(t)$ is continuous on $[a,b]$ satisfying (A.5) on (a,b) with $y(a) \ge x(a)$, then $y(t) \ge x(t)$ for all t in $[a,b]$.*

Proof. For $m = 1, 2, \ldots$, let $x_m(t)$ be a solution of the initial value problem

$$x_m' = f(t, x_m) + (1/m)e, \quad x_m(a) = x(a) + (1/m)e$$

where $e = (1, 1, \ldots, 1)$. By Lemma 3.1, Chapt. 1 of [**90**], $x_m(t)$ is defined on $[a,b]$ for all sufficiently large m and $x_m(t) \to x(t)$ as $t \to \infty$, uniformly on $[a,b]$. We show that $z(t) < x_m(t)$, $a < t < b$, for all large m, from which the first assertion follows by taking limits as $m \to \infty$. The second assertion is proved in a similar manner.

Let $m \ge 1$ be fixed such that $x_m(t)$ is defined on $[a,b]$. As $z(a) < x_m(a)$, continuity implies that $z(t) < x_m(t)$ for $t > a$ and $t - a$ small. Consequently, if $z(t) < x_m(t)$ is violated for some $t \in (a,b)$, then there exists j and $t_0 \in (a,b)$ such that $z_i(t) < x_{mi}(t)$, $a < t < t_0$, $1 \le i \le n$ and $z_j(t_0) = x_{mj}(t_0)$. Therefore,

$$\begin{aligned} f_j(t_0, z(t_0)) \ge & z_j'(t_0) \ge x_{mj}'(t_0) \\ = & f_j(t_0, x_m(t_0)) + (1/m) > f_j(t_0, x_m(t_0)). \end{aligned}$$

But $z(t_0) \le x_m(t_0)$ and $z_j(t_0) = x_{mj}(t_0)$ implies, by the quasimonotone condition, that $f_j(t_0, z(t_0)) \le f_j(t_0, x_m(t_0))$. This contradiction proves the theorem. □

See [**34**] for a more general result. Theorem A.19 is traditionally used when the solution of (A.3) is known, or a bound on it is known, and $z(t)$ or $y(t)$ arises from some more complicated differential equation whose right-hand side can be compared to f.

As a simple but useful example, suppose that $x(t)$ satisfies

$$x'(t) \le p - qx(t), \ t \ge t_0$$

for some p and $q \ne 0$. Then we may conclude from Theorem A.19 (or by elementary calculus), that

$$x(t) \le x(t_0)e^{-q(t-t_0)} + \frac{p}{q}\left[1 - e^{-q(t-t_0)}\right], \ t \ge t_0.$$

If $q > 0$, this implies that

$$x(t) \le \max\left\{x(t_0), \frac{p}{q}\right\} \quad \text{and} \quad x^\infty \le \frac{p}{q}.$$

The quasimonotone condition can be expressed in terms of the partial derivatives of f with respect to its second variable on suitable domains.

Definition A.20. We say that D is *p-convex* if $sx + (1-s)y \in D$ for all $s \in [0,1]$ whenever $x, y \in D$ and $x \leq y$.

Lemma A.21. *Let $f(t,x)$ and $\frac{\partial f}{\partial x}(t,x)$ be continuous on $\mathbb{R} \times D$, where D is a p-convex subset of $\mathbb{R}^n$. Let*

$$\frac{\partial f_i}{\partial x_j}(t,x) \geq 0, \qquad i \neq j, \quad (t,x) \in \mathbb{R} \times D, \tag{A.7}$$

hold. Then f satisfies the quasimonotone condition on D.

Proof. If $a \leq b$ and $a_i = b_i$, then

$$f_i(t,b) - f_i(t,a) = \int_0^1 \sum_{j \neq i} \frac{\partial f_i}{\partial x_j}(t, a + r(b-a))(b_j - a_j)dr \geq 0,$$

by (A.7). □

Definition A.22. The system (A.3) is said to be a *cooperative system* if the hypotheses of Lemma A.21 hold.

As a special case of Theorem A.19, note that if $x(t)$ and $y(t)$ are two solutions of the cooperative system (A.3) on $[a,b]$ with $x(a) \leq y(a)$, then $x(t) \leq y(t)$, $a \leq t \leq b$.

A.4.2. PDEs. There are comparison theorems analogous to (A.19) for systems of partial differential equations, but we will need only the scalar version here.

Let L be the second order differential operator given by

$$L = \sum_{i,j=1}^{n} a_{ij}(t,x)\frac{\partial^2}{\partial x_i \partial x_j} + \sum_{i=1}^{n} b_i(t,x)\frac{\partial}{\partial x_i}$$

where we may as well assume that $a_{ij} = a_{ji}$.

We assume that L is uniformly elliptic. That is, there exist $A_1 \geq A_0 > 0$ such that:

$$A_0|\eta|^2 \leq \sum_{i,j=1}^{n} a_{ij}(t,x)\eta_i\eta_j \leq A_1|\eta|^2$$

for all $\eta \in \mathbb{R}^n$ and all $(t,x) \in \overline{\Omega} \times (0,T]$ described below.

The most important example is the Laplacian

$$L = \sum_{i=1}^{n} \frac{\partial^2}{\partial x_i^2}$$

where we may take $A_i = 1$, $i = 0, 1$.

The following result may be found in [**25, 147**].

Theorem A.23. *Let $\Omega \subset \mathbb{R}^n$ be a bounded open connected set whose boundary, $\partial\Omega$ is C^2 with outer normal $n = n(x)$, $T > 0$, and a_{ij}, b_i bounded functions on $(0,T]\times\Omega$. Suppose that $f(t,x,u), \partial f(t,x,u)/\partial u \in C([0,T]\times\overline{\Omega}\times\mathbb{R})$. If $u, v \in C^{2,1}((0,T]\times\overline{\Omega}) \cap C([0,T]\times\overline{\Omega})$ with*

$$\begin{aligned} u_t - Lu &\geq f(t,x,u),\ (t,x) \in (0,T]\times\Omega, \\ v_t - Lv &\leq f(t,x,v),\ (t,x) \in (0,T]\times\Omega, \end{aligned}$$

$u(0,x) \geq v(0,x)$ on Ω, and either $u(t,x) \geq v(t,x)$ on $(0,T]\times\partial\Omega$ or

$$\gamma(t,x)u + \beta(t,x)\frac{\partial u}{\partial n} \geq \gamma(t,x)v + \beta(t,x)\frac{\partial v}{\partial n} \tag{A.8}$$

on $(0,T]\times\partial\Omega$, where $\gamma \geq 0$ and $\beta > 0$ on $(0,T]\times\partial\Omega$. Then either $u \equiv v$ or $u > v$ on $(0,T]\times\Omega$; if (A.8) holds, then these conclusions hold on $(0,T]\times\overline{\Omega}$.

Proof. We give only the proof that $u \geq v$.

Let $w = v - u$ and observe that it satisfies

$$\begin{aligned} w_t &\leq Lw + f(t,x,v) - f(t,x,u) \\ &= Lw + \int_0^1 f_u(t,x,sv+(1-s)u)ds\ w \\ &= Lw + c(t,x)w, \end{aligned}$$

$w(0,x) \leq 0$ on Ω, and either $w(t,x) \leq 0$ or

$$\gamma(t,x)w + \beta(t,x)\frac{\partial w}{\partial n} \leq 0$$

on $(0,T]\times\partial\Omega$, where

$$c(t,x) = \int_0^1 f_u(t,x,su(t,x)+(1-s)v(t,x))ds.$$

Now let $W = e^{\lambda t}w$ where λ is chosen such that $c(t,x)+\lambda \leq 0$ on $[0,T]\times\overline{\Omega}$. Then W satisfies

$$W_t \leq LW + (c+\lambda)W$$

and the same boundary and initial conditions as w. Let M be the maximum value attained by W on $[0,T]\times\overline{\Omega}$. Suppose that $M > 0$. Then, by the Strong Parabolic Maximum Principle (see e.g., [**147, 166, 178**]), if $W(t_0,x_0) = M$ for some $(t_0,x_0) \in (0,T]\times\Omega$ we may conclude that $W(t,x) = M$ for all $x \in \Omega$ and $t \leq t_0$. In particular, $W(0,x) = M > 0$ contradicting that $W(0,x) \leq 0$. If $W(t_0,x_0) = M$ for some $(t_0,x_0) \in (0,T]\times\partial\Omega$, then we have an immediate contradiction in case the hypothesis $w(t_0,x_0) \leq 0$ holds since it implies $W(t_0,x_0) \leq 0$. If, instead,

$$\gamma(t,x)W + \beta(t,x)\frac{\partial W}{\partial n} \leq 0$$

holds at (t_0, x_0) with $t_0 \leq T$, then we apply the Strong Parabolic Maximum Principle, see especially [**147**], again to conclude that $\frac{\partial W}{\partial n}(t_0, x_0) > 0$, which together with $W(t_0, x_0) = M > 0$ and the conditions on γ and β, imply a contradiction to the boundary inequality. Therefore, we conclude that our assumption $M > 0$ is untenable. Hence $W(t, x) \leq M \leq 0$ so $w(t, x) \leq 0$ and $v - u \leq 0$ in $(0, T] \times \overline{\Omega}$. □

See [**166, 207, 147, 25**] for more on differential inequalities for partial differential equations.

A.5. Perron-Frobenius theory

A square matrix A is called *positive*, $A > 0$, if all entries are nonnegative numbers and the matrix is not the zero matrix. It is called *quasipositive* if it is not the zero matrix and all off-diagonal entries are nonnegative numbers. It is called *strictly positive*, $A \gg 0$, if all entries are strictly positive. The analogous notation holds for vectors.

If $n \geq 2$, an $n \times n$ matrix $A = (a_{ik})$ is called *irreducible* if the following holds: For any proper nonempty subset P of $\{1, \ldots, n\}$ there are $k \in P, j \notin P$ such that $a_{jk} \neq 0$. A 1×1 matrix is called irreducible if it is not the 0 matrix.

Equivalently, A is irreducible if and only if, for all $i, k = 1, \ldots, n$, there exist numbers $j_1, \ldots, j_r \in \{1, \ldots, n\}$ such that $i = j_1, k = j_r$ and $a_{j_l j_{l+1}} \neq 0$ for all $l = 1, \ldots, r-1$. A nonnegative matrix A is irreducible if and only if the matrix exponential e^A is strictly positive.

A nonnegative square matrix A is called *primitive* if one of its powers, A^k, has strictly positive entries. It is easily seen that a nonnegative matrix is primitive if it is irreducible and all entries in its main diagonal are strictly positive.

If A is a complex square matrix, we denote by $\sigma(A)$ its eigenvalues. The *spectral radius* of the matrix A, $\mathbf{r}(A)$, is defined as

$$\mathbf{r}(A) = \max\{|\lambda|; \lambda \in \sigma(A)\},$$

while the spectral bound or stability modulus of A, $\mathbf{s}(A)$ is defined as

$$\mathbf{s}(A) = \max\{\Re\lambda; \lambda \in \sigma(A)\}.$$

Obviously, $\mathbf{s}(A) \leq \mathbf{r}(A)$. The following formula does not only hold for matrices, but bounded linear operators,

$$\mathbf{r}(A) = \lim_{n\to\infty} \|A^n\|^{1/n} = \inf_{n\in\mathbb{N}} \|A^n\|^{1/n}, \tag{A.9}$$

where $\|A\|$ is any matrix norm (or operator norm) ([**236**] VIII.2, Thm.3 and its proof). A norm $|\cdot|$ on $\mathbb{R}^m$ is called monotone if $|x| \leq |y|$ whenever $0 \leq x \leq y$.

Proposition A.24. *For every $r > \mathbf{r}(A)$, there exists a norm $|\cdot|'$ on $\mathbb{R}^m$ such that $|Ax|' \leq r|x|'$ for all $x \in \mathbb{R}^m$. If A is nonnegative, the norm $|\cdot|'$ is monotone.*

This result also holds for any bounded linear operator on a Banach space with the new norm being equivalent to the original one. If the Banach space is ordered, the operator is positive and the original norm is monotone, then also the new norm is monotone. The norms are related by

$$|x|' = |x| + \sum_{j=1}^{k} |A^j x|/r^j, \tag{A.10}$$

with k sufficiently large such that $\|A^{k+1}\| < r^{k+1}$ which is possible by (A.24) and $r > \mathbf{r}(A)$. For details see [**128**, 2.5.2].

Theorem A.25. *Let A be a positive matrix. Then its spectral radius, $\mathbf{r}(A)$, is an eigenvalue associated both with a positive eigenvector v of A and a positive eigenvector w of the transposed matrix A^*. In particular, $\mathbf{s}(A) = \mathbf{r}(A)$.*

If, in addition, A is irreducible, then $\mathbf{r}(A)$ is an algebraically simple eigenvalue of A and $v \gg 0$, $w \gg 0$. Moreover, any eigenvector u of A satisfying $u \geq 0$ is a multiple of v; a similar conclusion holds for A^*.

If A is primitive, then it is irreducible and $|\lambda| < \mathbf{r}(A)$ for every $\lambda \in \sigma(A)$ distinct from $\mathbf{r}(A)$.

These results have immediate consequences for quasipositive matrices because A is quasipositive if and only if $A + sI = B$ is a positive matrix for all large s. As $\sigma(B) = \sigma(A) + s$ we see that $\mathbf{s}(B) = \mathbf{r}(B) = \mathbf{s}(A) + s$.

Theorem A.26. *Let A be a quasipositive matrix. Then its stability modulus, $\mathbf{s}(A)$, is an eigenvalue associated both with a positive eigenvector v of A and a positive eigenvector w of the transposed matrix A^*.*

If, in addition, A is irreducible, then $\mathbf{s}(A)$ is an algebraically simple eigenvalue of A and $v \gg 0$, $w \gg 0$. Moreover, any eigenvector u of A satisfying $u > 0$ is a multiple of v; a similar conclusion holds for A^*.

The following estimate will be used often.

Corollary A.27. *Let $|\cdot|$ be a norm on $\mathbb{R}^m$, A an irreducible quasipositive matrix, and $\lambda = \mathbf{s}(A)$ be the spectral bound of A. Let $b \in (0, \infty]$ and $x : [0, b) \to \mathbb{R}^m$ be differentiable and satisfy $x' \geq Ax$ on $[0, b)$, $x(0) \in \mathbb{R}^m_+$. Then* *there exists some $\xi > 0$ which only depends on the choice of the norm such that*

$$|x(t)| \geq \xi e^{\lambda t}|x(0)|, \qquad t \in [0, b).$$

Proof. By Proposition A.17, $x(t) \geq 0$ for all $t \in [0, b)$. Let v be the strictly positive eigenvector of A^* associated with λ. Then

$$\langle x, v\rangle' = \langle x', v\rangle \geq \langle Ax, v\rangle = \langle x, A^*v\rangle = \langle x, \lambda v\rangle = \lambda\langle x, v\rangle.$$

We integrate this inequality,

$$\langle x(t), v\rangle \geq e^{\lambda t}\langle x(0), v\rangle, \qquad t \in [0, b).$$

Define the norm $|\cdot|_\diamond$ by

$$|x|_\diamond = \sum_{j=1}^{m} |x_j| v_j.$$

This is a norm, indeed, because $v \gg 0$, and

$$|x(t)|_\diamond \geq e^{\lambda t}|x(0)|_\diamond, \qquad t \in [0, b).$$

Since all norms are equivalent on $\mathbb{R}^m$, there exist $c_2 \geq 1 \geq c_1 > 0$ such that $c_2|x|_\diamond \geq |x| \geq c_1|x|_\diamond$. Thus

$$|x(t)| \geq c_1|x(t)|_\diamond \geq c_1 e^{\lambda t}|x(0)|_\diamond \geq c_2^{-1}c_1 e^{\lambda t}|x(0)|. \qquad \square$$

An excellent reference for this material is [**15**].

Consider the linear system of differential equations

$$x' = A(t)x \tag{A.11}$$

where $A(t)$ is an $n \times n$ continuous matrix on an interval $[t_0, b)$.

Proposition A.28. *If $A(t)$ is quasipositive for $[t_0, b)$ and $x(t)$ is a nonzero solution of (A.11) satisfying $x(t_0) \geq 0$, then $x(t) \geq 0$, $t_0 \leq t < b$, and $x(t) \gg 0$, $t_0 < t < b$, if $A(t_0)$ is irreducible.*

Proof. The first assertion follows from Proposition A.17. Observe that, for each i, $x_i' \geq a_{ii}(t)x_i$ and so

$$x_i(t) \geq x_i(s)\exp\Big(\int_s^t a_{ii}(\eta)d\eta\Big), \qquad t > s.$$

Once a component becomes positive, it remains so in the future.

Define $I = \{i : x_i(t) > 0, b > t > t_0\}$. I is nonempty. If I coincides with $\{1, 2, \ldots, n\}$, then our proof is complete; so we assume I is a proper subset. By the definition of irreducibility of $A(t_0)$, there exists $j \in I$ and $k \notin I$ such that $a_{kj}(t_0) > 0$, and by continuity, it is positive near t_0. As k is not in I,

$x_k(s) = 0$ for all $s > t_0$ sufficiently near t_0 by the above paragraph. Hence for $s > t_0$ sufficiently near t_0 that $x_k(s) = 0$ and $a_{kj}(s) > 0$, we have

$$x'_k(s) \geq a_{kj}(s)x_j(s) > 0,$$

a contradiction. □

An important special case of Proposition A.28 is that if A is a quasipositive and irreducible matrix, then

$$e^{At} \gg 0, \ t > 0. \tag{A.12}$$

A.6. Exercises

Exercise A.1. Let $f, g : [0, b) \to \mathbb{R}$ be continuous and $f'_+(t) \leq \alpha f(t) + g(t)$ for all $t \in (0, b)$. Then $f(t) \leq f(0)e^{\alpha t} + \int_0^t e^{\alpha(t-s)}g(s)ds$ for all $t \in [0, b)$.

Exercise A.2 (chain rule)**.** Let I be an interval, $f : [0, b) \to I$ be continuous and $g : I \to \mathbb{R}$ be differentiable with $g'(x) > 0$ for all $x \in I$. Show that

$$(g \circ f)'_+(t) = g'(f(t))f'_+(t), \qquad t \in [0, b).$$

This also holds if lim inf is replaced by lim sup in the definition of f'_+.

Hint: Imitate the proof of the classical chain-rule.

Exercise A.3. Prove Lemma A.12.

Hint: Use Proposition A.8.

Exercise A.4. Suppose that $x(t) = (x_1(t), x_2(t))$ satisfies

$$\begin{aligned} x'_1 &\leq a + bx_1 + cx_2, \\ x'_2 &\leq d + ex_1 + fx_2, \end{aligned}$$

for constants a, b, c, d, e, f. What restrictions must be made on these constants such that the right-hand side satisfies the quasimonotone condition in $\mathbb{R}^2$? Why does Theorem A.19 fail for (A.6)? If $a = 1,\ b = -2,\ c = 1,\ d = 0,\ e = 1,\ f = -2$ and $x(0) = (0, 1)^*$, give an explicit estimate of $x(t)$ for $t \geq 0$.

Exercise A.5. Consider a single population model

$$\begin{aligned} u_t &= d\Delta u + ru(1 - u/K), \quad x \in \Omega,\ t > 0, \\ \frac{\partial u}{\partial n}(t, x) &= 0, \quad x \in \partial\Omega,\ t > 0, \\ u(0, x) &= u_0(x) \geq 0, \quad x \in \Omega, \end{aligned}$$

where $r, K, d > 0$, Ω satisfy the assumptions of the comparison theorem and n denotes the outward normal to its boundary. Show that if $u_0 \geq 0$ then $u \geq 0$ for all $x \in \overline{\Omega}$ and $t > 0$ and $u > 0$ if $u_0 \neq 0$. Show that every solution with u_0 continuous and positive converges to the carrying capacity

K as $t \to \infty$, uniformly in $x \in \overline{\Omega}$. You may assume the solutions are smooth enough to apply the comparison theorem.

Hint: Solutions of the logistic ordinary differential equation are solutions of the equation with constant u_0.

Exercise A.6. Let A be a quasipositive matrix with $\mathbf{s}(A) < 0$. Show that $A^{-1} = -\int_0^\infty e^{At} dt$ and that $-A^{-1} \geq 0$. If, in addition, A is irreducible, show that $-A^{-1} \gg 0$. Hint: Use (A.12).

Exercise A.7. Prove Proposition A.24.

Appendix B

Tools from Functional Analysis and Integral Equations

B.1. Compact sets in $L^p(\mathbb{R}_+)$

This is the Fréchet-Kolmogorov criterion for the compactness of sets in $L^p(\mathbb{R})$ [**236**, X.1].

Theorem B.1. *Let K be a subset of $L^p(\mathbb{R})$, $1 \le p < \infty$. Then K has compact closure if and only if the following three conditions hold:*

(i) $\sup_{f \in K} \int_{-\infty}^{\infty} |f(s)|^p ds < \infty$,

(ii) $\lim_{t \to \infty} \int_{|s|>t} |f(s)|^p ds \to 0$ *uniformly in* $f \in K$,

(iii) $\lim_{t \to 0} \int_{-\infty}^{\infty} |f(s+t) - f(s)|^p ds \to 0$ *uniformly in* $f \in K$.

We now easily obtain a compactness criterion by identifying $L^p(\mathbb{R}_+)$ with the space of those functions in $L^p(\mathbb{R})$ which are 0 on the negative half-line.

Theorem B.2. *Let K be a subset of $L^p(\mathbb{R}_+)$, $1 \le p < \infty$. Then K has compact closure if and only if the following three conditions hold:*

(i) $\sup_{f \in K} \int_{0}^{\infty} |f(a)|^p da < \infty$,

(ii) $\lim_{r\to\infty}\int_r^\infty |f(a)|^p da \to 0$ *uniformly in* $f \in K$,

(iii) $\lim_{h\searrow 0}\int_0^\infty |f(a+h)-f(a)|^p da \to 0$ *uniformly in* $f \in K$,

(iv) $\lim_{h\searrow 0}\int_0^h |f(a)|^p da \to 0$ *uniformly in* $f \in K$.

B.2. Volterra integral equations

Theorem B.3. *Let $k : \mathbb{R}_+ \to \mathbb{R}$ be locally integrable and nonnegative. Then there exists a unique locally integrable nonnegative function $r : \mathbb{R}_+ \to \mathbb{R}_+$ (called the resolvent of k) such that*

$$r(t) = \int_0^t k(s)r(t-s)ds + k(t).$$

If k is not zero a.e., then there exists some $t_0 \geq 0$ such that $r(t) - k(t) > 0$ for all $t > t_0$. Actually, if $t_0 < t_1 < t_2 < \infty$, there exists some $\epsilon > 0$ such that $r(t) \geq \epsilon$ if $t \in [t_1, t_2]$.

In the following it is convenient to use the *convolution notation*

$$(k * r)(t) = \int_0^t k(s)r(t-s)ds. \tag{B.1}$$

Lemma B.4. *Let $f, g, h : \mathbb{R}_+ \to \mathbb{R}_+$ be locally integrable. Then*

$$f*(g*h) = (f*g)*h, \qquad f*(g\pm h) = f*g \pm f*h, \qquad f*g = g*h.$$

Further, if $g \leq h$, then $f * g \leq f * h$.

Proof. The associative law can be seen by applying the Laplace transform which converts the convolution into a product and using the uniqueness properties of the Laplace transform. The other properties are obvious. □

Proof of Theorem B.3. Existence and uniqueness of r follows from Banach's fixed point theorem which also provides that r is integrable on every bounded interval in $\mathbb{R}_+$. Since the fixed point is obtained by successive approximations,

$$r(t) = \sum_{j=1}^\infty k_n(t) \tag{B.2}$$

where $k_{n+1} = k * k_n$, $k_1 = k$. If $0 \leq \tilde{k} \leq k$, then $\tilde{k}_n \leq k_n$ for all $n \in N$ by induction, and the respective resolvents also satisfy $0 \leq \tilde{r} \leq r$ and $r - k \geq \tilde{r} - \tilde{k}$. So we can assume without restriction of generality that k is still not zero a.e., but also bounded and with compact support. Then

k_n is continuous for $n \geq 2$. Further, the convergence in (B.2) is uniform on bounded subintervals of $\mathbb{R}_+$. This implies that $r - k$ is continuous. Taking Laplace transforms, we find that

$$\widehat{k_n} = (\hat{k})^n > 0,$$

so each k_n is not zero a.e. In particular, for $n \geq 2$, there are $0 \leq a_n < b_n$ such that $k_n(t) > 0$ if $a_n < t < b_n$. Consider two nonnegative functions f and g such that $f(t) > 0$ if $0 \leq a < s < b$ and $g(s) > 0$ if $0 \leq c < t < d$. Then $f(s)g(t-s) > 0$ if $a < s < b$ and $t - d < s < t - c$. So $(f * g)(t) > 0$ if the intervals (a, b) and $(t - d, t - c)$ have nonempty intersection. This is the case if $t - c > a$ and $t - d < b$, i.e. $a + c < t < b + d$. Since k_2 is continuous and not zero a.e., there exist $0 \leq a < b$ such that $k_2(t) > 0$ for all $t \in (a, b)$. By our previous consideration $k_{2n}(t) > 0$ if $t \in (na, nb)$. There exists some $\ell \in \mathbb{N}$ such that $\ell b > (\ell + 1)a$ and so $k_{2n}(t) > 0$ for all $t \in \big(na, (n+1)a\big]$ and $n \geq \ell$. So

$$r(t) - k(t) \geq \sum_{j=\ell}^{\infty} k_{2j}(t) > 0, \qquad t > \ell a.$$

Further, the right-hand side is continuous. □

Theorem B.5. *Let k and r be as before and*

$$w(t) \geq \int_0^t k(s)w(t-s)ds + g(t).$$

Then

$$w(t) \geq \int_0^t r(s)g(t-s)ds + g(t).$$

Proof. By Theorem B.3, in convolution notation, $k = r - r * k$. So, by assumption,

$$w \geq k * w + g = (r - r * k) * w + g = r * (w - k * w) + g \geq r * g + g.$$

□

Corollary B.6. *Let w, r and g be nonnegative and locally integrable,*

$$w(t) \geq \int_0^t k(s)w(t-s)ds + g(t).$$

Assume that k is not zero a.e. Then there exists some $b > 0$, which only depends on k and not on g, such that $w(t) > 0$ for all $t > b$ with $\int_0^{t-b} g(s)ds > 0$.

In particular, if g is continuous at 0 and $g(0) > 0$, then $w(t) > 0$ for all $t > b$.

Proof. By our previous results,

$$w(t) \geq \int_0^t r(s)g(t-s)ds$$

and there is some $b > 0$ such that $r(s) > \delta_t > 0$ for all $s \in [b, t]$. Thus, for $t > b$,

$$w(t) \geq \delta_t \int_b^t g(t-s)ds = \delta_t \int_0^{t-b} g(s)ds. \qquad \square$$

B.3. Fourier transform methods for integro-differential equations

A general reference is [**85**] and the literature mentioned therein. We will present and prove a relatively crude result here in order to give the reader a flavor of the method (Parseval's relation for Fourier transforms and the conversion of convolutions into products under the Fourier transform). This result will be completely sufficient for our purpose and many other applications [**208, 223, 224**]; relaxing the main condition would come at the price of heavy technicalities that may obscure the essential ideas.

Consider a bounded differentiable function $u : \mathbb{R} \to \mathbb{R}$ which satisfies the integro-differential inequality

$$\text{(B.3)} \qquad \frac{d}{dt} f(u(t)) \geq g(u(t)) \int_0^\infty g(u(t-s))m(ds), \qquad t \in \mathbb{R},$$

with a continuous function $g : \mathbb{R} \to \mathbb{R}$, $g(0) = 0$, $g(u) \neq 0$ if $u \neq 0$, a differentiable function $f : \mathbb{R} \to \mathbb{R}$, and a signed Borel measure m on $\mathbb{R}_+$.

Assumption B.1. *$\int_{\mathbb{R}_+}(1+s)|m|(ds) < \infty$ where $|m|$ is the (total) variation of m.*

We aim at finding suitable extra conditions that make u vanish identically on $\mathbb{R}$.

Notice that all translations u_r, $u_r(t) = u(r+t)$, $r \in \mathbb{R}$, satisfy the same inequality. Let U be the set of translations of u. Let $v \in U$. For the time being, we only consider the differential inequality on $\mathbb{R}_+$,

$$\text{(B.4)} \qquad \left. \begin{aligned} \frac{d}{dt} f(v(t)) \geq {} & g(v(t)) \int_0^t g(v(t-s))m(ds) + w(t) \\ w(t) = {} & g(v(t)) \int_t^\infty g(v(t-s))m(ds) \end{aligned} \right\} \qquad t \geq 0.$$

We integrate this equation from 0 to t,

$$\text{(B.5)} \qquad \begin{aligned} & f(v(t)) - f(v(0)) \\ & \geq \int_0^t g(v(r))dr \int_0^r g(v(r-s))m(ds) + \int_0^t w(r)dr. \end{aligned}$$

By (B.4),

$$\int_0^t w(r)dr = \int_0^t g(v(r))dr \int_r^\infty g(v(r-s))m(ds).$$

We change the order of integration,

$$\int_0^t w(r)dr = \int_{\mathbb{R}_+} m(ds) \int_0^{t\wedge s} g(v(r))g(v(r-s))dr$$

with $t \wedge s = \min\{t, s\}$. Hence

$$\begin{aligned}\Big|\int_0^t w(r)dr\Big| &\le \int_{\mathbb{R}_+} |m|(ds) \int_0^s |g(v(r))|\,|g(v(r-s))|dr \\ &\le c^2 \int_{\mathbb{R}_+} s|m|(ds)\end{aligned}$$

where $c = \sup_{s\in\mathbb{R}} |g(u(s))|$ and the nonnegative measure $|m|$ is the variation of m. By (B.5),

$$\int_0^t g(v(r))dr \int_0^r g(v(r-s))m(ds) \le \text{const}$$

with a common constant for all $t \ge 0$ and $v \in U$. We fix an arbitrary $t > 0$ and set

$$\phi(r) = \begin{cases} g(v(r)), & 0 \le r \le t, \\ 0, & \text{otherwise.}\end{cases} \tag{B.6}$$

Then

$$\int_{\mathbb{R}} \phi(r)dr \int_{\mathbb{R}_+} \phi(r-s)m(ds) \le \text{const} \tag{B.7}$$

with a common constant for all $t \ge 0$ and $v \in U$. We extend m to all Borel sets in $\mathbb{R}$ in an even way,

$$m(\Omega) = m(\Omega \cap \mathbb{R}_+) + m((-\Omega)\cap\mathbb{R}_+) \tag{B.8}$$

for a Borel set Ω in $\mathbb{R}$.

Lemma B.7. *If $\phi \in L^2(\mathbb{R})$ and $m(\{0\}) \ge 0$, then*

$$\int_{\mathbb{R}} \phi(r)dr \int_{\mathbb{R}} \phi(r-s)m(ds) \le 2\int_{\mathbb{R}} \phi(r)dr \int_{\mathbb{R}_+} \phi(r-s)m(ds).$$

Proof. If $\phi \in L^2(\mathbb{R})$, then the convolutions ψ_1, ψ_2 defined by

$$\psi_1(r) = \int_{\mathbb{R}} \phi(r-s)m(ds), \qquad \psi_2(r) = \int_{\mathbb{R}_+} \phi(r-s)m(ds)$$

are also functions in $L^2(\mathbb{R})$ and

$$\|\psi_1\|_2 \le \|\phi\|_2\,|m|(\mathbb{R}), \qquad \|\psi_2\|_2 \le \|\phi\|_2\,|m|(\mathbb{R}_+).$$

Indeed, by Tonelli's theorem [**164**, Thm.4.16], the ψ_i can be shown to be bounded linear functionals on L^2 and can thus be identified with elements in L^2 by the Riesz representation theorem [**164**, Thm.9.8].

It is sufficient to show the statement for continuous functions ϕ with compact support because the set of such functions is dense in $L^2(\mathbb{R})$. This will allow us to change the order of integration freely. We first notice that

$$\begin{aligned}\int_{\mathbb{R}} \phi(r-s)m(ds) &= \int_{[0,\infty)} \phi(r-s)m(ds) + \int_{(-\infty,0)} \phi(r-s)m(ds)\\ &\overset{\text{(B.8)}}{=} \int_{[0,\infty)} \phi(r-s)m(ds) + \int_{(0,\infty)} \phi(r+s)m(ds).\end{aligned}$$

So,

$$\begin{aligned}&\int_{\mathbb{R}} \phi(r)dr \int_{\mathbb{R}} \phi(r-s)m(ds)\\ =&\int_{\mathbb{R}} \phi(r)dr \int_{[0,\infty)} \phi(r-s)m(ds) + \int_{\mathbb{R}} \phi(r)dr \int_{(0,\infty)} \phi(r+s)m(ds).\end{aligned}$$

Since ϕ is continuous and has compact support, we can change the order of integration forth and back,

$$\begin{aligned}&\int_{\mathbb{R}} \phi(r)dr \int_{(0,\infty)} \phi(r+s)m(ds) = \int_{(0,\infty)} m(ds) \int_{\mathbb{R}} \phi(r)\phi(r+s)dr\\ =&\int_{(0,\infty)} m(ds) \int_{\mathbb{R}} \phi(r-s)\phi(r)dr = \int_{\mathbb{R}} \phi(r)dr \int_{(0,\infty)} \phi(r-s)m(ds).\end{aligned}$$

We combine the two equations,

$$\begin{aligned}&\int_{\mathbb{R}} \phi(r)dr \int_{\mathbb{R}} \phi(r-s)m(ds) + m(\{0\}) \int_{\mathbb{R}} (\phi(r))^2 dr\\ =&2 \int_{\mathbb{R}} \phi(r)dr \int_{\mathbb{R}_+} \phi(r-s)m(ds).\end{aligned}$$

Since $m(\{0\}) \geq 0$ by assumption, this implies the assertion. □

Let $\hat{m}$ be the Fourier transform of the extended measure m,

$$\hat{m}(s) = \frac{1}{\sqrt{2\pi}} \int_{\mathbb{R}} e^{-its} m(dt). \tag{B.9}$$

Since m has been obtained by an even extension,

$$\hat{m}(s) = \frac{2}{\sqrt{2\pi}} \int_{\mathbb{R}_+} \cos(ts)m(dt) = \frac{2}{\sqrt{2\pi}} \Re \int_{\mathbb{R}_+} e^{-its} m(dt). \tag{B.10}$$

Notice that $\hat{m}(-s) = \hat{m}(s)$.

Lemma B.8. *Assume that* $\hat{m}(s) \geq 0$ *for all* $s > 0$. *Then*

$$\int_{\mathbb{R}} \phi(r)dr \int_{\mathbb{R}} \phi(r-s)m(ds) \geq \inf_{s\in\mathbb{R}} \hat{m}(s)\sqrt{2\pi} \int_{\mathbb{R}} |\phi(s)|^2 ds$$

for all $\phi \in L^2(\mathbb{R})$.

The proof is based on Parseval's relation for Fourier transforms. Let $\hat{\phi}$ denote the Fourier transform of $\phi \in L^1(\mathbb{R}, \mathbb{C})$,

$$\hat{\phi}(s) = \frac{1}{\sqrt{2\pi}} \int_{\mathbb{R}} \phi(t)e^{-ist}dt.$$

Lemma B.9. *Let* $\phi, \psi \in L^1(\mathbb{R}, \mathbb{C}) \cap L^2(\mathbb{R}, \mathbb{C})$. *Then* $\hat{\phi}, \hat{\psi} \in L^2(\mathbb{R})$ *and*

$$\int_{\mathbb{R}} \hat{\phi}(s)\overline{\hat{\psi}(s)}ds = \int_{\mathbb{R}} \phi(s)\overline{\psi(s)}ds.$$

This relation also holds for $\phi, \psi \in L^2(\mathbb{R}, \mathbb{C})$ if the Fourier transforms are understood in a generalized sense [**164**, Sec.11.5].

Proof of Lemma B.8. If $\phi \in L^2(\mathbb{R})$, then $L^2(\mathbb{R}) \ni \psi$ defined by

$$\psi(t) = \int_{\mathbb{R}} \phi(t-s)m(ds) \tag{B.11}$$

and $\|\psi\|_2 \leq \|\phi\|_2 |m|(\mathbb{R})$. Since the continuous functions with compact support are dense in $L^2(\mathbb{R})$, it is sufficient to show the statement for a continuous function ϕ with compact support in $\mathbb{R}$. By Parseval's relation,

$$\int_{\mathbb{R}} \overline{\phi(s)}\psi(s)ds = \int_{\mathbb{R}} \overline{\hat{\phi}(s)}\hat{\psi}(s)ds. \tag{B.12}$$

Since the Fourier transform essentially converts a convolution into a product of Fourier transforms, by (B.11),

$$\hat{\psi}(s) = \sqrt{2\pi}\hat{\phi}(s)\hat{m}(s).$$

Again by Parseval's relation,

$$\int_{\mathbb{R}} \phi(s)\psi(s)ds = \sqrt{2\pi} \int_{\mathbb{R}} |\hat{\phi}(s)|^2 \hat{m}(s)ds \geq \inf |\hat{m}(s)|\sqrt{2\pi} \int_{\mathbb{R}} |\phi(s)|^2 ds. \quad \square$$

We assume that $\inf_{s\in\mathbb{R}} \hat{m}(s) > 0$ and $m(\{0\}) \geq 0$ and combine (B.7), Lemma B.7, and Lemma B.8,

$$\int_{\mathbb{R}} |\phi(s)|^2 ds \leq \text{ const}$$

with a constant that does not depend on t and $v \in U$. We recall the definition of ϕ in (B.6),

$$\int_0^t (g(v(s)))^2 ds \leq \text{ const}$$

and that the constant is the same for all $t > 0$ and $v \in U$. By the monotone convergence theorem, $g \circ v \in L^2(\mathbb{R}_+)$ and

$$\int_{\mathbb{R}_+} (g(v(s)))^2 ds \leq \text{ const.}$$

This holds for all $v \in U$ (recall that U is the set of right translations of u) with the same constant, so

$$\int_r^\infty (g(u(s)))^2 ds \leq \text{ const}$$

with the constant being the same for all $r \in \mathbb{R}$. We let r tend to $-\infty$. Again by the monotone convergence theorem, $g \circ u \in L^2(\mathbb{R})$ and

$$\int_{\mathbb{R}} (g(u(s)))^2 ds \leq \text{ const.}$$

There exist positive sequence (r_n) and (t_n) such that $r_n, t_n \to \infty$ and $g(u(-r_n)) \to 0$ and $g(u(t_n)) \to 0$ as $n \to \infty$. Since $g(u) \neq 0$ for $u \neq 0$ and u is bounded, $u(-r_n) \to 0$ and $u(t_n) \to 0$ as $n \to \infty$.

We return to (B.3) which we integrate from $-r_n$ to t_n,

$$f(u(t_n)) - f(u(-r_n)) \geq \int_{-r_n}^{t_n} g(u(r))dr \int_{\mathbb{R}_+} g(u(r-s))m(ds).$$

Since $g \circ u \in L^2(\mathbb{R})$, $r \mapsto g(u(r)) \int_{\mathbb{R}_+} g(u(r-s))m(ds)$ is a function in $L^1(\mathbb{R})$ as the product of two functions in $L^2(\mathbb{R})$. By the dominated convergence theorem, we can take the limit as $n \to \infty$ on both sides of the inequality and obtain

$$\int_{\mathbb{R}} g(u(r))dr \int_{\mathbb{R}_+} g(u(r-s))m(ds) \leq f(0) - f(0) = 0.$$

We apply the combination of our lemmata to $\phi = g \circ u$ and obtain $g(u(r)) = 0$ and $u(r) = 0$ for all $r \in \mathbb{R}$.

We summarize.

Theorem B.10. *Let $f, g : \mathbb{R} \to \mathbb{R}$, f differentiable, g continuous, $g(u) = 0$ if and only if $u = 0$. Further, let m be a signed Borel measure on $\mathbb{R}_+$, $m(\{0\}) \geq 0$, whose variation $|m|$ satisfies*

$$\int_{\mathbb{R}_+} (1+s)|m|(ds) < \infty$$

and

$$\inf_{s \geq 0} \Re \int_{\mathbb{R}_+} e^{-ist} m(dt) > 0.$$

Then every bounded differentiable solution u of the integro-differential inequality (B.3) satisfies $u(t) = 0$ for all $t \in \mathbb{R}$,

An integro-differential equation. Consider a bounded real-valued solution of the scalar integro-differential equation

$$-u'(t) = \int_0^\infty g(u(t-s))m(ds), \qquad t \in \mathbb{R}, \tag{B.13}$$

with a signed Borel measure m on $\mathbb{R}_+$ and a continuous function $g : \mathbb{R} \to \mathbb{R}$, $g(u) = 0$ if and only if $u = 0$. We multiply both sides of the equation with $g(u(t))$,

$$-u'(t)g(u(t)) = g(u(t)) \int_0^\infty g(u(t-s))m(ds).$$

Set $f(u) = -\int_0^u g(x)dx$. Then u satisfies (B.3) and we have the following result from Theorem B.10.

Theorem B.11. *Let $g : \mathbb{R} \to \mathbb{R}$, g continuous, $g(u) = 0$ if and only if $u = 0$. Further, let m be a signed Borel measure on $\mathbb{R}_+$, $m(\{0\}) \geq 0$, whose variation $|m|$ satisfies*

$$\int_{\mathbb{R}_+} (1+s)|m|(ds) < \infty$$

and

$$\inf_{s\geq 0} \Re \int_{\mathbb{R}_+} e^{-ist} m(dt) > 0.$$

Then every bounded differentiable solution u of the integro-differential equation (B.13) satisfies $u(t) = 0$ for all $t \in \mathbb{R}$.

For generalizations of this result which relax the two assumptions for the measure m see the proof of Theorem 2 in [**151**].

B.4. Closed linear operators

Definition B.12. Let X, Y be normed linear spaces. A mapping A from a subset $D(A)$ of X into Y is called an *operator from X to Y*.

It is called a *closed operator* if the following holds in addition:

If (x_m) is a sequence in $D(A)$ and $x \in X, y \in Y$ such that $x_m \to x$ and $Ax_m \to y$ as $m \to \infty$, then $x \in D(A)$ and $Ax = y$.

A is called a *linear operator*, if $D(A)$ is a linear subspace of X and A is a linear mapping from $D(A)$ to Y.

A is called a bounded linear operator from X to Y if in addition $D(A) = X$ and there is some $c > 0$ such that $\|Ax\| \leq c\|x\|$ for all $x \in X$.

Notice that an operator A is closed if and only if its graph

$$G(A) = \{(x, Ax); x \in D(A)\}$$

is a closed subset of the product space $X \times Y$ with the norm $\|(x,y)\| = \|x\| + \|y\|$. Obviously, any bounded linear operator is closed.

B.4.1. Duality. If X is a normed vector space over $\mathbb{K} = \mathbb{R}$ or $\mathbb{K} = \mathbb{C}$, then X^* is the Banach space of bounded linear operators (functionals) $x^* : X \to \mathbb{K}$ with norm

$$\|x^*\| = \sup\{|x^*x|; \|x\| \leq 1\}.$$

Notice that $\|x^*\|$ is nothing else than the operator norm of x^*.

We recall the following corollaries of the Hahn-Banach Extension Theorem ([**164**, Thm. 10.3], [**166**, Sec.6.1]). They allow us to reduce statements for normed linear spaces to their counterparts for $\mathbb{R}$ or $\mathbb{C}$.

Lemma B.13. *For any $x \in X$ there exists some $x^* \in X^*$ with $x^*x = \|x\|$ and $\|x^*\| = 1$.*

Lemma B.14. *Let Z be a closed linear subspace of X and $x \in X \setminus Z$. Then there exists some $x^* \in X^*$ such that $x^*x \neq 0$, but $x^*z = 0$ for all $z \in Z$.*

Definition B.15. (a) A subset G of $X \times Y$ is called a *relation* from X to Y. G is called a *linear relation* if G is a linear subspace of $X \times Y$. G is called a *closed relation* if G is a closed subset of the product space $X \times Y$.

(b) The *dual relation* of G is the subset G^* of $Y^* \times X^*$ such that, for a pair (y^*, x^*),

$$(y^*, x^*) \in G^* \iff y^*y = x^*x \qquad \forall (x, y) \in G.$$

If G is the graph of a linear operator A, $G = G(A) = \{(x, Ax); x \in D(A)\}$, then

$$(y^*, x^*) \in G^*(A) \iff y^*(Ax) = x^*x \qquad \forall x \in D(A). \tag{B.14}$$

Remark B.16. (a) The dual relation G^* is a closed linear subspace of $Y^* \times X^*$ even if the relation G is not linear or closed.

(b) If A is a linear operator, then $G^*(A)$ is the graph of a linear operator from Y^* to X^* (denoted by A^*) if and only if A is densely defined (i.e., $D(A)$ is dense in X).

A^* is called the *dual operator* associated with A.

Proposition B.17. *Let G be a closed linear relation from X to Y and $x_0 \in X$ and $y_0 \in Y$. Then $(x_0, y_0) \in G$ if and only if $y^*y_0 = x^*x_0$ for all $(y^*, x^*) \in G^*$.*

Proof. 'Only if' immediately follows from the definition of G^* in Definition B.15. To prove the 'if' direction, we show the contraposition. Assume that $(x_0, y_0) \notin G$. By Lemma B.14, there exist a continuous linear functional f on $X \times Y$ such that $f(x_0, y_0) \neq 0$, but $f(x, y) = 0$ for all $(x, y) \in G$. We define elements $x^* \in X^*$ and $y^* \in Y^*$ by $x^*x = f(x,0)$ and $y^*y = -f(0,y)$. Then $f(x,y) = x^*x - y^*y$. For all $(x,y) \in G$ we have

$$0 = f(x,y) = x^*x - y^*y,$$

so $(y^*, x^*) \in G^*$. But

$$0 \neq f(x_0, y_0) = x^*x_0 - y^*y_0. \qquad \square$$

Corollary B.18. *Let A be a closed linear operator from X to Y and $x_0 \in X$ and $y_0 \in Y$. Then $x_0 \in D(A)$ and $y_0 = Ax_0$ if and only if $y^*y_0 = x^*x_0$ for all $(y^*, x^*) \in G^*(A)$.*

B.4.2. Inhomeogeneous Cauchy problems. Let A be a closed operator in the Banach space X, $x \in X$. Let I be an interval $[0,b)$ with $b \in (0,\infty]$ or an interval $[0,b]$ with $b \in (0,\infty)$, and $f : I \to X$.

We consider the inhomogeneous Cauchy problem

$$u' = Au + f(t), \qquad t \in I, \quad u(0) = x. \tag{B.15}$$

The function $u : I \to X$ is called a *classical solution* of (B.15) if u is continuous on I, differentiable on $I \setminus \{0\}$, $u(t) \in D(A)$ for all $t \in I \setminus \{0\}$, and $u' = Au + f$ on $I \setminus \{0\}$, $u(0) = x$. Assume in addition that f is continuous. Then $u : I \to X$ is called an *integral solution* of (B.15) if u is continuous on I, $\int_0^t u(s)ds \in D(A)$ for all $t \in I$ and

$$u(t) = x + A\int_0^t u(s)ds + \int_0^t f(s)ds, \qquad t \in I. \tag{B.16}$$

Finally, $u : I \to X$ is called a *weak solution* of (B.15) if u is continuous on I, $u(0) = x$ and, for all $(y^*, x^*) \in G^*(A)$, x^*u is differentiable on I and

$$\frac{d}{dt}y^*u(t) = x^*u(t) + y^*f(t), \qquad t \in I \setminus \{0\}. \tag{B.17}$$

Notice that $\frac{d}{dt}y^*u(t)$ is automatically continuous.

It follows immediately from the definition of $G^*(A)$ that every classical solution is a weak solution.

Theorem B.19. *$u : I \to X$ is an integral solution of (B.15) if and only if it is a weak solution.*

Proof. Let u be an integral solution and $(y^*, x^*) \in G^*(A)$. Then

$$y^*u(t) = y^*x + x^*\int_0^t u(s)ds + y^*\int_0^t f(s)ds.$$

Since x^* and y^* are bounded linear functionals, they commute with the integrals and $x^*u(t)$, and $y^*f(t)$ are continuous functions of t. So $y^*u(t)$ can be differentiated in $t \in I$, and u is a weak solution.

In turn, let u be a weak solution and $(y^*, x^*) \in G^*(A)$. See Definition B.15. Then x^*u is differentiable on $I \setminus \{0\}$ and

$$\frac{d}{dt} y^* u(t) = x^* u(t) + y^* f(t), \qquad t \in I \setminus \{0\}.$$

Since u is continuous, $y^*u(t)$ is continuously differentiable on $I \setminus \{0\}$. By the fundamental theorem of calculus,

$$\int_0^t x^* u(s) ds = y^* u(t) - y^* x - \int_0^t y^* f(s) ds.$$

Since x^* and y^* are linear bounded functionals,

$$x^* \int_0^t u(s) ds = y^* \Big(u(t) - x - \int_0^t f(s) ds \Big).$$

By Corollary B.18 , this implies that $\int_0^t u(s)ds \in D(A)$ and $A \int_0^t u(s)ds = u(t) - x - \int_0^t f(s)ds$. □

Corollary B.20. *Every classical solution of (B.15) is an integral solution. An integral solution $u : I \to X$ is a classical solution if and only if u is differentiable on $I \setminus \{0\}$.*

Proof. The first statement follows from Theorem B.19 because every classical solution is a weak solution. Now assume that u is an integral solution that is differentiable on $I \setminus \{0\}$. Then, for all (y^*, x^*) in $G^*(A)$ and $t \in I$, $t > 0$,

$$y^* u'(t) = x^* u(t) + y^* f(t).$$

By Corollary B.18, $u(t) \in D(A)$ and $u'(t) = Au(t) + f(t)$. □

Now we assume that A is the generator of a C_0-semigroup $\{S(t); t \geq 0\}$. The function $u : I \to X$ is called a *mild solution* of (B.15) if

$$u(t) = S(t)x + \int_0^t S(t-s) f(s) ds, \qquad t \in I. \tag{B.18}$$

Since A is the generator of a C_0-semigroup, $D(A)$ is dense in X.

Lemma B.21. *Let $x \in X$ and $(y^*, x^*) \in G^*(A)$. Then $y^*S(t)x$ is a differentiable function of $t \geq 0$ and $\frac{d}{dt} y^* S(t)x = x^* S(t)x$.*

Proof. We first assume that $x \in D(A)$. Then $S(t)x$ is differentiable in $t \geq 0$, $S(t)x \in D(A)$, and $\frac{d}{dt} S(t)x = AS(t)x = S(t)Ax$. So the statement holds. Now let $x \in X$. Since $D(A)$ is dense in X, there exists a sequence (x_n) in $D(A)$ such that $x_n \to x$ as $n \to \infty$. Further,

$$\frac{d}{dt} y^* S(t) x_n = x^* S(t) x_n \to x^* S(t) x, \qquad n \to \infty,$$

uniformly on finite subintervals of $\mathbb{R}_+$. Since $y^*x_n \to y^*x$, the assertion follows. □

Theorem B.22. *$u : I \to X$ is an integral (equivalently, weak) solution of (B.15) if and only if it is a mild solution.*

Proof. Let u be a mild solution and $(y^*, x^*) \in G*(A)$. Then

$$y^*u(t) = y^*S(t)x + \int_0^t y^*S(t-s)f(s)ds.$$

By Lemma B.21,

$$y^*u(t) = y^*S(t)x + \int_0^t \Big(\int_0^{t-s} x^*S(r)f(s)dr - y^*f(s)\Big)ds.$$

Since x^* is a bounded linear functional, it commutes with the integrals,

$$y^*u(t) = y^*S(t)x + x^* \int_0^t \Big(\int_0^{t-s} S(r)f(s)dr\Big)ds + \int_0^t y^*f(s)ds.$$

We make a substitution and change the order of integration,

$$y^*u(t) = y^*S(t)x + x^* \int_0^t \Big(\int_0^{r} S(r-s)f(s)ds\Big)dr + \int_0^t y^*f(s)ds.$$

So $y^*u(t)$ is differentiable and

$$\begin{aligned}\frac{d}{dt}y^*u(t) =& x^*S(t)x + x^*\int_0^t S(t-s)f(s)ds + y^*f(t)\\ =& x^*u(t) + y^*f(t).\end{aligned}$$

This implies that u is a weak solution.

In turn let u be a weak solution. Let v be the right-hand side of (B.18). Then v is also a weak solution as we have just shown. Set $w(t) = u(t) - v(t)$. Since A is linear, w is a weak solution of $w' = Aw$, $w(0) = 0$. By Theorem B.19, w is an integral solution of this equation, $w(t) = A\int_0^t w(s)ds$. Set $\phi(t) = \int_0^t w(s)ds$. Then ϕ is a classical solution of $\phi' = A\phi$, $\phi(0) = 0$. Since $\phi(s) \in D(A)$, for $0 < s < t$,

$$\begin{aligned}\frac{d}{ds}S(t-s)\phi(s) =& -\frac{d}{dt}S(t-s)\phi(s) + S(t-s)\phi'(s)\\ =& -S(t-s)A\phi(s) + S(t-s)A\phi(s) = 0.\end{aligned}$$

We integrate this equation from 0 to t,

$$\phi(t) = S(t)\phi(0) = 0.$$

This implies $w(t) = 0$ which in turn implies $u(t) = v(t)$. □

B.5. Exercises

Exercise B.1. Prove Remark B.16.

Bibliography

[1] R. Abraham, and J.E. Marsden, *Foundations of Mechanics*, 2nd edition, The Benjamin/Cummings Publishing Co. Inc., 1978.

[2] A.S. Ackleh, B.G. Fitzpatrick, H.R. Thieme, *Rate distributions and survival of the fittest: a formulation on the space of measures*, Disc. Cont. Dyn. Sys. B **5** (2005), 917–928.

[3] A.S. Ackleh, K. Ito, *Measure-valued solutions for a hierarchically size-structured population*, J. Differential Equations **217** (2005), 431–455.

[4] R.M. Anderson, R.M. May, *Infectious Diseases of Humans. Dynamics and Control*, Oxford University Press, 1991.

[5] W. Arendt, C.J.K. Batty, M. Hieber, F. Neubrander, *Vector-valued Laplace Transforms and Cauchy Problems*, Birkhäuser, 2001.

[6] J. Arino, C.C. McCluskey, P. van den Driessche, *Global results for an epidemic model with vaccination that exhibits backward bifurcation*, SIAM J. Appl. Math. **64** (2003), 260–276.

[7] L. Arnold, *Random Dynamical Systems*, Springer-Verlag, 1998.

[8] P. Ashwin, J. Buescu, I. Stewart, *From attractor to chaotic saddle: a tale of transverse instability*, Nonlinearity **9** (1996), 703–737.

[9] I. Barbalat, *Systèmes d'équations différentielles d'oscillations nonlinéaires*, Rev. Roumaine Math. Pures Appl. **4** (1959), 267–270.

[10] M. Barnsley, *Fractals Everywhere*, 2nd edition, Academic Press, 1993.

[11] M. Benaim, J. Hofbauer, W.H. Sandholm, *Robust permanence and impermanence for stochastic replicator dynamics*, J. Biol. Dynamics **2** (2008), 180–195.

[12] E. Beretta, T. Hara, W. Ma, Y. Takeuchi, *Global asymptotic stability of an SIR epidemic model with distributed time delay*, Nonlinear Analysis **47** (2001), 4107-4115.

[13] E. Beretta, F. Solimano, Y. Tang, *Analysis of a chemostat model for bacteria and virulent bacteriophage*, Discrete and Continuous Dynamical Systems-B, **2** (2002), 495–520.

[14] H.C. Berg, *Random Walks in Biology*, Princeton Univ. Press, 1993.

[15] A. Berman and R. Plemmons, *Nonnegative Matrices in the Mathematical Sciences*, Academic Press, 1979.

[16] N.P. Bhatia and G.P. Szegő, *Stability Theory of Dynamical Systems*, Springer, 2002.

[17] G.A. Bürger, *Wunderbare Reisen zu Wasser und Lande*, Feldzüge und lustige Abenteuer des Freiherrn von Münchhausen nach der Ausgabe von 1788 mit einem Anhang älterer Lügengeschichten - (I. Ruttmann, ed.) Philipp Reclam jun., 1969.

[18] T.A. Burton, V. Hutson, *Permanence for nonautonomous predator-prey systems*, Differential Integral Equations **4** (1991), 1269–1280.

[19] S.N. Busenberg and K.L. Cooke, *The effect of integral conditions in certain equations modelling epidemics and population growth*, J. Math. Biol. **10** (1980), 13–32.

[20] S.N. Busenberg and K.L. Cooke, *Vertically Transmitted Diseases*, Springer, 1993.

[21] S.N. Busenberg, K.L. Cooke, M.A. Pozio, *Analysis of a model of a vertically transmitted disease*, J. Math. Biol. **17** (1983), 305–329.

[22] G. Butler, H.I. Freedman, P. Waltman, *Uniformly persistent systems*, Proc. Amer. Math. Soc. **96** (1986), 425–430.

[23] G. Butler and P. Waltman, *Persistence in dynamical systems*, J. Diff. Eqns. **63** (1986), 255–263.

[24] R.S. Cantrell and C. Cosner, *Practical persistence in ecological models via comparison methods*, Proc. Royal Soc. Edinburgh A **126** (1996), 247–272.

[25] R.S. Cantrell and C. Cosner, *Spatial Ecology via Reaction-Diffusion Equations*, Wiley, 2003.

[26] Y. Cao and T.C. Gard, *Uniform persistence for population models with time delay using multiple Lyapunov functions*, Diff. Int. Equations **6** (1993), 883-898.

[27] Y. Cao and T.C. Gard, *Practical persistence for differential delay models of population interactions*, *Differential Equations and Computational Simulations III* (J. Graef, R. Shivaji, B. Soni, J. Zhu, eds.), *Electronic J. Differential Equations*, Conference 01, 1997, pp. 41–53.

[28] Y. Cao, J. Fan, T.C. Gard, *Persistence in Kolmogorov population models with time delay*, *Proc. First World Congress of Nonlinear Analysis* (V. Lakshmikantham, ed.), 3741-3751, de Gruyter, 1996

[29] H. Caswell, *Matrix Population Models*, Sinauer Assoc. Inc., 1989.

[30] C.-H. Chiu and S.-B. Hsu, *Extinction of the top predator in a three-level food chain*, J. Math. Biol. **37** (1998), 372–380.

[31] S.-N. Chow and J.K. Hale, *Methods of Bifurcation Theory*, Springer, 1982.

[32] C.W. Clarke, *Mathematical Bioeconomics: The Optimal Management of Renewable Resources*, Wiley, 1976.

[33] C. Conley, *Isolated Invariant Sets and the Morse Index*, CBMS 38, Amer. Math. Soc., 1978.

[34] W. Coppel, *Stability and Asymptotic Behavior of Differential Equations*, Heath, 1965.

[35] C. Cosner, *Persistence (permanence), compressibility, and practical persistence in some reaction-diffusion models from ecology*, *Comparison Methods and Stability Theory* (X. Liu, D. Siegel, eds.), Lecture Notes in Pure and Applied Mathematics, 162, Marcel Dekker, Inc., 1994.

[36] R.F. Costantino, J.M. Cushing, B. Dennis, R.A. Desharnais, *Experimentally induced transitions in the dynamic behavior of insect populations*, Nature **375** (1995), 227–230.

[37] P. Cull, *Local and global stability for population models*, Biol. Cybernetics **54** (1986), 141–149.

[38] J.M. Cushing, *Nonlinear Matrix Models and Population Dynamics*, Natural Resource Modeling **2** (1988), 539–580.

[39] J.M. Cushing, *An Introduction to Structured Population Dynamics*, CBMS-NSF Regional Conf. Series in Applied Math. vol. 77, 1998.

[40] J.M. Cushing, *Cycle chains and the LPA model*, J. Difference Equations Appl. **9** (2003), 655–670.

[41] J.M. Cushing, *Nonlinear semelparous Leslie models*, Math. Biosci. Eng. **3** (2006), 17–36.

[42] J.M. Cushing, *Three stage semelparous Leslie models*, J. Math. Biol. **59** (2009), 75–104.

[43] J.M. Cushing, R.F. Costantino, B. Dennis, R.A. Desharnais, S.M.Henson, *Nonlinear population dynamics: Models, experiments and data*, J. Theor. Biology **194** (1998), 1–9.

[44] J.M. Cushing, R.F. Costantino, B. Dennis, R.A. Desharnais, S.M.Henson, *Chaos in Ecology, experimental nonlinear dynamics*, Academic Press, 2003.

[45] J.M. Cushing and J. Li, *On Ebenman's model for the dynamics of population with competing juveniles and adults*, Bull. Math. Biol. **51** (1989), 687–713.

[46] E.M.C. D'Agata, P. Magal, S. Ruan, G.F. Webb, *Asymptotic behavior in nosocomial epidemic models with antibiotic resistance*, Diff. Integral Equations **19** (2006), 573–600.

[47] N. Davydova, *Old and Young. Can they coexist?* Thesis, University of Utrecht, 2004.

[48] N.V. Davydova, O. Diekmann, S.A. van Gils, *Year class coexistence or competitive exclusion for strict biennials?*, J. Math. Biol. **46** (2003), 95-131.

[49] K. Deimling, *Nonlinear Functional Analysis*, Springer, 1985.

[50] W. Desch and W. Schappacher, *Linearized stability for nonlinear semigroups*, *Differential Equations in Banach Spaces* (A. Favini, E. Obrecht, eds.), 61-67, Lecture Notes in Mathematics **1223**, Springer, 1986.

[51] T. Dhirasakdanon, *A Model of Infectious Diseases in Amphibian Populations with Ephemeral Habitat*, Ph.D. Thesis, Arizona State University, 2010.

[52] T. Dhirasakdanon and H.R. Thieme, *Persistence of vertically transmitted parasite strains which protect against more virulent horizontally transmitted strains*, *Modeling and Dynamics of Infectious Diseases* (Z. Ma, Y. Zhou, J. Wu, eds.), 187–215, World Scientific, 2009.

[53] T. Dhirasakdanon and H.R. Thieme, *Stability of the endemic coexistence equilibrium for one host and two parasites*, Mathematical Modelling of Natural Phenomena **5** (2010), 109-138.

[54] T. Dhirasakdanon, H.R. Thieme, P. van den Driessche, *A sharp threshold for disease persistence in host metapopulations*, J. Biol. Dynamics **1** (2007), 363-378

[55] O. Diekmann and Ph. Getto, *Boundedness, global existence and continuous dependence for nonlinear dynamical systems describing physiologically structured populations*, J. Diff. Equations **215** (2005), 268–319.

[56] O. Diekmann, M. Gyllenberg, H. Huang, M. Kirkilionis, J.A.J. Metz, H.R. Thieme, *On the formulation and analysis of general deterministic structured population models. II. Nonlinear theory*, J. Math. Biol. **43** (2001), 157–189.

[57] O. Diekmann, M. Gyllenberg, J.A.J. Metz, *Steady-state analysis of structured population models*, Theor. Pop. Biol. **63** (2003), 309–338.

[58] O. Diekmann, M. Gyllenberg, J.A.J. Metz, H.R. Thieme, *On the formulation and analysis of general deterministic structured population models. I. Linear theory*, J. Math. Biol. **36** (1998), 349–388.

[59] O. Diekmann and S.A. van Gils, *On the cyclic replicator equation and the dynamics of semelparous populations*, SIAM J. Appl. Dyn. Sys. **8** (2009), 1160–1189.

[60] O. Diekmann, S.A. van Gils, S.M. Verduyn Lunel, H.-O. Walter, *Delay Equations. Functional, Complex, and Nonlinear Analysis*, Springer, 1995.

[61] J. Dugundji, *Topology*, Allyn and Bacon, Boston 1966, Prenctice Hall of India, New Delhi 1975.

[62] N. Dunford and J.T. Schwartz, *Linear Operators, Part I: General Theory*, John Wiley & Sons, 1988.

[63] B. Ebenman, *Niche difference between age classes and intraspecific competition in age-structured populations*, J. Theor. Biol. **124** (1987), 25–33.

[64] B. Ebenman, *Competition between age-classes and population dynamics*, J. Theor. Biol. **131** (1988), 389–400.

[65] G. Edgar, *Measure, Topology, and Fractal Geometry*, Springer-Verlag, 1990.

[66] S.F. Ellermeyer, S.S. Pilyugin, R. Redheffer, *Persistence criteria for a chemostat with variable nutrient input*, J. Differential Equations **171** (2001), 132–147.

[67] K.-J. Engel and R. Nagel, *One-Parameter Semigroups for Linear Evolution Equations*, Springer, 2000.

[68] A. Fall, A. Iggidr, G. Sallet, J.J. Tewa, *Epidemiological models and Lyapunov functions*, Math. Model. Nat. Phenom. **2** (2007), 62-83.

[69] Z. Feng and H.R. Thieme, *Endemic models with arbitrarily distributed periods of infection. I. General theory*, SIAM J. Appl. Math. **61** (2000), 803–833.

[70] R. Ferriere and M. Gatto, *Lyapunov exponents and the mathematics of invasion in oscillatory or chaotic populations*, Theor. Population Biol. **48** (1995), 126–171.

[71] A. Fonda, *Uniformly persistent semi-dynamical systems*, Proc. Amer. Math. Soc. **104** (1988), 111–116.

[72] H. Freedman and P. Moson, *Persistence definitions and their connections*, Proc. Amer. Math. Soc. **109** (1990), 1025–1033.

[73] H. Freedman, S. Ruan, M. Tang, *Uniform persistence and flows near a closed positively invariant set*, J. Dyn. Diff. Eqn. **6** (1994), 583-600

[74] H. Freedman and J.W.-H. So, *Persistence in discrete semi-dynamical systems*, SIAM J. Math. Anal. **20** (1989), 930–938.

[75] H. Freedman and P. Waltman, *Mathematical analysis of some three species food-chain models*, Math. Biosci. **33** (1977), 257–276.

[76] H. Freedman and P. Waltman, *Persistence in models of three interacting predator-prey populations*, Math. Biosci. **68** (1984), 213–231.

[77] L.Q. Gao, J. Mena-Lorca, H.W. Hethcote, *Variations on a theme of SEI endemic models*, *Differential Equations and Applications to Biology and Industry* (M. Martelli, C.L. Cooke, E. Cumberbatch, B. Tang, H.R. Thieme, eds.), 191-207, World Scientific, 1996.

[78] B.M. Garay, *Uniform persistence and chain recurrence*, J. Math. Anal. Appl. **139** (1989), 372–381.

[79] B.M. Garay and J. Hofbauer, *Robust persistence for ecological differential equations, minimax, and discretizations*, SIAM J. Math. Anal. **34** (2003), 1007–1039.

[80] T.C. Gard, *Persistence in food webs: Holling-type food chains*, Math. Biosci. **49** (1980), 61–67.

[81] T.C. Gard, *Persistence in food chains with general interactions*, Math. Biosci. **51** (1980), 165-174

[82] T.C. Gard and T.G. Hallam, *Persistence in food webs. I. Lotka-Volterra food chains*, Bull. Math. Biol. **41** (1979), 877–891.

[83] B.S. Goh, *Global stability in many species systems*, Amer. Nat. **111** (1977), 135–142.

[84] M. Gobbino and M. Sardella, *On the connectedness of attractors for dynamical systems*, J. Diff. Eqn. **133**, 1-14.

[85] G. Gripenberg, S.-O. Londen, O. Staffans, *Volterra Integral Functional Equations*, Cambridge University Press, 1990.

[86] R.B. Guenther and J.W. Lee, *Partial Differential Equations of Mathematical Physics and Integral Equations*, Dover Publications Inc., 1988.

[87] H. Guo and M.Y. Li, *Global dynamics of a staged progression model for infectious diseases*, Math. Biosci. Engin. **3** (2006), 513–525.

[88] M. Gyllenberg, F. Jacobs, J. Metz, *On the concept of attractor for community-dynamical processes II: the case of structured populations*, J. Math. Biol. **47** (2003), 235–248.

[89] K.P. Hadeler, *Selection in a cyclically changing environment, Population Biology and Evolution* (K. Wöhrmann, V. Loeschcke, eds.), 102-114, Springer, 1984.

[90] J.K. Hale, *Ordinary Differential Equations*, Krieger, 1980.

[91] J.K. Hale, *Asymptotic Behavior of Dissipative Systems*, American Mathematical Society, 1988.

[92] J.K. Hale, *Dissipation and attractors*, Proceedings of the International Conference on Differential Equations (B. Fiedler, K. Groeger, and J. Sprekels, eds.), 622-637, World Scientific, 2000.

[93] J.K. Hale, *Dissipation and compact attractors*, J. Dyn. Diff. Eqn. **18** (2006), 485-523.

[94] J.K. Hale and S.M. Verduyn Lunel, *Introduction to Functional Differential Equations*, Springer, 1993.

[95] J.K. Hale and P. Waltman, *Persistence in infinite dimensional systems*, SIAM J. Math. Anal. **20** (1989), 388–395.

[96] T.G. Hallam, *Structural sensitivity of grazing formulations in nutrient controlled plankton models* J. Math. Biol. **5** (1978), 269–280.

[97] A. Hastings and T. Powell, *Chaos in a three-species food chain*, Ecology **72**(1991), 896–903.

[98] H.W. Hethcote, *The mathematics of infectious diseases*, SIAM Review **42** (2000), 599–653.

[99] E. Hewitt and K. Stromberg, *Real and Abstract Analysis*, Springer, 1965.

[100] E. Hille and R.S. Phillips, *Functional Analysis and Operator Semigroups*, American Mathematical Society, 1957.

[101] M.W. Hirsch, H.L. Smith, and X.-Q. Zhao, *Chain transitivity, attractivity and strong repellors for semidynamical systems*, J. Dynamics and Diff. Eqns. **13** (2001), 107–131.

[102] W.M. Hirsch, H. Hanisch, J.-P. Gabriel, *Differential equation models for some parasitic infections: methods for the study of asymptotic behavior*, Comm. Pure Appl. Math. **38** (1985), 733–753.

[103] J. Hofbauer, *A general cooperation theorem for hypercycles*, Monatshefte für Math. **91** (1981), 233–240.

[104] J. Hofbauer, *A unified approach to persistence*, Acta Applicandae Math. **14** (1989), 11–22.

[105] J. Hofbauer and S. Schreiber, *Robust permanence for structured populations*, J. Diff. Eqns. **248** (2010), 1955–1971.

[106] J. Hofbauer and K. Sigmund, *Evolutionary Games and Population Dynamics*, Cambridge Univ. Press, 1998.

[107] J. Hofbauer and J.W.-H. So, *Uniform persistence and repellors for maps*, Proceedings Amer. Math. Soc. **107** (1989), 1137–1142.

[108] J. Hofbauer, J. Mallet-Paret, H.L. Smith, *Stable periodic solutions for the hypercycle system*, Journal of Dynamics and Differential Equations **3** (1991), 423–436.

[109] V. Hutson, *A theorem on average Liapunov functions*, Monatshefte für Math. **98** (1984), 267–275.

[110] V. Hutson and K. Schmitt, *Permanence and the dynamics of biological systems*, Math. Biosciences **111** (1992), 293–326.

[111] M. Iannelli, M. Martcheva, X.-Z. Li, *Strain replacement in an epidemic model with super-infection and perfect vaccination*, Math. Biosci. **195** (2005), 23–46.

[112] A. Iggidr, J. Mbang, G. Sallet, *Stability analysis of within-host parasite models with delays*, Math. Biosci. **209** (2007), 51–75.

[113] F.J.A. Jacobs and J.A.J. Metz, *On the concept of attractor for community-dynamical processes I: the case of unstructured populations*, J.Math. Biol. **47** (2003), 222–234.

[114] A. Katok and B. Hasselblatt, *Introduction to the Modern Theory of Dynamical Systems*, Cambridge University Press, 1995.

[115] W.O. Kermack and A.G. McKendrick, *A contribution to the mathematical theory of epidemics*, Proc. Roy. Soc., A, **115** (1927), 700–721, reprinted in Bull. Math Biol. **53** (1991), 33–55.

[116] W.O. Kermack and A.G. McKendrick, *Contribution to the mathematical theory of epidemics. II. The problem of endemicity*, Proc. Roy. Soc., A, **138** (1932), 55–85, Reprinted in Bull. Math Biol. **53** (1991), 57–87.

[117] W.O. Kermack and A.G. McKendrick, *Contribution to the mathematical theory of epidemics. III. Further studies on the problem of endemicity*, Proc. Roy. Soc., A, **141** (1933), 94–122, reprinted in Bull. Math Biol. **53** (1991), 89–118.

[118] J.R. Kirkwood, *An Introduction to Analysis*, PWS Publishing Company 1995.

[119] A. Klebanoff and A. Hastings, *Chaos in three species food chains*, J. Math. Biol. **32** (1994), 427–451.

[120] R. Kon, *Multiple attractors in host-parasitoid interactions: Coexistence and extinction*, Math. Biosci. **201** (2006), 172–183.

[121] R. Kon, *Nonexistence of synchronous orbits and class coexistence in matrix population models*, SIAM J. Applied Math. **66** (2006), 616–626.

[122] R. Kon and Y. Iwasa, *Single-class orbits in nonlinear Leslie matrix models for semelparous populations*, J. Math. Biol. **55** (2007), 781-802.

[123] R. Kon, Y. Saito, Y. Takeuchi, *Permanence of single-species stage structured models*, J. Math. Biol. **48** (2004), 515–528.

[124] R. Kon and Y. Takeuchi, *Permanence of host-parasitoid systems*, Nonlinear Anal., **47** (2001), 1383–1393.

[125] A. Korobeinikov, *Lyapunov functions and global properties for SEIR and SEIS epidemic models*, Mathematical Medicine and Biology **21** (2004), 75–83.

[126] A. Korobeinikov, *Lyapunov functions and global stability for SIR and SIRS epidemiological models with non-linear transmission*, Bull. Math. Biol. **30** (2006), 615–626.

[127] A. Korobeinikov and G.C. Wake, *Lyapunov functions and global stability for SIR, SIRS and SIS epidemiological models*, Appl. Math. Letters **15** (2002), 955–961.

[128] M.A. Krasnosel'skii, *Positive Solutions of Operator Equations*, Noordhoff, 1964.

[129] U. Krause and R.D. Nussbaum, *A limit set trichotomy for self-mappings of normal cones in Banach spaces*, Nonlinear Analysis, TMA **20** (1993), 855–1993.

[130] C. Kribs-Zaleta and J. Velasco-Hernández, *A simple vaccination model with multiple endemic states*, Math. Biosci. **164** (2000), 183–201.

[131] Y. Kuznetsov, O. Feo, S. Rinaldi, *Belyakov Homoclinic bifurcations in a tri-trophic food chain model*, SIAM J. Appl. Math. **62** (2001), 462–487.

[132] Y. Kuang, *Global stability of Gause-type predator-prey systems*, J. Math. Biol. **28** (1990), 463–474.

[133] Y. Kuang and H. Freedman, *Uniqueness of limit cycles in Gause-type predator-prey systems*, Math. Biosci. **88** (1988), 67–84.

[134] O. Ladyzhenskaya, *Attractors for Semigroups and Evolution Equations*, Cambridge University Press, 1991.

[135] V. Lakshmikantham and S. Leela, *Nonlinear Differential Equations in Abstract Spaces*, Pergamon Press, 1981.

[136] P. Lancaster and M. Tismenetsky, *The Theory of Matrices*, Academic Press, 1985.

[137] J.P. LaSalle, *The Stability of Dynamical Systems*, SIAM, 1976.

[138] D. Le and H.L. Smith, *A parabolic system modeling microbial competition in an unmixed bio-reactor*, Journal of Differential Equations **130** (1996), 59–91.

[139] P. De Leenheer and H.L. Smith, *Virus Dynamics: a global analysis*, SIAM J. Appl. Math. **63**(2003), 1313–1327.

[140] R. Lenski and B. Levin, *Constraints on the evolution of bacteria and virulent phage: a model, some experiments, and predictions for natural communities*, Amer. Naturalist **125** (1985), 585–602.

[141] P.H. Leslie, *The use of matrices in certain population mathematics*, Biometrika, **33** (1945), 183-212.

[142] C.-K. Li and H. Schneider, *Applications of Perron-Frobenius theory to population dynamics*, J. Math. Biol. **44** (2002), 450–462.

[143] J. Li, *Simple stage-structured models for wild and transgenic mosquito populations*, J. Difference Eqn. Appl. **15** (2009), 327-347.

[144] M.Y. Li and J.S. Muldowney, *Global stability for the SEIR model in epidemiology*, Math. Biosci. **125** (1995), 155–164.

[145] M.Y. Li, Z. Shuai, *Global stability problem for coupled systems of differential equations on networks*, J. Diff. Eqn. **248** (2010), 1-20.

[146] M.Y. Li, Z. Shuai, C. Wang, *Global stability of multi-group epidemic models with distibuted delays*, J. Math. Anal. Appl. **361** (2010), 38-47.

[147] G.M. Lieberman, *Second Order Parabolic Differential Equations*, World Scientific, 1996.

[148] L.-P. Liou and K.-S. Cheng, *On the uniqueness of a limit cycle for a predator-prey system*, SIAM J. Math. Anal. **19** (1988), 867–878.

[149] S. Liu, L. Wang, *Global stability of an HIV-1 model with distributed intracellular delays and a combination therapy*, Math. Biosci. Eng. **7** (2010), 675-685.

[150] S.-O. Londen, *On the variation of the solution of a nonlinear integral equation*, J. Math. Anal. Appl. **52** (1975), 430–449.

[151] S.-O. Londen, *On some integral equations with locally finite measures and L^∞-perturbations*, SIAM J. Math. Anal. **14** (1983), 1187–1203.

[152] A.L. Lotka, *Relation between birth rates and death rates*, Science **26** (1907), 21–22.

[153] A. Lunardi, *Analytic Semigroups and Optimal Regularity in Parabolic Problems*, Progress in Nonlinear Differential Equations and Their Applications 16, Birkhäuser, 1995.

[154] P. Magal, *Perturbation of a globally stable steady state and uniform persistence*, J. Dyn. Diff. Eqn. **21** (2009), 1-20

[155] P. Magal, C.C. McCluskey, G.F. Webb, *Liapunov functional and global asymptotic stability for an infection-age model*, Applicable Analysis, **89** (2010), 1109–1140.

[156] P. Magal and S. Ruan, *On integrated semigroups and age structured models in L^p spaces*, Diff. Integr. Eqn. **20** (2007), 197–239.

[157] P. Magal and S. Ruan, *Center manifolds for semilinear equations with non-dense domain and applications to Hopf bifurcation in age structured models*, Memoirs of the AMS vol. 202, (2009), no. 951.

[158] P. Magal and X.-Q. Zhao, *Global attractors and steady states for uniformly persistent dynamical systems*, SIAM J. Math. Anal. **37** (2005), 251–275.

[159] R.H. Martin, *Nonlinear Operators and Differential Equations in Banach Spaces*, Wiley and Sons, 1976.

[160] R.H. Martin Jr. and H.L. Smith, *Abstract functional-differential equations and reaction-diffusion systems*, Trans. Amer. Math. Soc. **321** (1990), 1–44.

[161] M. Martcheva and H.R. Thieme, *Progression age enhanced backward bifurcation in an epidemic model with super-infection*, J. Math. Biol. **46** (2003), 385–424.

[162] R.M. May and W.J. Leonard, *Nonlinear aspects of competition between three species*, SIAM J. Appl. Math. **29** (1975), 243–253.

[163] C.C. McCluskey, *Global stability for an SEIR epidemiological model with varying infectivity and infinite delay*, Math. Biosci. and Eng. **6**(2009) 603–610.

[164] J.N. McDonald and N.A. Weiss, *A Course in Real Analysis*, Academic Press, 1999.

[165] A.G. McKendrick, *Applications of mathematics to medical problems*, Proc. Edinb. Math. Soc. **14** (1926), 98–130.

[166] R.C. McOwen, *Partial Differential Equations, methods and applications*, Prentice-Hall, 2003.

[167] J.A.J. Metz, (2008) *Fitness*, Evol. Ecol. **2** (2008), 1599–1612.

[168] J.A.J. Metz, R.M. Nisbet, S.A.H. Geritz, *How should we define "fitness" for general ecological scenarios?*, Tree **7** (1992), 198–202.

[169] J. Mierczynski, W. Shen, X.-Q. Zhao, *Uniform persistence for nonautonomous and random parabolic Kolmogorov systems*, J. Diff. Equations, **204** (2004), 471–510.

[170] K. Mischaikow, H.L. Smith, and H.R. Thieme, *Asymptotically autonomous semiflows: Chain recurrence and Lyapunov functions*, Trans. Amer. Math. Soc. **347** (1995), 1669–1685.

[171] S. Muratori and S. Rinaldi, *Low- and high-frequency oscillations in three-dimensional food chains*, SIAM J. Appl. Math. **52** (1992) 1688–1706.

[172] M.A. Nowak and R.M. May, *Virus Dynamics*, Oxford Univ. Press, 2000.

[173] L. Oksanen, S.D. Fretwell, J. Arruda, P. Niemela, *Exploitation ecosystems in gradients of primary productivity*, American Naturalist **118** (1981), 240–261.

[174] C. V. Pao, *Nonlinear Parabolic and Elliptic Equations*, Plenum Press, 1992.

[175] A. Pazy, *Semigroups of Linear Operators and Applications to Partial Differential Equation*, Springer, 1983.

[176] A.S. Perelson and P.W. Nelson, *Mathematical analysis of HIV-1 dynamics in vivo*, SIAM Review **41** (1999), 3–44.

[177] L. Perko, *Differential Equations and Dynamical Systems*, Springer-Verlag, (1991).

[178] M. Protter and H.F. Weinberger, *Maximum Principles in Differential Equations*, Prentice Hall, 1967.

[179] J. Prüss, L. Pujo-Menjouet, G.F. Webb, *Analysis of a model for the dynamics of prions*, Discrete Contin. Dyn. Syst. **6** (2006), 225–235.

[180] D.A. Rand, H.B. Wilson, J.M. McGlade, *Dynamics and evolution: evolutionarily stable attractors, invasion exponents and phenotype dynamics*, Philosophical Transactions: Biological Sciences **343** (1994), 261–283.

[181] G. Raugel, *Global attractors in partial differential equations*, *Handbook of Dynamical Systems* **2**, 885–982, North-Holland, 2002.

[182] M.G. Roberts and J.A.P. Heesterbeek, *A simple parasite model with complicated dynamics*, J. Math. Biol. **37** (1998), 272–290.

[183] C. Robinson, *Dynamical Systems, Stability, Symbolic Dynamics, and Chaos*, 2nd ed., CRC Press, 1999.

[184] G. Röst and J. Wu, *SEIR epidemiological model with varying infectivity and infinite delay*, Math. Biosci. Engin. **5** (2008), 389–401.

[185] M. Salmani and P. van den Driessche, *A model for disease transmission in a patchy environment*, Disc. Cont. Dyn. Sys. -B **6** (2006), 185–202.

[186] P.L. Salceanu, *Lyapunov Exponents and Persistence in Dynamical Systems, Ph.D. Thesis*, Arizona State University, 2009.

[187] P.L. Salceanu and H.L. Smith, *Persistence in a discrete-time, stage-structured epidemic model*, J. Difference Equ. & Appl. **16** (2010), 73–103.

[188] P.L. Salceanu and H.L. Smith, *Persistence in a discrete-time stage-structured fungal disease model*, J. Biol. Dynamics **3** (2009), 271–285.

[189] P.L. Salceanu and H.L. Smith, *Lyapunov exponents and persistence in discrete dynamical systems*, Discrete & Continuous Dynamical Systems-B **12** (2009), 187–203.

[190] P.L. Salceanu and H.L. Smith, *Lyapunov exponents and uniformly weak normally repelling invariant sets*, *Positive Systems* (R. Bru and S. Romero-Vivo, eds.), Lecture Notes in Control and Information Sciences 389, Springer, 2009.

[191] S.J. Schreiber, *Persistence despite perturbations for interacting populations*, J. of Theor. Biol. **242** (2006), 844–852.

[192] S.J. Schreiber, *On growth rates of subadditive functions for semiflows*, J. Differential Equations **148** (1998), 334–350.

[193] S.J. Schreiber, *Criteria for C^r robust permanence*, J. Differential Equations **162** (2000), 400–426.

[194] S.J. Schreiber, *On persistence and extinction for randomly perturbed dynamical systems*, Discrete Contin. Dyn. Syst. Ser. B **7** (2007), 457–463.

[195] J. Schröder, *Anwendung von Fixpunktsätzen bei der numerischen Behandlung nichtlinearer Gleichungen in halbgeordneten Räumen*, Arch. Rat. Mech. Analysis **4** (1959/60), 177–192.

[196] P. Schuster, K. Sigmund, R. Wolff, *Dynamical systems under constant organization. III. Cooperative and competitive behavior of hypercycles*, J. Diff. Eqns. **32** (1979), 357–368.

[197] G.R. Sell and Y. You, *Dynamics of Evolutionary Equations*, Springer, 2002.

[198] L.A. Segel, *Modeling dynamic phenomena in molecular and cellular biology*, Cambridge Univ. Press, 1987.

[199] A. Singh and R.M. Nisbet, *Semi-discrete host-parasitoid models*, J. Theor. Biol. **247** (2007), 733–742.

[200] H.L. Smith, *Monotone Dynamical Systems: an introduction to the theory of competitive and cooperative systems*, Amer. Math. Soc. Surveys and Monographs, vol. 41, 1995.

[201] H.L. Smith, *The discrete dynamics of monotonically decomposable maps*, J. Math. Biol. **53** (2006), 747–758.

[202] H.L Smith, *Models of virulent phage growth with application to phage therapy*, SIAM J. Appl. Math. **68** (2008), 1717–1737.

[203] H.L. Smith and P. Waltman, *The Theory of the Chemostat*, Cambridge Studies in Mathematical Biology 13, Cambridge University Press 1995.

[204] H.L. Smith and P. Waltman, *Perturbation of a globally stable steady state*, Proc. Amer. Math. Soc. **127** (1999), 447–453.

[205] H.L. Smith and H. R. Thieme, *Persistence of bacteria and phages in a chemostat*, preprint.

[206] H.L. Smith and X.-Q. Zhao, *Robust persistence for semidynamical systems*, Nonlinear Analysis **47** (2001), 6169–6179.

[207] J. Smoller, *Shock Waves and Reaction Diffusion Equations*, Springer-Verlag, 1983.

[208] H.W. Stech and M. Williams, *Stability in a class of cyclic epidemic models with delay*, J. Math. Biol. **11** (1981), 95–103.

[209] P. Takáč, *Convergence in the part metric for discrete dynamical systems in ordered topological cones*, Nonlinear Analysis, TMA, **26** (1996), 1753–1777.

[210] R. Temam, *Infinite Dimensional Dynamical Systems in Mechanics and Physics*, Springer, 1988

[211] H.R. Thieme, *Density-dependent regulation of spatially distributed populations and their asymptotic speed of spread*, J. Math. Biology **8** (1979), 173–187.

[212] H.R. Thieme, *On a class of Hammerstein integral equations*, Manuscripta mathematica **29** (1979), 49–84.

[213] H.R. Thieme, *Well-posedness of physiologically structured population models for Daphnia magna (How biological concepts can benefit by abstract mathematical analysis)*, J. Math. Biology **26** (1988), 299–317.

[214] H.R. Thieme, *Semiflows generated by Lipschitz perturbations of non-densely defined operators*, Differential and Integral Equations **3** (1990), 1035–1066.

[215] H.R. Thieme, *Persistence under relaxed point-dissipativity (with applications to an endemic model)*, SIAM J. Math. Anal. **24** (1993), 407–435.

[216] H.R. Thieme, *Uniform persistence and permanence for non-autonomous semiflows in population biology*, Math. Biosci. **166** (2000), 173–201.

[217] H.R. Thieme, *Mathematics in Population Biology*, Princeton University Press, 2003.

[218] H.R. Thieme, *Pathogen competition and coexistence and the evolution of virulence*, *Mathematics for Life Sciences and Medicine*, (Y. Takeuchi, Y. Iwasa, K. Sato, eds.), 123-153, Springer, 2007.

[219] H.R. Thieme, *Differentiability of convolutions, integrated semigroups of bounded semi-variation, and the non-homogeneous Cauchy problem*, J. Evolution Equations **8** (2008), 283–305.

[220] H.R. Thieme, *Spectral bound and reproduction number for infinite dimensional population structure and time-heterogeneity*, SIAM J. Appl. Math. **70** (2009), 188–211.

[221] H.R. Thieme and C. Castillo-Chavez, *How may infection-age-dependent infectivity affect the dynamics of HIV/AIDS?*, SIAM J. Appl. Math. **53** (1993), 1447–1479.

[222] H.R. Thieme, A. Tridane, Y. Kuang, *An epidemic model with post-contact prophylaxis of distributed length. I. Thresholds for disease persistence and extinction*, J. Biological Dynamics **2** (2008), 221–239.

[223] H.R. Thieme, A. Tridane, Y. Kuang, *An epidemic model with post-contact prophylaxis of distributed length. II. Stability and oscillations if treatment is fully effective*, Mathematical Modelling of Natural Phenomena, **3** (2008), 267-293

[224] H.R. Thieme and P. van den Driessche, *Global stability in cyclic epidemic models with disease fatalities* (Proceedings of the conference on Differential Equations with Applications to Biology), Fields Inst. Comm. **21** (1999), 459–472.

[225] H. R. Thieme and H. Voßeler, *Semilinear perturbations of Hille Yosida operators. Mathematical Modelling of Population Dynamics* (R. Rudnicki, ed.), 87-122. Banach Center Publ. **63**, Polish Academy of Sciences, 2004.

[226] H.R. Thieme and I.I. Vrabie, *Relatively compact orbits and compact attractors for a class of nonlinear evolution equations*, J. Dyn. Diff. Equ. **15** (2003), 731–750.

[227] H.R. Thieme and J. Yang, *An endemic model with variable re-infection rate and application to influenza*, Math. Biosci. **180** (2002), 207–235.

[228] A.C. Thompson, *On certain contraction mappings in a partially ordered space*, Proc. Amer. Math. Soc. **14** (1963), 438–443.

[229] H. Thunberg, *Periodicity versus chaos in one-dimensional dynamics*, SIAM Review **43** (2001), 3–30.

[230] P. Turchin, *Complex Population Dynamics*, Monographs in Population Biology, Princeton University Press, 2003.

[231] P. Waltman, *A brief survey of persistence in dynamical systems*, *Delay Differential Equations and Dynamical Systems* (S. Busenberg, M. Martelli, eds.), 31–40, Lecture Notes in Math., 1475, Springer, 1991.

[232] P. Waltman, *A Second Course in Elementary Differential Equations*, Dover Publications, 2004.

[233] G.F. Webb, *Theory of Nonlinear Age-dependent Population Dynamics*, Marcel Dekker, 1985.

[234] L. Xu, B. Chen, Y. Zhao, and Y. Cao, *Normal Lyapunov exponents and asymptotically stable attractors*, Dynamical Systems **23** (2008), 207–218.

[235] F. Yang and S. Ruan, *A generalization of the Butler-McGehee lemma and its applications in persistence theory*, Diff. Int. Eqn. **9** (1996), 1321-1330

[236] K. Yosida, *Functional Analysis*, second ed., Springer, 1968.

[237] M.L. Zeeman, *Hopf bifurcations in competitive three-dimensional Lotka-Volterra systems*, Dynamics and Stability of Systems **8** (1993), 189–217.

[238] X.-Q. Zhao, *Dynamical Systems in Population Biology*, CMS Books in Mathematics, Springer, 2003.

[239] Y. Zhou and J.M. Cushing, *The net reproductive value and stability in matrix population models*, Nat. Res. Mod. **8** (1994), 297–333.

Index